AF737093

Advancement of Melatonin Research in Plants

Melatonin (*N-acetyl*-5-methoxytryptamine) is a widely studied biomolecule, and its function has been investigated in bacteria, mammals, birds, amphibians, reptiles, fish and plants. Melatonin research in plants the last five years has attained an exponential growth phase and the number of publications related to phytomelatonin has significantly increased in present times. All this research is directed towards establishing multifarious roles of melatonin in plants, including the promotion of seed germination and seedling growth, influencing plant development and senescence, alteration of flowering and fruiting time and grain yield, modulating circadian rhythms, improving resistance against biotic stress as well as tolerance to abiotic stress, reducing oxidative stress, and stimulating plant immune response. However, to date, an exhaustive book, solely and exclusively dedicated to melatonin research on various aspects of plant biology, has not been published by any renowned publisher or is available in the market. Keeping all these facts in mind, this book aims to compile the most recent developments in the arena of melatonin-mediated regulation of plant physiology under optimal and sub-optimal conditions. This book will initiate and introduce the readers to the state-of-the-art developments and trends in this field of study and will largely cater to the scientific need of scientists engaged in melatonin research across the globe. The book will be unique of its kind and will draw the attention of scientific and research communities engaged in the field of melatonin research.

The key features of this book:

- Authors will address the topics which would provide advances in the understanding of diverse functions of melatonin in plants, encompassing almost all the aspects that have been investigated so far.
- Integrate the overall mechanistic function of melatonin, genomics and proteomics tools, and genetic engineering to manipulate endogenous melatonin levels, thus giving a holistic view of plant-level studies under normal or challenging situations.
- Largely help the readers to design future research programs in exploiting melatonin as a protective agent to develop climate-resilient and pathogen-resistant plants.
- Benefit the crop protection program and ensure food security.
- Chapters will be supplemented with ample illustrations and figures.

This book will be beneficial to research scholars and Ph.D. students involved in melatonin research and working on stress physiology, stress proteins, genomics, proteomics and genetic engineering in different universities and research institutes all over the world. The book can also be used as an advanced reference book for the coursework of research and master's level students. It will be of use to people involved in improving plant growth and yield, as well as the performance of plant species in the face of climate change and pathogen infection, using melatonin as the prime protective agent and signaling molecule. Although the book will directly benefit the academic and scientific community, it will also indirectly draw the attention of industries or companies that manufacture or sell chemical products, since the book will highlight in detail the antioxidative potential of melatonin. This will create impetus at the industrial level for the isolation and synthesis of melatonin as a beneficial chemical agent. Applying melatonin as an ameliorating agent against different stresses at the field level will also increase the demand for this chemical among the farmers and will promote large-scale sale and utilization of melatonin for ensuring crop protection and increasing yield.

Advancement of Melatonin Research in Plants

Multi-Faceted Role in Regulating Development and Stress Protection

Edited by
Aryadeep Roychoudhury

CRC Press
Taylor & Francis Group
Boca Raton London New York

CRC Press is an imprint of the
Taylor & Francis Group, an **informa** business

Cover image: © shutterstock

First edition published 2024
by CRC Press
2385 NW Executive Center Drive, Suite 320, Boca Raton FL 33431

and by CRC Press
4 Park Square, Milton Park, Abingdon, Oxon, OX14 4RN

CRC Press is an imprint of Taylor & Francis Group, LLC

© 2024 selection and editorial matter, Aryadeep Roychoudhury; individual chapters, the contributors

Reasonable efforts have been made to publish reliable data and information, but the author and publisher cannot assume responsibility for the validity of all materials or the consequences of their use. The authors and publishers have attempted to trace the copyright holders of all material reproduced in this publication and apologize to copyright holders if permission to publish in this form has not been obtained. If any copyright material has not been acknowledged please write and let us know so we may rectify in any future reprint.

Except as permitted under U.S. Copyright Law, no part of this book may be reprinted, reproduced, transmitted, or utilized in any form by any electronic, mechanical, or other means, now known or hereafter invented, including photocopying, microfilming, and recording, or in any information storage or retrieval system, without written permission from the publishers.

For permission to photocopy or use material electronically from this work, access www.copyright.com or contact the Copyright Clearance Center, Inc. (CCC), 222 Rosewood Drive, Danvers, MA 01923, 978–750–8400. For works that are not available on CCC please contact mpkbookspermissions@tandf.co.uk

Trademark notice: Product or corporate names may be trademarks or registered trademarks and are used only for identification and explanation without intent to infringe.

ISBN: 978-1-032-38155-8 (hbk)
ISBN: 978-1-032-38156-5 (pbk)
ISBN: 978-1-003-34375-2 (ebk)

DOI: 10.1201/9781003343752

Typeset in Times
by Apex CoVantage, LLC

Contents

Preface

Since its discovery in bovine pineal gland in 1958, melatonin (*N-acetyl*-5-methoxytryptamine) was considered exclusively as an animal neurohormone, participating in diverse functions like regulation of sleep, mood, body temperature and bone metabolism, modulation of circadian rhythms, enhancement of immunity, as a multifunctional oncostatic agent, etc., while retaining its ability to reduce oxidative stress. The presence of melatonin in a number of edible plants was unequivocally detected in 1995. Most of the plant species where melatonin has been reported belong to the families like Rosaceae, Vitaceae, Poaceae, Apiaceae and Brassicaceae. Melatonin is now regarded as a low molecular weight, ubiquitous biologically active compound that acts as a multifunctional molecule and regulates key physiological and biochemical processes in plants. The major role of melatonin as an antioxidant and redox homeostasis manager during various forms of abiotic and biotic stresses is widely reported in recent investigations by several research groups. Melatonin can efficiently scavenge ROS, thereby providing a strong defense mechanism for the plants. Melatonin has also been observed to exert influence on several phytohormones and signaling molecules, activating or down regulating the genes involved in their biosynthetic pathways. The genes involved in melatonin biosynthesis have been genetically manipulated through transgenic approaches to enhance environmental stress tolerance and pathogen resistance. Researchers all over the globe are actively engaged in the present time in exploring and understanding the in-depth mechanism of melatonin-mediated regulation of various aspects of plant biology, through biochemical and molecular approaches. There has been a major boom, since the last five years, in the number of valuable and informative research papers involving melatonin. All these researches are directed towards establishing multifarious roles of melatonin in plants, including promotion of seed germination and seedling growth, influencing plant development and senescence, alteration of flowering and fruiting time and grain yield, modulating circadian rhythms, improving resistance against biotic stress as well as tolerance to abiotic stress, reducing oxidative stress, and stimulating plant immune response. Melatonin biosynthesis pathway has been characterized in plants and the accumulated melatonin has been estimated using different methods like radioimmunoassay (RIA), enzyme-linked immunosorbent assay (ELISA), gas chromatography–mass spectrometry (GC–MS), and high-performance liquid chromatography (HPLC) with electrochemical detection (HPLC-ECD), fluorescence detection (HPLC-FD), or HPLC-MS.

Exogenous application of melatonin at appropriate concentration can improve tolerance against salinity, cold, heat, drought, UV radiation, heavy metals and chemical stresses. Melatonin also serves as an important therapeutic alternative to fight against bacterial, viral and fungal diseases. In all these cases, melatonin acts as a powerful antioxidant and membrane stabilizer. Melatonin improves the redox state of cells, thereby decreasing reactive oxygen species (ROS) and reactive nitrogen species (RNS) levels. It can also increase the effective quantum yield of photosystem II and reduce damage to photosynthetic electron transport chain during stress conditions. Melatonin scavenges ROS efficiently during oxidative stress induced by the herbicide called butafenacil. Melatonin treatment can hasten the development of color and flavor of the tomato fruit via its action on ethylene synthesis in the post-harvest state, thereby improving fruit quality. Melatonin has also been found to alter the levels of phytohormones like auxin, gibberellins and abscisic acid. Moreover, integrative roles of signaling molecules like nitric oxide and hydrogen sulfide in melatonin-induced tolerance is also reported from several plant species. Transgenic plants, where genes encoding enzymes for melatonin biosynthesis have been overexpressed, have shown desirable results in improving growth and mitigation of various forms of stress. Consumption of melatonin-rich plant products like cereals, fruits (grapes, strawberries and cherries), vegetables (tomatoes and peppers), nuts, medical herbs (*Scutellaria*, *Hypericum*, etc.) can significantly increase melatonin concentration in human serum and show immense bioactivities within the human system.

While tracing the footprints of research on different aspects of melatonin in the context of vegetative and reproductive development, and its role in stress mitigation in plant systems, I found that there is dearth of reliable books as well as a loophole of proper knowledge between book content and ongoing research trends. There are quite a number of books available in the market dealing with the functioning of melatonin in human and animal systems. Till date, no such dedicated book only on plant melatonin is available. There is not a single exclusive book in the market that compiles together in a comprehensive and elaborate way the multiple functions of melatonin in plant science, so as to provide a direct testimonial to highlight melatonin as a potent biomolecule and disseminate knowledge regarding its immense application in agriculture to improve plant performance. Being quite an inexpensive metabolite, widely distributed among all the organisms, exogenously added melatonin and its accumulation within the plants are environment-friendly events and are not going to cause serious biosafety issues.

Considering the vital roles played by melatonin in plants, I feel the extreme necessity to have a compiled book on this aspect that will enhance proper knowledge to the scientists, researchers and academicians working in the related discipline and further motivate them for future research. Thus, I feel the strong urge to edit this valuable book containing chapters from renowned scientists, researchers and global leaders in this field. I hope that this book will serve as a proper documentation of researches undertaken so far and future trends to be followed to focus on this biomolecule for improvement in plant performances under both optimal and sub-optimal conditions. This book will provide a more holistic view of melatonin research exclusively in plant science. The book will be unique of its kind in broadly overemphasizing the recent advances in melatonin research in plant species and their role in plant growth, yield and stress protection.

Prof. Aryadeep Roychoudhury
Professor
Discipline of Life Sciences
School of Sciences
Indira Gandhi National Open University
Maidan Garhi, New Delhi

Editor Biography

Aryadeep Roychoudhury, Discipline of Life Sciences, School of Sciences, Indira Gandhi National Open University, Maidan Garhi, New Delhi-110068, India

Prof. Aryadeep Roychoudhury is currently working as Professor in the Discipline of Life Sciences, Indira Gandhi National Open University, New Delhi, India. Earlier, he served as Assistant Professor at the Postgraduate Department of Biotechnology, St. Xavier's College (Autonomous), Kolkata, West Bengal, India. He received his B.Sc. (Hons.) in Botany from Presidency College, Kolkata, and M.Sc. in Biophysics and Molecular Biology, University of Calcutta, West Bengal, India. He did his Ph.D. from Bose Institute, Kolkata, under Jadavpur University, Kolkata. His Ph.D. Thesis was based on molecular characterization of salt-inducible genes and transcription factors in rice. Following his Ph.D. work, he joined as Research Associate (Postdoctorate) at the University of Calcutta, continuing with translational research on transgenic rice. He is presently involved in active research in the field of abiotic stress responses in plants with perspectives to the physiology, molecular biology and cell signaling under diverse stress conditions. He holds 23 years of research experience in the concerned discipline. Prof. Roychoudhury has handled several government-funded projects as principal investigator and supervised five (5) Ph.D. students as principal investigator. To date, he has published more than 250 articles in peer-reviewed journals and chapters in books of international and national repute. He has already edited twelve books from reputed publishers till date and has also handled as Guest Editor some Special Issues in renowned international journals. He is a regular reviewer of articles in high-impact, international journals, Life Member of different scientific associations and societies, and the recipient of the Young Scientist Award 2019, conferred upon him by International Foundation for Environment and Ecology, at University of Allahabad, Prayagraj, Uttar Pradesh. His name is included in the Stanford University's List of World's Top 2% Most Influential Scientists.

Contributors

Krishnendu Acharya
Molecular and Applied Mycology and Plant Pathology Laboratory, Department of Botany, University of Calcutta, Kolkata 700019, India

Iffat Zareen Ahmad
Natural Products Laboratory, Department of Bioengineering and Biosciences, Integral University, Dasauli, Kursi Road, Lucknow-226026, Uttar Pradesh, India

Haider Ali
Department of Environmental Sciences, Quaid-e-Azam University, Islamabad

Muhammad Ahsan Altaf
College of Horticulture, Hainan University, Haikou, 570228, PR China

Phaniendra Alugoju
Natural Products for Neuroprotection and Anti-Ageing Research Unit, Chulalongkorn University, Bangkok 10330, Thailand; phaniendra.a@gmail.com
Department of Clinical Chemistry, Faculty of Allied Health Sciences, Chulalongkorn University, Bangkok 10330, Thailand

Kousik Atta
Department of Plant Physiology, Bidhan Chandra Krishi Viswavidyalaya, Mohanpur 741252, Nadia, West Bengal, India

Kakan Ball
Plant Molecular Biology Laboratory, Department of Botany, Raiganj University, Raiganj-733134, Uttar Dinajpur, West Bengal, India

Purabi Banerjee
Department of Agronomy, Bidhan Chandra Krishi Viswavidyalaya, Mohanpur 741252, Nadia, West Bengal, India

L. Barbanti
Department of Agricultural and Food Sciences, University of Bologna, Bologna 40127, Italy

Biswaranjan Behera
ICAR-Indian Institute of Water Management, Bhubaneswar, Odisha, India

Kuntal Bera
Plant Molecular Biology Laboratory, Department of Botany, Raiganj University, Raiganj-733134, Uttar Dinajpur, West Bengal, India
Department of Seed Science and Technology, Uttar Banga Krishi Viswavidyalaya, Pundibari-736165, Cooch Behar, West Bengal, India

Pooja Bhadrecha
Department of Biotechnology, UIBT, Chandigarh University, Mohali, Punjab, India

Rajan Bhatt
Regional Research Station, Kapurthala, Punjab Agricultural University, Ludhiana, Punjab, India, 144601

Parijat Bhattacharya
Department of Agricultural Chemistry and Soil Science, Bidhan Chandra Krishi Viswavidyalaya, Mohanpur 741252, Nadia, West Bengal, India

Paramita Bhowmik
Malda Krishi Vigyan Kendra, Ratua, Uttar banga Krishi Viswavidyalaya, West Bengal-732205
School of Agriculture, Swami Vivekananda University, Barrackpore-700121, North 24 Parganas, West Bengal, India

R. Calone
Council for Agricultural Research and Economics, Research Centre for Agriculture and Environment, I-40128 Bologna, Italy

Nilanjan Chakraborty
Department of Botany, Scottish Church College, Kolkata 700006, India

Chandini
Department of Agronomy and Agroforestry, Centurion University of Technology and Management, Odisha-761211

Kshouni Das
Malda Krishi Vigyan Kendra, Ratua, Uttar banga Krishi Viswavidyala, West Bengal-732205

Suddhasuchi Das
Malda Krishi Vigyan Kendra, Ratua, Uttar banga Krishi Viswavidyala, West Bengal-732205

Puspendu Dutta
Department of Seed Science and Technology, Uttar Banga Krishi Viswavidyalaya, Pundibari-736165, Cooch Behar, West Bengal, India

Dinkar J. Gaikwad
Department of Biochemistry and Plant Physiology, Centurion University of Technology and Management, Odisha-761211

Sreeramaiah N. Gangappa
Department of Biological Sciences, Indian Institute of Science Education and Research Kolkata, Mohanpur 741246 West Bengal, India

Retwika Ganguly
Department of Botany, Scottish Church College, Kolkata 700006, India

Pabitra Kumar Ghosh
Department of Seed Science and Technology, Bidhan Chandra Krishi Viswavidyalaya, Mohanpur 741252, Nadia, West Bengal, India

Subir Ghosh
Plant Molecular Biology Laboratory, Department of Botany, Raiganj University, Raiganj-733134, Uttar Dinajpur, West Bengal, India

Alvina Gul
Department of Plant Biotechnology, Atta ur Rahman School of Applied Biosciences, National University of Sciences & Technology, Islamabad

Partha Halder
Department of Botany, Scottish Church College, Kolkata 700006, India

Akbar Hossain
Division of Agronomy, Bangladesh Wheat and Maize Research Institute, Dinajpur 5200, Bangladesh

Noor ul huda
Department of Plant Biotechnology, Atta ur Rahman School of Applied Biosciences, National University of Sciences & Technology, Islamabad

M. Ijaz
College of Agriculture, Bahauddin Zakariya University, Bahadur Sub Campus Layyah 31200, Pakistan

Husnain Jawad
Agronomic Research Institute, Ayub Agricultural Research Institute, Faisalabad, Pakistan

Hafiz Muhammad Jhanzab
Department of Agronomy, University of Agriculture, Dera Ismail Khan, 29111, Dera Ismail Khan, Pakistan

Meera Kumari Kar
ICAR-National Rice Research Institute, Cuttack

Bushra Hafeez Kiani
Department of Biological Sciences (Female Campus), International Islamic University, Islamabad, 44000, Pakistan

G.K. Krishna
Department of Plant Physiology, KAU-College of Agriculture, Thrissur 680656, Kerala

Awadhesh Kumar
ICAR-National Rice Research Institute, Cuttack, Odisha, India

Ravinder Kumar
ICAR-Central Potato Research Institute, Shimla, Himachal Pradesh, India

K.K Kumaree
Natural Products for Neuroprotection and Anti-Ageing Research Unit, Chulalongkorn University, Bangkok 10330, Thailand
College of Public Health Sciences, Chulalongkorn University, Bangkok 10330, Thailand

Aradhna Kumari
Department of Plant Physiology, College of Agriculture, Jawaharlal Nehru Krishi Vishwa Vidyalaya, Ganj Basoda 464221, Madhya Pradesh

Milan Kumar Lal
ICAR-Central Potato Research Institute, Shimla, Himachal Pradesh, India

Sagar Lalichetti
Department of Agronomy and Agroforestry, Centurion University of Technology and Management, Odisha-761211

Biswajit Lenka
Indira Gandhi Agricultural University, Raipur-492012, Chhattisgarh, India

Sagar Maitra
Department of Agronomy and Agroforestry, Centurion University of Technology and Management, Odisha-761211

Debjyoti Majumder
Malda Krishi Vigyan Kendra, Ratua, Uttar Banga Krishi Viswavidyala, West Bengal-732205

Oksana Mandal
Department of Plant Physiology, Bidhan Chandra Krishi Viswavidyalaya, Mohanpur 741252, Nadia, West Bengal, India

Vikas Mangal
ICAR-Central Potato Research Institute, Shimla, Himachal Pradesh, India

M.S. Mikhina
Department of Plant Physiology, KAU-College of Agriculture, Thrissur 680656, Kerala

Subhasis Mondal
Department of Plant Physiology, Bidhan Chandra Krishi Viswavidyalaya, Mohanpur 741252, Nadia, West Bengal, India

Triptesh Mondal
Department of Agronomy and Agroforestry, Centurion University of Technology and Management, Odisha-761211

Debojyoti Moulick
Department of Environmental Science, University of Kalyani, Nadia-741235, West Bengal, India

Arup Kumar Mukherjee
ICAR-National Rice Research Institute, Cuttack

Faisal Nadeem
Department of Agronomy, University of Agriculture, Dera Ismail Khan, 29111, Dera Ismail Khan, Pakistan

Rajib Nath
Department of Agronomy, Bidhan Chandra Krishi Viswavidyalaya, Mohanpur 741252, Nadia, West Bengal, India

Babita Patni
High Altitude Plant Physiology Research Centre, H.N.B. Garhwal University (A Central University), Srinagar Garhwal 246174, Uttarakhand

Anamika Paul
Scottish Church College, Kolkata 700006, India

Aman Prakash
Department of Molecular and Cellular Engineering, Jacob Institute of Biotechnology and Bioengineering, Sam Higginbottom University of Agriculture, Technology and Sciences, Naini, Pragraj-211007

Anchalee Prasansuklab
Natural Products for Neuroprotection and Anti-Ageing Research Unit, Chulalongkorn University, Bangkok 10330, Thailand
College of Public Health Sciences, Chulalongkorn University, Bangkok 10330, Thailand

Abdul Qayyum
Department of Agronomy, The University of Haripur, Haripur, Pakistan

Shashi Ranjan
Department of Molecular and Cellular Engineering, Jacob Institute of Biotechnology and Bioengineering, Sam Higginbottom University of Agriculture, Technology and Sciences, Naini, Pragraj-211007
ICAR-Central Potato Research Institute, Shimla, Himachal Pradesh, India

Sayanika Roy
Scottish Church College, Kolkata 700006, India

Aryadeep Roychoudhury
Discipline of Life Sciences, School of Sciences, Indira Gandhi National Open University, Maidan Garhi, New Delhi-110068, India

Muhammad Saad
Department of Agronomy, University of Agriculture, Dera Ismail Khan, 29111, Dera Ismail Khan, Pakistan

Sanjoy Sadhukhan
Plant Molecular Biology Laboratory, Department of Botany, Raiganj University, Raiganj-733134, Uttar Dinajpur, West Bengal, India

Soumya Kumar Sahoo
Institute of Agricultural Sciences, SOA (Deemed to be) University, Bhubaneswar-751030, Odisha, India

Upasana Sahoo
Department of Agronomy and Agroforestry, Centurion University of Technology and Management, Odisha-761211

Jyoti Saini
Noida International University, Gautam Budh Nagar, 203201, U.P., India

Masina Sairam
Department of Agronomy and Agroforestry, Centurion University of Technology and Management, Odisha-761211

Pankajini Samal
ICAR-National Rice Research Institute, Cuttack

Muhammad Saqib
Department of Horticulture, University of Agriculture, Dera Ismail Khan, 29111, Dera Ismail Khan, Pakistan

Haram Sarfraz
Natural Products Laboratory, Department of Bioengineering and Biosciences, Integral University, Dasauli, Kursi Road, Lucknow-226026, Uttar Pradesh, India

Anup Kumar Sarkar
Department of Botany, Dukhulal Nibaran Chandra College, Aurangabad-742201, Murshidabad, West Bengal, India
Plant Molecular Biology Laboratory, Department of Botany, Raiganj University, Raiganj-733134, Uttar Dinajpur, West Bengal, India

Anik Sarkar
Molecular and Applied Mycology and Plant Pathology Laboratory, Department of Botany, University of Calcutta, Kolkata 700019, India

B. Sarwar
College of Agriculture, Bahauddin Zakariya University, Bahadur Sub Campus Layyah 31200, Pakistan

A. Sattar
College of Agriculture, Bahauddin Zakariya University, Bahadur Sub Campus Layyah 31200, Pakistan

Vishmita Sethi
Department of Biological Sciences, Indian Institute of Science Education and Research Kolkata, Mohanpur 741246 West Bengal, India

Tanmoy Shankar
Department of Agronomy and Agroforestry, Centurion University of Technology and Management, Odisha-761211

A. Sher
College of Agriculture, Bahauddin Zakariya University, Bahadur Sub Campus Layyah 31200, Pakistan; Department of Agricultural and Food Sciences, University of Bologna, Bologna 40127, Italy

Ankur Singh
Department of Biotechnology, St. Xavier's College (Autonomous), 30, Mother Teresa Sarani, Kolkata-700016, West Bengal, India

Santosh Kumar Singh
Department of Soil Science, Dr. Rajendra Prasad Central Agriculture University, Pusa, Samastipur 848125, Bihar

Shilpy Singh
Noida International University, Gautam Budh Nagar, 203201, U.P., India

Tewin Tencomnao
Natural Products for Neuroprotection and Anti-Ageing Research Unit, Chulalongkorn University, Bangkok 10330, Thailand; phaniendra.a@gmail.com
Department of Clinical Chemistry, Faculty of Allied Health Sciences, Chulalongkorn University, Bangkok 10330, Thailand

Rahul Kumar Tiwari
ICAR-Central Potato Research Institute, Shimla, Himachal Pradesh, India

S. Ul-Allah
College of Agriculture, Bahauddin Zakariya University, Bahadur Sub Campus Layyah 31200, Pakistan

Gouranga Upadhyaya
Department of Biological Sciences, Indian Institute of Science Education and Research Kolkata, Mohanpur 741246 West Bengal, India

Muhammad Zubair
Directorate of Oilseeds, Ayub Agricultural Research Institute, Faisalabad, Pakistan

1 Versatile Role of Melatonin in Plants

Debjyoti Majumder, Paramita Bhowmik, Kshouni Das, Suddhasuchi Das, Akbar Hossain, and Rajan Bhatt

1.1 INTRODUCTION

Melatonin (MEL) is a biogenic indole amine that shares structural similarities with tryptophan, serotonin, indole-3-acetic acid (IAA), and other significant compounds (Figure 1.1). It regulates numerous physiological reactions including improvement of photosynthesis activity, plant development, carbon acquisition, root growth, seed germination, and defence mechanisms against biotic and abiotic stress.

The scavenging ability (antioxidant activity) of MEL against biological free radicals, such as reactive oxygen and nitrogen species, including the hydroxyl radical, singlet oxygen, peroxyl radical, hydrogen peroxide, peroxynitrite anion, and nitric oxide is another recent and significant action that has been attributed to it (Samanta et al., 2021). In this regard, it is intriguing to note that MEL differs from other antioxidants in that the MEL molecule itself has no pro-oxidative effects, whereas MEL-intermediates exhibit antioxidant properties and have a significant synergistic effect with other antioxidants like ascorbic acid and glutathione (Tan, 1993; Reiter and Tan, 2000; Cano et al., 2003). MEL was initially discovered in vascular plants in 1995, when multiple groups discovered it mostly in the monocotyledon and dicotyledon plant families (Dubbels et al., 1995; Hattori et al., 1995; Kolar et al., 1995). According to Arnao and Hernández-Ruiz (2014, 2021); Nawaz et al. (2016), it is engaged in a variety of physiological processes in plants, including growth and development, seed germination, root development, blooming, branching, photosynthesis, and ripening. According to Chan and Shi (2015), Reiter et al. (2015) and Arnao and Hernández-Ruiz (2019a), the most important role of MEL in plants is as a protective agent against various biotic and abiotic stresses, such as cold, drought, PEG-induced osmotic stress, heat, salinity, chemical pollutants, herbicides, and UV irradiation. In recent years, it has also been shown that MEL promotes plant nutrient absorption in addition to protecting plants from a variety of biotic and abiotic stressors. In research on the *Malus* plant, it was shown that MEL could control the reactive oxygen species (ROS) and activate the CBL1-CIPK23 pathway to control the production of a potassium channel protein gene, which in turn enhanced potassium absorption in salinity- or nutrient-deficient environments (Li et al., 2016). According to Sarafi et al. (2017), MEL may counteract the harmful effects of elevated boron levels in pepper plants. MEL applied to the rhizosphere of drought-stressed alfalfa might encourage N metabolism and proline balance (Antoniou et al., 2017). Exogenous MEL treatment controls a multitude of genes and physiological processes involved in plant resistance to salt stress under salt environment (Murch et al., 2004; Simopoulos et al., 2005; Ljung et al., 2002; Mir et al., 2020). These MEL effects, however, differ based on the application technique and plant type. N, one of the most crucial nutrients for plant development, has also been shown to be controlled by MEL in some plant species. According to Zhang et al. (2017a), exogenous MEL treatment increased resistance of cucumber seedlings to high nitrate stress by increasing the activity of N metabolism-related enzymes. Exogenous MEL substantially reduced the leaf NH_4 + N concentration in sub-low temperature melon seedlings by enhancing the activities of N metabolism-related enzymes such as glutamine synthetases

DOI: 10.1201/9781003343752-1

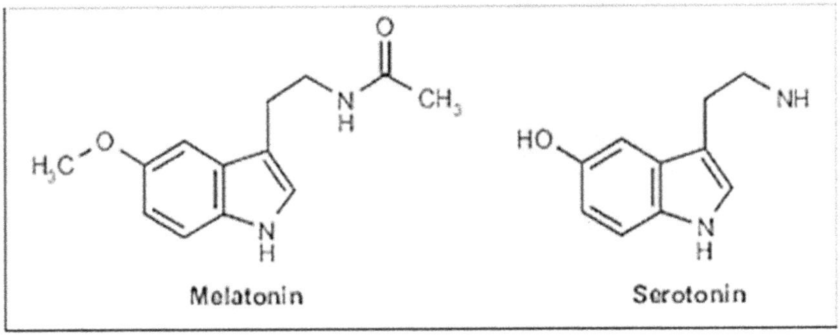

FIGURE 1.1 Structural compositions of MEL

(GS) and glutamic acid synthase (GOGAT) (Gao et al., 2016). In drought-stressed *Medicago sativa*, Antoniou et al. (2016) discovered that MEL might preserve nitro-oxidative equilibrium by controlling N species at the enzymatic and/or transcript level (Van Tassel, 1995). Additionally, by interacting with reactive oxygen species and reactive nitrogen species in a very effective free radical scavenging cascade, MEL may protect organisms from oxidative stress (Tan et al., 2007). According to reports, MEL has a very comparable impact to IAA on the development of lateral roots in lupine (Arnao and Hernández-Ruiz, 2019n). In these investigations, auxin-induced root and cytokinin-induced shoot organogenesis were blocked by variations in the endogenous concentration of MEL and inhibitors of the transfer of serotonin and MEL (Murch et al., 2001). According to research, MEL controls leaf senescence, root, shoot and explant development, as well as seed germination (Arnao and Hernández-Ruiz, 2014). Tan et al. (2012) also noted that high MEL concentration in plants boosted the germination rate of seeds under challenging conditions, extending seed germination time and enhancing plant output quality. MEL boosted plant root and shoot biomass in lupin with effects comparable to those of IAA for root biomass at the utilised doses (Arnao and Hernández-Ruiz, 2006). The hormone MEL is regarded as a master plant regulator since it promotes plant growth and development by serving as a signalling molecule linked to defence mechanisms against various biotic and abiotic challenges including cold, drought, salt, and nutritional deficiencies (Posmyk and Janas, 2009; Kabiri et al., 2021). Additionally, it has a role in several physiological processes including photosynthesis and seed germination in plants (Sadak et al., 2020). In response to biotic and abiotic challenges, plants experience losses in yield, growth, senescence, and occasionally even death, but they also mount a variety of defence mechanisms.

MEL is important in the control of roots and shoots because it shares structural similarities with auxins, especially IAA. To promote vegetative development in a range of plant species, auxin plays a major function (Kolář and Macháčková, 2005). In vivo impacts on the existence of internal levels of MEL and other regulating hormones are found to promote plant development. Depending on the plant species, MEL accumulates in various amounts (Fan et al., 2018). Byeon and Back (2014) produced transgenic rice by overexpressing the sheep *SNAcT* gene in rice, which resulted in excessive MEL build-up. Transgenic rice had a much better growth potential than wild rice and exhibited traits including delayed blooming and decreased yield. Other research demonstrated that 50 mol L^{-1} of MEL solution may greatly boost soybean seedling development and yield (Wei et al., 2015). According to Liu et al. (2019), spraying 100 mol L^{-1} of MEL solution on to the pear tree at the late stage of fruit development might encourage the accumulation of endogenous MEL in pear fruit. Boosting the maximal quantum efficiency of photosystem II and the net photosynthetic rate, MEL expanded the size of pear fruit. Research into the significance and uses of MEL in relation to its role in plant growth and development is becoming more in-depth, and this might lead to the emergence of a new area of study.

1.2 BIOSYNTHESIS OF PLANT MEL

The main site of MEL biosynthesis is the chloroplasts, but if the chloroplast pathway is disrupted, the mitochondrial pathway will be energized for MEL generation to preserve homeostasis (Tan and Russel, 2020).

In both plants and animals, MEL is an indole compound synthesized from serotonin, by a minor difference in the two synthetic pathways. Synthesis of MEL occurs from tryptophan (an essential aromatic amino acid) through four-step reactions catalyzed by six enzymes, that is, tryptophan decarboxylase (TDC), tryptophan hydroxylase (TPH), tryptamine 5-hydroxylase (T5H), serotonin N-acetyl transferase (SNAT), N-acetyl serotonin O-methyl transferase (ASMT), and caffeic acid O-methyl transferase (COMT) (Figure 1.2).

For the determination of MEL content in plant samples, several detection methods and assays have been reported, including enzyme-linked immuno sorbent assay (ELISA), radioimmunoassay (RIA), gas chromatography–mass spectrometry (GC–MS), and high-performance liquid chromatography (HPLC) with electrochemical detection (HPLC-ECD), fluorescence detection (HPLC-FD), or high-performance liquid chromatography-mass spectrometry HPLC-MS (Feng et al., 2014).

In plants, tryptophan is initially decarboxylated to form tryptamine by tryptophan decarboxylase (TDC) and then tryptamine is hydroxylated at the 5 positions of the indole ring to form 5-hydroxytryptamine (serotonin) by tryptamine 5-hydroxylase (T5H) (Tan and Russel, 2020).

MEL formation in plants begins with tryptophan and involves four enzymatic steps catalyzed by at least six enzymes: tryptophan decarboxylase (TDC), tryptophan hydroxylase (TPH), tryptamine 5-hydroxylase (T5H), serotonin N-acetyl transferase (SNAT), N-acetyl serotonin methyl transferase (ASMT), and caffeic acid O-methyl transferase (COMT) (Back et al., 2016). The first two enzymes that contribute to tryptophan hydroxylation and decarboxylation in herbivores have been identified in plants (Park et al., 2012).

MEL biosynthesis must be appropriately regulated among different plant species, tissues, developmental stages, and growth conditions. The MEL content varies depending on the degree of fruit maturation in tomatoes and grapes (Murch et al., 2010). Cherry fruit is rich in MEL, and contents are particularly high during endocarp lignification and embryonic development; the MEL content varies among different varieties (Zhao et al., 2013). At the tips of mulberry leaves, the MEL content is the highest and is higher in juvenile leaves than in older leaves (Pothinuch and Tongchitpakdee, 2011).

The biosynthesis of MEL in plants is far more complex than in animals (Park and Back, 2012; Back et al., 2016). In plants, MT synthesis begins with tryptophan, and the intermediate steps involve six enzymes and at least four different routes (Tan et al., 2015).

Exogenous N-acetyl serotonin can be converted to serotonin in rice seedlings by N-acetyl serotonin deacetylase (ASDAC), which may result in a reduction in the content of MEL studied by Lee et al. (2018). In various sub-cellular locations, MEL intermediates are produced, such as the cytoplasm, endoplasmic reticulum, and chloroplasts. Depending on the pathway, cytoplasm or the

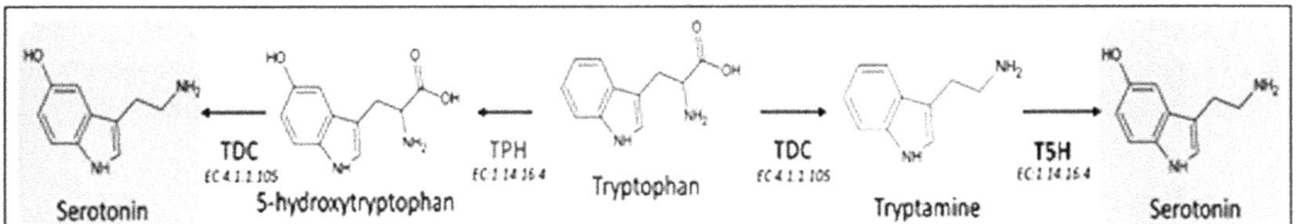

FIGURE 1.2 First two reactions of the MEL biosynthetic pathway leading to the formation of the essential intermediate serotonin. (TDC: L-tryptophan decarboxylase; TPH: tryptophan hydroxylase; T5H: tryptamine 5-hydroxylase; EC: enzyme commission number).

TABLE 1.1
Content of MEL in Particular Plants

Species	Common name	Organ	Content [ng g^{-1} FW (DW)]	Reference
Allium cepa L.	Onion	Bulbs	0.03	Hattori et al. (1995)
Ananas comosus L.	Pineapple	Fruits	0.30	Sae-Teaw et al. (2013)
Brassica nigra L. Black	mustard	Seeds	129 (DW)	Manchester et al. (2000)
Brassica oleracea L.	Cabbage	Leaves	0.107	Hattori et al. (1995)
Capsicum annuum L.	Green pepper	Seeds	0.52 (DW)	Huang and Mazza (2011)
Cucumis sativus L	Cucumber	Seeds	11–80	Kołodziejczyk et al. (2015)
Daucus carota L.	Carrot	Roots	0.06	Hattori et al. (1995)
Elettaria cardamomum White et Maton	Green cardamom	Seeds	15 (DW)	Manchester et al. (2000)
Foeniculum vulgare Gilib.	Fennel	Seeds	28 (DW)	Manchester et al. (2000)
Helianthus annuus L.	Sunflower	Seeds	29 (DW)	Manchester et al. (2000)

chloroplasts can be the final sub-cellular site of MEL synthesis, which may differentially affect the mode of action of MEL in plants (Wei et al., 2017, 2018; Tan et al., 2016).

It is found in almost all plant species, varying in amount according to the plant tissues in a light-depending manner (Zhao et al., 2018) (Table 1.1), though the level of MEL is higher in aromatic plants and leaves than that of seeds. Moreover, the majority of the plants containing MEL are from families Brassicaceae, Vitaceae, Poaceae, Rosaceae, and Apiaceae. During the last decade, several researchers revealed that depending on location, growth stage, organ, and harvest timing, the concentration of MEL also differs among the varieties of the same species (Murch et al., 2009). In horticulture plants, Japanese morning glory (*Pharbitis nil*) MEL was first observed in 1993. MEL levels, synthetic pathways, and roles in plant growth and postharvest storage have attracted increasing attention in the last 20 years.

The cellular compartments with the highest MEL levels in plants are mitochondria and chloroplasts (Kanwar et al., 2018). The whole MEL biosynthetic pathway has been discovered in several plant species, except for one putative gene encoding for a tryptophan hydroxylase (TPH), which catalyzes the conversion of tryptophan into 5-hydroxytryptophan (Vigliante et al., 2019). MEL biosynthesis begins with the amino acid tryptophan, a compound that plants can synthesize de novo via the shikimate pathway. This pathway consists of seven different steps that allow for the biosynthesis of all aromatic amino acids in plants, including tryptophan (Mannino et al., 2021).

1.3 ROLE OF MEL IN PLANT GROWTH

MEL exhibits different levels of expression in different plant species (Fan et al., 2018). It helps in the germination of seed, biomass production, increase in photosynthesis, delay in senescence, flowering and ripening of fruit (Lee et al., 2017; Shi et al., 2017; Wang et al., 2017) (Figure 1.3). According to Nawaz et al. (2020), MEL is involved in the enhancement of seed germination, flower formation and fruit development, with improved survival rates under unfavourable conditions of plants.

1.3.1 Impact on Vegetative Growth

MEL, as a growth-promoting molecule, plays a dual nature in promoting plant growth by increasing growth parameters both under in vitro and in vivo conditions (Sadak et al., 2020). Seed germination, seedling growth, morphology, and nutrient element content in cucumber and wheat were promoted by MEL (Zhang et al., 2013; Qiao et al., 2019). The exogenous application of MEL during the

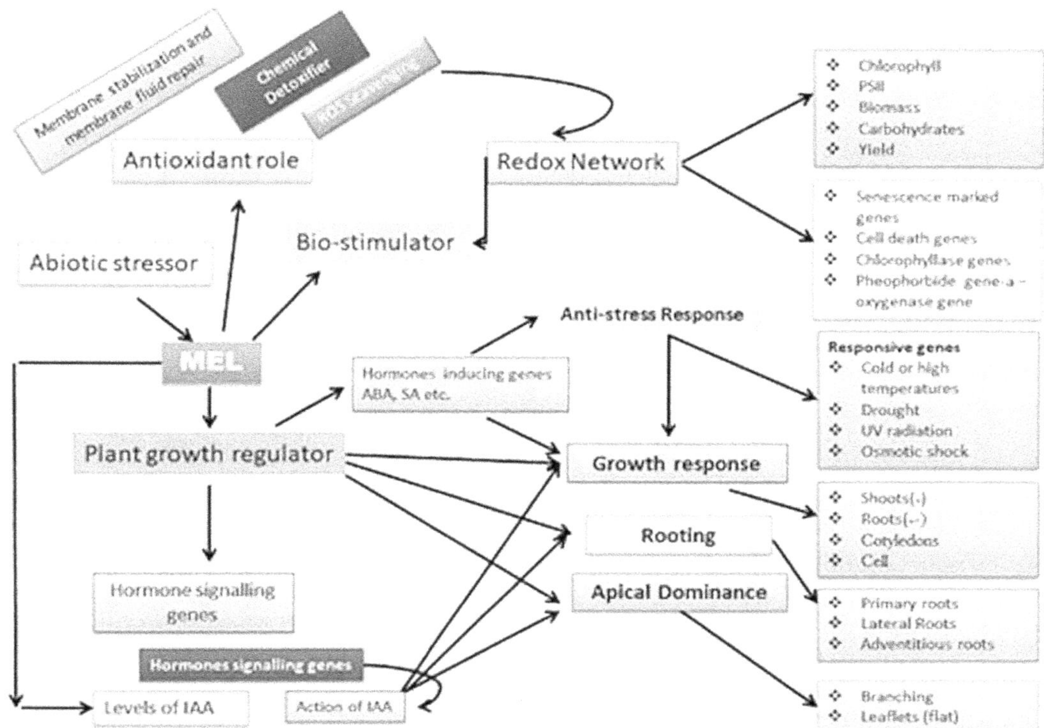

FIGURE 1.3 Roles of MEL in plants (authors' unpublished material).

flowering stage of the short-day plant *Chenopodium rubrum* (Kolář et al., 2003) provoked neither toxic effects nor changes in shape, colour, or number of leaves of this plant. MEL treatment significantly improved the quality and quantity of chickpea seed; leaf size, plant height, pod and seed numbers of soybean; seed germination of red cabbage and postharvest leaf quality of Chinese flowering cabbage (Dawood, 2017; Tan et al., 2021). MEL played an integral role to regulate the shoot and root growth in *Prunus cerasus* L., *A. thaliana*, and *Lupinus albus* L. (Arnao and Hernández-Ruiz, 2007; Sarropoulou et al., 2012a; Zia et al., 2019). The fresh weight, primary root length, and shoot height were significantly higher after MEL treatment in *Arabidopsis* plants (Bajwa et al., 2014). In cucumber, MEL also provoked root development and vitality and enhanced the ratio of root to shoot (Zhang et al., 2013). The germination rate of chilling stressed cucumber was influenced by MEL when seeds were pretreated (Posmyk et al., 2009). It was also reported that there was a clear increase in the seed germination rate and root growth in MEL-treated water-stressed cucumber plants (Zhang et al., 2013). Melatonin significantly improved plant height, leaf growth, and the number of seeds and increased vegetative growth regeneration of adventitious and lateral roots of soybean when the seeds were coated with MEL (Wei et al., 2015). Chen et al. (2009) observed that 0.1 μM MEL exhibited a root growth-enhancing effect in wild leaf mustard, *Brassica juncea* (L.) Czern, whereas 100 μM MEL showed an inhibitory effect, while a stimulatory effect was only detectable in younger seedlings. The usefulness of MEL for the long-term storage of germplasm for plant cell culture was demonstrated by Uchendu et al. (2013) during the enhancement of the growth of frozen shoot explants of American elm (*Ulmus americana*) in both pre-culture and growth media of MEL. There was also a significant increase in fresh weight, primary root length, and shoot height in MEL-treated *Arabidopsis* plants grown at 4°C temperature (Bajwa et al., 2014).

A maximum concentration of 240–550 pg/g fresh weight was recorded in cultivated 15-day-old plants of *Chenopodium rubrum* L. when the levels of MEL was low or undetectable during the light period and the level was increased considerably in darkness (Kolář et al., 1997). Cucumber seed treatment with 1 μM MEL significantly enhanced the growth and germination rate which proved to be efficient in a fivefold increase of the activities of antioxidant enzymes like catalase and superoxide dismutase (Nawaz et al., 2020). When *Arabidopsis thaliana* plants were treated with 10–30 μM of MEL, it resulted in the enhancement of root length, fresh weight, and height of the plant (Bajwa et al., 2014).

1.3.2 Impact on Reproductive Growth

In plants, MEL produces dark and light signals, which coincide with the photoperiodic ecology that influences the daylight responses of flowering within a short period of time (Kolář et al., 1997). Exogenous application of MEL at 100 and 500 μM concentration (much above physiological levels) had a weak inhibitory effect on flowering, but not on the timing and rhythm (phase, period) of the process (Reiter et al., 1980). Brzezinski (1997) suggested that there was an unclear effect of MEL on flowering because H-MEL remained practically unmetabolized and was only minimally transported to other regions after 37 h when it was applied to the apex and cotyledons of young plants (Kolář et al., 2003; Machácčková and Krekule, 2002). It was concluded by Van Tassel et al. (2001) that there was no relationship between MEL levels and photoperiod or ripening. Exogenous application of MEL regulates the transition of flowering in *Chenopodium rubrum* L., *Pterygpra californica* Ruprecht, and *Arabidopsis thaliana* (Wolf et al., 2001). It played an important role in influencing reproductive physiology and flower development of *Hypericum perforatum* L. (St John's wort). High concentrations of indole and MEL during the uninucleate microsporogenesis stage led to an increased regenerative potential of isolated anthers (Kolář et al., 2003). Murch et al. (2001) suggested that the increased rates of MEL in juicy fruits indicated its impact on the ripening of the fruits. As a gene regulator of ethylene-related factors, it also played a role in fruit ripening and post-harvest processes (Abd El-Naby et al., 2019; Arnao and Hernández-Ruiz, 2007; Koyama et al., 2013).

1.3.3 Impact of MEL on Bacteria

The application of exogenous MEL and plant growth-promoting rhizobacteria (PGPR) in agriculture enhanced plant growth under abiotic and biotic stress without disturbing the natural soil ecosystem. They both are non-toxic substances that increase and stimulate the processes involved in plant growth and development (Asif et al., 2019). The application of exogenous MEL and *Bacillus licheniformis* to *Spinacia oleracea* L. seedlings revealed that MEL application increased seedling growth by 14% (Asif et al., 2020).

1.4 IMPACT OF MEL IN HEAVY METAL STRESS

When MEL was applied to 25 ppm cadmium-treated *Spinacia oleracea* L. plants, a greater increase of 13% seedling growth was observed, and a 17% increase in 75 ppm arsenic-stressed plants. MEL-treated plants exhibited a 19% increase in the number of leaves at 25 ppm arsenic-stressed condition. MEL application and *Bacillus licheniformis* inoculation alleviated arsenic and cadmium toxicity by significantly reducing the negative impacts of stresses and increasing the fresh and dry weight as well as preventing the damage to the chlorophyll content of *S. oleracea* L. Supplementation of MEL and *B. licheniformis* enhanced the activities of antioxidant enzymes, superoxide dismutase, peroxidase, and catalase, thus acting as a line of defence against arsenic and cadmium stresses. Lipid peroxidation was also inhibited by exogenous MEL and *B. licheniformis* inoculation. Exogenous application of MEL and inoculating roots of *S. oleracea* L. with *B. licheniformis* were found to ameliorate the harmful effects of arsenic and cadmium contamination. Asif et al. (2020) also found that there is a strong negative correlation between the concentration of arsenic and cadmium in the presence of MEL treatment and chlorophyll content in plants because MEL preserves chlorophyll by down-regulation of the expression of chlorophyllase (CLH1), pheide, and oxygenase (PAO) enzymes that play a role in the degradation of chlorophyll (Zhang et al., 2017b). MEL and plant-growth-promoting rhizobium enhance antioxidant activity under cadmium and arsenic stress. Plants treated with MEL under arsenic (at 75 ppm) stress condition had the highest SOD content (143.6 units/g FW) and under cadmium (at 75 ppm cadmium) stress condition had a significantly high CAT amount (44.1 units/g FW). MEL enhances growth parameters more efficiently at low metal concentrations (Asif et al., 2020).

1.5 IMPACT OF MEL ON PLANT HORMONES

MEL, along with other phytohormones, enhances plant growth (Liu et al., 2016). The indole derivative structure and biosynthetic pathway (from tryptophan) suggested that MEL might be related to IAA (Murch and Saxena, 2002a; Kolář et al., 2003; Van Tassel and O'Neill, 2001; Kolář and Macháčková, 2005; Arnao and Hernández-Ruiz, 2007; Fan et al., 2018; Murch et al., 2000). The optimum degree of growth promotion (100%) was obtained by MEL with IAA, while the optimum effect of MEL alone ranged between 22 to 63% at 10 μM MEL. The similar concentration gradient of MEL and IAA coexist in the tissues of different zones (apical, central, and basal) of the hypocotyls (Hernández-Ruiz et al., 2004). The concentration of MEL was higher than IAA in etiolated coleoptiles of barley and oat (Tanimoto, 2005). Similar to IAA, MEL was also responsible for slight acidification of the media during coleoptile growth promotion (Hernández-Ruiz et al., 2005), thus indicating the possible co-participation of MEL with IAA (Hager, 2003). Increased formation of root primordia in adventitious and lateral roots after treatment with MEL at physiological concentrations affected the length, number, and rate of growth of roots in a similar way to IAA (Murch et al., 2001). MEL plays an important role in modulating the gene expression of auxin carrier proteins and helps in the metabolism of auxin (IAA), cytokinin, abscisic acid (ABA), gibberellins (GAs), and ethylene (Arnao and Hernández-Ruiz, 2018; Samanta et al., 2021). It was observed that IAA stimulated root organogenesis, and cytokinin-induced shoot organogenesis was repressed with the endogenous application of MEL and repressors of the transport of serotonin and MEL (Murch et al., 2001). It was observed that incubation of the etiolated hypocotyls from *Lupinus albus* L. (lupin) with MEL and IAA at similar concentration gradient produced an active elevation of growth at concentrations in the micro-molar range and exhibited an inhibitory growth effect at higher concentrations in de-rooted lupins. The reversal effects on the application of MEL and IAA were observed due to the removal of meristematic regions (Hernández-Ruiz et al., 2004).

1.6 IN VIVO IMPACT OF MEL

MEL, in combination with IAA, played a functional role in hypocotyl growth to cell expansion in tissues. They stimulated the formation of root primordia from pericyclic cells, resulting in a modification in adventitious or lateral root spreading pattern and increasing the number and length of adventitious roots and number of lateral roots (Arnao and Hernández-Ruiz, 2006, 2018). It was stated that MEL formed the maximum number of hypocotyls with analogous values of IAA for root length at an entire range of concentrations, and the inhibitory analogous effects of MEL at high concentrations showed an active elevation of growth and development probably due to auxin-induced ethylene biosynthesis in monocotyledons, for example, wheat, oat, and canary grass. The optimum level of growth enhancement attained in coleoptiles with IAA was 100%, and the optimum growth-elevation effect of MEL was 10% for oat, 20% for wheat, 32% for canary grass, and 55% for barley coleoptiles (Hernández-Ruiz et al., 2005). In the same way, the root growth ranged between 56%

to 86% for canary grass and wheat, respectively, with the growth-enhancing effect of MEL in lupin tissues up to 63% in different bioassays (Hernández-Ruiz et al., 2004, 2005). MEL also promoted the extension of etiolated cotyledons of *Lupinus albus* L. to the same level which was noticed with IAA (Hernández-Ruiz and Arnao, 2008).

1.7 IN VITRO IMPACT OF MEL

The amount of MEL in *Glycyrrhiza uralensis* elevated the plant growth as it grew older, with values that were four times higher for a 6-month-older plant than for a 3-month-old plant (Afreen et al., 2006). It was also proposed that the inhibitory effect of MEL on root elongation and growth is irrelevant to the mechanism of IAA (Chen et al., 2009). High concentrations of serotonin and MEL was noted in thidiazuron (TDZ)-treated leaf explants with auxin-transported action suppressors; the augmentation of the TDZ medium with lidocaine, a sodium channel blocker, elevated the levels of serotonin and MEL but not auxin and considerably reduced the rate of TDZ-induced regeneration in the explants (Jones et al., 2007). Thus, the levels of serotonin and MEL as well as auxin elevated by exposure to TDZ stimulated the regeneration of explant, indicating that MEL acts either as a hormone autonomously or in combination with auxin and its own precursors and metabolites (Posmyk et al., 2008). It is also responsible for the protein kinase Ca^{2+} dependent inhibition, via the interaction of calmodulin—kinase being related to reorganizations of the cytoskeleton, which denotes some of the most primitive effects of MEL (Kolář and Macháčková, 2005; Hardeland, 2008). So it was reported that activation of calcium channels (that change polarity of cells) enhanced the levels of MEL, along with inhibition in TDZ-induced callus induction in *Echinacea purpurea* (L.) explants (Jones et al., 2007). This describes the in vitro growth-enhancing effects of MEL at low concentrations and inhibitory effects at higher concentrations (Paredes et al., 2009).

1.8 ROLE OF MEL IN PLANT PHYSIOLOGY

MEL can easily cross the cell and nuclear membrane due to its amphiphilic nature through receptor-independent actions (Tarocco et al., 2019). MEL was absorbed through epidermal cells and transported through a long distance via the vascular system in plant tissues (Erland et al., 2022; Li et al., 2017a). Thus, MEL exerts an impact on physiological processes throughout the tissues at cellular and molecular levels (Zheng et al., 2017b; Tan et al., 2015; Fleta-Soriano et al., 2017). MEL interferes with numerous physiological responses through the Ca^{2+}-calmodulin mechanism of signalling, considering the Ca^{2+}-dependent action of auxin. MEL has a significant impact on the cytoskeleton in plants, and it has a great affinity for calmodulin (Hardeland, 1997). MEL-treated plants under stress conditions showed a lower ROS burst, electrolyte leakage, and cell damage and higher plant height/weight, high levels of amino acids, organic acids, sugars, and sugar alcohols, thus affecting the carbohydrate and nitrogen metabolism, mainly solutes involved in the osmotic-stress response (Shi et al., 2014; Samanta et al., 2022). A similar type of activity was also noticed in *Citrus aurantium* seedlings (Kostopoulou et al., 2015); roots, seedlings, and cotyledons of sunflower (Mukherjee et al., 2014); cucumber seeds (Zhang et al., 2014); and *Chara australis* cell cultures (Beilby et al., 2015).

1.8.1 Impact on Photosynthesis and Senescence

MEL has an impact on photosynthesis and senescence suspension (Arnao and Hernández-Ruiz, 2018; Sadak et al., 2020). Higher levels of MEL delay leaf senescence, prevent chlorophyll degradation, and also increase photosynthetic activity by maintaining redox homeostasis levels (Sharif et al., 2018; Nawaz et al., 2020). MEL can detoxify free radicals which are generated by photosynthesis (Galano and Reiter, 2018). It enhances the efficiency of the electron transport chain in mitochondria, and thus electron leakage is reduced (Reiter et al., 2007; Zhao et al., 2019). MEL provokes higher nitrogen and chlorophyll content, a higher ability of photosynthesis, and higher content of soluble proteins and RuBisCo. It also increases the up-regulation of photosynthetic protein during leaf senescence, can uptake the regulation of three chloroplast ATP synthases enzymes, and can improve the efficiency of photosystem I and photosystem II during light and dark cycles (Arnao and Hernández-Ruiz, 2015). According to Parida and Das (2005), salt stress has a severe negative influence on plant life throughout the life cycle, from seed germination through the process of senescence. However, 0.1 M MEL significantly reduced this stress, demonstrating the importance of MEL in preserving plant photosynthetic capacity (Arnao and Hernández-Ruiz, 2014). It led to a suspension in the senescence process by restricting the chlorophyll-degrading enzyme, chlorophyllase; pheophorbide, which are oxygenase; or red chlorophyll catabolite reductase (Murch et al., 2001). MEL helps in the up-regulation of three chloroplastic ATP synthases, two RuBisCo small subunits, one RuBisCo-interacting protein, and one photosystem I reaction centre subunit which correlates with better photosynthetic activity and down-regulation of PaO (related to chlorophyll degradation), one senescence dehydration-associated protein, four dehydrins, and two heat-shock proteins (related to abiotic stress), two ά-amylases, one pullulanase (a debranching starch enzyme), two ά-glucans, and one ά-glucano-transferase. All of these are related to the breakdown of starch granules, thus reducing photosynthesis and starch degradation during leaf senescence. MEL also down regulated several proteins involved in protein folding and post-translational modifications, such as some MAP kinases, serine/threonine, and protein kinases/phosphatases, many of which are involved in senescence and stress responses (Wang et al., 2014).

1.9 IMPACT ON THE FIELD AND HORTICULTURAL CROPS

Higher concentrations of MEL protect wheat, barley, sweet, cherry and many other crops by preventing damage in the photosynthetic apparatus (Sharma and Zheng, 2019; Wen et al., 2016). A reduction in CO_2 concentration was observed through spraying of 25–100 µM MEL in cucumber seedlings under both normal and heat stress conditions leading to increased photosynthesis (Xu, 2010; Wu et al., 2021). Higher photosynthetic rates, stomatal conductance, and transpiration rates in MEL-treated maize seedlings were noted after foliar spraying of MEL during drought stress (Ye et al., 2016). Arnao and Hernández-Ruiz (2009) reported the inhibitory effect of exogenous MEL on increased senescence in barley leaves, thus delaying the loss of chlorophyll. This effect of MEL was in contrast with the inductive effect of the hormone ABA and the retardant effect of kinetin (a synthetic cytokinin with plant hormone activity) on foliar senescence. This result was confirmed in apple (Wang et al., 2012, 2013a), *Arabidopsis* (Weeda et al., 2014), cucumber (Zhang et al., 2013), rice (Byeon et al., 2012, 2013), and cherry (Sarropoulou et al., 2012b). Exogenous application of MEL delayed dark-induced senescence in apple leaves (*Malus domestica*) through increasing ROS scavenging enzyme activities, which are involved in the elimination of the excess H_2O_2 generated in stressed leaves while maintaining the higher levels of ascorbic acid and glutathione content (Wang et al., 2012). In *Arabidopsis*, a direct relationship between MEL and natural leaf senescence was observed through down-regulation of ubiquitin-mediated degradation of auxin/indole-3-acetic acid (AUX/IAA) protein, the auxin resistant 3/indole-3-acetic acid inducible 17 by exogenous MEL (Shi et al., 2015a). MEL participated in the better efficiency of photosystem II under dark and light conditions in apple trees, alleviated the inhibition in photosynthesis caused by drought stress, and also allowed the leaves to maintain a higher capacity for CO_2 assimilation and stomatal conductance (Wang et al., 2013b). Exogenous application of MEL at low concentrations took part in plant stress metabolism by slightly enhancing the content of photosynthetic pigments, total biomass, and total carbohydrates and reducing the proline content of roots in shoot tip explants of cherry rootstock PHL-C, *Prunus avium*, and *Prunus cerasus* (Sarropoulou et al., 2012b).

1.10 IMPACT ON TRANSGENIC CROPS

The effects of MEL on increased photosynthetic activity of transgenic horticultural crops were also evaluated (Zheng et al., 2017b; Byeon and Back, 2014; Meloni et al., 2003). Increased production of endogenous MEL in the chloroplasts of transgenic lines of *A. thaliana* exhibited an improved redox state with lower ROS levels under salt-stressed conditions, resulting in the transgenic plants being more tolerant to salt stress due to enhanced photosynthetic efficiency and higher biomass of the transgenic plants under salt-stressed conditions (Zheng et al., 2017a). MEL-treated transgenic Bermuda grass exhibited increased photochemical efficiency of PSII and total chlorophyll content by enhancing the biosynthesis of chlorophyll and decreasing the rate of its decomposition by up-regulating the expression of photosynthesis-related genes and thus protecting PSII and ameliorating the decrease of chlorophyll content under salt stress condition (Shi et al., 2014).

1.11 IMPACT ON CHLOROPHYLL PRESERVATION

Environmental stress factors such as temperature and UV radiation damage chlorophyll by generating ROS in plants and triggering damage in reaction centre D1 protein binding (Yamamoto et al., 2008). Leaf senescence also causes chlorophyll damage and plant death by changing plant hormone levels, damaging molecular integrity, and degenerating the cell wall. Being an antioxidant, MEL protects chlorophyll against degradation. It also protects plants from senescence (loss of chlorophyll) which occurs due to ROS/RNS and regulates SAGs. Twofold higher chlorophyll content was noted in 1 mM MEL treated barley leaves for 48 h (Arnao and Hernández-Ruiz, 2009). Indoleamine characteristics of MEL not only preserve chlorophyll content by radical scavenging activity through exogenous application (Tal et al., 2011) under stress condition but also enhances the efficiency of photosystem II and induces the levels of ascorbic acid (Wang et al., 2012). The senescence condition of leaves was delayed under abiotic stress conditions in both monocots and dicots due to MEL (Arnao and Hernández-Ruiz, 2019b).

1.12 IMPACT ON DORMANCY

MEL helps in the regulation of growing stages that may lead to the conservation of dormancy germs and in the preservation of different states in fruit tissues (Murch et al., 2001). Hardeland and Rodríguez (1995) and Balzer and Hardeland (1996) hypothesized the possible role of MEL as a dormancy maintenance factor in seeds due to high concentration.

1.13 IMPACT OF PROTECTIVE MECHANISM OF MEL

MEL significantly induces gene expression of plant hormones like auxin, ABA, and cytokinin (Arnao and Hernández-Ruiz, 2018) which have a protective mechanism against heavy metal toxicity in plants. ABA retards the uptake of toxic metals via their translocation from root to shoot through directing stomatal closure (Bücker-Neto et al., 2017). MEL is also responsible for plant protection from free radical damage (Reiter and Tan, 2000; Reiter et al., 2001, 2002; Tan et al., 2000, 2003; Rodriguez et al., 2004; Cano et al., 2003). Dubbels et al. (1995) reported the different MEL content of two cultivars of *Nicotiana tabacum* and their susceptibility to ozone. Manchester et al. (2000) proposed a germ tissue protection role for MEL while analyzing 15 seeds of edible plants. Murch and Saxena (2002b) studied the possible protective role of MEL during the flower development stage of *Hypericum perforatum* L. Serotonin and MEL present at higher concentrations throughout the flower development stage and the higher MEL level coincided with the maximum regeneration of potentially isolated anthers. MEL also acts as a membrane integrity factor (in nuclear and plasmatic membranes). The effect of MEL on polyamine levels may be related to some photoperiodic events. Application of MEL has been found to exert a protective role and defense mechanisms against cold, heat, drought, heavy metal, chemicals, and ultraviolet radiation stress (Arnao and Hernández-Ruiz, 2015; Raza, 2020). MEL can curb negative impacts due to salinity stress and improves plant growth and seed germination (Nawaz et al., 2016). It is also responsible for the enhancement of tolerance levels to Fe deficiency and salinity stress by increasing active Fe, K^+, endogenous NO, and H_2S as well as activities of antioxidant enzymes along with reduced levels of H_2O_2 and MDA (Kaya et al., 2020; Ahammed et al., 2020). Under arsenic-induced stress, MEL works synergistically with Ca^{2+} to promote resilience through reactive oxygen species (ROS) metabolism (Siddiqui et al., 2020). It was reported that MEL-induced stomatal closure was mediated by its receptor CAND2/PMTR1 which activates the G protein α subunit-regulated H_2O_2 and Ca^{2+} signal transduction cascade in *Arabidopsis thaliana* (Wei et al., 2018).

1.14 IMPACT OF MEL AS AN ANTIOXIDANT

MEL has a scavenging capacity (antioxidant activity) against biological free radicals, such as reactive oxygen and nitrogen species, including the hydroxyl radical, singlet oxygen, peroxyl radical, hydrogen peroxide, peroxynitrite anion, and nitric oxide. It also exhibits pleiotropic nature, which can also act as an antioxidant in the control of reactive oxygen and nitrogen species (ROS, RNS) in plants (Fan et al., 2018; Tan et al., 2003). MEL molecule exerts no pro-oxidative effects, while MEL-intermediate products have antioxidant properties and an important synergistic action with ascorbic acid, glutathione, etc. (Reiter and Tan, 2000; Reiter et al., 2001, 2002; Tan et al., 2000, 2003; Rodriguez et al., 2004; Cano et al., 2003). MEL and its metabolites also play a role in free radical scavenging which directly scavenges ROS and enhances mitochondrial oxidative phosphorylation efficiency. It also increases the activities of antioxidant enzymes like SOD, POD, CAT, and GR and up-regulates the anti-stress-related genes (Zhang et al., 2013; Asif et al., 2019). Being an antioxidant, MEL has a highly efficient scavenging capability of free radicals against hazardous and deleterious reactive molecules like nitric oxide, hydrogen peroxide, singlet oxygen, and hypochlorous acid (Nawaz et al., 2020). The role of MEL as an antioxidant against the oxidation of lipid vesicles of *Gonyaulax polyedra* cysts during photo-periodically induced encystment has also been proposed by Fuhrberg et al. (1995, 1997).

1.15 IMPACT ON CATABOLISM AND BIOLOGICAL SYNTHESIS

MEL is an independent plant growth regulator that can regulate the biosynthesis of its own and other natural and synthetic plant growth regulators, viz., auxin, abscisic acid, gibberellins, cytokinins, ethylene, polyamines, jasmonic acid, and salicylic acid. It is a modulator of gene expression of plant growth regulators and can also influence their activities (Nawaz et al., 2016, 2020). MEL helps in catabolism and biological synthesis of gibberellic acid and abscisic acid and also in down-regulation of ABA biosynthetic genes and up-regulation of ABA catabolism genes and biosynthetic genes of GAs in early germination stages, resulting in better plant growth and seed germination during initial crop growth stages (Zhang et al., 2014). MEL is efficient in inducing metabolic activity due to the over-expression of genes involved in hormones, nitrogen and secondary metabolism, redox, transport, and transformation of the tricarboxylic acid (Nawaz et al., 2020). MEL up regulates ribosome biosynthesis, lipid and carbohydrate metabolism, as well as storage protein degradation during seed germination under salt stress (Zhang et al., 2017a). Exogenous application of MEL altered glucosinolate profiles and contents by stimulating the expression of the primary metabolism genes for S and N assimilation, as well as the secondary metabolism genes for glucosinolate biosynthesis, thereby contributing to *Sclerotinia sclerotiorum* resistance in *Brassica rapa* (Teng et al., 2021, 2022). MEL up regulates the expression of genes related to photosynthesis, carbohydrate/fatty acid metabolism, and ascorbate biosynthesis and also up regulates the genes including two subunits of photosystem I (PsaK and PsaG), two elements (PsbO and PsbP) related to the oxygen-evolving complex of photosystem II (oxygen-evolving enhancer proteins), the ferredoxin gene PetF, and the VTC4 gene, which encodes the L-galactose 1-P phosphatase involved in ascorbate biosynthesis (Wei et al., 2015).

1.16 ROLE OF MEL IN CROPS IN OVERCOMING MINERAL STRESS

Nitrogen is one of the most important macronutrients for optimum growth and development of plants so that the application of nitrogenous fertilizer significantly influences the yield and quality of the crop (Mader et al., 2011; Mehmood et al., 2021). Apart from nitrogen, other macronutrients like phosphorous, potassium, and micronutrients like boron, molybdenum, etc. also play a vital role in plant growth and development. In recent years, applications of chemical nitrogenous fertilizers have been increased drastically to meet the growing demand of the vast population throughout the world (Signor et al., 2013; Qiao et al., 2019). Besides increasing yield and productivity, enhanced use of nitrogenous fertilizer reduced nitrogen use efficiency of plants, and excessive use of chemical or synthetic nitrogenous fertilizers resulted in adverse effects on soil health, environmental pollution, particularly water pollution that causes eutrophication, enhanced nitrate content in the groundwater level, and higher emission of greenhouse gases like nitrous oxide (Qiao et al., 2019). Therefore, optimum utilization of organic or sustainable sources of nutrients is an urgent necessity under these consequences which may contribute in a better way to develop sustainable agriculture.

MEL (*N-acetyl*-5-methoxy tryptamine) as a natural bio-stimulant and plant growth regulator, besides being active participation in several physiological actions like seed germination, photosynthesis, flowering, root development, ripening, has also been found to be actively involved in several plant protective actions under different abiotic stress conditions such as cold, drought, salinity, heat, UV radiation, etc. and plant nutrient uptake under nutrient deficit environment (Arnao and Hernández-Ruiz, 2018; Qiao et al., 2019).

Research revealed that MEL may directly or indirectly stimulate nutrient uptake metabolism under nutrient deficiency conditions. MEL may control the ROS signal and thereby stimulate the CBL1-CIPK23 pathway to control the expression of a gene involved in the potassium channel that actually triggers potassium absorption under salinity or nutrient-deficient conditions in *Malus* (Li et al., 2016). MEL may also reduce the noxious effect of excess boron contents in *Capsicum annuum* (Sarafi et al., 2017). MEL application in root rhizosphere of alfalfa could stimulate nitrogen metabolism and proline homeostasis under drought-stressed conditions (Antoniou et al., 2017). It was also found that MEL may control the horizontal root development which actually increases the root/shoot ratio (Arnao and Hernández-Ruiz, 2006; Chen et al., 2018) and thereby develop a vigorous root system that would enhance the nutrient uptake. Under drought and salinity stress conditions, MEL alleviates the effects of stresses or hindrance of nutrient uptake metabolism that usually occurr under stressed conditions. MEL application was also found to enhance the yield of the crops in some experimental studies which may be attributed to the enhancement of nutrient uptake metabolism in the plants because of the application of MEL (Qiao et al., 2019).

Several research studies revealed that MEL may regulate nitrogen metabolism in some plant species. Exogenous application of MEL under high nitrate stress conditions was found to increase high nitrate stress tolerance in cucumber seedlings due to enhanced enzymatic activities essential for nitrogen metabolism (Zhang et al., 2017b). Exogenous MEL application under sub-low temperature was also found to reduce ammonium nitrogen content (NH_4^+-N) of melon seedlings by stimulating enzymatic activities that are involved in nitrogen metabolism such as glutamine synthetase and glutamic acid synthase (Gao et al., 2016). MEL was found to sustain nitro-oxidative homeostasis by regulating enzymatic activities of nitrogen metabolism in alfalfa plants under drought-stressed conditions (Antoniou et al., 2016). Still, the role of MEL on plant nitrogen uptake and its metabolism inside the plant bodies remains undefined.

Previous research work on exogenous application of MEL in wheat under nitrogen deficit conditions revealed that MEL enhanced the enzymatic activities of glutamine synthetase and glutamic acid synthase which are directly involved in nitrogen metabolism, and the root/shoot ratio was also increased due to application of MEL, which ultimately enhanced nitrogen absorption, nitrogen metabolism, and its assimilation (Qiao et al., 2019). Wang et al. (2021) suggested that exogenous

MEL application under nitrogen deficit conditions could enhance nitrogen metabolism in soybean seedlings by regulating the enzymatic reactions involved in nitrogen metabolism which increased total nitrogen content in plants.

Sulphur is one of the macronutrients essential for optimum growth and development of crop plants. Sulphur deficiency induces oxidative stress that leads to changes in cellular structure and may cause damage to chloroplast DNA molecules that inhibit photosynthesis so that chlorophyll content is also reduced. However, the research studies as explained by Hassan et al. (2017) indicated that application of MEL in sulphur-deficient tomato reduced accumulation of ROS and thus sulphur-deficiency-induced symptoms were alleviated (Zhou, 2018). Furthermore, MEL encouraged sulphur uptake by plants and its absorption inside the plant cells by regulating the genes involved in enzymatic reactions of sulphur transportation and its metabolism. MEL also decreased ROS-induced damage by up-regulation of 2-cysteine peroxiredoxin and sulphur biosynthetic compounds. These findings suggested that MEL can increase the stress tolerance of tomato plants under sulphur-deficiency-induced stress conditions by stimulating sulphur uptake metabolism and redox homeostasis (Hassan et al., 2020).

The environmental conditions which favour plant growth and development greatly influence the melatonin content of the plants. Plants exposed to sunlight have more melatonin than those exposed to artificial light (Tan et al., 2007; Arnao and Hernández-Ruiz, 2013).

Drought, which alters physiological, biochemical, and molecular processes in plants, is also a major stress (Sharma and Zheng, 2019; Ahmed et al., 2021; Sehar et al., 2021). Plants grown in arid and semi-arid regions face more intense and longer periods of drought (Shao et al., 2008). Adaptive agricultural strategies are promoted in these changing environmental conditions because numerous levels of sensitivity to water stress are shown by different organs (Ahmed et al., 2021). Application of melatonin in roots causes a reduction in the leaf senescence, along with significant improvement in the efficiency of PSII and enhancement in the antioxidative eenzymes that led authors to propose the use of melatonin for agricultural purposes (Wang et al., 2012, 2014). During drought stress, two genes controlling phenylalanine ammonia-lyase (PAL) expression were shown to be down regulated, but up regulated after treatment with exogenous melatonin (Shi et al., 2016; Wan et al., 2018).

Salt stress also significantly limits crop growth and production and threatens global food production. It mainly controls ionic and nutritional imbalance, osmotic stress, and ROS, resulting in a significant loss in plant growth (Abbasi et al., 2016; Dustgeer et al., 2021; Sultan et al., 2021; Seleiman et al., 2022). Melatonin application considerably improves the effectiveness of PS-II for photochemical and non-photochemical quenching, which favours increased photosynthetic efficacy under salt stress (Li et al., 2017b). The role of melatonin in reducing the adverse effects of salt stress in plants is through the effect on the salt-tolerant genes. In rice seedlings, melatonin treatment induced salt tolerance by delaying apoptosis and cellular senescence, reducing chlorophyll degradation, down regulating senescence-related genes, enhancing antioxidant enzyme activity, and decreasing H_2O_2 accumulation.

Crop senescence in extreme environment decreases crop productivity and may even cause death accelerated by heat stress (Raza et al., 2021; Giordano et al., 2021; Farooq et al., 2022; Haider et al., 2022). Melatonin plays an essential role in maintaining many physio-biochemical and molecular processes to cope with heat stress conditions and acts as a plant growth regulator. Cold stress also has negative impacts on plants and considerably limits crop growth and development (Mishra et al., 2011). It induces substantial changes in physiological, molecular, and metabolic activity, altering the membrane permeability and antioxidant activity (Bajwa et al., 2014; Hu et al., 2016) (Table 1.2). The cold-responsive genes (COR15a) and antioxidant genes (ZAT10 and ZAT12), which increases cold tolerance, are up regulated through the application of melatonin (Bajwa et al., 2014).

Another severe environmental problem for all organisms and particularly in plants is heavy metal contamination. However, the application of melatonin has a beneficial effect on plant tolerance to

TABLE 1.2
Selected Examples of the Melatonin Application-Induced Enhancement in Photosynthesis and Associated Traits under Abiotic Stress Conditions

Stress	Plant species	Response of photosynthesis and associated traits in melatonin-treated plants	References
Salinity	*Glycine max*	Increased chlorophyll synthesis and PSII activity	Alharbi et al. (2021)
Drought	*Zea mays*	Increased photosynthetic activity	Ahmad et al. (2019)
Cold	*Pistacia vera*	Reduced chlorophyll degradation and increased photosynthesis	Barand et al. (2020)
High temperature	*Oryza sativa*	Increased photosynthetic activity and PSII efficiency	Han et al. (2017)

different metal stresses. Cucumber seedlings pre-treated with exogenous melatonin exhibited an improved endogenous melatonin concentration and also showed a positive response in alleviating nitrate-induced growth retardation (Zhang et al., 2017b).

Ten free radicals are potentially scavenged by one molecule of melatonin, which is higher than that for antioxidant enzymes at one radical per molecule (Ye et al., 2016). Therefore, melatonin was initially thought to be an antioxidant molecule that alleviated oxidative stress by scavenging ROS and reactive nitrogen species (RNS) species and providing protection to plants (Weeda et al., 2014). Melatonin affects the expression of several genes involved in plant stress defense. For example, chlorophyll degradation or delayed leaf senescence in apple plants is prevented by melatonin through down regulating chlorophyllase (*CLH1*) enzyme activity, an enzyme involved in chlorophyll breakdown (Weeda et al., 2014).

1.17 CONCLUSION

MEL controls several physiological processes in plants, such as fruit ripening and taste, shoot and root development, blooming period, and circadian rhythm. It enhances plant growth and development while augmenting the functions of mineral nutrition under various stress circumstances. MEL is always present in plants. Even though a plant homologue of the classic arylalkylamine N-acetyltransferase (AANAT) has not yet been found, evidence has shown that a large amount of MEL is produced by plants themselves. As a result, the sequence and structure of the serotonin N-acetylating enzyme in plants may be very different from that of animal AANAT. It is believed that plants, which cannot move as mammals do, have significantly greater amounts of MEL as a way of compensating for the harsh surroundings that they must endure. The efficacy of MEL to boost agricultural output is also supported by evidence. Increased MEL levels in transgenic plants may result in innovations that boost agricultural crop output and benefit humans in general.

REFERENCES

Abbasi, H., Jamil, M., Haq, A., Ali, S., Ahmad, R., and Malik, Z. (2016). Salt stress manifestation on plants, mechanism of salt tolerance and potassium role in alleviating it: A review. *Zemdirbyste Agriculture*, *103*, 229–238.

Abd El-Naby, S. K. M., Mohamed, A. A. A., and El-Naggar, Y. I. M. (2019). Effect of melatonin, GA3 and NAA on vegetative growth, yield and quality of 'canino' apricot fruits. *Scientiarum Polonorum Hortorum Cultus*, *18*, 167–174.

Afreen, F., Zobayed, S. M., and Kozai, T. (2006). Melatonin in glycyrrhizauralensis: Response of plant roots to spectral quality of light and UV-B radiation. *Journal of Pineal Research*, *41*(2), 108–115.

Ahammed, G. J., Wu, M., Wang, Y., Yan, Y., Mao, Q., Ren, J., and Chen, S. (2020). Melatonin alleviates iron stress by improving iron homeostasis, antioxidant defense and secondary metabolism in cucumber. *Scientia Horticulturae*, *265*, 109205.

Ahmad, S., Kamran, M., Ding, R., Meng, X., Wang, H., Ahmad, I., Fahad, S., and Han, Q. (2019). Exogenous melatonin confers drought stress by promoting plant growth, photosynthetic capacity and antioxidant defense system of maize seedlings. *Peer J*, *7*, e7793.

Ahmed, S., Kouser, S., Asgher, M., and Gandhi, S. G. (2021). Plant aquaporins: A frontward to make crop plants drought resistant. *Physiologia Plantarum*, *172*, 1089–1105.

Alharbi, B. M., Elhakem, A. H., Alnusairi, G. S., Soliman, M. H., Hakeem, K. R., Hasan, M. M., and Abdelhamid, M. T. (2021). Exogenous application of melatonin alleviates salt stress-induced decline in growth and photosynthesis in *Glycine max* (L.) seedlings by improving mineral uptake, antioxidant and glyoxalase system. *Plant, Soil and Environment*, *67*, 208–220.

Antoniou, C., Chatzimichail, G., Xenofontos, R., Pavlou, J. J., Panagiotou, E., Christou, A., and Fotopoulos, V. (2017). Melatonin systemically ameliorates drought stress-induced damage in *Medicago sativa* plants by modulating nitro-oxidative homeostasis and proline metabolism. *Journal of Pineal Research*, *62*(4), e12401.

Antoniou, C., Savvides, A., Christou, A., and Fotopoulos, V. (2016). Unravelling chemical priming machinery in plants: The role of reactive oxygen-nitrogen-sulfur species in abiotic stress tolerance enhancement. *Current Opinion in Plant Biology*, *33*, 101–107.

Arnao, M. B., and Hernández-Ruiz, J. (2006). The physiological function of MEL in plants. *Plant Signaling & Behavior*, *1*(3), 89–95.

Arnao, M. B., and Hernández-Ruiz, J. (2007). Inhibition of ACC oxidase activity by MEL and indole-3-acetic acid in etiolated lupin hypocotyls. *Advances in Plant Ethylene Research* (pp. 101–103). Springer, Dordrecht.

Arnao, M. B., and Hernández-Ruiz, J. (2009). Protective effect of melatonin against chlorophyll degradation during the senescence of barley leaves. *Journal of Pineal Research*, *46*(1), 58–63.

Arnao, M. B., and Hernández-Ruiz, J. (2013). Growth conditions determine different melatonin levels in *Lupinus albus* L. *Journal of Pineal Research*, *55*, 149–155.

Arnao, M. B., and Hernández-Ruiz, J. (2014). Melatonin: Plant growth regulator and/or biostimulator during stress? *Trends in Plant Science*, *19*(12), 789–797.

Arnao, M. B., and Hernández-Ruiz, J. (2015). Phytomelatonin: Searching for plants with high levels for use as a natural nutraceutical. *Studies in Natural Products Chemistry*, *46*, 519–545.

Arnao, M. B., and Hernández-Ruiz, J. (2017). Growth activity, rooting capacity, and tropism: Three auxinic precepts fulfilled by melatonin. *Acta Physiologiae Plantarum*, *39*(6), 1–9.

Arnao, M. B., and Hernández-Ruiz, J. (2018). Melatonin and its relationship to plant hormones. *Annals of Botany*, *121*, 195–207.

Arnao, M. B., and Hernández-Ruiz, J. (2019a). Melatonin: A new plant hormone and/or a plant master regulator? *Trends in Plant Science*, *24*(1), 38–48.

Arnao, M. B., and Hernández-Ruiz, J. (2019b). Role of melatonin to enhance phytoremediation capacity. *Applied Sciences*, *9*(24), 5293.

Arnao, M. B., and Hernández-Ruiz, J. (2021). Melatonin as a regulatory hub of plant hormone levels and action in stress situations. *Plant Biology*, *23*, 7–19.

Asif, M., Pervez, A., and Ahmad, R. (2019). Role of melatonin and plant-growth-promoting rhizobacteria in the growth and development of plants. *Clean: Soil, Air, Water*, *47*(6), 1800459.

Asif, M., Pervez, A., Irshad, U., Mehmood, Q., and Ahmad, R. (2020). Melatonin and plant growth-promoting rhizobacteria alleviate the cadmium and arsenic stresses and increase the growth of *Spinacia oleracea* L. *Plant, Soil and Environment*, *66*(5), 234–241.

Back, K., Tan, D. X., and Reiter, R. J. (2016). Melatonin biosynthesis in plants: Multiple pathways catalyze tryptophan to melatonin in the cytoplasm orchloroplasts. *Journal of Pineal Research*, *61*, 426–437.

Bajwa, V. S., Shukla, M. R., Sherif, S. M., Murch, S. J., and Saxena, P. K. (2014). Role of melatonin in alleviating cold stress in *Arabidopsis thaliana*. *Journal of Pineal Research*, *56*, 238–245. doi:10.1111/jpi.12115.

Balzer, I., and Hardeland, R. (1996). Melatonin in algae and higher plants-possible new roles as a phytohormone and antioxidant. *Botanica Acta*, *109*(3), 180–183.

Barand, A., Nasibi, F., Kalantari, K. M., and Moradi, M. (2020). The effects of foliar application of melatonin on some physiological and biochemical characteristics and expression of fatty acid desaturase gene in pistachio seedlings (*Pistacia vera* L.) under freezing stress. *Journal of Plant Interactions*, *15*, 257–265.

Beilby, M. J., Al Khazaaly, S., and Bisson, M. A. (2015). Salinity-induced noise in membrane potential of characeae *Chara australis*: Effect of exogenous melatonin. *The Journal of Membrane Biology*, *248*(1), 93–102.

Brzezinski, A. (1997). Melatonin in humans. *New England Journal of Medicine*, *336*(3), 186–195.
Bücker-Neto, L., Paiva, A. L. S., Machado, R. D., Arenhart, R. A., and Margis-Pinheiro, M. (2017). Interactions between plant hormones and heavy metals responses. *Genetics and Molecular Biology*, *40*, 373–386.
Byeon, Y., and Back, K. (2014). An increase in melatonin in transgenic rice causes pleiotropic phenotypes, including enhanced seedling growth, delayed flowering, and low grain yield. *Journal of Pineal Research*, *56*(4), 408–414.
Byeon, Y., Park, S., Kim, Y. S., and Back, K. (2013). Microarray analysis of genes differentially expressed in melatonin-rich transgenic rice expressing a sheep serotonin N-acetyltransferase. *Journal of Pineal Research*, *55*(4), 357–363.
Byeon, Y., Park, S., Kim, Y. S., Park, D. H., Lee, S., and Back, K. (2012). Light-regulated melatonin biosynthesis in rice during the senescence process in detached leaves. *Journal of Pineal Research*, *53*(1), 107–111.
Cano, A., Alcaraz, O., and Arnao, M. B. (2003). Free radical-scavenging activity of indolic compounds in aqueous and ethanolic media. *Analytical and Bioanalytical Chemistry*, *376*(1), 33–37.
Chan, Z., and Shi, H. (2015). Improved abiotic stress tolerance of bermudagrass by exogenous small molecules. *Plant Signaling & Behavior*, *10*(3), e991577. https://doi:10.4161/15592324.2014.991577
Chen, Q., Qi, W. B., Reiter, R. J., Wei, W., and Wang, B. M. (2009). Exogenously applied MEL stimulates root growth and raises endogenous indoleacetic acid in roots of etiolated seedlings of *Brassica juncea*. *Journal of Plant Physiology*, *166*(3), 324–328.
Chen, Z., Gu, Q., Yu, X., Huang, L., Xu, S., Wang, R., and Shen, W. (2018). Hydrogen peroxide acts downstream of MEL to induce lateral root formation. *Annals of Botany*, *121*(6), 1127–1136.
Dawood, M. G. (2017). Physiological effect of MEL, IAA and their precursor on quality and quantity of chickpea plants grown under sandy soil conditions. *Agricultural Engineering International: CIGR Journal*, 35–44.
Dubbels, R., Reiter, R. J., Klenke, E., Goebel, A., Schnakenberg, E., Ehlers, C., and Schloot, W. (1995). Melatonin in edible plants identified by radioimmunoassay and by high performance liquid chromatography-mass spectrometry. *Journal of Pineal Research*, *18*(1), 28–31.
Dustgeer, Z., Seleiman, M. F., Imran, K., Chattha, M., Alhammad, B. A., and Jalal, R. S. (2021). Glycine-betaine induced salinity tolerance in maize by regulating the physiological attributes, antioxidant defense system and ionic homeostasis. *Notulae Botanicae Horti Agrobotanici Cluj-Napoca*, *49*, 12248. doi:10.15835/nbha49112248.
Erland, L. A., Dumigan, C. R., Forsyth, J. A., Frolova, L., Yasunaga, A. B., Pun, W., and Murch, S. J. (2022). Mammalian MEL agonist pharmaceuticals stimulate rhomboid proteins in plants. *Biomolecules*, *12*(7), 882.
Fan, J., Xie, Y., Zhang, Z., and Chen, L. (2018). Melatonin: A multifunctional factor in plants. *International Journal of Molecular Sciences*, *19*(5), 1528.
Farooq, M. S., Uzaiir, M., Raza, A., Habib, M., Xu, Y., Yousuf, M., Yang, S. H., and Ramzan Khan, M. (2022). Uncovering the research gaps to alleviate the negative impacts of climate change on food security: A review. *Frontiers in Plant Science*, *13*, 927535. doi:10.3389/fpls.2022.927535.
Feng, X., Wang, M., Zhao, Y., Han, P., and Dai, Y. (2014). Melatonin from different fruit sources, functional roles, and analytical methods. *Trends in Food Science & Technology*, *37*(1), 21–31.
Fleta-Soriano, E., Díaz, L., Bonet, E., and Munné-Bosch, S. (2017). Melatonin may exert a protective role against drought stress in maize. *Journal of Agronomy and Crop Science*, *203*(4), 286–294.
Fuhrberg, B., Hardeland, R., Poeggeler, B., and Behrmann, C. (1997). Dramatic rises of melatonin and 5-methoxytryptamine in gonyaulax exposed to decreased temperature. *Biological Rhythm Research*, *28*(1), 144–150.
Fuhrberg, B., Hardeland, R., Poeggeler, B., and Behrmann, G. (1995). Melatonin rises dramatically in *Gonyaulax polyedra* exposed to decreased temperature: Implications for photoperiodic cyst induction. *Biological Rhythm Research*, *26*, 391–396.
Galano, A., and Reiter, R. J. (2018). Melatonin and its metabolites vs oxidative stress: From individual actions to collective protection. *Journal of Pineal Research*, *65*(1), e12514.
Gao, Q. H., Jia, S. S., Miao, Y. M., Lu, X. M., and Li, H. M. (2016). Effects of exogenous melatonin on nitrogen metabolism and osmotic adjustment substances of melon seedlings under sub-low temperature. *Ying Yong Sheng taixuebao= The Journal of Applied Ecology*, *27*(2), 519–524.
Giordano, M., Petropoulos, S. A., and Rouphael, Y. (2021). Response and defence mechanisms of vegetable crops against drought, heat and salinity stress. *Agriculture*, *11*(5), 463. doi:10.3390/agriculture11050463.

Hager, A. (2003). Role of the plasma membrane H+-ATPase in auxin-induced elongation growth: Historical and new aspects. *Journal of Plant Research*, *116*(6), 483–505.

Haider, S., Raza, A., Iqbal, J., Shaukat, M., and Mahmood, T. (2022). Analyzing the regulatory role of heat shock transcription factors in plant heat stress tolerance: A brief appraisal. *Molecular Biology Reports*, 1–15. doi:10.1007/s11033-022-07190-x.

Han, Q., Huang, B., and Ding, C. (2017). Effects of melatonin on anti-oxidative systems and photosystem II in cold-stressed rice seedlings. *Frontiers in Plant Science*, *8*, 785.

Hardeland, R. (1997). New actions of MEL and their relevance to biometeorology. *International Journal of Biometeorology*, *41*(2), 47–57.

Hardeland, R. (2008). Melatonin, hormone of darkness and more: Occurrence, control mechanisms, actions and bioactive metabolites. *Cellular and Molecular Life Sciences*, *65*(13), 2001–2018.

Hardeland, R., and Rodríguez, C. (1995). Minireview: Versatile MEL: A pervasive molecule serves various functions in signaling and protection. *Chronobiology International*, *12*(3), 157–165.

Hassan, M. U., Aamer, M., Chattha, M. U., Ullah, M. A., Sulaman, S., Nawaz, M., and Guoqin, H. (2017). The role of potassium in plants under drought stress: Mini review. *Journal of Basic and Applied Sciences*, *13*, 268–271. doi:10.6000/1927-5129.2017.13.44.

Hassan, M. U., Aamer, M., Umer Chattha, M., Haiying, T., Shahzad, B., and Barbanti, L. (2020). The critical role of zinc in plants facing the drought stress. *Agriculture*, *10*, 396.

Hattori, A., Migitaka, H., Iigo, M., Itoh, M., Yamamoto, K., Ohtani-Kaneko, R., Hara, M., Suzuki, T., and Reiter, R. J. (1995). Identification of melatonin in plants and its effects on plasma melatonin levels and binding to melatonin receptors in vertebrates. *Biochemistry and Molecular Biology International*, *35*(3), 627–634.

Hernández-Ruiz, J., and Arnao, M. B. (2008). Melatonin stimulates the expansion of etiolated lupin cotyledons. *Plant Growth Regulation*, *55*(1), 29–34.

Hernández-Ruiz, J., Cano, A., and Arnao, M. B. (2004). Melatonin: A growth-stimulating compound present in lupin tissues. *Planta*, *220*(1), 140–144.

Hernández-Ruiz, J., Cano, A., and Arnao, M. B. (2005). Melatonin acts as a growth-stimulating compound in some monocot species. *Journal of Pineal Research*, *39*(2), 137–142.

Hu, Z., Fan, J., Xie, Y., Amombo, E., Liu, A., Gitau, M. M., and Fu, J. (2016). Comparative photosynthetic and metabolic analyses reveal mechanism of improved cold stress tolerance in bermudagrass by exogenous melatonin. *Plant Physiology and Biochemistry*, *100*, 94–104. doi:10.1016/j.plaphy.2016.01.008.

Huang, X., and Mazza, G. (2011). Simultaneous analysis of serotonin, melatonin, piceid and resveratrol in fruits using liquid chromatography tandem mass spectrometry. *Journal of Chromatography A*, *1218*(25), 3890–3899. https://doi.org/10.1016/j.chroma.2011.04.049.

Jones, M., Cao, J., O'Brien, R., Murch, S. J., and Saxena, P. K. (2007). The mode of action of thidiazuron: Auxins, indoleamines, and ion channels in the regeneration of *Echinacea purpurea* L. *Plant Cell Reports*, *26*(9), 1481–1490.

Kabiri, R., Hatami, A., Oloumi, H., Naghizadeh, M., Nasibi, F., and Tahmasebi, Z. (2018). Foliar application of melatonin induces tolerance to drought stress in moldavian balm plants (*Dracocephalum moldavica*) through regulating the antioxidant system. *Folia Horticultarae*, *30*, 155. doi:10.2478/fhort-2018-0016.

Kanwar, M. K., Yu, J., and Zhou, J. (2018). Phytomelatonin: Recent advances and future prospects. *Journal of Pineal Research*, *65*, e12526.

Kaya, C., Higgs, D., Ashraf, M., Alyemeni, M. N., and Ahmad, P. (2020). Integrative roles of nitric oxide and hydrogen sulfide in MEL-induced tolerance of pepper (*Capsicum annuum* L.) plants to iron deficiency and salt stress alone or in combination. *Physiologia Plantarum*, *168*(2), 256–277.

Kolář, J., Johnson, C. H., and Macháčková, I. (2003). Exogenously applied melatonin (N-acetyl-5-methoxytryptamine) affects flowering of the short-day plant *Chenopodium rubrum*. *Physiologia Plantarum*, *118*(4), 605–612.

Kolář, J., and Macháčková, I. (2005). Melatonin in higher plants: Occurrence and possible functions. *Journal of Pineal Research*, *39*(4), 333–341.

Kolář, J., Macháčková, I., Eder, J., Prinsen, E., Van Dongen, W., Van Onckelen, H., and Illnerová, H. (1997). Melatonin: Occurrence and daily rhythm in *Chenopodium rubrum*. *Phytochemistry*, *44*(8), 1407–1413.

Kolar, J., Machackova, I., Illnerova, H., Prinsen, E., Van Dongen, W., and Van Onckelen, H. (1995). Melatonin in higher plant determined by radioimmunoassay and liquid chromatography-mass spectrometry. *Biological Rhythm Research*, *26*, 406–409.

Kołodziejczyk, I., Bałabusta, M., Szewczyk, R., and Posmyk, M. (2015). The levels of melatonin and its metabolites in conditioned corn (*Zea mays* L.) and cucumber (*Cucumis sativus* L.) seeds during storage. *Acta Physiologiae Plantarum*, *37*, 105. https://doi.org/10.1007/s11738-015-1850-7.

Kostopoulou, Z., Therios, I., Roumeliotis, E., Kanellis, A. K., and Molassiotis, A. (2015). Melatonin combined with ascorbic acid provides salt adaptation in *Citrus aurantium* L. seedlings. *Plant Physiology and Biochemistry*, *86*, 155–165.

Koyama, F. C., Carvalho, T. L., Alves, E., Da Silva, H. B., De Azevedo, M. F., Hemerly, A. S., and Garcia, C. R. (2013). The structurally related auxin and MEL tryptophan-derivatives and their roles in *Arabidopsis thaliana* and in the human malaria parasite *Plasmodium falciparum*. *Journal of Eukaryotic Microbiology*, *60*(6), 646–651.

Lee, K., Choi, G. H., and Back, K. (2017). Cadmium-induced MEL synthesis in rice requires light, hydrogen peroxide, and nitric oxide: Key regulatory roles for tryptophan decarboxylase and caffeic acid o-methyltransferase. *Journal of Pineal Research*, *63*(4), e12441.

Lee, K., Lee, H. Y., and Back, K. (2018). Rice histone deacetylase 10 and arabidopsis histone deacetylase 14 genes encode N-acetylserotonin deacetylase, which catalyzes conversion of N-acetylserotonin into serotonin, a reverse reaction for melatonin biosynthesis in plants. *Journal of Pineal Research*, *64*(2), e12460. https://doi.org/10.1111/jpi.12460.

Li, C., Liang, B., Chang, C., Wei, Z., Zhou, S., and Ma, F. (2016). Exogenous melatonin improved potassium content in malus under different stress conditions. *Journal of Pineal Research*, *61*(2), 218–229.

Li, H., Chang, J., Chen, H., Wang, Z., Gu, X., and Wei, C. (2017a). Exogenous melatonin confers salt stress tolerance to watermelon by improving photosynthesis and redox homeostasis. *Frontiers in Plant Science*, *8*, 295. doi:10.3389/fpls. 2017.00295

Li, H., Chang, J., Zheng, J., Dong, Y., Liu, Q., Yang, X., and Zhang, X. (2017b). Local melatonin application induces cold tolerance in distant organs of *Citrullus lanatus* L. via long distance transport. *Scientific Reports*, *7*(1), 1–15.

Liu, J., Yue, R., Si, M., Wu, M., Cong, L., Zhai, R., and Xu, L. (2019). Effects of exogenous application of MEL on quality and sugar metabolism in 'Zaosu' pear fruit. *Journal of Plant Growth Regulation*, *38*(3), 1161–1169.

Liu, J., Zhang, R., Sun, Y., Liu, Z., Jin, W., and Sun, Y. (2016). The beneficial effects of exogenous MEL on tomato fruit properties. *Scientia Horticulturae*, *207*, 14–20.

Ljung, K., Hull, A. K., Kowalczyk, M., Marchant, A., Celenza, J., Cohen, J. D., and Sandberg, G. (2002). Biosynthesis, conjugation, catabolism and homeostasis of indole-3-acetic acid in *Arabidopsis thaliana*. *Plant Molecular Biology*, *49*(3), 249–272.

Macháčková, I., and Krekule, J. (2002). Sixty-five years of searching for the signals that trigger flowering. *Russian Journal of Plant Physiology*, *49*(4), 451–459.

Mader, P., Kaiser, F., Adholeya, A., Singh, R., Uppal, H. S., Sharma, A. K., and Fried, P. M. (2011). Inoculation of root microorganisms for sustainable wheat-rice and wheat-black gram rotations in India. *Soil Biology and Biochemistry*, *43*(3), 609–619.

Manchester, L. C., Tan, D. X., Reiter, R. J., Park, W., Monis, K., and Qi, W. (2000). High levels of MEL in the seeds of edible plants: Possible function in germ tissue protection. *Life Sciences*, *67*(25), 3023–3029. https://doi.org/10.1016/S0024-3205(00)00896-1.

Mannino, G., Gentile, C., Ertani, A., Serio, G., and Bertea, C. M. (2021). Anthocyanins: Biosynthesis, distribution, ecological role, and use of biostimulants to increase their content in plant foods: A review. *Agriculture*, *11*, 212.

Mehmood, M., Khan, I., Chattha, M. U., Hussain, S., Ahmad, N., and Aslam, M. T. (2021). Thiourea application protects maize from drought stress by regulating growth and physiological traits. *Pakistan Journal of Science*, *73*, 355–363.

Meloni, D. A., Oliva, M. A., Martinez, C. A., and Cambraia, J. (2003). Photosynthesis and activity of superoxide dismutase, peroxidase and glutathione reductase in cotton under salt stress. *Environmental and Experimental Botany*, *49*(1), 69–76.

Mir, A. R., Faizan, M., Bajguz, A., Sami, F., Siddiqui, H., and Hayat, S. (2020). Occurrence and biosynthesis of melatonin and its exogenous effect on plants. *Acta Societatis Botanicorum Poloniae*, *89*(2).

Mishra, A., Mishra, K. B., Höermiller, I. I., Heyer, A. G., and Nedbal, L. (2011). Chlorophyll fluorescence emission as a reporter on cold tolerance in *Arabidopsis thaliana* accessions. *Plant Signal*, *6*, 301–310. https:/doi:10.4161/psb.6.2.15278.

Mukherjee, S., David, A., Yadav, S., Baluška, F., and Bhatla, S. C. (2014). Salt stress-induced seedling growth inhibition coincides with differential distribution of serotonin and MEL in sunflower seedling roots and cotyledons. *Physiologia Plantarum*, *152*(4), 714–728.

Murch, S. J., Alan, A. R., Cao, J., and Saxena, P. K. (2009). Melatonin and serotonin in flowers and fruits of *Datura metel* L. *Journal of Pineal Research*, *47*, 277–283.

Murch, S. J., Campbell, S. S., and Saxena, P. K. (2001). The role of serotonin and melatonin in plant morphogenesis: Regulation of auxin-induced root organogenesis in in vitro-cultured explants of St. John's wort (*Hypericum perforatum* L.). *In Vitro Cellular & Developmental Biology-Plant*, *37*(6), 786–793.

Murch, S. J., Hall, B. A., and Le, C. H. (2010). Changes in the levels of indoleamine phytochemicals during vḗraison and ripening of wine grapes. *Journal of Pineal Research*, *49*(1), 95–100.

Murch, S. J., Krishna Raj, S., and Saxena, P. K. (2000). Tryptophan is a precursor for melatonin and serotonin biosynthesis in in vitro regenerated St. John's wort (*Hypericum perforatum* L. cv. Anthos) plants. *Plant Cell Reports*, *19*(7), 698–704.

Murch, S. J., Rupasinghe, H. P., Goodenowe, D., and Saxena, P. K. (2004). A metabolomic analysis of medicinal diversity in Huang-qin (*Scutellaria baicalensis* georgi) genotypes: Discovery of novel compounds. *Plant Cell Reports*, *23*(6), 419–425.

Murch, S. J., and Saxena, P. K. (2002a). Mammalian neurohormones: Potential significance in reproductive physiology of St. John's wort (*Hypericum perforatum* L.)? *Naturwissenschaften*, *89*(12), 555–560.

Murch, S. J., and Saxena, P. K. (2002b). Melatonin: A potential regulator of plant growth and development? *In Vitro Cellular & Developmental Biology-Plant*, *38*(6), 531–536.

Nawaz, K., Chaudhary, R., Sarwar, A., Ahmad, B., Gul, A., Hano, C., and Anjum, S. (2020). Melatonin as master regulator in plant growth, development and stress alleviator for sustainable agricultural production: Current status and future perspectives. *Sustainability*, *13*(1), 294. https://doi.org/10.3390/su13010294.

Nawaz, M. A., Huang, Y., Bie, Z., Ahmed, W., Reiter, R. J., Niu, M., and Hameed, S. (2016). Corrigendum: Melatonin: Current status and future perspectives in plant science. *Frontiers in Plant Science*, *7*, 714. https://doi:10.3389/fpls.2015.01230.

Paredes, S. D., Korkmaz, A., Manchester, L. C., Tan, D. X., and Reiter, R. J. (2009). Phytomelatonin: A review. *Journal of Experimental Botany*, *60*(1), 57–69.

Parida, A. K., and Das, A. B. (2005). Salt tolerance and salinity effects on plants: A review. *Ecotoxicology and Environmental Safety*, *60*(3), 324–349. https://doi.org/10.1016/j.ecoenv.2004.06.010.

Park, S., and Back, K. (2012). Melatonin promotes seminal root elongation and root growth in transgenic rice after germination. *Journal of Pineal Research*, *53*(4), 385–389.

Park, S., Lee, K., Kim, Y. S., and Back, K. (2012). Tryptamine 5-hydroxylasedeficient sekiguchi rice induces synthesis of 5-hydroxytryptophan and N-acetyltryptamine but decreases melatonin biosynthesis during senescence process of detached leaves. *Journal of Pineal Research*, *52*, 211–216.

Posmyk, M. M., Bałabusta, M., Wieczorek, M., Sliwinska, E., and Janas, K. M. (2009). Melatonin applied to cucumber (*Cucumis sativus* L.) seeds improves germination during chilling stress. *Journal of Pineal Research*, *46*(2), 214–223.

Posmyk, M. M., and Janas, K. M. (2009). Melatonin in plants. *Acta Physiologiae Plantarum*, *31*(1), 1–11.

Posmyk, M. M., Kuran, H., Marciniak, K., and Janas, K. M. (2008). Presowing seed treatment with MEL protects red cabbage seedlings against toxic copper ion concentrations. *Journal of Pineal Research*, *45*(1), 24–31.

Pothinuch, P., and Tongchitpakdee, S. (2011). Melatonin contents in mulberry (Morus spp.) leaves: Effects of sample preparation, cultivar, leaf age and tea processing. *Food Chemistry*, *128*(2), 415–419.

Qiao, Y., Yin, L., Wang, B., Ke, Q., Deng, X., and Wang, S. (2019). Melatonin promotes plant growth by increasing nitrogen uptake and assimilation under nitrogen deficient condition in winter wheat. *Plant Physiology and Biochemistry*, *139*, 342–349.

Raza, A. (2020). Metabolomics: A systems biology approach for enhancing heat stress tolerance in plants. *Plant Cell Reports*, 1–23. doi:10.1007/s00299-020-02635-8.

Raza, A., Tabassum, J., Kudapa, H., and Varshney, R. K. (2021). Can omics deliver temperature resilient ready-to-grow crops? *Critical Reviews in Biotechnology*, *41*(8), 1209–1232. doi:10.1080/07388551.2021.1898332.

Reiter, R. J., Burkhardt, S., Cabrera, J., and Garcia, J. J. (2002). Beneficial neurobiological effects of MEL under conditions of increased oxidative stress. *Current Medicinal Chemistry-Central Nervous System Agents*, *2*(1), 45–58.

Reiter, R. J., Richardson, B. A., Johnson, L. Y., Ferguson, B. N., and Dinh, D. T. (1980). Pineal melatonin rhythm: Reduction in aging syrian hamsters. *Science*, *210*(4476), 1372–1373.

Reiter, R. J., and Tan, D. X. (2000). Melatonin: Mechanisms and actions as an antioxidant. *Current Topics in Biophysics*, *24*, 171–183.

Reiter, R. J., Tan, D. X., Manchester, L. C., and Qi, W. (2001). Biochemical reactivity of melatonin with reactive oxygen and nitrogen species. *Cell Biochemistry and Biophysics*, *34*(2), 237–256.

Reiter, R. J., Tan, D. X., Manchester, L. C., Simopoulos, A. P., Maldonado, M. D., Flores, L. J., and Terron, M. P. (2007). Melatonin in edible plants (phytomelatonin): Identification, concentrations, bioavailability and proposed functions. *World Review of Nutrition and Dietetics*, *97*, 211–230.

Reiter, R. J., Tan, D. X., Zhou, Z., Cruz, M. H. C., Fuentes-Broto, L., and Galano, A. (2015). PhytoMEL: Assisting plants to survive and thrive. *Molecules*, *20*(4), 7396–7437.

Rodriguez, C., Mayo, J. C., Sainz, R. M., Antolín, I., Herrera, F., Martín, V., and Reiter, R. J. (2004). Regulation of antioxidant enzymes: A significant role for melatonin. *Journal of Pineal Research*, *36*(1), 1–9.

Sadak, M. S., Abdalla, A. M., Abd Elhamid, E. M., and Ezzo, M. I. (2020). Role of MEL in improving growth, yield quantity and quality of *Moringa oleifera* L. plant under drought stress. *Bulletin of the National Research Centre*, *44*(1), 1–13.

Sae-Teaw, M., Johns, J., Johns, N. P., and Subongkot, S. (2013). Serum melatonin levels and antioxidant capacities after consumption of pineapple, orange, or banana by healthy male volunteers. *Journal of Pineal Research*, *55*(1), 58–64. https://doi.org/10.1111/jpi.12025.

Samanta, S., Banerjee, A., and Roychoudhury, A. (2021). Melatonin application differentially modulates the enzymes associated with antioxidative machinery and ascorbate glutathione cycle during arsenate exposure in indica rice varieties. *Plant Biology*, *23*(S1), 193–201.

Samanta, S., Banerjee, A., and Roychoudhury, A. (2022). Arsenic toxicity is counteracted by exogenous application of melatonin to different extents in arsenic-susceptible and arsenic-tolerant rice cultivars. *Journal of Plant Growth Regulation*, *41*, 2210–2231.

Sarafi, E., Tsouvaltzis, P., Chatzissavvidis, C., Siomos, A., and Therios, I. (2017). Melatonin and resveratrol reverse the toxic effect of high boron (B) and modulate biochemical parameters in pepper plants (*Capsicum annuum* L.). *Plant Physiology and Biochemistry*, *112*, 173–182.

Sarropoulou, V. N., Dimassi-Theriou, K., Therios, I., and Koukourikou-Petridou, M. (2012a). Melatonin enhances root regeneration, photosynthetic pigments, biomass, total carbohydrates and proline content in the cherry rootstock PHL-C (Prunusaviumx Prunuscerasus). *Plant Physiology and Biochemistry*, *61*, 162–168.

Sarropoulou, V. N., Therios, I. N., and Dimassi-Theriou, K. N. (2012b). Melatonin promotes adventitious root regeneration in in vitro shoot tip explants of the commercial sweet cherry rootstocks CAB-6P (Prunus cerasus L.), Gisela 6 (P. cerasus× P. canescens), and MxM 60 (P. avium× P. mahaleb). *Journal of Pineal Research*, *52*(1), 38–46.

Sehar, Z., Jahan, B., Masood, A., Anjum, N. A., and Khan, N. A. (2021). Hydrogen peroxide potentiates defense system in presence of sulfur to protect chloroplast damage and photosynthesis of wheat under drought stress. *Physiologia Plantarum*, *172*(2), 922–934.

Seleiman, M. F., Aslam, M. T., Alhammad, B. A., Hassan, M. U., Maqbool, R., Chattha, M. U., and Battaglia, M. L. (2022). Salinity stress in wheat: Effects, mechanisms and management strategies. *Phyton (0031-9457)*, *91*(4).

Shao, H. B., Chu, L. Y., Jaleel, C. A., and Zhao, C. X. (2008). Water-deficit stress-induced anatomical changes in higher plants. *Comptesrendus Biologies*, *331*(3), 215–225.

Sharif, R., Xie, C., Zhang, H., Arnao, M. B., Ali, M., Ali, Q., Muhammad, I., Shalmani, A., Nawaz, M. A., Chen, P., and Li, Y. (2018). Melatonin and its effects on plant systems. *Molecules*, *23*(9), 2352.

Sharma, A., and Zheng, B. (2019). Melatonin mediated regulation of drought stress: Physiological and molecular aspects. *Plants*, *8*(7), 190. https://doi.org/10.3390/plants8070190.

Shi, H. T., Chen, K., and Wei, Y. (2016). Fundamental issues of melatonin-mediated stress signaling in plants. *Frontiers in Plant Science*, *7*, 1124.

Shi, H. T., Jiang, C., Ye, T., Tan, D. X., Reiter, R. J., Zhang, H., Liu, R., and Chan, Z. (2014). Comparative physiological, metabolomic, and transcriptomic analyses reveal mechanisms of improved abiotic stress resistance in bermudagrass [*Cynodon dactylon* (L). Pers.] by exogenous melatonin. *Journal of Experimental Botany*, *66*, 681–694.

Shi, H. T., Love, J., and Hu, W. (2017). Melatonin in development and stress responses: An emerging research area in plant science. *Frontiers in Plant Science*, *8*, 1666.

Shi, H. T., Reiter, R. J., Tan, D. X., and Chan, Z. (2015a). Indole-3-acetic acid inducible 17 positively modulates natural leaf senescence through melatonin-mediated pathway in Arabidopsis. *Journal of Pineal Research*, *58*(1), 26–33.

Shi, H. T., Wang, X., Tan, D. X., Reiter, R. J., and Chan, Z. (2015b). Comparative physiological and proteomic analyses reveal the actions of melatonin in the reduction of oxidative stress in Bermuda grass (*Cynodon dactylon* (L.) Pers.). *Journal of Pineal Research*, *59*(1), 120–131.

Siddiqui, M. H., Alamri, S., Khan, M. N., Corpas, F. J., Al-Amri, A. A., Alsubaie, Q. D., Alia, H. M., Kalajid, H. M., and Ahmad, P. (2020). Melatonin and calcium function synergistically to promote the resilience through ROS metabolism under arsenic-induced stress. *Journal Hazardous Material*, *398*, 122882.

Signor, D., Cerri, C. E. P., and Conant, R. (2013). N_2O emissions due to nitrogen fertilizer applications in two regions of sugarcane cultivation in Brazil. *Environmental Research Letters*, *8*(1), 015013.

Simopoulos, A. P., Tan, D. X., Manchester, L. C., and Reiter, R. J. (2005). Purslane: A plant source of omega-3 fatty acids and melatonin. *Journal of Pineal Research*, *39*(3), 331–332.

Sultan, I., Khan, I., Chattha, M. U., Hassan, M. U., Barbanti, L., and Calone, R. (2021). Improved salinity tolerance in early growth stage of maize through salicylic acid foliar application. *Italian Journal of Agronomy*, *16*, 1810.

Tal, O., Haim, A., Harel, O., and Gerchman, Y. (2011). Melatonin as an antioxidant and its semi-lunar rhythm in green macroalga Ulva sp. *Journal of Experimental Botany*, *62*(6), 1903–1910.

Tan, D. X. (1993). Melatonin: A potent, endogenous hydroxyl radical scavenger. *Endocrine Journal*, *1*, 57–60.

Tan, D. X., Hardeland, R., Back, K., Manchester, L. C., Alatorre-Jimenez, M. A., and Reiter, R. J. (2016). On the significance of an alternate pathway of melatonin synthesis via 5-methoxytryptamine: Comparisons across species. *Journal of Pineal Research*, *61*(1), 27–40. https://doi.org/10.1111/jpi.12336.

Tan, D. X., Hardeland, R., Manchester, L. C., Korkmaz, A., Ma, S., Rosales-Corral, S., and Reiter, R. J. (2012). Functional roles of MEL in plants, and perspectives in nutritional and agricultural science. *Journal of Experimental Botany*, *63*(2), 577–597. https://doi:10.1093/jxb/err256.

Tan, D. X., Hardeland, R., Manchester, L. C., Poeggeler, B., Lopez-Burillo, S., Mayo, J. C., and Reiter, R. J. (2003). Mechanistic and comparative studies of MEL and classic antioxidants in terms of their interactions with the ABTS cation radical. *Journal of Pineal Research*, *34*(4), 249–259.

Tan, D. X., Manchester, L. C., Di Mascio, P., Martinez, G. R., Prado, F. M., and Reiter, R. J. (2007). Novel rhythms of N1-acetyl-N2-formyl-5-methoxykynuramine and its precursor melatonin in water hyacinth: Importance for phytoremediation. *The FASEB Journal*, *21*(8), 1724–1729. doi:10.1093/jxb/eraa235.

Tan, D. X., Manchester, L. C., Esteban-Zubero, E., Zhou, Z., and Reiter, R. J. (2015). Melatonin as a potent and inducible endogenous antioxidant: Synthesis and metabolism. *Molecules*, *20*(10), 18886–18906.

Tan, D. X., Manchester, L. C., Hardeland, R., Lopez-Burillo, S., Mayo, J. C., Sainz, R. M., and Reiter, R. J. (2003). Melatonin: A hormone, a tissue factor, an autocoid, a paracoid, and an antioxidant vitamin. *Journal of Pineal Research*, *34*(1), 75–78.

Tan, D. X., Manchester, L. C., Reiter, R. J., Plummer, B. F., Limson, J., Weintraub, S. T., and Qi, W. (2000). Melatonin directly scavenges hydrogen peroxide: A potentially new metabolic pathway of MEL biotransformation. *Free Radical Biology and Medicine*, *29*(11), 1177–1185.

Tan, D. X., Manchester, L. C., Terron, M. P., Flores, L. J., and Reiter, R. J. (2007). One molecule, many derivatives: A never-ending interaction of melatonin with reactive oxygen and nitrogen species? *Journal of Pineal Research*, *42*(1), 28–42.

Tan, D. X., and Russel, J. R. (2020). An evolutionary view of melatonin synthesis and metabolism related to its biological functions in plants. *Journal of Experimental Botany*, *71*(16), 4677–4689.

Tan, X. L., Fan, Z. Q., Zeng, Z. X., Shan, W., Kuang, J. F., Lu, W. J., and Zhao, Y. T. (2021). Exogenous MEL maintains leaf quality of postharvest Chinese flowering cabbage by modulating respiratory metabolism and energy status. *Postharvest Biology and Technology*, *177*, 111524.

Tanimoto, E. (2005). Regulation of root growth by plant hormones: Roles for auxin and gibberellin. *Critical Reviews in Plant Sciences*, *24*(4), 249–265.

Tarocco, A., Caroccia, N., Morciano, G., Wieckowski, M. R., Ancora, G., Garani, G., and Pinton, P. (2019). Melatonin as a master regulator of cell death and inflammation: Molecular mechanisms and clinical implications for newborn care. *Cell Death & Disease*, *10*(4), 1–12.

Teng, Z., Yu, Y., Zhu, Z., Hong, S. B., Yang, B., and Zang, Y. (2021). Melatonin elevated sclerotiniasclerotiorum resistance via modulation of ATP and glucosinolate biosynthesis in *Brassica rapa* ssp. pekinensis. *Journal of Proteomics*, *243*, 104264.

Teng, Z., Zheng, W., Yu, Y., Hong, S. B., Zhu, Z., and Zang, Y. (2022). Melatonin regulated glucosinolate profile via modulation of genes related with sulfur and nitrogen metabolism in *Brassica rapa* ssp. pekinensis. *Industrial Crops and Products*, *177*, 114538.

Uchendu, E. E., Shukla, M. R., Reed, B. M., and Saxena, P. K. (2013). Melatonin enhances the recovery of cryopreserved shoot tips of a merican elm (*Ulmus americana* L.). *Journal of Pineal Research*, *55*(4), 435–442.

Van Tassel, D. L., and O'Neill, S. D. (2001). Putative regulatory molecules in plants: Evaluating MEL. *Journal of Pineal Research*, *31*, 1–7.

Van Tassel, D. L., Roberts, N. J., Lewy, A., and O'Neill, S. D. (2001). Melatonin in plant organs. *Journal of Pineal Research*, *31*(1), 8–15.

Van Tassel, D. L., Roberts, N. J., and Oenill, S. D. (1995, June). Melatonin from higher-plants-isolation and identification of N-acetyl 5-methoxytryptamine. *Plant Physiology*, *108*(2), 10.

Vigliante, I., Mannino, G., and Maffei, M. E. (2019). Chemical characterization and DNA fingerprinting of *Griffonia simplicifolia* baill. *Molecules*, *24*, 1032.

Wan, J., Zhang, P., Wang, R., Sun, L., Ju, Q., and Xu, J. (2018). Comparative physiological responses and transcriptome analysis reveal the roles of melatonin and serotonin in regulating growth and metabolism in Arabidopsis. *BMC Plant Biology*, *18*(1), 1–14.

Wang, H., Ren, C., Cao, L., Jin, X., Wang, M., Zhang, M., and Wu, H. (2021). The mechanisms underlying MEL improved soybean seedling growth at different nitrogen levels. *Functional Plant Biology*, *48*(12), 1225–1240.

Wang, L., Feng, C., Zheng, X., Guo, Y., Zhou, F., Shan, D., and Kong, J. (2017). Plant mitochondria synthesize MEL and enhance the tolerance of plants to drought stress. *Journal of Pineal Research*, *63*(3), e12429.

Wang, P., Sun, X., Chang, C., Feng, F., Liang, D., Cheng, L., and Ma, F. (2013a). Delay in leaf senescence of *Malus hupehensis* by long-term MEL application is associated with its regulation of metabolic status and protein degradation. *Journal of Pineal Research*, *55*(4), 424–434.

Wang, P., Sun, X., Li, C., Wei, Z., Liang, D., and Ma, F. (2013b). Long-term exogenous application of MEL delays drought-induced leaf senescence in apple. *Journal of Pineal Research*, *54*(3), 292–302.

Wang, P., Sun, X., Xie, Y., Li, M., Chen, W., Zhang, S., and Ma, F. (2014). MEL regulates proteomic changes during leaf senescence in *Malus hupehensis*. *Journal of Pineal Research*, *57*(3), 291–307.

Wang, P., Yin, L., Liang, D., Li, C., Ma, F., and Yue, Z. (2012). Delayed senescence of apple leaves by exogenous MEL treatment: Toward regulating the ascorbate: Glutathione cycle. *Journal of Pineal Research*, *53*(1), 11–20.

Weeda, S., Zhang, N., Zhao, X., Ndip, G., Guo, Y., Buck, G. A., Fu, C., and Ren, S. (2014). Arabidopsis transcriptome analysis reveals key roles of melatonin in plant defense systems. *PLoS ONE*, *9*(3), e93462.

Wei, J., Li, D. X., Zhang, J. R., Shan, C., Rengel, Z., Song, Z. B., and Chen, Q. (2018). PhytoMEL receptor PMTR 1-mediated signaling regulates stomatal closure in *Arabidopsis thaliana*. *Journal of Pineal Research*, *65*(2), e12500.

Wei, W., Li, Q. T., Chu, Y. N., Reiter, R. J., Yu, X. M., Zhu, D. H., and Chen, S. Y. (2015). MEL enhances plant growth and abiotic stress tolerance in soybean plants. *Journal of Experimental Botany*, *66*(3), 695–707.

Wei, Y. X., Liu, G. Y., Bai, Y. J., Xia, F. Y., He, C. Z., and Shi, H. T. (2017). Two transcriptional activators of N-acetylserotonin O-methyltransferase 2 and melatonin biosynthesis in cassava. *Journal of Experimental Botany*, *68*(17), 4997–5006. https://doi.org/10.1093/jxb/erx305.

Wen, D., Gong, B., Sun, S., Liu, S., Wang, X., Wei, M., Yang, F., Li, Y., and Shi, Q. (2016). Promoting roles of melatonin in adventitious root development of *Solanum lycopersicum* L. by regulating auxin and nitric oxide signaling. *Frontiers in Plant Science*, *7*, 718. https://doi.org/10.3389/fpls.2016.00718.

Wolf, K., Kolář, J., Witters, E., van Dongen, W., van Onckelen, H., and Macháčková, I. (2001). Daily profile of MEL levels in *Chenopodium rubrum* L. depends on photoperiod. *Journal of Plant Physiology*, *158*(11), 1491–1493.

Wu, X., Ren, J., Huang, X., Zheng, X., Tian, Y., Shi, L., and Li, Z. (2021). Melatonin: Biosynthesis, content, and function in horticultural plants and potential application. *Scientia Horticulturae*, *288*, 110392.

Xu, X. D. (2010). Effects of exogenous MEL on physiological response of cucumber seedlings under high temperature stress. *Master's degree thesis, Northwest A and F University*.

Yamamoto, Y., Aminaka, R., Yoshioka, M., Khatoon, M., Komayama, K., Takenaka, D., and Yamamoto, Y. (2008). Quality control of photosystem II: Impact of light and heat stresses. *Photosynthesis Research*, *98*(1), 589–608.

Ye, J., Wang, S., Deng, X., Yin, L., Xiong, B., and Wang, X. (2016). Melatonin increased maize (*Zea mays* L.) seedling drought tolerance by alleviating drought-induced photosynthetic inhibition and oxidative damage. *Acta Physiologiae Plantarum*, *38*(2), 1–13.

Zhang, H. J., Zhang, N. A., Yang, R. C., Wang, L., Sun, Q. Q., Li, D. B., and Guo, Y. D. (2014). Melatonin promotes seed germination under high salinity by regulating antioxidant systems, ABA and GA 4 interaction in cucumber (*Cucumis sativus* L.). *Journal of Pineal Research*, *57*(3), 269–279.

Zhang, N., Zhang, H. J., Sun, Q. Q., Cao, Y. Y., Li, X., Zhao, B., and Guo, Y. D. (2017a). Proteomic analysis reveals a role of melatonin in promoting cucumber seed germination under high salinity by regulating energy production. *Scientific Reports*, *7*(1), 1–15.

Zhang, N., Zhao, B., Zhang, H. J., Weeda, S., Yang, C., Yang, Z. C., and Guo, Y. D. (2013). Melatonin promotes water-stress tolerance, lateral root formation, and seed germination in cucumber (*Cucumis sativus* L.). *Journal of Pineal Research*, *54*(1), 15–23. https://doi.org/10.1111/j.1600-079X.2012.01015.x.

Zhang, R., Sun, Y., Liu, Z., Jin, W., and Sun, Y. (2017b). Effects of melatonin on seedling growth, mineral nutrition, and nitrogen metabolism in cucumber under nitrate stress. *Journal of Pineal Research*, *62*(4), e12403.

Zhao, D., Wang, R., Liu, D., Wu, Y., Sun, J., and Tao, J. (2018). Melatonin and expression of tryptophan decarboxylase gene (TDC) in herbaceous peony (*Paeonia lactiflora* Pall.) flowers. *Molecules*, *23*(5), 1164.

Zhao, D., Yu, Y., Shen, Y., Liu, Q., Zhao, Z., Sharma, R., and Reiter, R. J. (2019). Melatonin synthesis and function: Evolutionary history in animals and plants. *Frontiers in Endocrinology*, *10*, 249.

Zhao, Y., Tan, D. X., Lei, Q., Chen, H., Wang, L., Li, Q. T., and Kong, J. (2013). Melatonin and its potential biological functions in the fruits of sweet cherry. *Journal of Pineal Research*, *55*(1), 79–88.

Zheng, X., Tan, D. X., Allan, A. C., Zuo, B., Zhao, Y., Reiter, R. J., and Kong, J. (2017a). Chloroplastic biosynthesis of melatonin and its involvement in protection of plants from salt stress. *Scientific Reports*, *7*(1), 1–12.

Zheng, X., Zhou, J., Tan, D. X., Wang, N., Wang, L., Shan, D., and Kong, J. (2017b). Melatonin improves waterlogging tolerance of *Malus baccata* (Linn.) Borkh seedlings by maintaining aerobic respiration, photosynthesis and ROS migration. *Frontiers in Plant Science*, *8*, 483.

Zhou, J. (2018). Melatonin alleviates low-sulphur stress by promoting sulphur homeostasis in tomato plants. *Scientific Reports*, *8*, 10182.

Zia, S. F., Berkowitz, O., Bedon, F., Whelan, J., Franks, A. E., and Plummer, K. M. (2019). Direct comparison of arabidopsis gene expression reveals different responses to melatonin versus auxin. *BMC Plant Biology*, *19*(1), 1–18.

2 Melatonin as a Key Regulator of Plant Growth and Development

*Aradhna Kumari, Santosh Kumar Singh,
M.S. Mikhina, Babita Patni and G.K. Krishna*

2.1 INTRODUCTION

Melatonin is an indoleamine that was first isolated and identified as a neurotransmitter in animals and proved to be very important in several indispensable biological processes. It was recognized in animals mainly as a hormone secreted at night by the pineal gland and has been found to play a key role in controlling sleep and waking cycles (Faraone and Antshel, 2014; Auld et al., 2017). Melatonin helps in seasonal rhythmicity in vertebrates, which includes reproduction, fattening, moulting, and hibernation, as well as their circadian rhythms, which regulate blood pressure and sleep-wake timing (Altun and Ugur-Altun, 2007). Most of its functions are due to the activation of melatonin receptors, and some owe to their role as an antioxidant (Boutin et al., 2005; Reiter et al., 2001).

Melatonin was discovered in 1958 by Aaron B. Lerner (a dermatology professor) and his co-workers (Lerner et al., 1958; Auld et al., 2017) at Yale University. He believed that there may be a substance from the pineal gland that can be effective in treating skin problems. He subsequently extracted the hormone from bovine pineal gland extracts, named it as melatonin, and identified it as *N-acetyl*-5-methoxy-tryptamine (Lerner et al., 1960). Melatonin was given that name because it can lighten the skin of some fish, reptiles, and amphibians (Lerner et al., 1958; Carlson, 1994). Research by Lynch et al. (1975) in the middle of the 1970s showed that melatonin controls the circadian rhythm.

In animals, melatonin participates in the regulation of the circadian rhythm, which includes the sleep and awakening cycle, and blood pressure control, besides controlling seasonal rhythms comprising fertility, obesity, constipation and sleep. Most of its outcomes are due to the stimulation of melatonin receptors and scavenging of antioxidants (Boutin et al., 2005; Reiter et al., 2001). Antioxidant properties of melatonin were discovered in evolutionarily diverse organisms, viz., rats and algae (Poeggeler et al., 1993). In plants, it mediates several signalling pathways and helps to protect against oxidative stress (Tan et al., 2012). Melatonin is also found in many of the foods which we consume in our daily life.

2.2 ROLES OF MELATONIN IN PLANT GROWTH AND DEVELOPMENT

Even though the function of melatonin as a plant hormone has not been firmly recognized yet, an association of it with processes such as growth, development and photosynthesis are confirmed. It is a multi-regulatory molecule in plants. Melatonin performs many hormone-like functions. It has important roles in plant growth and development, like enhancement of photosynthesis and carbon fixation, modulation of reproductive development, seed germination, root growth and development, leaf senescence, circadian rhythms, post-harvest fruit ripening, maintenance of plant tissues, delay

 DOI: 10.1201/9781003343752-2

of senescence and responses to biotic and abiotic stresses (Samanta et al., 2020). Various scientists have shown evidence which indicates that melatonin has the capacity to improve the productivity of most crops. The mechanisms of enhancing crop productivity may involve the involvement of melatonin in the protection of chlorophyll, enhancement of photosynthesis and other metabolic processes and induction of root development. Transgenic plants with improved melatonin concentration might be able to improve crop productivity of agricultural crops and betterment in the general health of humans (Tan et al., 2012). Melatonin performs important tasks in plants as a growth regulator, and sometimes its role is overlapping with different plant hormones like auxin, gibberellic acids (GAs), strigolactones and ABA. It is synthesized in plants when they are exposed to stress, both biotic and abiotic, for example, fungal infection and abiotic stresses like drastic temperature, toxicity, high soil salinity and drought (Arnao and Hernández-Ruiz, 2015; Hardeland, 2015; Reiter et al., 2015; Banerjee and Roychoudhury, 2017). The following section describes the physiological and molecular actions of melatonin.

2.2.1 Regulation of Seed Germination

Seed germination is thought of as the starting of the new plant life cycle, and it is vital in deciding how the plant will continue to grow and develop. Reports indicate that melatonin can stimulate seed germination, especially under various abiotic stresses. It can be applied either by seed treatment or seed priming, which improves seed germination in various crops. For meeting this purpose, optimization of dose is essential. Xiao et al. (2019) reported that melatonin in low concentration can promote germination, while at high concentrations, it proved to be unworthy to stimulate germination and even inhibited germination in case of cotton. Thus, a lower dose is recommended for improving the germination rate of cotton seeds.

Li et al. (2019) reported that in *Limonium bicolour* (a halophyte), seed germination was suppressed considerably under 200 mM NaCl treatment. Pre-treatment of seeds with melatonin enhanced seed germination significantly under stress. Hormone profiling indicated that exogenous melatonin application spiked the levels of endogenous melatonin and gibberellic acid and attenuated the levels of abscisic acid. A high level of amylase and alpha-amylase enzyme activities was also recorded. Upon melatonin treatment, there is an up regulated expression of the key genes involved in GA biosynthesis (*GA20ox* and *GA3ox*), down regulation of key genes involved in ABA biosynthesis (*LbNCED1* and *LbNCED3*) and up regulation of ABA 8′-hydroxylase genes (*LbCYP707A1* and *LbCYP707A2*).

Zhang et al. (2017) examined whether there is any involvement of melatonin in proteomic changes under salt stress imposed on cucumber seeds. Findings indicated that several pathways were noticeably up regulated by melatonin. These regulate pathways include ribosome biosynthesis, lipid and carbohydrate metabolism and storage protein degradation. Proteins involved in primary metabolism such as enzymes involved in glycolysis, citric acid cycle and glyoxylate cycle were also up regulated. Rajora et al. (2022) reviewed that melatonin seed priming leads to resistance in different crops under elevated drought, salinity, heat, and heavy metal stress. They illustrated that primed seeds hold a greater content of various osmoprotectants, up regulation of energy-producing pathways and an upsurge in protein content, which may lead to augmenting abiotic stress response in plants. They pointed out that the primary function ascribed to melatonin is reactive oxygen species scavenging, which therefore results in greater plant stress resistance against multiple stresses (Rajora et al., 2022).

2.2.2 Regulation of Vegetative Growth

Arnao and Hernández-Ruiz (2018) concluded from literature surveys that melatonin induced a three- to four-fold upsurge in growth over untreated plants as observed in *Lupinus*, *Phalaris*, *Triticum*, *Hordeum*, *Arabidopsis* and *Cucumis*. Hernández-Ruiz et al. (2005) reported through their experiments in some monocots that melatonin had a growth-promoting activity in coleoptiles of oat, wheat, canary grass and barley. Tanimoto (2005) reported that the effect of melatonin in promoting growth in aerial organs (coleoptile, hypocotyl) and inhibiting growth in roots was similar as shown by IAA. The mechanism of action of IAA might be mediated by cell wall proton extrusion through the activation and/or *de novo* formation of plasma membrane H^+-ATPases. Arnao and Hernández-Ruiz (2006) showed that modest acidification of the media at the time of coleoptile development, caused by melatonin, is similar to the mechanism of action by auxin. Coating seeds with melatonin (50 μmol L^{-1}) appreciably promoted the vegetative growth of soybean crops by enhancing leaf size and plant height (Wei et al., 2015). El-Naby et al. (2019) used a 25 ppm foliar spray of melatonin on 'Canino' apricot along with GA_3 and NAA and found that melatonin was most efficient in enhancing the vegetative parameters such as leaf area and total leaf chlorophyll. The effects were attributed to hormone metabolism, enhanced cell division, elongation and expansion of cells.

2.2.3 Regulation of Reproductive Growth Stages and Floral Transition

Melatonin has a profound role in the processes of flowering. Macháčková and Krekule (2002) and Kolář et al. (2003) performed studies regarding the possible role of melatonin in the short-day plant *Chenopodium rubrum*. They found that high concentrations of melatonin impeded flower induction up to 40–50% when applied exogenously at 2 h before and 2 h after the inductive dark period. They also reported that melatonin treatment had no effect on the flowering phase, and suggested that melatonin controls any process before floral transition. Kolář (2003) performed a similar study in the long-day plant *Arabidopsis thaliana* and found that flowering was slightly delayed in plants treated with 100 or 500 μM melatonin, which affected leaves and rosette numbers and the process of anthesis. Murch and Saxena (2002) observed the appearance of melatonin peak in the intermediate stages of flower development while studying floral development of the plant St John's wort (*Hypericum perforatum*). They observed melatonin accumulation in specific tissues and stages that acted as a signal in development of gametophyte, which turned into available microspore. Murch et al. (2009) probed a narcotic plant, devil's trumpet (*Datura metel*), used in natural medicine and noted that melatonin concentration peaked at the time of flower bud maturation. This is suggestive of melatonin playing a protective function in the reproductive parts of the plant.

Zhao et al. (2018) analyzed the melatonin content during the different flower developmental stages of herbaceous peony (*Paeonia lactifora*). They observed that melatonin content gradually increased from flower bud to initial bloom stage, which later started declining till withering. Exposure to sunlight and a blue beam of light stimulated melatonin production, whereas lower values were recorded in shade and white and green lights (Zhao et al., 2018). Park et al. (2013) analyzed the melatonin contents of the field-grown rice plant at vegetative and different reproductive stages (pre-flowering, flowering, and post-flowering). They found that melatonin production in the panicle (flower) was six times more as compared to flag leaves. This stimulation coincided with the induction of melatonin biosynthetic enzymes (TDC, T5H and COMT), indicating that melatonin may be involved in flower induction and development. Kotchoni et al. (2009) reported that melatonin may help in regulating flowering time in an identical fashion to that of ascorbic acid in Arabidopsis.

Studies in Arabidopsis provided insight into the function of melatonin on floral transition. A unique role of DELLAs and FLC in melatonin-mediated flowering in Arabidopsis similar to the action of gibberellins was observed (Li et al., 2016; Shi et al., 2016). They showed that melatonin-treated plants (500 μM) delayed flowering by five days with a higher number of rosette leaves as compared to control. This indicated that higher concentration of exogenous melatonin inhibits flowering. Melatonin has been reported to increase the protein stability of two DELLA proteins, without affecting their transcript levels (Shi et al., 2016). It was also observed that exogenous melatonin does not alter endogenous GA levels, indicating that melatonin impacts DELLA proteins independent of GA. Melatonin stimulates endogenous nitric oxide (NO) levels and may also participate in melatonin-mediated signalling pathways (Shi et al., 2015a; Mukherjee, 2019). Thus NO might have participated in melatonin-mediated DELLA stabilization, which resulted in delayed flowering (Shi et al., 2016). Melatonin up regulated GA biosynthesis genes such as *GA20ox* and *GA3ox* in cucumber seedlings under salinity, which leads to a higher level of GAs like GA3 and GA4, promoting inhibition of germination under salinity (Zhang et al., 2014). In cotton seedlings, exogenous melatonin application showed an upsurge in the GA concentration (Xiao et al., 2019), whereas in pear fruits, it stimulated parthenocarpy (Liu et al., 2018a). Zhang et al. (2019) reported that the application of strigolactone (SL) suppresses melatonin signalling and/or biosynthesis. Thus an antagonistic regulation was exhibited in the cross-talk of the flowering process.

2.2.4 Protective Role of Melatonin

A 'dual peak' of melatonin appeared in peony flowers during a 24-h light/dark cycle, coinciding with temperature and light interception patterns and with a reduction in the malondialdehyde (MDA) leading to reduced membrane lipid peroxidation (Zhao et al., 2018). In cut flowers of anthurium (*Anthurium andraeanum*), this protective function of melatonin was also observed at the time of storage under low temperature (Aghdam et al., 2019). Hardeland and Rodríguez (1995) postulated that melatonin acts as a dormancy-maintaining agent in seeds since a high content of melatonin was noted in seeds. Lei et al. (2004) demonstrated a significant function of melatonin as a membrane integrity factor of both nuclear and plasma membranes in carrot suspension cell culture.

2.2.5 Plant Metabolism and Physiological Processes

Melatonin is involved in multiple facets of cellular and physiological functions, *viz.*, changes in intracellular Ca^{2+} level and in the permeability of membranes via various ion transporters (Li et al., 2016). It has significant role in the opening and closing of stomata, carbohydrate, lipid and nitrogen metabolisms, as well as in the building up of osmoprotectants inside the cell (Shi et al., 2015b; Wei et al., 2015). It influences the processes such as growth promotion, rooting induction, gravitropism (Arnao and Hernández-Ruiz, 2017), seed germination enhancement (Zhang et al., 2017), photosynthesis and leaf water/CO_2 exchange (Li et al., 2017). Reports on the regulation of processes such as ripening, senescence, circadian cycle (the internal biological clock), parthenocarpy and senescence (Arnao and Hernández-Ruiz, 2018; Liang et al., 2018; Liu et al., 2018a) are also observed for melatonin. Arnao and Hernández-Ruiz (2009) reported an increase in photosynthesis by exogenous melatonin owing to delayed loss of chlorophylls as compared to untreated plants.

Coating seeds with melatonin (50 μmol L^{-1}) leads to enhancement in the fatty acid content of the soybean as well as the yield of the crop at maturity (Wei et al., 2015). The up regulation of genes involved in cell division, photosynthesis, carbohydrate metabolism, fatty acid biosynthesis and ascorbate metabolism were observed in treated samples. Meng et al. (2015) observed that spraying 100 mol L^{-1} of melatonin solution to the fruits of young grapes led to synchronicity by the reduction of underripe and overripe grape berries.

2.2.6 Circadian Cycle Regulation

In plants, the circadian rhythm can maintain the coincidence of most of the physiological processes with daily and/or seasonal cycles. Various researchers had reported that circadian rhythm regulates photosynthesis and growth rates, may affect flowering and seed set in crops and biotic/abiotic stress responses (Dodd et al., 2005; Atkins and Dodd, 2014; Dixon et al., 2014; Dodd et al., 2014).

Researchers visualized the probable task of melatonin in plants as a chrono-regulator of circadian rhythms, in an identical fashion as in animals (Kolář et al., 1995, 1997; Wolf et al., 2001). Kolář et al. (1997) showed the existence of melatonin in a 15-day-old seedling of *Chenopodium rubrum* L. through liquid chromatography/tandem mass spectrometry and radioimmunoassay. They performed a study to show the possibility of involvement of melatonin in circadian rhythm and probable changes in the melatonin levels in light/dark cycles of 12 h. They showed an oscillating behaviour of melatonin. They observed a very minute concentration of melatonin during the light period and a substantial upsurge in darkness. In the dark period, a concentration of 240–550 pg/g fresh weight was noted. In another experiment, plants were exposed to different photoperiodic time lengths (6, 12 and 18 h of darkness). No change in the duration of the melatonin enhancement was noted with the different photoperiodic time length applications, though the maximum concentration of melatonin always appeared after switching off the lights (Wolf et al., 2001).

2.2.7 Shoot Growth and Differentiation

Reports on the effects on differentiation are meagre in plants for melatonin. Huang et al. (2022) reported improvement in the number of shoot buds per explant in the two species of *Cymbidium* orchids (*C. goeringii* and *C. faberi*), with 1.0 μM melatonin addition in tissue culture media as compared to their control. They also reported improved seedling height with the incorporation of melatonin in specific concentrations in the culture.

2.2.8 Biomass Accumulation and Yield Enhancement

Transgenic rice overexpressing sheep SNAT (serotonin N-acetyltransferase) showed an upsurge of endogenous melatonin content up to three-fold as compared to wild type. Enhanced seedling growth and two times increase in biomass were observed. Also, melatonin-rich lines exhibited a one-week delay in flowering with a reduced grain yield by 33% (Byeonand Back, 2014). Coating seeds with melatonin (50 μmol L^{-1}) leads to enhancement in soybean biomass and production (Wei et al., 2015). They observed that melatonin increased leaf size, plant height, pod number and seed number, but not 100-seed weight. Liu et al. (2019) experimented on the effect of exogenous melatonin (100 μmol L^{-1}) on pear. They observed that it promoted the deposition of endogenous melatonin in pear fruit. During the advanced stage of pear fruit development, net photosynthetic rate and maximal quantum efficiency of photosystem II was enhanced, which led to an increase in fruit weight by 47.85% over untreated plants. Soluble sugars, namely sucrose and sorbitol, were also elevated in fruit. All these factors contributed to increasing yield. They recommended that the use of melatonin in pear trees can be utilized to produce bigger and sweeter fruit of higher economic value (Liu et al., 2019). Meng et al. (2015) observed that application of 100 mol L^{-1} of melatonin solution through spray on the fruits of young grapes might help in promoting the build-up of endogenous melatonin contents in berries of grapes and had a substantial influence on enhancing the yield of the berries and their weight.

2.2.9 Root Organogenesis and Lateral Root Development

Most of the experiments done have proved that melatonin shows auxin-like activity, in the manner that melatonin can induce root and shoot growth along with stimulation of root generation in plants. Arnao and Hernández-Ruiz (2007) showed that melatonin induced growth in aerial parts and promoted or inhibited growth in roots, depending on the concentration assayed. Generally, only at high melatonin concentrations (100 μM) does root growth inhibition takes place, although in one recent study using *Zea mays*, melatonin was seen not to have any effect on growth-promoting/inhibiting activities. The strong induction of roots by melatonin was comparable with the effect of IAA. Melatonin induced the root primordia formation from pericycle cells and affected the occurrence of both adventitious and lateral roots. Therefore, melatonin changed the time course and pattern of distribution, number and length of adventitious roots and number of new lateral roots. Studies in *Arabidopsis thaliana* revealed that melatonin improves the emergence of adventitious roots two-fold and the emergence of lateral roots by up to three-fold but has no effect on root hair density (Pelagio-Flores et al., 2012). Additionally, three transgenic lines of

Arabidopsis thaliana that overproduce melatonin induced numerous lateral roots than wild type (Zuo et al., 2014). Thus it may function parallel to IAA in the induction of lateral and adventitious roots (Arnao and Hernández-Ruiz, 2014).

Huang et al. (2022) performed an experiment to understand the effect of melatonin on rhizome proliferation, differentiation and seedling rooting in the two species of *Cymbidium* (*C. goeringii* and *C. faberi*). They found a substantial enhancement in rhizome proliferation in both the species, with 0.5 to 1 μM dosage in rhizome culture. Its application also enhanced the survival rate of rhizomes. They also reported that incorporation of melatonin in culture medium in optimized concentrations had improved seedling height and root number and length in both the species. Thus melatonin application is effective in promoting rhizome proliferation, differentiation, and seedling rooting in the rapid propagation of *C. goeringii* and *C. faberi*. Concomitantly, rooting was inhibited with the application of higher concentrations of melatonin, and the requirement of melatonin concentration varied significantly in all propagation stages of the two species for root initiation and differentiation.

2.2.10 Effect of Melatonin on Root Tropism

Arnao and Hernández-Ruiz (2018) experimented on lupin primary root responses 24 h after the localized treatment with agar blocks containing IAA or melatonin (10 and 100 μM). They analyzed the growth trajectories of roots containing agar blocks with IAA and melatonin at 10 and 100 μM, respectively, and found that both trajectories are similar in shape and time, and both affect the normal tropic response of the root. This confirmed the role of melatonin in the gravitropic response of the lupin seedlings, which was frequently attributed to compounds with auxin activity. Weeda et al. (2014) reported that melatonin may influence the action of auxin through changes in auxin carriers by changing the local IAA gradients. It was suggested that melatonin interacts directly with auxin carriers or alters the local IAA gradients.

2.2.11 Effect of Melatonin on Growth Stimulation

Melatonin is reported to play pivotal roles in plant growth and development (Chen et al., 2019). Plant produces melatonin endogenously during various stages of development (Sharif et al., 2018). The vegetative growth promotion by phytohormone is hypothetically attributed to its similarity in structure and biosynthesis with IAA (Kolář et al., 2003). In addition, exogenous application of melatonin up regulated the synthesis of IAA in plants (Wang et al., 2016). Studies reported both the promotion and inhibitory actions in species such as wheat, oat, canary grass and barley (Hernández-Ruiz et al., 2005). Depending upon the crop and dosage, melatonin induces growth in shoot, while it either induces or inhibits the growth in roots (Arnao and Hernández-Ruiz, 2017). Roots are the major plant parts in which most of the melatonin-related studies were conducted. The first report was by Murch et al. (2001) that the supplement of melatonin (100 μM) led to faster root formation in John Wort. In Arabidopsis, it enhanced adventitious and lateral root growth by two- and three-folds, respectively (Pelagio-Flores et al., 2012). In higher plants, the connection between carbon and nitrogen metabolism underlies the growth process (Duan et al., 2018). Scientists have reported that the application of melatonin hastened the root elongation, plant height, leaf surface area, increased total soluble protein content, total soluble carbohydrate content and total chlorophyll content by enhancing the biosynthetic reactions inside plants. Melatonin-treated maize seedlings exhibited a high amount of shoot fresh and dry biomass accumulation (Ahmad et al., 2019). An inhibitory growth was found in cherry rootstocks grown at a concentration of 10 μM (Sarropoulou et al., 2012). Ye et al. (2020) reported that wheat seeds presoaked in melatonin (10, 100, 500 μM) showed better root system in 100 μM. Melatonin is also considered as a bio-stimulator due to the stimulation of plant life (Kolodziejczyk and Posmyk, 2016) and a modulator of gene expression related to various plant growth regulators and their activities (Nawaz et al., 2020).

2.2.12 Melatonin as Antioxidant

In plants, melatonin is involved in various stress responses (Szafrańska et al., 2016). It is known as a broad spectrum antioxidant because of its dual nature (hydrophilic and hydrophobic) and small size, which enable easy migration between cell compartments and protects from excessive ROS (Bałabusta et al., 2016). In rice, exogenous application of melatonin induced salt stress tolerance by reducing membrane lipid peroxidation and alleviated cell injury. It also improved nitrogen and silicon accumulation in leaf and root (Wei et al., 2022). Melatonin-treated (100 μM) cucumber seedlings showed a maximum number of lateral root generation under water stress conditions. It also imparted a positive effect on root number, root diameter and viability (Zhang et al., 2013). In kiwi seedlings, it showed more dry matter accumulation and stronger stem under water deficit condition (Liang et al., 2019). Similarly, it increased the root length and diameter in rape seedlings under salinity stress (Liu et al., 2018b). Melatonin acts as primary line of defence against internal and environmental oxidative stressors (Samanta et al., 2022). Oxidative stress such as UV irradiation probably consumes a large quantity of melatonin in plants (Tan et al., 2012). Studies documented that melatonin directly scavenges ROS in plants and decreases the levels of malondialdehyde (MDA) in leaves of mung bean exposed to UV-B radiation (Wang et al., 2009). Under heat stress, foliar spray of melatonin on cucumber seedlings significantly reduced the MDA, super oxide anion and hydrogen peroxide. Apart from the down regulation of ROS, it elevated the endogenous antioxidant enzyme activity of catalase, glutathione reductase, ascorbate peroxidase (Xu et al., 2010). Mode of supplement of melatonin was considerable in case of maize seedlings because root irrigation with 100 μM allowed better growth, higher chlorophyll content, plant biomass and protection from cell death by maintaining the membrane stability and lipid peroxidation (Huang et al., 2019).

2.2.13 Regulation of Autophagy

Autophagy refers to the intracellular process in which cytoplasmic components are destructed via vacuole (Su et al., 2020) and is required for the normal homeostasis condition of a cell (Todde et al., 2009). Both melatonin and autophagy have their own place in plant growth, development and stress response (Wei et al., 2020). When we look into the stress regulation by melatonin, Arabidopsis seedlings exhibited an induced autophagy with melatonin (10 μM) pre-treatment under simulated oxidative stress by using methyl viologen (Wang et al., 2015). Chilling stress tolerance was accomplished in the eggplant by the exogenous application of melatonin (5 μM) via soil drench after being subjected to chilling stress at 5/10° C for three days, along with increase of antioxidant enzymes and photosynthetic activity and decreased membrane damage (Yakuboğlu et al., 2022). The ability of melatonin in protein quality control was noted by Lee et al. (2022) through autophagy process and by inducing autophagosome genes. The genes responsible for autophagy (MeATG8b, 8c and 8e) were induced by melatonin treatment in cassava (Wei et al., 2020). In tomato, irrigation treatment of melatonin (20 μM) elevated high temperature-induced inactivation of pollen and its germination (Qi et al., 2018). Exogenous application of melatonin under flood condition was able to ameliorate effects such as cell death, lignification, with reduced RNA and protein degradation (Wang et al., 2021).

2.2.14 Modulation of Leaf Senescence and Post-Harvest Fruit Ripening

Melatonin has an inevitable role in delaying senescence by various means. It delays post-harvest leaf senescence of Chinese cabbage by maintaining ROS homeostasis through antioxidant enzymes and AsA-GSH cycle (Tan et al., 2020). It is well known that chlorophyll degradation and reduced photosynthesis are collateral with senescence. Melatonin control senescence by down regulating the senescence associated genes such as *SAG12* in cucumber (Liu et al., 2022) and apple

(Wang et al., 2014). Study of melatonin in tomato at dark condition revealed that it helps in delaying leaf starch degradation and inhibiting the starch-degrading enzymes at dusk and inhibiting dark-induced senescence by miR171b expression in tomato (Wang et al., 2022). Melatonin can be considered as a blooming biomolecule in post-harvest management of perishable fruits and vegetables (Jayarajan and Sharma, 2021) and can be used for post-harvest treatment against biotic and abiotic stress. A study conducted by Zhai et al. (2018) in pear showed that melatonin treated fruits showed a significant reduction in ethylene production that maintained the firmness of fruits. Melatonin was also used in ripening of strawberry (Mansouri et al., 2021) by the activation of GAMYB gene and lower expression of SnRK 2.6 gene, resulting from endogenous accumulation of melatonin by injecting 1000 μM of melatonin exogenously. In peach fruits, a slow senescence (Gao et al., 2016) was observed by treatment with melatonin along with reduced weight loss, less decay and low respiration rate as well as maintained firmness, total soluble solids and ascorbic acid content. A study conducted on the effect of light (red, blue and mix of red and blue and white as control) on endogenous melatonin by Yan et al. (2021) showed that the combined red and blue light promoted fruit ripening and quality improvement of tomato through a cross-talk with other molecules by increased production and signalling.

2.2.15 Intracellular and Genomic Actions of Melatonin

In plants, mitochondria and chloroplast are the primary sites of melatonin synthesis, and these organelles are highly protected from free radicals and oxidative stress (Tan et al., 2012). It was reported that five genes are implicated in the synthesis of melatonin, viz., TDC, TPH, T5H, SNAT, ASMT and COMT in plants (Back et al., 2016). Melatonin regulates the transcription factors, namely phosphorylation of cAMP responsive element binding protein and expression of c-Fos, and also second messengers such as Ca^{2+}, DAG and cAMP (Vanecek, 1998). The regulation of stress-specific genes and activation of pathogenesis-related proteins and antioxidant enzymes under various stress conditions make this molecule a more versatile type (Sharif et al., 2018). Soybean root tip treated with melatonin ameliorated the retarded growth and cell death under flood (Wang et al., 2021). Genome wide profiling conducted by Shi et al. (2015b) for melatonin pre-treated Bermuda grass identified 3933 transcripts which were differentially expressed in treated over control along with overrepresentation of nitrogen metabolism, carbohydrate metabolism, hormone metabolism and secondary metabolism. In sweet potato, it improved K^+/Na^+ homeostasis and tolerance to 150 mM NaCl (Yu et al., 2018) by exogenous application of melatonin through root and leaf. Wei et al. (2018) identified for the first time that melatonin governing receptor-dependent stomatal closure in Arabidopsis by heteromeric G protein alpha sub unit regulated H_2O_2 and Ca^{2+} signals. Zhang et al. (2014) observed that peroxidase-related genes are the candidates for melatonin response. The RNA-sequencing analysis conducted in cotton for its salt stress tolerance showed that exogenous applications of melatonin regulated the transcription factors, plant hormones, signal molecules and Ca^{2+} (Zhang et al., 2021). Chang et al. (2021) found melatonin mediated induction of C-Repeat Binding Factor pathway under cold stress through a positive interaction between H_2O_2 and Ca^{2+}.

2.3 CONCLUSION

Melatonin can be considered as a new member of next-generation phytohormones. Its exogenous as well as endogenous application promotes various physiological, biochemical, molecular and developmental effects in plants. Melatonin is universally found in plants and hence must be an integral part of plant signalling. The developmental attributes regulated by melatonin include seed germination, biomass development, delayed flowering, circadian cycle, organ differentiation, root development and fruit ripening. Melatonin also exhibits a protective role on tissues by reducing the ROS levels.

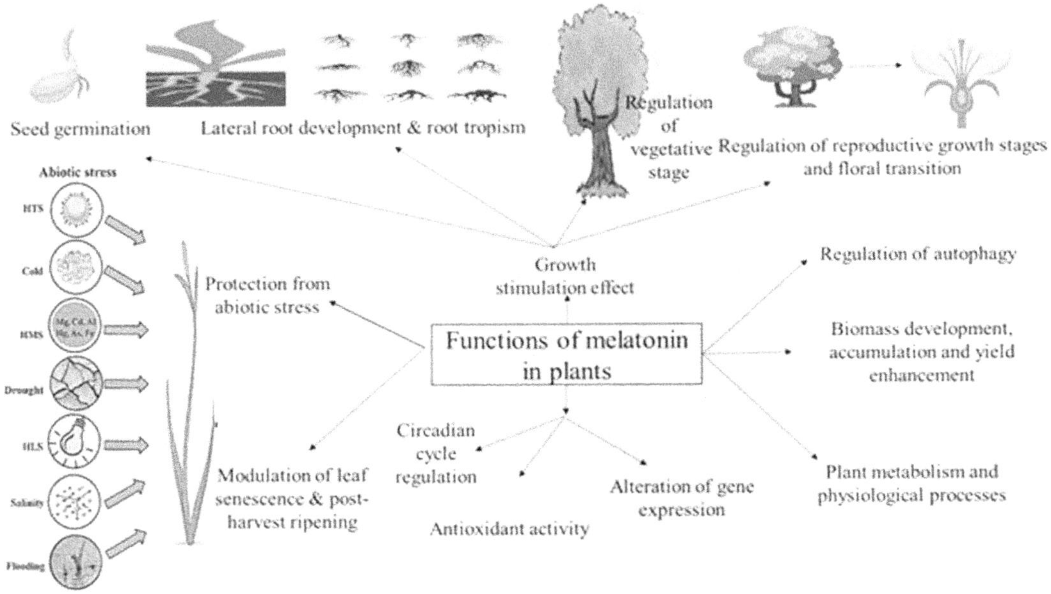

FIGURE 2.1 Physiological and biochemical roles of melatonin in the growth, development and survival of plants.

Plant metabolism, soluble sugar accumulation, autophagy and senescence are also regulated by the phytohormone (Figure 2.1). More insights into the applicability of melatonin in stress tolerance, tissue culture and yield enhancement are expected with progressing research in the coming decades.

REFERENCES

Aghdam, Morteza Soleimani, Abbasali Jannatizadeh, Mohsen Sabzi Nojadeh, and Asghar Ebrahimzadeh. "Exogenous melatonin ameliorates chilling injury in cut anthurium flowers during low temperature storage." Postharvest Biology and Technology 148 (2019): 184–191. https://doi.org/10.1016/j.postharvbio.2018.11.008

Ahmad, Shakeel, Muhammad Kamran, Ruixia Ding, Xiangping Meng, Haiqi Wang, Irshad Ahmad, Shah Fahad, and Qingfang Han. "Exogenous melatonin confers drought stress by promoting plant growth, photosynthetic capacity and antioxidant defense system of maize seedlings." PeerJ 7 (2019): e7793. https://doi.org/10.7717/peerj.7793

Altun, Armagan, and B. Ugur-Altun. "Melatonin: Therapeutic and clinical utilization." International Journal of Clinical Practice 61, no. 5 (2007): 835–845. https://doi.org/10.1111/j.1742-1241.2006.01191.x

Arnao, M. B., and J. Hernández-Ruiz. "Growth activity, rooting capacity, and tropism: Three auxinic precepts fulfilled by melatonin." Acta Physiologiae Plantarum 39, no. 6 (2017): 1–9. https://doi.org/10.1007/s11738-017-2428-3

Arnao, M. B., and J. Hernández-Ruiz. "Melatonin: Plant growth regulator and/or biostimulator during stress?" Trends in Plant Science 19, no. 12 (2014): 789–797. https://doi.org/10.1016/j.tplants.2014.07.006

Arnao, M. B., and J. Hernández-Ruiz. "Melatonin promotes adventitious-and lateral root regeneration in etiolated hypocotyls of *Lupinus albus* L." Journal of Pineal Research 42, no. 2 (2007): 147–152. https://doi.org/10.1111/j.1600-079X.2006.00396.x

Arnao, M. B., and J. Hernández-Ruiz. "The multi-regulatory properties of melatonin in plants." In: Roshchina V.V., Ramakrishna A. (Eds.) Neurotransmitters in Plants, CRC Press, United States, 2018, pp. 71–101. https://doi.org/10.1201/b22467-5

Arnao, M. B., and J. Hernández-Ruiz. "The physiological function of melatonin in plants." Plant Signaling & Behavior 1, no. 3 (2006): 89–95. https://doi.org/10.4161/psb.1.3.2640

Arnao, M. B., and J. Hernández-Ruiz. "Protective effect of melatonin against chlorophyll degradation during the senescence of barley leaves." Journal of Pineal Research 46, no. 1 (2009): 58–63. https://doi.org/10.1111/j.1600-079X.2008.00625.x

Arnao, M. B., and J. Hernández-Ruiz. "Studies in natural products chemistry." Bioactive Natural Products 46 (2015): 519–545. https://doi.org/10.1016/B978-0-444-63462-7.00011-7

Atkins, Kelly A., and Antony N. Dodd. "Circadian regulation of chloroplasts." Current Opinion in Plant Biology 21 (2014): 43–50. https://doi.org/10.1016/j.pbi.2014.06.008

Auld, Fiona, Emily L. Maschauer, Ian Morrison, Debra J. Skene, and Renata L. Riha. "Evidence for the efficacy of melatonin in the treatment of primary adult sleep disorders." Sleep Medicine Reviews 34 (2017): 10–22. https://doi.org/10.1016/j.smrv.2016.06.005

Back, Kyoungwhan, Dun-Xian Tan, and Russel J. Reiter. "Melatonin biosynthesis in plants: Multiple pathways catalyze tryptophan to melatonin in the cytoplasm or chloroplasts." Journal of Pineal Research 61, no. 4 (2016): 426–437. https://doi.org/10.1111/jpi.12364

Bałabusta, M., K. Szafrańska, and M. M. Posmyk. "Exogenous melatonin improves antioxidant defense in cucumber seeds (*Cucumis sativus* L.) germinated under chilling stress." Frontiers in Plant Science 7 (2016): 575. https://doi.org/10.3389/fpls.2016.00575

Banerjee, A., and A. Roychoudhury. "Melatonin as a regulator of abiotic stress tolerance in plants." In: Singh V. P., Singh S., Prasad S. M. (Eds.) Mechanisms Behind Phytohormonal Signaling and Crop Abiotic Stress Tolerance, New York, Nova Science Publishers, 2017, pp. 47–60.

Boutin, Jean A., Valérie Audinot, Gilles Ferry, and Philippe Delagrange. "Recently reported effects of melatonin." Trends in Pharmacological Sciences 8, no. 26 (2005): 412–419. https://doi.org/10.1016/j.tips.2005.06.006

Byeon, Yeong, and Kyoungwhan Back. "An increase in melatonin in transgenic rice causes pleiotropic phenotypes, including enhanced seedling growth, delayed flowering, and low grain yield." Journal of Pineal Research 56, no. 4 (2014): 408–414. https://doi.org/10.1111/jpi.12129

Carlson, N. Physiology of Behavior. 5th ed., Needham Heights, MA, Paramount Publishing, 1994.

Chang, Jingjing, Yanliang Guo, Jiayue Li, Zhuangzhuang Su, Chunxia Wang, Ruimin Zhang, Chunhua Wei, Jianxiang Ma, Xian Zhang, and Hao Li. "Positive interaction between H2O2 and Ca2+ mediates melatonin-induced CBF pathway and cold tolerance in watermelon (*Citrullus lanatus* L.)." Antioxidants 10, no. 9 (2021): 1457. https://doi.org/10.3390/antiox10091457

Chen, Jian, Hui Li, Kang Yang, Yongzhu Wang, Lifei Yang, Liangbin Hu, Ruixian Liu, and Zhiqi Shi. "Melatonin facilitates lateral root development by coordinating PAO-derived hydrogen peroxide and Rboh-derived superoxide radical." Free Radical Biology and Medicine 143 (2019): 534–544. https://doi.org/10.1016/j.freeradbiomed.2019.09.011

Dixon, Laura E., Sarah K. Hodge, Gerben van Ooijen, Carl Troein, Ozgur E. Akman, and Andrew J. Millar. "Light and circadian regulation of clock components aids flexible responses to environmental signals." New Phytologist 203, no. 2 (2014): 568–577. https://doi.org/10.1111/nph.12853

Dodd, Antony N., Neeraj Salathia, Anthony Hall, Eva Kévei, Réka Tóth, Ferenc Nagy, Julian M. Hibberd, Andrew J. Millar, and Alex A. R. Webb. "Plant circadian clocks increase photosynthesis, growth, survival, and competitive advantage." Science 309, no. 5734 (2005): 630–633. https://doi.org/10.1126/science.1115581

Dodd, Antony N., Neil Dalchau, Michael J. Gardner, Seong-Jin Baek, and Alex A. R. Webb. "The circadian clock has transient plasticity of period and is required for timing of nocturnal processes in a rabidopsis." New Phytologist 201, no. 1 (2014): 168–179. https://doi.org/10.1111/nph.12489

Duan, W., Q. Wang, H. Zhang, B. Xie, A. Li, F. Hou, S. Dong, B. Wang, Z. Qin, and L. Zhang. "Comparative study on carbon-nitrogen metabolism and endogenous hormone contents in normal and overgrown sweetpotato." South African Journal of Botany 115 (2018): 199–207. https://doi.org/10.1016/j.sajb.2017.11.016

El-Naby, Saied Kamel Mohamed Abd, Amr Abdelkhalek Ahmed Mohamed, and Yahia Ibrahim Mohamed El-Naggar. "Effect of Melatonin, GA3 and NAA on vegetative growth, yield and quality of 'canino' apricot fruits." Acta Scientiarum Polonorum Hortorum Cultus 18, no. 3 (2019): 167–174. https://doi.org/10.24326/asphc.2019.3.16

Faraone, Stephen V., and Kevin M. Antshel. "ADHD: Non-pharmacologic interventions." Child and Adolescent Psychiatric Clinics 23, no. 4 (2014): 13–14. https://doi.org/10.1016/j.chc.2014.06.004

Gao, Hui, Zheng Ke Zhang, Hong Kang Chai, Ni Cheng, Yue Yang, Dan Na Wang, Ting Yang, and Wei Cao. "Melatonin treatment delays postharvest senescence and regulates reactive oxygen species metabolism in peach fruit." Postharvest Biology and Technology 118 (2016): 103–110. https://doi.org/10.1016/j.postharvbio.2016.03.006

Hardeland, Rüdiger. "Melatonin in plants and other phototrophs: Advances and gaps concerning the diversity of functions." Journal of Experimental Botany 66, no. 3 (2015): 627–646. https://doi.org/10.1093/jxb/eru386

Hardeland, Rüdiger, and Carmen Rodríguez. "Minireview: Versatile melatonin: A pervasive molecule serves various functions in signaling and protection." Chronobiology International 12, no. 3 (1995): 157–165. https://doi.org/10.3109/07420529509057261

Hernández-Ruiz, Josefa, Antonio Cano, and Marino B. Arnao. "Melatonin acts as a growth-stimulating compound in some monocot species." Journal of Pineal Research 39, no. 2 (2005): 137–142. https://doi.org/10.1111/j.1600-079X.2005.00226.x

Huang, Bo, Yang-Er Chen, Yu-Qing Zhao, Chun-Bang Ding, Jin-Qiu Liao, Chao Hu, Li-Jun Zhou, Zhong-Wei Zhang, Shu Yuan, and Ming Yuan. "Exogenous melatonin alleviates oxidative damages and protects photosystem II in maize seedlings under drought stress." Frontiers in Plant Science 10 (2019): 677. https://doi.org/10.3389/fpls.2019.00677

Huang, Weiting, Xiaoyan Huang, and Zhongming Fang. "Melatonin-induced rhizome proliferation, differentiation, and rooting during rapid propagation of cymbidium goeringii and cymbidium faberi." Hort Science 57, no. 9 (2022): 1072–1077. https://doi.org/10.21273/HORTSCI16704-22

Jayarajan, Smruthi, and R. R. Sharma. "Melatonin: A blooming biomolecule for postharvest management of perishable fruits and vegetables." Trends in Food Science & Technology 116 (2021): 318–328. https://doi.org/10.1016/j.tifs.2021.07.034

Kolár, J. "Effects of melatonin on circadian rhythms and photoperiodism in higher plants." PhD diss., Faculty of Natural Sciences, Charles University, (2003).

Kolář, Jan, Carl Hirschie Johnson, and Ivana Macháčková. "Exogenously applied melatonin (N-acetyl-5-methoxytryptamine) affects flowering of the short-day plant Chenopodium rubrum." Physiologia Plantarum 118, no. 4 (2003): 605–612. https://doi.org/10.1034/j.1399-3054.2003.00114.x

Kolář, Jan, Ivana Macháčková, Josef Eder, Els Prinsen, Walter Van Dongen, Harry Van Onckelen, and Helena Illnerová. "Melatonin: Occurrence and daily rhythm in Chenopodium rubrum." Phytochemistry 44, no. 8 (1997): 1407–1413. https://doi.org/10.1016/S0031-9422(96)00568-7

Kolář, Jan, Ivana Machackova, H. Illnerova, E. Prinsen, W. Van Dongen, and H. Van Onckelen. "Melatonin in higher plant determined by radioimmunoassay and liquid chromatography-mass spectrometry." Biological Rhythm Research 26 (1995): 406–409.

Kołodziejczyk, I., and Malgorzata M. Posmyk. "Melatonin-a new plant biostimulator?" Journal of Elementology 21, no. 4 (2016): 1187–1198. https://doi.org/10.5601/jelem.2015.20.3.1012

Kotchoni, S. O., K. E. Larrimore, M. Mukherjee, C. F. Kempinski, and C. Barth. "Alterations in the endogenous ascorbic acid content affect flowering time in Arabidopsis." Plant Physiology, 149, no. 2 (2009), pp. 803–815. https://doi.org/10.1104/pp.108.132324

Lee, Hyoung Yool, Ok Jin Hwang, and Kyoungwhan Back. "Phytomelatonin as a signaling molecule for protein quality control via chaperone, autophagy, and ubiquitin-proteasome systems in plants." Journal of Experimental Botany (2022). https://doi.org/10.1093/jxb/erac002

Lei, Xiao-Yong, Rui-Yu Zhu, Gui-You Zhang, and Yao-Ren Dai. "Attenuation of cold-induced apoptosis by exogenous melatonin in carrot suspension cells: The possible involvement of polyamines." Journal of Pineal Research 36, no. 2 (2004): 126–131. https://doi.org/10.1046/j.1600-079X.2003.00106.x

Lerner, Aaron B., James D. Case, and Yoshiyata Takahashi. "Isolation of melatonin and 5-methoxyindole-3-acetic acid from bovine pineal glands." Journal of Biological Chemistry 235, no. 7 (1960): 1992–1997. https://doi.org/10.1016/S0021-9258(18)69351-2

Lerner, Aaron B., James D. Case, Yoshiyata Takahashi, Teh H. Lee, and Wataru Mori. "Isolation of melatonin, the pineal gland factor that lightens melanocyteS1." Journal of the American Chemical Society 80, no. 10 (1958): 2587. https://doi.org/10.1021/ja01543a060

Li, Hao, Jingjing Chang, Hejie Chen, Zhongyuan Wang, Xiurong Gu, Chunhua Wei, Yong Zhang, Jianxiang Ma, Jianqiang Yang, and Xian Zhang. "Exogenous melatonin confers salt stress tolerance to watermelon by improving photosynthesis and redox homeostasis." Frontiers in Plant Science 8 (2017): 295. https://doi.org/10.3389/fpls.2017.00295

Li, Junpeng, Chen Zhao, Mingjing Zhang, Fang Yuan, and Min Chen. "Exogenous melatonin improves seed germination in Limonium bicolor under salt stress." Plant Signaling & Behavior 14, no. 11 (2019): 1659705. https://doi.org/10.1080/15592324.2019.1659705

Li, Mingzhe, Fengying An, Wenyang Li, Mengdi Ma, Ying Feng, Xing Zhang, and Hongwei Guo. "Della proteins interact with FLC to repress flowering transition." Journal of Integrative Plant Biology 58, no. 7 (2016): 642–655. https://doi.org/10.1111/jipb.12451

Liang, Dong, Zhiyou Ni, Hui Xia, Yue Xie, Xiulan Lv, Jin Wang, Lijin Lin, Qunxian Deng, and Xian Luo. "Exogenous melatonin promotes biomass accumulation and photosynthesis of kiwifruit seedlings under drought stress." Scientia Horticulturae 246 (2019): 34–43. https://doi.org/10.1016/j.scienta.2018.10.058

Liang, Dong, Yanqiu Shen, Zhiyou Ni, Qin Wang, Zhi Lei, Nengqin Xu, Qunxian Deng et al. "Exogenous melatonin application delays senescence of kiwifruit leaves by regulating the antioxidant capacity and biosynthesis of flavonoids." Frontiers in Plant Science 9 (2018): 426. https://doi.org/10.3389/fpls.2018.00426

Liu, Jianlong, Rongrong Yue, Min Si, Meng Wu, Liu Cong, Rui Zhai, Chengquan Yang, Zhigang Wang, Fengwang Ma, and Lingfei Xu. "Effects of exogenous application of melatonin on quality and sugar metabolism in 'Zaosu' pear fruit." Journal of Plant Growth Regulation 38, no. 3 (2019): 1161–1169. https://doi.org/10.1007/s00344-019-09921-0

Liu, Jianlong, Rui Zhai, Fengxia Liu, Yingxiao Zhao, Huibin Wang, Lulu Liu, Chengquan Yang, Zhigang Wang, Fengwang Ma, and Lingfei Xu. "Melatonin induces parthenocarpy by regulating genes in gibberellin pathways of 'Starkrimson' pear (Pyrus communis L.)." Frontiers in Plant Science 9 (2018a): 946. https://doi.org/10.3389/fpls.2018.00946

Liu, Kun, Tongtong Jing, Yanan Wang, Xizhen Ai, and Huangai Bi. "Melatonin delays leaf senescence and improves cucumber yield by modulating chlorophyll degradation and photoinhibition of PSII and PSI." Environmental and Experimental Botany (2022): 104915. https://doi.org/10.1016/j.envexpbot.2022.104915

Liu, Zeng, Jun-song Cai, Jing-jing Li, Guang-yuan Lu, Chun-sheng Li, Gui-ping Fu, Xue-kun Zhang, Qing-yun Liu, Xi-ling Zou, and Yong Cheng. "Exogenous application of a low concentration of melatonin enhances salt tolerance in rapeseed (*Brassica napus* L.) seedlings." Journal of Integrative Agriculture 17, no. 2 (2018b): 328–335. https://doi.org/10.1016/S2095-3119(17)61757-X

Lynch, H. J., R. J. Wurtman, M. A. Moskowitz, M. C. Archer, and M. H. Ho. "Daily rhythm in human urinary melatonin." Science 187, no. 4172 (1975): 169–171. https://doi.org/10.1126/science.1167425

Macháčková, I., and J. Krekule. "Sixty-five years of searching for the signals that trigger flowering." Russian Journal of Plant Physiology 49, no. 4 (2002): 451–459. https://doi.org/10.1023/A:1016395405884

Mansouri, Sirvan, Hassan Sarikhani, Mohammad Sayyari, and Morteza Soleimani Aghdam. "Melatonin accelerates strawberry fruit ripening by triggering GAMYB gene expression and promoting ABA accumulation." Scientia Horticulturae 281 (2021): 109919. https://doi.org/10.1016/j.scienta.2021.109919

Meng, Jiang-Fei, Teng-Fei Xu, Chang-Zheng Song, Yong Yu, Fan Hu, Li Zhang, Zhen-Wen Zhang, and Zhu-Mei Xi. "Melatonin treatment of pre-veraison grape berries to increase size and synchronicity of berries and modify wine aroma components." Food Chemistry 185 (2015): 127–134. https://doi.org/10.1016/j.foodchem.2015.03.140

Mukherjee, Soumya. "Recent advancements in the mechanism of nitric oxide signaling associated with hydrogen sulfide and melatonin crosstalk during ethylene-induced fruit ripening in plants." Nitric Oxide 82 (2019): 25–34. https://doi.org/10.1016/j.niox.2018.11.003

Murch, S. J., and Saxena, P. K. "Melatonin: A potential regulator of plant growth and development?" In Vitro Cellular Developmental Biology-Plant 38 (2002): 531–536. https://doi.org/10.1079/IVP2002333

Murch, Susan J., Ali R. Alan, Jin Cao, and Praveen K. Saxena. "Melatonin and serotonin in flowers and fruits of Datura metel L." Journal of Pineal Research 47, no. 3 (2009): 277–283. https://doi.org/10.1111/j.1600-079X.2009.00711.x

Murch, Susan J., Skye S. B. Campbell, and Praveen K. Saxena. "The role of serotonin and melatonin in plant morphogenesis: Regulation of auxin-induced root organogenesis in in vitro-cultured explants of St. John's wort (Hypericum perforatum L.)." Vitro Cellular & Developmental Biology-Plant 37, no. 6 (2001): 786–793. https://doi.org/10.1007/s11627-001-0130-y

Nawaz, Khadija, Rimsha Chaudhary, Ayesha Sarwar, Bushra Ahmad, Asma Gul, Christophe Hano, Bilal Haider Abbasi, and Sumaira Anjum. "Melatonin as master regulator in plant growth, development and stress alleviator for sustainable agricultural production: Current status and future perspectives." Sustainability 13, no. 1 (2020): 294. https://doi.org/10.3390/su13010294

Park, Sangkyu, Tieu-Ngoc Nguyen Le, Yeong Byeon, Young Soon Kim, and Kyoungwhan Back. "Transient induction of melatonin biosynthesis in rice (*Oryza sativa* L.) during the reproductive stage." Journal of Pineal Research 55, no. 1 (2013): 40–45. https://doi.org/10.1111/jpi.12021

Pelagio-Flores, Ramón, Edith Muñoz-Parra, Randy Ortiz-Castro, and José López-Bucio. "Melatonin regulates Arabidopsis root system architecture likely acting independently of auxin signaling." Journal of Pineal Research 53, no. 3 (2012): 279–288. https://doi.org/10.1111/j.1600-079X.2012.00996.x

Poeggeler, Burkhard, Russel J. Reiter, Dun-Xian Tan, Li-Dun Chen, and Lucien C. Manchester. "Melatonin, hydroxyl radical-mediated oxidative damage, and aging: A hypothesis." Journal of Pineal Research 14, no. 4 (1993): 151–168. https://doi.org/10.1111/j.1600-079X.1993.tb00498.x

Qi, Zhen-Yu, Kai-Xin Wang, Meng-Yu Yan, Mukesh Kumar Kanwar, Dao-Yi Li, Leonard Wijaya, Mohammed Nasser Alyemeni, Parvaiz Ahmad, and Jie Zhou. "Melatonin alleviates high temperature-induced pollen abortion in Solanum lycopersicum." Molecules 23, no. 2 (2018): 386. https://doi.org/10.3390/molecules23020386

Rajora, Nitika, Sanskriti Vats, Gaurav Raturi, Vandana Thakral, Sukhmandeep Kaur, Vinaykumar Rachappanavar, Manish Kumar et al. "Seed priming with melatonin: A promising approach to combat abiotic stress in plants." Plant Stress 4 (2022): 100071. https://doi.org/10.1016/j.stress.2022.100071

Reiter, Russel J., Dario Acuña-Castroviejo, Dun-Xian Tan, and Susanne Burkhardt. "Free radical-mediated molecular damage: Mechanisms for the protective actions of melatonin in the central nervous system." Annals of the New York Academy of Sciences 939, no. 1 (2001): 200–215. https://doi.org/10.1111/j.1749-6632.2001.tb03627.x

Reiter, Russel J., Dun-Xian Tan, Zhou Zhou, Maria Helena Coelho Cruz, Lorena Fuentes-Broto, and Annia Galano. "Phytomelatonin: Assisting plants to survive and thrive." Molecules 20, no. 4 (2015): 7396–7437. https://doi.org/10.3390/molecules20047396

Samanta, S., A. Banerjee, and A. Roychoudhury. "Arsenic toxicity is counteracted by exogenous application of melatonin to different extents in arsenic-susceptible and arsenic-tolerant rice cultivars." Journal of Plant Growth Regulation 41 (2022): 2210–2231.

Samanta, S., A. Singh, A. Banerjee, and A. Roychoudhury. "Exogenous supplementation of melatonin alters representative organic acids and enzymes of respiratory cycle as well as sugar metabolism during arsenic stress in two contrasting indica rice cultivars." Journal of Biotechnology 324 (2020): 220–232.

Sarropoulou, Virginia, Kortessa Dimassi-Theriou, Ioannis Therios, and Magdalene Koukourikou-Petridou. "Melatonin enhances root regeneration, photosynthetic pigments, biomass, total carbohydrates and proline content in the cherry rootstock PHL-C (Prunus avium× Prunus cerasus)." Plant Physiology and Biochemistry 61 (2012): 162–168. https://doi.org/10.1016/j.plaphy.2012.10.001

Sharif, Rahat, Chen Xie, Haiqiang Zhang, Marino B. Arnao, Muhammad Ali, Qasid Ali, Izhar Muhammad et al. "Melatonin and its effects on plant systems." Molecules 23, no. 9 (2018): 2352. https://doi.org/10.3390/molecules23092352

Shi, Haitao, Yinhua Chen, Dun-Xian Tan, Russel J. Reiter, Zhulong Chan, and Chaozu He. "Melatonin induces nitric oxide and the potential mechanisms relate to innate immunity against bacterial pathogen infection in Arabidopsis." Journal of Pineal Research 59, no. 1 (2015a): 102–108. https://doi.org/10.1111/jpi.12244

Shi, Haitao, Chuan Jiang, Tiantian Ye, Dun-Xian Tan, Russel J. Reiter, Heng Zhang, Renyi Liu, and Zhulong Chan. "Comparative physiological, metabolomic, and transcriptomic analyses reveal mechanisms of improved abiotic stress resistance in bermudagrass [*Cynodon dactylon* (L). Pers.] by exogenous melatonin." Journal of Experimental Botany 66, no. 3 (2015b): 681–694. https://doi.org/10.1093/jxb/eru373

Shi, Haitao, Yunxie Wei, Qiannan Wang, Russel J. Reiter, and Chaozu He. "Melatonin mediates the stabilization of DELLA proteins to repress the floral transition in A rabidopsis." Journal of Pineal Research 60, no. 3 (2016): 373–379. https://doi.org/10.1111/jpi.12320

Su, Tong, Xuezhi Li, Mingyue Yang, Qun Shao, Yanxiu Zhao, Changle Ma, and Pingping Wang. "Autophagy: An intracellular degradation pathway regulating plant survival and stress response." Frontiers in Plant Science 11 (2020): 164. https://doi.org/10.3389/fpls.2020.00164

Szafrańska, Katarzyna, Russel J. Reiter, and Małgorzata M. Posmyk. "Melatonin application to *Pisum sativum* L. seeds positively influences the function of the photosynthetic apparatus in growing seedlings during paraquat-induced oxidative stress." Frontiers in Plant Science 7 (2016): 1663. https://doi.org/10.3389/fpls.2016.01663

Tan, Dun-Xian, Rudiger Hardeland, Lucien C. Manchester, Ahmet Korkmaz, Shuran Ma, Sergio Rosales-Corral, and Russel J. Reiter. "Functional roles of melatonin in plants, and perspectives in nutritional and agricultural science." Journal of Experimental Botany 63, no. 2 (2012): 577–597. https://doi.org/10.1093/jxb/err256

Tan, Xiao-li, Ya-ting Zhao, Wei Shan, Jian-fei Kuang, Wang-jin Lu, Xin-guo Su, Neng-guo Tao, Prakash Lakshmanan, and Jian-ye Chen. "Melatonin delays leaf senescence of postharvest Chinese flowering cabbage through ROS homeostasis." Food Research International 138 (2020): 109790. https://doi.org/10.1016/j.foodres.2020.109790

Tanimoto, Eiichi. "Regulation of root growth by plant hormones-roles for auxin and gibberellin." Critical Reviews in Plant Sciences 24, no. 4 (2005): 249–265. https://doi.org/10.1080/07352680500196108

Todde, Virginia, Marten Veenhuis, and Ida J. van der Klei. "Autophagy: Principles and significance in health and disease." Biochimica et Biophysica Acta (BBA)-Molecular Basis of Disease 1792, no. 1 (2009): 3–13. https://doi.org/10.1016/j.bbadis.2008.10.016

Vanecek, Jiri. "Cellular mechanisms of melatonin action." Physiological Reviews 78, no. 3 (1998): 687–721. https://doi.org/10.1152/physrev.1998.78.3.687

Wang, K., S. Cai, Q. Xing, Z. Qi, V. Fotopoulos, and J. Yu. "Melatonin delays dark-induced leaf senescence by inducing miR171b expression in tomato." Journal of Pineal Research 72, no. 3 (2022): e12792. https://doi.org/10.1111/jpi.12792

Wang, Ping, Xun Sun, Na Wang, Dun-Xian Tan, and Fengwang Ma. "Melatonin enhances the occurrence of autophagy induced by oxidative stress in Arabidopsis seedlings." Journal of Pineal Research 58, no. 4 (2015): 479–489. https://doi.org/10.1111/jpi.12233

Wang, Ping, Xun Sun, Yinpeng Xie, Mingjun Li, Wei Chen, Sheng Zhang, Dong Liang, and Fengwang Ma. "Melatonin regulates proteomic changes during leaf senescence in Malus hupehensis." Journal of Pineal Research 57, no. 3 (2014): 291–307. https://doi.org/10.1111/jpi.12169

Wang, Qiannan, Bang An, Yunxie Wei, Russel J. Reiter, Haitao Shi, Hongli Luo, and Chaozu He. "Melatonin regulates root meristem by repressing auxin synthesis and polar auxin transport in Arabidopsis." Frontiers in Plant Science 7 (2016): 1882. https://doi.org/10.3389/fpls.2016.01882

Wang, Xin, Fang Li, Zhenyuan Chen, Bingxian Yang, Setsuko Komatsu, and Shunli Zhou. "Proteomic analysis reveals the effects of melatonin on soybean root tips under flooding stress." Journal of Proteomics 232 (2021): 104064. https://doi.org/10.1016/j.jprot.2020.104064

Wang, Y., Y. J. Wang, J. G. Hao, and J. F. Jia. "Defend effects of melatonin on mung bean UV-B irradiation." Acta Photonica Sinica 38 (2009): 2629–2633.

Weeda, Sarah, Na Zhang, Xiaolei Zhao, Grace Ndip, Yangdong Guo, Gregory A. Buck, Conggui Fu, and Shuxin Ren. "Arabidopsis transcriptome analysis reveals key roles of melatonin in plant defense systems." PloS One 9, no. 3 (2014): e93462.

Wei, Jian, Dong-Xu Li, Jia-Rong Zhang, Chi Shan, Zed Rengel, Zhong-Bang Song, and Qi Chen. "Phytomelatonin receptor PMTR 1-mediated signaling regulates stomatal closure in Arabidopsis thaliana." Journal of Pineal Research 65, no. 2 (2018): e12500. https://doi.org/10.1111/jpi.12500

Wei, Long, Haiyan Zhao, Baoxiang Wang, Xinyi Wu, Rujia Lan, Xu Huang, Bo Chen et al. "Exogenous melatonin improves the growth of rice seedlings by regulating redox balance and ion homeostasis under salt stress." Journal of Plant Growth Regulation 41, no. 6 (2022): 2108–2121. https://doi.org/10.1007/s00344-021-10417-z

Wei, Wei, Qing-Tian Li, Ya-Nan Chu, Russel J. Reiter, Xiao-Min Yu, Dan-Hua Zhu, Wan-Ke Zhang et al. "Melatonin enhances plant growth and abiotic stress tolerance in soybean plants." Journal of Experimental Botany 66, no. 3 (2015): 695–707. https://doi.org/10.1093/jxb/eru392

Wei, Yunxie, Yujing Bai, Xiao Cheng, Binbin Zhu, Russel J. Reiter, and Haitao Shi. "The dual roles of melatonin biosynthesis enzymes in the coordination of melatonin biosynthesis and autophagy in cassava." Journal of Pineal Research 69, no. 1 (2020): e12652. https://doi.org/10.1111/jpi.12652

Wolf, Karel, Jan Kolář, Erwin Witters, Walter van Dongen, Harry van Onckelen, and Ivana Macháčková. "Daily profile of melatonin levels in Chenopodium rubrum L. depends on photoperiod." Journal of Plant Physiology 158, no. 11 (2001): 1491–1493. https://doi.org/10.1078/0176-1617-00561

Xiao, Shuang, Liantao Liu, Hao Wang, Dongxiao Li, Zhiying Bai, Yongjiang Zhang, Hongchun Sun, Ke Zhang, and Cundong Li. "Exogenous melatonin accelerates seed germination in cotton (*Gossypium hirsutum* L.)." PloS One 14, no. 6 (2019): e0216575. https://doi.org/10.1371/journal.pone.0216575

Xu, X. D., Y. Sun, X. Q. Guo, B. Sun, and J. Zhang. "Effects of exogenous melatonin on ascorbate metabolism system in cucumber seedlings under high temperature stress." Yingyong Shengtai Xuebao 21, no. 10 (2010): 2580–2586.

Yakuboğlu, Gökçen, Şebnem Köklü Ardıç, Yakup Cuci, and Ahmet Korkmaz. "Mitigation of chilling stress effects on eggplant seedlings by exogenous application of melatonin." Turkish Journal of Agriculture-Food Science and Technology 10, no. 4 (2022): 562–568. https://doi.org/10.24925/turjaf.v10i4.562-568.4601

Yan, Li, Chang Liu, Qinghua Shi, Fengjuan Yang, and Min Wei. "Mixed red and blue light promotes ripening and improves quality of tomato fruit by influencing melatonin content." Environmental and Experimental Botany 185 (2021): 104407. https://doi.org/10.1016/j.envexpbot.2021.104407

Ye, Jun, Wenjia Yang, Yulin Li, Shiwen Wang, Lina Yin, and Xiping Deng. "Seed pre-soaking with melatonin improves wheat yield by delaying leaf senescence and promoting root development." Agronomy 10, no. 1 (2020): 84. https://doi.org/10.3390/agronomy10010084

Yu, Yicheng, Aimin Wang, Xiang Li, Meng Kou, Wenjun Wang, Xianyang Chen, Tao Xu et al. "Melatonin-stimulated triacylglycerol breakdown and energy turnover under salinity stress contributes to the maintenance of plasma membrane H+-ATPase activity and K+/Na+ homeostasis in sweet potato." Frontiers in Plant Science 9 (2018): 256. https://doi.org/10.3389/fpls.2018.00256

Zhai, Rui, Jianlong Liu, Fengxia Liu, Yingxiao Zhao, Lulu Liu, Chen Fang, Huibin Wang et al. "Melatonin limited ethylene production, softening and reduced physiology disorder in pear (*Pyrus communis* L.) fruit during senescence." Postharvest Biology and Technology 139 (2018): 38–46. https://doi.org/10.1016/j.postharvbio.2018.01.017

Zhang, Hai-Jun, N. A. Zhang, Rong-Chao Yang, Li Wang, Qian-Qian Sun, Dian-Bo Li, Yun-Yun Cao et al. "Melatonin promotes seed germination under high salinity by regulating antioxidant systems, ABA and GA 4 interaction in cucumber (*Cucumis sativus* L.)." Journal of Pineal Research 57, no. 3 (2014): 269–279. https://doi.org/10.1111/jpi.12167

Zhang, Na, Hai-Jun Zhang, Qian-Qian Sun, Yun-Yun Cao, Xingsheng Li, Bing Zhao, Ping Wu, and Yang-Dong Guo. "Proteomic analysis reveals a role of melatonin in promoting cucumber seed germination under high salinity by regulating energy production." Scientific Reports 7, no. 1 (2017): 1–15. https://doi.org/10.1038/s41598-017-00566-1

Zhang, Na, Bing Zhao, Hai-Jun Zhang, Sarah Weeda, Chen Yang, Zi-Cai Yang, Shuxin Ren, and Yang-Dong Guo. "Melatonin promotes water-stress tolerance, lateral root formation, and seed germination in cucumber (*Cucumis sativus* L.)." Journal of Pineal Research 54, no. 1 (2013): 15–23. https://doi.org/10.1111/j.1600-079X.2012.01015.x

Zhang, Yuexin, Yapeng Fan, Cun Rui, Hong Zhang, Nan Xu, Maohua Dai, Xiugui Chen et al. "Melatonin improves cotton salt tolerance by regulating ROS scavenging system and Ca 2+ signal transduction." Frontiers in Plant Science 12 (2021): 1239. https://doi.org/10.3389/fpls.2021.693690

Zhang, Z, Q. Hu, Y. Liu, P. Cheng, H. Cheng, W. Liu, X. Xing, Z. Guan, W. Fang, S. Chen, J. Jiang, Zhang Zixin, Qian Hu, Yanan Liu, Peilei Cheng, Hua Cheng, Weixin Liu, Xiaojuan Xing et al. "Strigolactone represses the synthesis of melatonin, thereby inducing floral transition in Arabidopsis thaliana in an FLC-dependent manner." Journal of Pineal Research 67, no. 2 (2019): e12582. https://doi.org/10.1111/jpi.12582

Zhao, Daqiu, Rong Wang, Ding Liu, Yanqing Wu, Jing Sun, and Jun Tao. "Melatonin and expression of tryptophan decarboxylase gene (TDC) in herbaceous peony (Paeonia lactiflora Pall.) flowers." Molecules 23, no. 5 (2018): 1164. https://doi.org/10.3390/molecules23051164

Zuo, Bixiao, Xiaodong Zheng, Pingli He, Lin Wang, Qiong Lei, Chao Feng, Jingzhe Zhou, Qingtian Li, Zhenhai Han, and Jin Kong. "Overexpression of MzASMT improves melatonin production and enhances drought tolerance in transgenic Arabidopsis thaliana plants." Journal of Pineal Research 57, no. 4 (2014): 408–417. https://doi.org/10.1111/jpi.12180

3 Melatonin

An Emerging Phytohormone with Multifaceted Functions in the Regulation of Plant Growth and Development

Vishmita Sethi, Gouranga Upadhyaya and Sreeramaiah N. Gangappa

3.1 PHYTOMELATONIN: ORIGIN, OCCURRENCE AND DIVERSE ROLES IN PLANTS

Melatonin (*N-acetyl*-5-methoxy-tryptamine) was discovered in the bovine pineal gland in 1958 by Lerner et al. (1958). Melatonin was considered a neurohormone for the longest time possible until its role was found in plants nearly forty years later (Reiter, 1991). It was discovered in Japanese morning glory (*Pharbitis nil*) (Vantassel et al., 1993) and subsequently in edible plant materials independently by Dubbels et al. (1995), Kolar et al. (1997) and Hattori et al. (1995). The past decade has seen research on melatonin exponentially. Since then, melatonin has been found to control various biological processes in plants. Consequently, with new roles assigned to melatonin in plants, the number of publications relating to these studies has also increased in the past decade (Reiter et al., 2015). Moreover, the structural similarity of melatonin to auxin with a common biosynthetic precursor, that is, tryptophan, makes it a vital molecule to investigate in plant growth and development. Melatonin can be considered a universal metabolite involved in diverse physiological processes in unicellular organisms, microorganisms, algae, plants and vertebrates (Banerjee and Roychoudhury, 2017). Melatonin occurrence can be traced back to the origin of life and is known to have evolved in bacteria. α-proteobacteria and photosynthetic cyanobacteria show accumulation of melatonin. According to the early endosymbiotic theory of evolution, it is known that primitive eukaryotes phagocytosed bacteria for nutrition, after which the bacteria eventually formed a symbiotic association with the host. The ingested α-proteobacteria developed into mitochondria, and the cyanobacteria evolved into the chloroplast. During evolution, both these organelles retained their ability to synthesize melatonin. Therefore, during the course of evolution, all the species that existed or are existing produce melatonin through these two organelles, that is, chloroplast and mitochondria in plants and only mitochondria in animals. Functions of melatonin, especially in animals, occur via receptor-mediated and receptor-independent pathways. The main physiological processes in animals controlled by melatonin but not restricted to these are the regulation of immunity, circadian clocks, sleep cycles, mood, reproduction, locomotion and food ingestion (Carrillo-Vico et al., 2013; Pandi-Perumal et al., 2008; Maronde and Stehle, 2007; Hardeland et al., 2012; Reiter et al., 2009).

Since its discovery, melatonin has been a potent antioxidant in response to biotic and abiotic stresses in plants. Increased research on the metabolite has contributed to discovering an array of

DOI: 10.1201/9781003343752-3

roles in plants. This includes numerous processes occurring in various developmental stages in plants ranging from seed germination, seedling growth, vegetative growth, modulating circadian rhythms, flowering control, stomatal movements and senescence (Xiao et al., 2019; Erdal, 2019; Castanares and Bouzo, 2019; Arnao and Hernandez-Ruiz, 2014a; Lee et al., 2019; Wei et al., 2018; Tan et al., 2020; Sharif et al., 2018). Therefore, considering the multiplicity of functions regulated by melatonin, engineering these for sustainable agriculture can provide new paradigms for phytomelatonin research in the future.

3.2 BIOSYNTHESIS, DYNAMICS AND REGULATION OF PHYTOMELATONIN

Melatonin (MT) is ubiquitously expressed in plant organs such as germinated seeds, shoots, roots, flowers, fruits, coleoptiles, bulbs and leaves, etc., in most plant species studied (Hardeland, 2015, 2016). The biosynthesis of melatonin is influenced by external factors and also varies with time and the developmental stages of the plant. As the name suggests, melatonin is *N-acetyl*-5-methoxytryptamine. It is an indolic derivative of serotonin, and the biogenic amines serotonin and melatonin are synthesized from a common precursor, the amino acid tryptophan. The biosynthesis pathway in vertebrates is well known (Zhao et al., 2019), and an extensive study on plant melatonin has also elucidated the pathway in plants (Tan et al., 2016; Back et al., 2016, 2021). In plants, tryptophan decarboxylase (TDC) mediated the conversion of tryptophan to tryptamine. Tryptamine is converted to 5-hydroxytryptamine (serotonin) by tryptamine 5-hydroxylase (T5H), and serotonin N-acetyl transferase (SNAT) further catalyzes the N-acetylation of serotonin. There are several SNATs known in plants as compared to only one known in animals, which accounts for more levels of phytomelatonin accumulation in plants as compared to animals (Wang et al., 2022b; Tan et al., 2016). Next, a hydroxyindole-O-methyltransferase, acetyl serotonin methyl transferase (ASMT) methylates N-acetylserotonin to generate melatonin. Other alternative pathways leading to the final product of melatonin exist and involve alternate intermediates. The availability and expression of enzymes in different cellular compartments such as cytosol, chloroplast and endoplasmic reticulum determines the intermediates generated and thus subsequent reactions leading to the production of the final product melatonin. One of the alternative enzymes causing the methylation of N-acetylserotonin in plants is caffeic acid O-methyltransferase (COMT) (Sun et al., 2021). COMT also acts on multiple substrates such as caffeic acid and quercetin. Along with ASMT, it can convert serotonin into 5-methoxy tryptamine, which is then acetylated to melatonin by SNAT (Sun et al., 2021; Wu et al., 2021). This forms an alternative pathway for the biosynthesis of melatonin. Another alternative route of melatonin synthesis is via the production of N-acetyl tryptamine from tryptamine through SNAT (Wu et al., 2021). T5H enzyme modulates the generation of N-acetyl serotonin, which eventually converts to melatonin by ASMT or COMT. Most enzymes mediating melatonin biosynthesis are well characterized in Arabidopsis and rice, except one, that is, tryptophan hydrolase, which is found in animals only. It is interesting to note that in addition to the intrinsic factors that control melatonin biosynthesis, external factors also significantly influence the process. Environmental changes majorly affect plant growth and development. Also, they are closely associated with biosynthesis and regulation of metabolites such as carotenoids, flavonoids and other nutrient and anti-nutrient factors. While mammalian systems have well-known receptor-mediated gene expression, the plant systems lack receptor-dependent responses having more chemical responses (Park, 2011). However, more recently, a receptor of phytomelatonin was discovered called the CAND2/PMTR1 by Wei et al. (2018). CAND2/PMTR1 localizes to the plasma membrane and interacts with GPA1 (G protein alpha subunit 1) to regulate stomatal movement in *Arabidopsis* via the NADPH oxidase-mediated reactive oxygen species (ROS) signaling pathway.

3.3 CROSSTALK OF MELATONIN WITH OTHER PLANT HORMONES AND ITS INFLUENCES ON PLANT GROWTH AND DEVELOPMENT

Melatonin, an emerging growth regulator, interacts with plant growth hormones at various levels and stages of plant growth and development (Arnao and Hernandez-Ruiz, 2018, 2020). Melatonin, like other growth hormones, is active at low concentrations. The exogenous application of melatonin, including the concentration, time and application method, has to be comprehensively selected to maximize the effects of application. Table 3.1 summarizes these parameters and potential growth and developmental impact in different plant species. The interactions with plant hormones are also a significant part of the defence strategy employed by plants when challenged by biotic and abiotic stresses (Samanta et al., 2021). It can be considered a signaling hub of interactions with auxin, cytokinin, abscisic acid, ethylene, salicylic acid and jasmonic acid. With the increasing evidence, it can be considered a master modulator of plant growth and defence responses (Arnao et al., 2017). In addition to these interactions, the intrinsic levels of phytohormones get altered owing to external applications of melatonin.

How these processes are modulated at the molecular levels is still an open question and is an upcoming field of research that, when explored, may significantly impact crop improvement (Sun et al., 2021). Melatonin is a growth regulator for both above-ground (aerial) and below-ground (roots) tissues. Auxin and melatonin share huge similarities, whether the same precursor, that is, tryptophan in this case or the structural similarities of melatonin to indole-3- acetic acid (Arnao and Hernández-Ruiz, 2006). All these make melatonin a predominant candidate for controlling growth and development responses and a mimic for functional auxin (Edgerton et al., 1994). During seedling development, melatonin promotes cotyledon expansion and hypocotyl growth, similar to IAA (Hernandez-Ruiz et al., 2004, 2005).

Melatonin also increases the generation of lateral and adventitious roots across plant species (Arnao and Hernández-Ruiz, 2007). There are abundant studies on root growth and development aspects in Arabidopsis (Pelagio-Flores et al., 2012; Koyama et al., 2013) and other plant species such as *Oryza, Triticum, Brassica, Hordeum, Avena*, tomato and maize etc. In tomato and mustard, low melatonin levels have been shown to increase endogenous IAA levels, resulting in increased root growth (Chen et al., 2009; Wen et al., 2016). Melatonin treatment reduced the expression of IAA biosynthesis genes, *YUCCA* and *TAA*, which are the key enzymes involved in auxin biosynthesis (Wang et al., 2016). Nitric oxide (NO) is an emerging pivotal player in auxin and melatonin

TABLE 3.1
Growth and Developmental Effects on the Tissue-Specific Treatment of Exogenous Melatonin Applications in Different Plant Species

Plant species	Tissue treatment	Effective melatonin concentration (μM)	Growth and developmental effects	References
Gossypium hirsutum	Soaking Seeds	20	Seed germination increased; antioxidant enzyme activity increased	(Xiao et al., 2019)
Solanum lycopersicum	Soaking Seeds	100	Fruit yield and quality improved	(Liu et al., 2016)
Cucumis melo L.	Soaking seeds and irrigating plants	50	Improved growth and enhanced chlorophyll	(Castanares and Bouzo, 2019)
Zea mays	Spraying seedlings	100–1000	Seedling growth promoted	(Erdal, 2019)
Brassica rapa	Immersing leaves	100	Leaf senescence delayed	(Tan et al., 2020)

TABLE 3.1 ***(Continued)***

Growth and Developmental Effects on the Tissue-Specific Treatment of Exogenous Melatonin Applications in Different Plant Species

Plant species	Tissue treatment	Effective melatonin concentration (µM)	Growth and developmental effects	References
Fragaria x ananassa	Spaying leaves	100	Fruit yields and quality increased	(Zahedi et al., 2020)
Hordeum vulgare	Immersing leaves	0.01–1	Dark-induced leaf senescence delayed; chlorophyll degradation decreased	(Arnao and Hernándes-Ruiz, 2009)
Malus domestica	Immersing leaves	1000–10000	Leaf senescence delayed	(Wang et al., 2013)
Prunus avium	Spraying leaves and fruits	50–100	Anthocyanin content, fruit yield and quality improved	(Xia et al., 2020)
Malus domestica	Immersing fruits	50	Fruit quality improved and delayed ripening	(Fan et al., 2022)
Mangifera indica	Immersing fruits	500	Fruit quality improved and delayed ripening	(Liu et al., 2020)
Pyrus communis	Immersing fruits	100	Fruit quality improved, delayed ethylene production	(Zhai et al., 2018)
Solanum lycopersicum	Immersing fruits	50	Fruit quality improved	(Sun et al., 2015)
Ziziphus jujuba	Immersing fruits	25	Improved antioxidative capacity, post-harvest fruit quality improved	(Tang et al., 2020)
Vitis vinifera	Soaking fruits	50–200	Fruit quality improved, delayed fruit senescence, ethylene biosynthesis increased	(Wang et al., 2022a; Ma et al., 2021)
Vitis vinifera	Spraying fruits	100–500	Fruit quality and yield increased, fruit coloration and sugar accumulation enhanced	(Meng et al., 2017; Xia et al., 2021)
Brassica oleracea	Immersing florets	1–100	Post-harvest senescence delayed, positive regulation of glucoraphanin-sulforaphane system	(Miao et al., 2020; Wei et al., 2020)
Solanum lycopersicum	Roots irrigated from flowering stage to maturity	100	Fruit yield and quality improved	(Liu et al., 2016)
Malus domestica	Roots irrigated	100	Leaf senescence delayed	(Wang et al., 2013)
Arabidopsis thaliana	Roots immersed	0.01	Primary root growth promoted	(Yang et al., 2021)
Solanum lycopersicum	Seedlings pruned off primary roots	50	Adventitious roots induced	(Wen et al., 2016)
Cucumis melo L.	Roots treated	300	Root vigour increased and antioxidant enzyme levels increased, proline and malondialdehyde (MDA) contents decreased in the presence of copper	(Hu et al., 2020)
Chenopodium rubrum	Cotyledon and plumules of seedlings treated	500	Reduced flowering in plants	(Kolar et al., 2003)

signaling crosstalk. NO regulates auxin-mediated processes such as adventitious root formation, root hair development and root gravitropism (Yu et al., 2014). These studies of auxin and melatonin crosstalk will be helpful in engineering plants with increased water and nutrient uptake capability.

Melatonin and cytokinin are known to retard the induced-senescence process. Anti-senescence function of melatonin is also known in several species, such as apple (Wang et al., 2013), rice (Liang et al., 2015), Arabidopsis (Shi et al., 2015) and barley (Arnao and Hernández-Ruiz, 2009). Barley leaves treated with melatonin delay dark-induced senescence and slow down chlorophyll loss in detached leaves in a concentration-dependent manner (Arnao and Hernández-Ruiz, 2009). Chlorophyll retention in treated samples compared to the control samples was proportional to the concentration of kinetin used. A report from Weeda et al. (2014) proposed that two chlorophyll-degrading enzymes, chlorophyllase and pheophorbide a oxygenase, were down regulated by melatonin in *Arabidopsis*. It is also reported that two cytokinin biosynthetic genes (LpIPT2 and LpOG1) were up regulated in stress conditions through melatonin. Two types of transcription factors active during cytokinin signaling, Type A-ARRs and Type B- ARRs, were differentially regulated during heat stress alleviation through interactions of melatonin and cytokinin (Zhang et al., 2017). Transgenic rice plants with increased melatonin levels produce larger seeds, indicative of increased cytokinin synthesis (Huangfu et al., 2022). Melatonin also plays an exciting role in alleviating photo-induced damage by protecting the pigments such as chlorophylls and carotenoids (Samanta et al., 2022). Melatonin and cytokinin crosstalk alleviate heat and salinity stress and act as buffer zones for plant systems to combat the ill effects of these stresses (Arnao and Hernandez-Ruiz, 2014a, 2014b).

Gibberellic acid (GA) promotes plant growth through cell elongation. It is also responsible for seed germination, flowering and fruit ripening responses. There are very few reports that suggest the crosstalk of melatonin with GA. A recent report suggests ROS generation through 2-hydroxymelatonin treatments that lead to increased seed germination, followed by an increase in GA biosynthesis and expression of GA-responsive genes in *Arabidopsis* (Lee and Back, 2022). GA also acts as an elicitor of melatonin synthesis in rice seedlings and immature seeds. Brassinosteroids (BR) promote GA biosynthesis and signaling, and these hormones cooperatively function in melatonin production (Wang et al., 2022b; Tong et al., 2014; Unterholzner et al., 2015; Banerjee and Roychoudhury, 2019). The DELLA proteins are negative regulators of GA-mediated flowering responses. It was shown that melatonin, when applied exogenously, delays floral transitions in *Arabidopsis* (Shi et al., 2016b). The exogenously applied melatonin stabilizes DELLA proteins (REPRESSOR of ga1–3 (RGA) and RGA-LIKE3 (RGL3) but does not affect their transcripts or the endogenous GA levels.

Jasmonic acid is another well-known lipid derivative hormone which has roles in plant growth and development (Huang et al., 2017). Crosstalk between melatonin and jasmonic acid occurs during developmental processes in plants. An interesting study suggested that melatonin suppresses jasmonate-induced anthocyanin production but involves the abolishment of jasmonic acid signaling (Ai et al., 2018). Another study highlights the role of melatonin in rootstock-mediated cold tolerance. Melatonin increases the accumulation of methyl jasmonic acid (MeJA) and H_2O_2. MeJA in a regulatory feedback loop increases melatonin accumulation, which increases H_2O_2 accumulation and cold tolerance (Li et al., 2021). In *Solanum lycopersicum*, melatonin and JA act synergistically in cold tolerance. During cold stress, JA and MT were substantially accumulated, and foliar application of methyl jasmonate and melatonin promoted cold tolerance and declined H_2O_2 accumulation in tomato plants (Ding et al., 2021). MeJA also increased the expression of two critical melatonin biosynthetic enzymes, namely *SlSNAT* and *SlAMST* (Ding et al., 2021). Thus, tomato varieties can be genetically manipulated for the JA and MT pathways to develop cold-resistant tomato cultivars, which can have significant agronomic potential (Ding et al., 2021).

An interesting study also highlights the effect of melatonin application on reduced ethylene production in apples through the down regulation of ethylene biosynthetic genes. The activity of three enzymes, ascorbate peroxidase (APX), superoxide dismutase (SOD) and catalase (CAT)

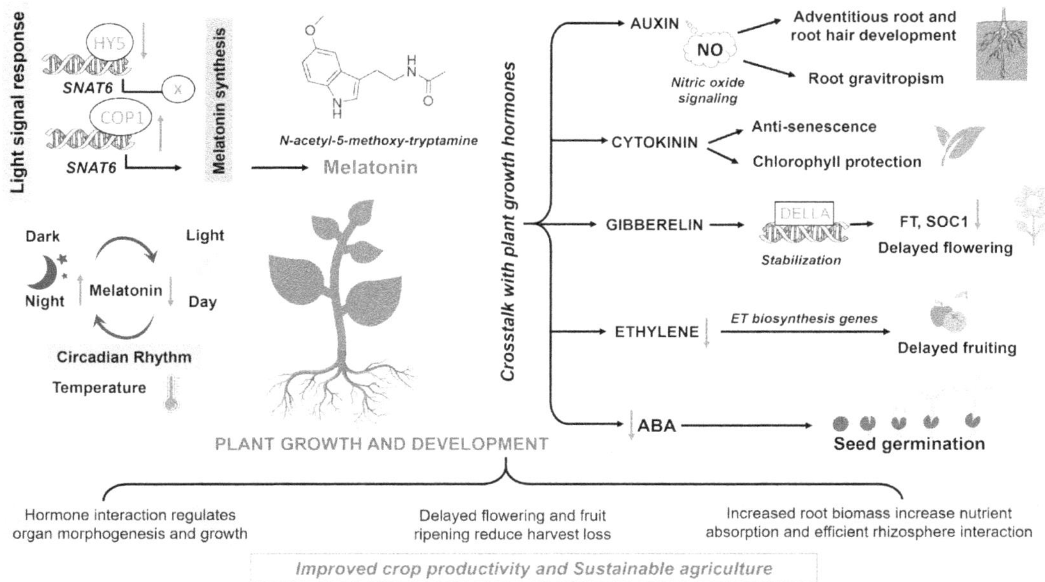

FIGURE 3.1 Roles of melatonin in controlling plant growth and developmental responses. Melatonin crosstalks with various plant growth hormones to regulate diverse aspects of plant growth responses such as adventitious root and root hair development that is controlled by auxin in a nitric oxide-dependent manner. MT crosstalks with cytokinin to control anti-senescence responses and chlorophyll protection. MT interacts with GA to stabilize DELLA, leading to a delayed flowering response. MT's interaction with ET lowers the gene expression of ethylene-responsive genes, leading to delayed fruiting. MT promotes seed germination by inhibiting ABA responses. In response to light signals, HY5 lowers the expression of the rate-limiting enzyme SNAT6, resulting in the down regulation of melatonin biosynthesis, whereas in the presence of COP1, the melatonin levels go up with an increased expression of SNAT6. Melatonin levels are also regulated by circadian rhythm and temperature. MT levels go up in dark and elevated temperature. The major agricultural implications from a larger perspective are given in purple boxes. All these functions of MT have a concerted effect on improving crop productivity and contributing to sustainable agriculture in the future.

was up regulated in melatonin-treated fruits, thus increasing the post-harvest quality of apples (Onik et al., 2021). If manipulated and exploited to their potential, these pathways can significantly improve the post-harvest quality and storage of fruits, thus leading to enhanced productivity and yield (**Figure 3.1**).

3.4 MELATONIN IN ACTION: CONTROL OF OTHER PHYSIOLOGICAL PROCESSES

3.4.1 Role of Melatonin in Regulating Light Responses in Plants

Light, as an environmental factor, is essential for the growth and development of plants. Light intensity, wavelength, direction and duration all determine plant growth. Photoreceptors mainly perceive light, and the downstream signaling components such as Elongated Hypocotyl 5 (HY5), Constitutive Photomorphogenic 1 (COP1) and Phytochrome Interacting Factor (PIFs) etc. are the major regulators of light signaling. In the dark, COP1, an E3-ubiquitin ligase, degrades positive regulators of light signaling such as HY5 in the nucleus, inhibiting their activity. In the light, it shuttles to the cytosol, thus leaving HY5 active in the nucleus to transcriptionally regulate its targets (Gangappa and Botto, 2016). Most studies focus on the physiological implications of these interactions, but an interesting study by Wang et al. (2022b) suggests the transcriptional control of melatonin biosynthesis via COP1 and HY5. They identified SNAT6 as the prime enzyme out of a total of nine homologous SNATs in *Arabidopsis* to be expressed in the dark.

Interestingly, the study also showed that HY5 specifically binds to the G-box of the *SNAT6* promoter and negatively regulates its expression. The *HY5* mutants showed elevated melatonin production in the dark. The *COP1* mutants showed decreased melatonin production and reduced *SNAT6* expression in the dark conditions. HY5 also participates in cotyledon opening by regulating melatonin biosynthesis (Wang et al., 2022b). However, light can also have detrimental effects on plants, such as causing photodamage to cellular components and the generation of ROS. UV light is harmful to plants, and melatonin secreted by plants is known to combat these stresses (Tan et al., 2007; Afreen et al., 2006; Wei et al., 2019).

Changes in light quantity and quality can have stress-like effects on plants. Recently, it was reported for the first time that melatonin levels transiently go up during high light stress (Lee et al., 2018). High light stress also induces the expression of melatonin biosynthetic genes, primarily serotonin N-acetyl transferase 1 (SNAT1). The step in melatonin biosynthesis mediated by SNAT is crucial, wherein serotonin gets converted to N-acetylserotonin in the chloroplast. The final steps of conversion of N-acetylserotonin to melatonin also occur in the chloroplast, mediated by N-acetyl serotonin O-methyltransferase (ASMT). Hence, even a low level of ROS accumulation in chloroplast can directly affect melatonin production and function. This phenomenon may also prevent the chloroplast from oxidative damage by melatonin accumulation in these organelles.

3.4.2 Melatonin and Seed Germination in Plants

Seed growth and development in plants is a complex process tightly regulated by the interplay of various regulatory molecules. Melatonin is a growth regulator and signaling molecule that regulates seed germination under stressful situations such as heavy metal pollution, salinity and temperature. Under all these circumstances, seed vitality and germination rates are greatly affected (Hernández et al., 2015; Marta et al., 2016). Melatonin has a potent antioxidant capacity and inhibits the accumulation of hydrogen peroxide, thereby also elevating the levels of SOD, CAT and APX (Zhang et al., 2015). In cotton (*Gossypium hirsutum*), low concentrations of melatonin can promote cotton seed germination by increasing the antioxidant enzyme activity and decreasing the accumulation of malondialdehyde (MDA) (Xiao et al., 2019).

Melatonin also interacts with phytohormones to control seed germination. Melatonin treatment leads to the down regulation of the ABA biosynthetic gene NCED (9-cis-epoxycarotenoid dioxygenase) and up regulation of the ABA catabolism gene (CYP707 monooxygenases), which leads to a decrease in ABA levels during seed germination under salt stress (Zhang et al., 2015). Melatonin is also found to increase the germination rates of *Limonium bicolor* seeds and melon seeds (*Cucumis melo*) (Li et al., 2021; Castanares and Bouzo, 2019) under salinity stress by synthesis of new proteins and soluble sugar utilization and also increasing amylase and alpha-amylase activity. Melatonin increases seed germination under salt stress by regulating ABA and GA3 in cotton. The ABA signaling gene GhABF2 was up regulated and GhDPBF2 was down regulated and the expression of GA signaling genes GhGID1C and GhGID1B was up regulated by melatonin. This study can be beneficial in studying regulatory mechanisms of cotton resistance and exploiting the interactions with the phytohormones for cultivating salt-resilient varieties (Chen et al., 2021).

3.4.3 Melatonin Controls Flowering in Plants

Flowering in plants is an intricate process which involves control by regulatory networks comprising transcription factors and other signaling molecules. It marks the transition from the vegetative growth phase to the reproductive growth phase. One of the critical regulators of flowering is Flowering Locus C (FLC), a MADS-box gene that represses flowering. Flowering control via FLC is dependent on vernalization. Before vernalization, the vegetative meristem sustains, but after vernalization, it gets transitioned to reproductive or floral meristem. FLC is repressed on vernalization, and plants can initiate flowering. Flowering Locus T (FT) is a positive regulator of flowering. It is a small mobile protein synthesized in the leaves and travels to the shoot apex on floral transition, where it interacts with the bZIP transcription factor FD. This complex further leads to the activation of floral identity genes in the shoot apex (Wigge et al., 2005). Melatonin as a growth regulator can have direct consequences on flowering in plants. One of the studies involved the phenotypic characterization of melatonin-rich rice transgenics overexpressing sheep serotonin N-acetyl transferase. It was observed that the transgenic plants exhibited delayed flowering compared to the wild type (Byeon and Back, 2014). Studies in *Chenopodium rubrum* also suggested a delay in flowering on melatonin application (Kolar et al., 2003). When melatonin was applied to the shoot apical meristems before the beginning of a 12-hour darkness period, the levels were maintained even after light exposure. This suggested that melatonin regulates both photoperiodic flowering induction and developmental processes in the shoot apical meristem. The delayed flowering response on melatonin application may have potential agronomic relevance, such as in the case of apple. A recent study highlights the crosstalk of melatonin and light signaling to control flowering in apples. On treatment with melatonin, there was up regulation of genes which inhibit flowering and down regulation of genes that promote flowering, leading to delayed flowering on melatonin treatments.

Further, blue and far-red light up regulated the gene expression of *MdSNAT5* and *MdASMT9*, which increased melatonin levels (Zhang et al., 2019). Exogenously applied melatonin retards flowering by elevating the DELLA proteins, the master regulators of GA production, suppressing the expression of flowering genes such as FT and SOC1 (SUPPRESSOR OF OVEREXPRESSION OF CONSTANT 1) (Shi et al., 2016a, 2016b; Lee et al., 2019). However, reports by Lee et al. (2019) also suggest that Arabidopsis *SNAT2*knockout mutant has low melatonin levels that delay flowering, which is contradictory to earlier reports of melatonin function in flowering. The role of GA in this process is also an area of active investigation. Melatonin engineering to delay flowering and reduce harvest losses in crop plants can be a direct implication of this research.

3.4.4 Role of Phytomelatonin in Regulating Circadian Rhythms in Plants

Melatonin is considered as one of the prime molecules controlling sleep cycles or circadian rhythms in vertebrates. Phytomelatonin also has similar functions in plants. Its levels fluctuate with circadian rhythms. Phytomelatonin content peaks at night and shows low levels during the daytime for the short-day plant *Chenopodium rubrum* (Kolar et al., 1997, 2004). Sometimes a "double peak" of melatonin levels may be seen, as shown by Zhao et al. (2013), corresponding to darkness, and a second peak corresponding to high temperature and high light intensity (stressful condition) in ripe cherry fruit. Through enzyme kinetic analysis, it was shown that SNAT and ASMT enzymes exhibited thermophilic features (Byeon et al., 2014). An interesting paper by Shi et al. (2016a) shows that in *Arabidopsis*, phytomelatonin content and also the gene expression of *AtSNAT1* (At1G32070), *AtCOmelatonin1* (At5G54160), *AtASmelatonin* (At4G35160) shows time-specific rhythmicity, and the maximum term is at zeitgeber time (ZT4). These rhythmic signals also control stomatal closure and antioxidant enzyme-related gene expression (Wei et al., 2018). Thus, phytomelatonin helps regulate the circadian clock in plants and is regulated by the light and temperature signals. How these environmental cues integrate to maintain circadian rhythmicity is an exciting area of research and can contribute to understanding these processes in model plants like Arabidopsis, which can be further extrapolated to various crop species.

3.4.5 Role of Melatonin in Regulating Root Growth in Plants

The melatonin levels in plants (ng/g or μg/g) are usually higher than in animals (in the range of pg/g in the bloodstream). Though in plants, in general, the melatonin levels are similar to IAA

(Indole-3-Acetic Acid), an essential endogenous auxin is required for root formation (Parkand Back, 2012; Shewry, 2001; Arnao et al., 2007). In one of the studies, when sheep SNAT was expressed in rice, the transgenics expressed more melatonin than wild-type plants (Byeon and Back, 2014). These elevated melatonin levels increase seminal root growth and root biomass of rice seedlings. Additional studies are underway to elucidate whether the enhanced root growth positively correlates with rice grain yield and biomass in field conditions (Park and Back, 2012).

Further, an interesting study shows that the adventitious root formation in *Solanum lycopersicum* results from crosstalk between melatonin, auxin and NO signaling. Externally applied melatonin has the same effects as auxin applications and induces adventitious root formation in a concentration-dependent manner. An intriguing study by Hu et al. (2020) shows that melatonin alleviates the copper stress responses and promotes lateral root development in melon seedlings. The stress-activated proline and malondialdehyde (MDA) content was reduced in these copper-stressed melon seedlings, accompanied by elevated antioxidant enzyme activity and root vigour. The transcriptomics and metabolomics analyses revealed the suppression of jasmonic acid biosynthesis and the involvement of linoleic acid and lecithin on the effect of melatonin on alleviating copper toxicity (Hu et al., 2020). In addition to its different roles in growth and development, one of the crucial functions of melatonin lies in determining root growth and architecture. It has been found to promote primary root growth in an IAA-dependent manner. This role of melatonin can be attributed to the fact that it has the same precursor as auxin. In the wild-type and *pin-formed* (*pin*) double and triple mutants, treatment with optimal concentrations of melatonin ranging from 10^{-9} to 10^{-6} M functions as IAA mimics to promote primary root growth. This study also included a large-scale transcriptomic analysis of the genes coregulated by auxin and melatonin, suggesting that a subset of genes coregulated by both these hormones largely overlap.

Interestingly, the effects of melatonin on primary root growth were abolished in the presence of high concentrations of auxin biosynthesis inhibitor or polar transport inhibitor. Further, the melatonin treatments could not rescue the root length defects of the auxin biosynthesis quintuple mutant *yucQ*, thus suggesting that melatonin promotes primary root growth in an IAA-dependent manner (Yang et al., 2021). It is also known that early response activation of auxin signaling by IAA is not mimicked by melatonin (Koyama et al., 2013; Pelagio-Flores et al., 2012). Though there are potentially similarities between auxin and melatonin functions, these studies are still in their infancy, and a lot more research is required to implement related strategies for crop improvement.

3.5 CONCLUSIONS AND FUTURE PERSPECTIVES

Melatonin is known to have pleiotropic roles during plants' biotic and abiotic stress responses. These roles can be attributed to the fact that it is a potent ROS scavenger and can alleviate these stresses through multiple pathways. The last decade has seen a surge in reports suggesting that melatonin is an essential factor in regulating plant growth and developmental functions. In plants, it controls seed development and germination, seedling growth, flowering, root growth, circadian rhythms, biomass production, senescence and cell death. Identification of a plasma membrane-localized putative receptor, CAND2/PMTR1, in plants strengthened the fact that it is a phytohormone (Wei et al., 2018). Further, research on the receptor and its downstream effectors may lead to deciphering novel mechanisms on how melatonin controls different physiological processes. More pathways via novel receptors, enzymes and transcription factors must be dissected to study the crosstalk of melatonin with other phytohormones, especially auxin, due to its structural similarities and the fact that both share a common precursor. The study of interaction of melatonin with auxin and cytokinin can be used to manipulate various biological processes such as shoot and root morphogenesis. Future studies should also focus on mechanisms that lead to melatonin degradation and how the levels are controlled at the molecular levels. This can answer questions about how phytomelatonin levels are maintained for different tissues and plant developmental stages. The regulation of both above-ground and under-ground tissues in the plant can directly affect nutrient absorption, water movement, vasculature development and interactions with the rhizosphere. Melatonin level modifications can also lead to customizing the chlorophyll levels, promoting photosynthesis and also prolonging the post-harvest shelf life. Melatonin levels in edible plant parts such as leaves, grains, fruits and vegetables can be enhanced for rich dietary supplies of melatonin. Human health improvement programs can focus on genetic manipulations that lead to enhanced phytomelatonin accumulation in these plant parts that can be used for human consumption and treatment of sleep disorders, where melatonin acts as a neuroprotectant and antioxidant agent (Mannino et al., 2021). Thus, breeding phytomelatonin-rich varieties of fruits and vegetables can have significant pharmacological implications. The crop varieties with elevated melatonin concentrations have better stress tolerance capabilities with better phenotypes in seed germination, flowering, fruiting, rooting and biomass production. Thus, melatonin engineering can be a powerful tool to get desired phenotypes under stress conditions (Zhang et al., 2015). Melatonin research so far mainly focuses on the physiological, molecular and metabolic aspects of plant growth and development. However, there is still much potential in the unexplored areas related to melatonin research. These challenges include the identification of additional receptor(s), transcriptional networks and chaperone involvement in normal and stress conditions. Underpinning these aspects of plant growth and development will open new paradigms of melatonin research for crop improvement and sustainable agriculture.

REFERENCES

Afreen, F., Sayed M.A. Zobayed, and Toyoki Kozai. "Melatonin in Glycyrrhiza uralensis: Response of plant roots to spectral quality of light and UV-B radiation." *Journal of Pineal Research* 41, no. 2 (2006): 108–115.

Ai, Yu, and Ziqiang Zhu. "Melatonin antagonizes jasmonate-triggered anthocyanin biosynthesis in Arabidopsis thaliana." *Journal of Agricultural and Food Chemistry* 66, no. 21 (2018): 5392–5400.

Arnao, Marino B., and Josefa Hernández-Ruiz. "The physiological function of melatonin in plants." *Plant Signaling & Behavior* 1, no. 3 (2006): 89–95.

Arnao, Marino B., and Josefa Hernández-Ruiz. "Melatonin promotes adventitious-and lateral root regeneration in etiolated hypocotyls of *Lupinus albus* L." *Journal of Pineal Research* 42, no. 2 (2007): 147–152.

Arnao, Marino B., and Josefa Hernández-Ruiz. "Protective effect of melatonin against chlorophyll degradation during the senescence of barley leaves." *Journal of Pineal Research* 46, no. 1 (2009): 58–63.

Arnao, Marino B., and Josefa Hernandez-Ruiz. "Melatonin: Plant growth regulator and/or biostimulator during stress?." *Trends in Plant Science* 19, no. 12 (2014a): 789–797.

Arnao, Marino B., and Josefa Hernandez-Ruiz. "Melatonin: Possible role as light-protector in plants." *UV Radiation: Properties, Effects, and Applications* (2014b): 79–92.

Arnao, Marino B., and Josefa Hernandez-Ruiz. "Growth activity, rooting capacity, and tropism: Three auxinic precepts fulfilled by melatonin." *Acta Physiologiae Plantarum* 39, no. 6 (2017): 1–9.

Arnao, Marino B., and Josefa Hernández-Ruiz. "Melatonin and its relationship to plant hormones." *Annals of Botany* 121, no. 2 (2018): 195–207.

Arnao, Marino B., and Josefa Hernández-Ruiz. "Is phytomelatonin a new plant hormone?." *Agronomy* 10, no. 1 (2020): 95.

Back, Kyoungwhan. "Melatonin metabolism, signaling and possible roles in plants." *The Plant Journal* 105, no. 2 (2021): 376–391.

Back, Kyoungwhan, Dun-Xian Tan, and Russel J. Reiter. "Melatonin biosynthesis in plants: Multiple pathways catalyze tryptophan to melatonin in the cytoplasm or chloroplasts." *Journal of Pineal Research* 61, no. 4 (2016): 426–437.

Banerjee, A., and A. Roychoudhury. "Melatonin as a regulator of abiotic stress tolerance in plants." In: Singh V.P., Singh S., Prasad S.M. (Eds.) *Mechanisms behind phytohormonal signaling and crop abiotic stress tolerance*, Nova Science Publishers, New York, 2017, Pp. 47–60.

Banerjee, A., and A. Roychoudhury. "Melatonin application reduces fluoride uptake and toxicity in rice seedlings by altering abscisic acid, gibberellin, auxin and antioxidant homeostasis." *Plant Physiology and Biochemistry* 145 (2019): 164–173.

Byeon, Yeong, and Kyoungwhan Back. "An increase in melatonin in transgenic rice causes pleiotropic phenotypes, including enhanced seedling growth, delayed flowering, and low grain yield." *Journal of Pineal Research* 56, no. 4 (2014): 408–414.

Byeon, Yeong, Hyoung Yool Lee, Kyungjin Lee, Sangkyu Park, and Kyoungwhan Back. "Cellular localization and kinetics of the rice melatonin biosynthetic enzymes SNAT and ASMT." *Journal of Pineal Research* 56, no. 1 (2014): 107–114.

Carrillo-Vico, Antonio, Patricia J. Lardone, Nuria Álvarez-Sánchez, Ana Rodríguez-Rodríguez, and Juan M. Guerrero. "Melatonin: Buffering the immune system." *International Journal of Molecular Sciences* 14, no. 4 (2013): 8638–8683.

Castanares, Jose Luis, and Carlos Alberto Bouzo. "Effect of exogenous melatonin on seed germination and seedling growth in melon (*Cucumis melo* L.) under salt stress." *Horticultural Plant Journal* 5, no. 2 (2019): 79–87.

Chen, Li, Bin Lu, Liantao Liu, Wenjing Duan, Dan Jiang, Jin Li, Ke Zhang et al. "Melatonin promotes seed germination under salt stress by regulating ABA and GA3 in cotton (*Gossypium hirsutum* L.)." *Plant Physiology and Biochemistry* 162 (2021): 506–516.

Chen, Qian, Wen-bo Qi, Russel J. Reiter, Wei Wei, and Bao-min Wang. "Exogenously applied melatonin stimulates root growth and raises endogenous indoleacetic acid in roots of etiolated seedlings of *Brassica juncea*." *Journal of Plant Physiology* 166, no. 3 (2009): 324–328.

Ding, Fei, Liming Ren, Fang Xie, Meiling Wang, and Shuoxin Zhang. "Jasmonate and melatonin act synergistically to potentiate cold tolerance in tomato plants." *Frontiers in Plant Science* 12 (2021).

Dubbels, R., R.J. Reiter, E. Klenke, A. Goebel, E. Schnakenberg, C. Ehlers, H.W. Schiwara, and W. Schloot. "Melatonin in edible plants identified by radioimmunoassay and by high performance liquid chromatography-mass spectrometry." *Journal of Pineal Research* 18, no. 1 (1995): 28–31.

Edgerton, Michael D., Alexander Tropsha, and Alan M. Jones. "Modelling the auxin-binding site of auxin-binding protein 1 of maize." *Phytochemistry* 35, no. 5 (1994): 1111–1123.

Erdal, Serkan. "Melatonin promotes plant growth by maintaining integration and coordination between carbon and nitrogen metabolisms." *Plant Cell Reports* 38, no. 8 (2019): 1001–1012.

Fan, Yiting, Canying Li, Jie Zhu, Lei Sun, Rui Huang, Mi Guo, Yushuo Wu, and Yonghong Ge. "Organic acids metabolism and GABA shunt involved in maintaining quality of Malus domestica by methyl jasmonate treatment." *Food Research International* (2022): 111741.

Gangappa, Sreeramaiah N., and Javier F. Botto. "The multifaceted roles of HY5 in plant growth and development." *Molecular Plant* 9, no. 10 (2016): 1353–1365.

Hardeland, Rüdiger. "Melatonin in plants and other phototrophs: Advances and gaps concerning the diversity of functions." *Journal of Experimental Botany* 66, no. 3 (2015): 627–646.

Hardeland, Rüdiger. "Melatonin in plants: Diversity of levels and multiplicity of functions." *Frontiers in Plant Science* 7 (2016): 198.

Hardeland, Rüdiger, Juan Antonio Madrid, Dun-Xian Tan, and Russel J. Reiter. "Melatonin, the circadian multioscillator system and health: The need for detailed analyses of peripheral melatonin signaling." *Journal of Pineal Research* 52, no. 2 (2012): 139–166.

Hattori, Atsuhiko, Hiro Migitaka, Masayuki Iigo, Masanori Itoh, Koji Yamamoto, Ritsuko Ohtani-Kaneko, Masayuki Hara, Takuro Suzuki, and Russel J. Reiter. "Identification of melatonin in plants and its effects on plasma melatonin levels and binding to melatonin receptors in vertebrates." *Biochemistry and Molecular Biology International* 35, no. 3 (1995): 627–634.

Hernández, Ismaél Gatica, Federico José Vicente Gomez, Soledad Cerutti, María Verónica Arana, and María Fernanda Silva. "Melatonin in *Arabidopsis thaliana* acts as plant growth regulator at low concentrations and preserves seed viability at high concentrations." *Plant Physiology and Biochemistry* 94 (2015): 191–196.

Hernandez-Ruiz, Josefa, Antonio Cano, and Marino B. Arnao. "Melatonin: A growth-stimulating compound present in lupin tissues." *Planta* 220, no. 1 (2004): 140–144.

Hernandez-Ruiz, Josefa, Antonio Cano, and Marino B. Arnao. "Melatonin acts as a growth-stimulating compound in some monocot species." *Journal of Pineal Research* 39, no. 2 (2005): 137–142.

Hu, Zhicheng, Qiushi Fu, Jing Zheng, Aiai Zhang, and Huaisong Wang. "Transcriptomic and metabolomic analyses reveal that melatonin promotes melon root development under copper stress by inhibiting jasmonic acid biosynthesis." *Horticulture Research* 7 (2020).

Huang, Huang, Bei Liu, Liangyu Liu, and Susheng Song. "Jasmonate action in plant growth and development." *Journal of Experimental Botany* 68, no. 6 (2017): 1349–1359.

Huangfu, Liexiang, Rujia Chen, Yue Lu, Enying Zhang, Jun Miao, Zhihao Zuo, Yu Zhao et al. "OsCOMT, encoding a caffeic acid O-methyltransferase in melatonin biosynthesis, increases rice grain yield through dual regulation of leaf senescence and vascular development." *Plant Biotechnology Journal* 20, no. 6 (2022): 1122–1139.

Kolar, Jan, Carl Hirschie Johnson, and Ivana Macháčková. "Exogenously applied melatonin (N-acetyl-5-methoxytryptamine) affects flowering of the short-day plant Chenopodium rubrum." *Physiologia Plantarum* 118, no. 4 (2003): 605–612.

Kolar, Jan, Carl Hirschie Johnson, and Ivana Machackova. "Effects of melatonin on circadian rhythms and photoperiodic flower induction in higher plants." *Acta Physiologiae Plantarum* 26, no. 3 suppl. (2004).

Kolar, Jan, Ivana Macháčková, Josef Eder, Els Prinsen, Walter Van Dongen, Harry Van Onckelen, and Helena Illnerová. "Melatonin: Occurrence and daily rhythm in Chenopodium rubrum." *Phytochemistry* 44, no. 8 (1997): 1407–1413.

Koyama, Fernanda C., Thais L.G. Carvalho, Eduardo Alves, Henrique B. Da Silva, Mauro F. De Azevedo, Adriana S. Hemerly, and Célia R.S. Garcia. "The structurally related auxin and melatonin tryptophan-derivatives and their roles in Arabidopsis thaliana and in the human malaria parasite Plasmodium falciparum." *Journal of Eukaryotic Microbiology* 60, no. 6 (2013): 646–651.

Lee, Hyoung Yool, and Kyoungwhan Back. "Melatonin induction and its role in high light stress tolerance in Arabidopsis thaliana." *Journal of Pineal Research* 65, no. 3 (2018): e12504.

Lee, Hyoung Yool, and Kyoungwhan Back. "2-hydroxymelatonin promotes seed germination by increasing reactive oxygen species production and gibberellin synthesis in arabidopsis thaliana." *Antioxidants* 11, no. 4 (2022): 737.

Lee, Hyoung Yool, Kyungjin Lee, and Kyoungwhan Back. "Knockout of Arabidopsis serotonin N-acetyltransferase-2 reduces melatonin levels and delays flowering." *Biomolecules* 9, no. 11 (2019): 712.

Lerner, Aaron B., James D. Case, Yoshiyata Takahashi, Teh H. Lee, and Wataru Mori. "Isolation of melatonin, the pineal gland factor that lightens melanocyteS1." *Journal of the American Chemical Society* 80, no. 10 (1958): 2587–2587.

Li, Hao, Yanliang Guo, Zhixiang Lan, Kai Xu, Jingjing Chang, Golam Jalal Ahammed, Jianxiang Ma, Chunhua Wei, and Xian Zhang. "Methyl jasmonate mediates melatonin-induced cold tolerance of grafted watermelon plants." *Horticulture Research* 8 (2021).

Liang, Chengzhen, Guangyong Zheng, Wenzhen Li, Yiqin Wang, Bin Hu, Hongru Wang, Hongkai Wu et al. "Melatonin delays leaf senescence and enhances salt stress tolerance in rice." *Journal of Pineal Research* 59, no. 1 (2015): 91–101.

Liu, Jianlong, Ruimin Zhang, Yunkuo Sun, Zeyu Liu, Wen Jin, and Yan Sun. "The beneficial effects of exogenous melatonin on tomato fruit properties." *Scientia Horticulturae* 207 (2016): 14–20.

Liu, Shuaimin, Hua Huang, Donald J. Huber, Yonggui Pan, Xuequn Shi, and Zhengke Zhang. "Delay of ripening and softening in 'Guifei' mango fruit by postharvest application of melatonin." *Postharvest Biology and Technology* 163 (2020): 111136.

Ma, Zonghuan, Shijin Yang, Juan Mao, Wenfang Li, Wei Li, Cunwu Zuo, Mingyu Chu, Xin Zhao, Qi Zhou, and Baihong Chen. "Effects of shading on the synthesis of volatile organic compounds in 'Marselan' grape berries (Vitis vinifera L.)." *Journal of Plant Growth Regulation* 40, no. 2 (2021): 679–693.

Mannino, Giuseppe, Carlo Pernici, Graziella Serio, Carla Gentile, and Cinzia M. Bertea. "Melatonin and phytomelatonin: Chemistry, biosynthesis, metabolism, distribution and bioactivity in plants and animals: An overview." *International Journal of Molecular Sciences* 22, no. 18 (2021): 9996.

Maronde, Erik, and Jörg H. Stehle. "The mammalian pineal gland: Known facts, unknown facets." *Trends in Endocrinology & Metabolism* 18, no. 4 (2007): 142–149.

Marta, Bałabusta, Katarzyna Szafrańska, and Małgorzata M. Posmyk. "Exogenous melatonin improves antioxidant defense in cucumber seeds (*Cucumis sativus* L.) germinated under chilling stress." *Frontiers in Plant Science* 7 (2016): 575.

Meng, Jiang-Fei, Tian-Ci Shi, Shuo Song, Zhen-Wen Zhang, and Yu-Lin Fang. "Melatonin in grapes and grape-related foodstuffs: A review." *Food Chemistry* 231 (2017): 185–191.

Miao, Huiying, Wei Zeng, Meng Zhao, Jiansheng Wang, and Qiaomei Wang. "Effect of melatonin treatment on visual quality and health-promoting properties of broccoli florets under room temperature." *Food Chemistry* 319 (2020): 126498.

Onik, Jakaria Chowdhury, Su Chit Wai, Ang Li, Qiong Lin, Qianqian Sun, Zhidong Wang, and Yuquan Duan. "Melatonin treatment reduces ethylene production and maintains fruit quality in apple during postharvest storage." *Food Chemistry* 337 (2021): 127753.

Pandi-Perumal, Seithikurippu R., Ilya Trakht, Venkataramanujan Srinivasan, D. Warren Spence, Georges J.M. Maestroni, Nava Zisapel, and Daniel P. Cardinali. "Physiological effects of melatonin: Role of melatonin receptors and signal transduction pathways." *Progress in Neurobiology* 85, no. 3 (2008): 335–353.

Park, Sangkyu, and Kyoungwhan Back. "Melatonin promotes seminal root elongation and root growth in transgenic rice after germination." *Journal of Pineal Research* 53, no. 4 (2012): 385–389.

Park, Woong June. "Melatonin as an endogenous plant regulatory signal: Debates and perspectives." *Journal of Plant Biology* 54, no. 3 (2011): 143–149.

Pelagio-Flores, Ramón, Edith Muñoz-Parra, Randy Ortiz-Castro, and José López-Bucio. "Melatonin regulates Arabidopsis root system architecture likely acting independently of auxin signaling." *Journal of Pineal Research* 53, no. 3 (2012): 279–288.

Reiter, Russel J. "Pineal melatonin: Cell biology of its synthesis and of its physiological interactions." *Endocrine Reviews* 12, no. 2 (1991): 151–180.

Reiter, Russel J., Dun-Xian Tan, Lucien C. Manchester, Sergio D. Paredes, Juan C. Mayo, and Rosa M. Sainz. "Melatonin and reproduction revisited." *Biology of Reproduction* 81, no. 3 (2009): 445–456.

Reiter, Russel J., Dun-Xian Tan, Zhou Zhou, Maria Helena Coelho Cruz, Lorena Fuentes-Broto, and Annia Galano. "Phytomelatonin: Assisting plants to survive and thrive." *Molecules* 20, no. 4 (2015): 7396–7437.

Samanta, S., A. Banerjee, and A. Roychoudhury. "Exogenous melatonin regulates endogenous phytohormone homeostasis and thiol-mediated detoxification in two indica rice cultivars under arsenic stress." *Plant Cell Reports* 40 (2021): 1585–1602.

Samanta, S., A. Banerjee, and A. Roychoudhury. "Arsenic toxicity is counteracted by exogenous application of melatonin to different extents in arsenic-susceptible and arsenic-tolerant rice cultivars." *Journal of Plant Growth Regulation* 41 (2022): 2210–2231.

Sharif, Rahat, Chen Xie, Haiqiang Zhang, Marino B. Arnao, Muhammad Ali, Qasid Ali, Izhar Muhammad et al. "Melatonin and its effects on plant systems." *Molecules* 23, no. 9 (2018): 2352.

Shewry, Peter R. "Biochemistry & molecular biology of plants." *Plant Growth Regulation* 35, no. 1 (2001): 105.

Shi, Haitao, Russel J. Reiter, Dun-Xian Tan, and Zhulong Chan. "Indole-3-acetic acid inducible 17 positively modulates natural leaf senescence through melatonin-mediated pathway in Arabidopsis." *Journal of Pineal Research* 58, no. 1 (2015): 26–33.

Shi, Haitao, Yunxie Wei, and Chaozu He. "Melatonin-induced CBF/DREB1s are essential for diurnal change of disease resistance and CCA1 expression in Arabidopsis." *Plant Physiology and Biochemistry* 100 (2016a): 150–155.

Shi, Haitao, Yunxie Wei, Qiannan Wang, Russel J. Reiter, and Chaozu He. "Melatonin mediates the stabilization of DELLA proteins to repress the floral transition in Arabidopsis." *Journal of Pineal Research* 60, no. 3 (2016b): 373–379.

Sun, Chengliang, Lijuan Liu, Luxuan Wang, Baohai Li, Chongwei Jin, and Xianyong Lin. "Melatonin: A master regulator of plant development and stress responses." *Journal of Integrative Plant Biology* 63, no. 1 (2021): 126–145.

Sun, Qianqian, Na Zhang, Jinfang Wang, Haijun Zhang, Dianbo Li, Jin Shi, Ren Li et al. "Melatonin promotes ripening and improves quality of tomato fruit during postharvest life." *Journal of Experimental Botany* 66, no. 3 (2015): 657–668.

Tan, Dun-Xian, Rüdiger Hardeland, Kyoungwhan Back, Lucien C. Manchester, Moises A. Alatorre-Jimenez, and Russel J. Reiter. "On the significance of an alternate pathway of melatonin synthesis via 5-methoxytryptamine: Comparisons across species." *Journal of Pineal Research* 61, no. 1 (2016): 27–40.

Tan, Dun-Xian, Lucien C. Manchester, Maria P. Terron, Luis J. Flores, and Russel J. Reiter. "One molecule, many derivatives: A never-ending interaction of melatonin with reactive oxygen and nitrogen species?." *Journal of Pineal Research* 42, no. 1 (2007): 28–42.

Tan, Dun-Xian, and Russel J. Reiter. "An evolutionary view of melatonin synthesis and metabolism related to its biological functions in plants." *Journal of Experimental Botany* 71, no. 16 (2020): 4677–4689.

Tang, Qi, Canying Li, Yonghong Ge, Xue Li, Yuan Cheng, Jiabao Hou, and Jianrong Li. "Exogenous application of melatonin maintains storage quality of jujubes by enhancing anti-oxidative ability and suppressing the activity of cell wall-degrading enzymes." *Lwt* 127 (2020): 109431.

Tong, Hongning, Yunhua Xiao, Dapu Liu, Shaopei Gao, Linchuan Liu, Yanhai Yin, Yun Jin, Qian Qian, and Chengcai Chu. "Brassinosteroid regulates cell elongation by modulating gibberellin metabolism in rice." *The Plant Cell* 26, no. 11 (2014): 4376–4393.

Unterholzner, Simon J., Wilfried Rozhon, Michael Papacek, Jennifer Ciomas, Theo Lange, Karl G. Kugler, Klaus F. Mayer, Tobias Sieberer, and Brigitte Poppenberger. "Brassinosteroids are master regulators of gibberellin biosynthesis in Arabidopsis." *The Plant Cell* 27, no. 8 (2015): 2261–2272.

Vantassel, D.L., J.A. Li, and S.D. Oneill. "Melatonin-identification of a potential dark signal in plants." *Plant Physiology* 102, no. 1 (1993): 117.

Wang, Lei, Mingyi Yang, Yingying Dong, Russel J. Reiter, Yanqun Xu, Xingyu Lin, Zisheng Luo, and Li Li. "Melatonin confers enhanced polyamine metabolism and cell tolerance in Vitis vinifera against oxidative damage: Quantitative proteomic evidence." *Postharvest Biology and Technology* 184 (2022a): 111756.

Wang, Lin, Fangfang Zhou, Xuan Liu, Haixia Zhang, Tianci Yan, Yanzhao Sun, Kun Shi et al. "ELONGATED HYPOCOTYL 5-mediated suppression of melatonin biosynthesis is alleviated by darkness and promotes cotyledon opening." *Journal of Experimental Botany* 73, no. 14 (2022b): 4941–4953.

Wang, Ping, Xun Sun, Chao Li, Zhiwei Wei, Dong Liang, and Fengwang Ma. "Long-term exogenous application of melatonin delays drought-induced leaf senescence in apple." *Journal of Pineal Research* 54, no. 3 (2013): 292–302.

Wang, Qiannan, Bang An, Yunxie Wei, Russel J. Reiter, Haitao Shi, Hongli Luo, and Chaozu He. "Melatonin regulates root meristem by repressing auxin synthesis and polar auxin transport in Arabidopsis." *Frontiers in Plant Science* 7 (2016): 1882.

Weeda, Sarah, Na Zhang, Xiaolei Zhao, Grace Ndip, Yangdong Guo, Gregory A. Buck, Conggui Fu, and Shuxin Ren. "Arabidopsis transcriptome analysis reveals key roles of melatonin in plant defense systems." *PloS One* 9, no. 3 (2014): e93462.

Wei, Jian, Dong-Xu Li, Jia-Rong Zhang, Chi Shan, Zed Rengel, Zhong-Bang Song, and Qi Chen. "Phytomelatonin receptor PMTR 1-mediated signaling regulates stomatal closure in Arabidopsis thaliana." *Journal of Pineal Research* 65, no. 2 (2018): e12500.

Wei, Liyang, Changhong Liu, Huanhuan Zheng, and Lei Zheng. "Melatonin treatment affects the glucoraphanin-sulforaphane system in postharvest fresh-cut broccoli (*Brassica oleracea* L.)." *Food Chemistry* 307 (2020): 125562.

Wei, Zhiwei, Chao Li, Tengteng Gao, Zhijun Zhang, Bowen Liang, Zuosen Lv, Yangjun Zou, and Fengwang Ma. "Melatonin increases the performance of Malus hupehensis after UV-B exposure." *Plant Physiology and Biochemistry* 139 (2019): 630–641.

Wen, Dan, Biao Gong, Shasha Sun, Shiqi Liu, Xiufeng Wang, Min Wei, Fengjuan Yang, Yan Li, and Qinghua Shi. "Promoting roles of melatonin in adventitious root development of *Solanum lycopersicum* L. by regulating auxin and nitric oxide signaling." *Frontiers in Plant Science* 7 (2016): 718.

Wigge, Philip A., Min Chul Kim, Katja E. Jaeger, Wolfgang Busch, Markus Schmid, Jan U. Lohmann, and Detlef Weigel. "Integration of spatial and temporal information during floral induction in Arabidopsis." *Science* 309, no. 5737 (2005): 1056–1059.

Wu, Xiao, Jie Ren, Xiaoqing Huang, Xianzhe Zheng, Yuchuan Tian, Lei Shi, Pan Dong, and Zhengguo Li. "Melatonin: Biosynthesis, content, and function in horticultural plants and potential application." *Scientia Horticulturae* 288 (2021): 110392.

Xia, Hui, Yanqiu Shen, Honghong Deng, Jin Wang, Lijin Lin, Qunxian Deng, Xiulan Lv et al. "Melatonin application improves berry coloration, sucrose synthesis, and nutrient absorption in 'Summer Black' grape." *Food Chemistry* 356 (2021): 129713.

Xia, Hui, Yanqiu Shen, Tian Shen, Xin Wang, Xuefeng Zhang, Peng Hu, Dong Liang et al. "Melatonin accumulation in sweet cherry and its influence on fruit quality and antioxidant properties." *Molecules* 25, no. 3 (2020): 753.

Xiao, Shuang, Liantao Liu, Hao Wang, Dongxiao Li, Zhiying Bai, Yongjiang Zhang, Hongchun Sun, Ke Zhang, and Cundong Li. "Exogenous melatonin accelerates seed germination in cotton (*Gossypium hirsutum* L.)." *PloS One* 14, no. 6 (2019): e0216575.

Yang, Li, Jun You, Jinzhu Li, Yanping Wang, and Zhulong Chan. "Melatonin promotes Arabidopsis primary root growth in an IAA-dependent manner." *Journal of Experimental Botany* 72, no. 15 (2021): 5599–5611.

Yu, Manda, Lorenzo Lamattina, Steven H. Spoel, and Gary J. Loake. "Nitric oxide function in plant biology: A redox cue in deconvolution." *New Phytologist* 202, no. 4 (2014): 1142–1156.

Zahedi, Seyed Morteza, Marjan Sadat Hosseini, Javier Abadía, and Mina Marjani. "Melatonin foliar sprays elicit salinity stress tolerance and enhance fruit yield and quality in strawberry (Fragaria× ananassa Duch.)." *Plant Physiology and Biochemistry* 149 (2020): 313–323.

Zhai, Rui, Jianlong Liu, Fengxia Liu, Yingxiao Zhao, Lulu Liu, Chen Fang, Huibin Wang et al. "Melatonin limited ethylene production, softening and reduced physiology disorder in pear (*Pyrus communis* L.) fruit during senescence." *Postharvest Biology and Technology* 139 (2018): 38–46.

Zhang, Haixia, Lin Wang, Kun Shi, Dongqian Shan, Yunpeng Zhu, Chanyu Wang, Yixue Bai, Tianci Yan, Xiaodong Zheng, and Jin Kong. "Apple tree flowering is mediated by low level of melatonin under the regulation of seasonal light signal." *Journal of Pineal Research* 66, no. 2 (2019): e12551.

Zhang, Jing, Yi Shi, Xunzhong Zhang, Hongmei Du, Bin Xu, and Bingru Huang. "Melatonin suppression of heat-induced leaf senescence involves changes in abscisic acid and cytokinin biosynthesis and signaling pathways in perennial ryegrass (*Lolium perenne* L.)." *Environmental and Experimental Botany* 138 (2017): 36–45.

Zhang, Na, Qianqian Sun, Haijun Zhang, Yunyun Cao, Sarah Weeda, Shuxin Ren, and Yang-Dong Guo. "Roles of melatonin in abiotic stress resistance in plants." *Journal of Experimental Botany* 66, no. 3 (2015): 647–656.

Zhao, Dake, Yang Yu, Yong Shen, Qin Liu, Zhiwei Zhao, Ramaswamy Sharma, and Russel J. Reiter. "Melatonin synthesis and function: Evolutionary history in animals and plants." *Frontiers in Endocrinology* 10 (2019): 249.

Zhao, Yu, Dun-Xian Tan, Qiong Lei, Hao Chen, Lin Wang, Qing-tian Li, Yinan Gao, and Jin Kong. "Melatonin and its potential biological functions in the fruits of sweet cherry." *Journal of Pineal Research* 55, no. 1 (2013): 79–88.

4 Melatonin in Flowering and Fruit Ripening Response

Bushra Hafeez Kiani

4.1 INTRODUCTION

Melatonin, also called melatol, was identified in plants in 1995 after being discovered as an animal hormone. Melatonin affects many physiological aspects in animals, including body temperature, immune system, mood, sexual behavior, endocrine hormone levels, and sleep (Arnao and Hernández-Ruiz 2015a). Extensive research has been carried out on melatonin in several plants, linking it to several important roles. Melatonin, in general, protects against both abiotic (salinity, high temperatures, UV radiation, chemical contamination, drought, etc.) and biotic (fungi, virus, and bacteria) stressors in stressful conditions (Arnao and Hernández-Ruiz 2014, 2015b).

Melatonin appears to have a protective function in biological mechanisms like photosynthesis, primary and secondary metabolism, vegetative development, germination, water economy, and growth. Melatonin has a direct regulatory effect in all of these processes, functioning as an antioxidant or indirectly affecting gene activity patterns of transcriptional factors particular to stress, as well as other variables, which include hormones found in plants like indole-3-acetic acid, cytokinins, gibberellic acid, ethene, abscisic acid, and regulators such as brassinosteroids, 2-hydroxybenzoic acid, polyamines, strigolactones, and jasmonates (Arnao and Hernández-Ruiz 2018a, 2018b, 2019c, 2020; Sharif et al. 2018; Samanta et al. 2021a). Less research has been done on reproductive development. Since the preliminary research on *Chenopodium rubrum* by Kolar and coworkers in 1995, a lot of progress has been made, and intriguing findings have been reported.

Another unique and significant function linked to melatonin is its antioxidative properties to scavenge biological oxidants such as oxidants, monooxygen, hydrogen superoxide, perhydrol, peroxynitrite, and nitrogen monoxide (Samanta et al. 2021b). In this regard, it is worth noting that melatonin is not a pro-oxidant, whereas its intermediary products exhibit antioxidative potential along with a significant synergy with antioxidants like glutathione, vitamin C, and others (Cano et al. 2003; Tan et al. 2003).

4.2 MELATONIN IN FLOWERING

Preliminary research on the potential impact of melatonin on flowering had been focused on the short-day plant *Chenopodium rubrum* (Machackova and Krekule 2002; Kolar et al. 2003). When applied two hours prior to and two hours following the start of the induced dark period, elevated levels of melatonin caused a hindrance in the induction of flowers by 40 to 50 percent on average. CGP-52608 (1-methyl-3-[(E)-(4-oxo-3-prop-2-enyl-1,3-thiazolidin-2-ylidene) amino] thiourea), a melatonin agonist being used in animal experiments, was successful in suppressing the induction of flowers.

Moreover, melatonin application had an almost negligible effect on the flowering stage, implying that it influenced some processes prior to flower transformation (Kolar and Machackova 2005). All such findings were a portion of further investigation into the potential of melatonin contribution in crops as a Chrono regulator of the circadian system, comparable to how it exists in animal life (Kolar et al. 1995, 1997; Wolf et al. 2001). Previous research using *Arabidopsis thaliana*, a long-day

DOI: 10.1201/9781003343752-4

plant, revealed that blooming was only slightly postponed in plants administered with melatonin in 100 or 500 micromolar concentrations. This influenced the flowering period of the plant and the number of leaves and rosettes. The aforementioned statistics were included in a doctoral search but do not seem to have been published in journals (Kolar 2003). Additionally, preliminary research on the flower development of *Hypericum perforatum* provided an explanation for the melatonin peaks that appeared during the intermediate phases of floral development.

According to researchers, this peak in melatonin corresponded to the enhanced regeneration abilities of separated anthers, thereby causing a buildup of melatonin in particular tissue cells. These stages were reported to be a development signal in gametophytes, which resulted in usable microspores (Murch and Saxena 2002). Working with the flowers of *Datura metal*, which is a sedative shrub utilized as a herbal therapeutic worldwide, the same authors discovered the highest concentration of melatonin during the maturation of the flower buds, implying the defensive function of melatonin in the reproductive tissues (Murch et al. 2009). Melatonin content in the *Paeonia lactiflora* is investigated in various developing phases of the flower. The concentration of melatonin is improved in the flower-bud stage (S1), reduced in the initiating bloom stage (S2), peaks in the bloom stage (S3), and the melatonin concentration is slightly greater in the wither stage (S4) than it was in the S2 phase. These differences are directly linked with tyrosine decarboxylase transcription. Researchers also investigated the effects of varying light spectra components on melatonin concentration throughout flowering, revealing that exposure to sunlight, as well as blue light, stimulates the synthesis of melatonin, whereas a shaded environment and white and green light produce lower levels. Furthermore, melatonin dual peaks were observed throughout the one-day light and dark cycle, coexisting with light and temperature trends as well as a decline in malonaldehyde concentration, which indicates cytomembrane lipid oxidation. The researchers stressed the complexity of reaching conclusions from this information on the involvement of melatonin in *Paeonia lactiflora* flower development. Nonetheless, these findings indicate that melatonin may protect these flowers from stresses caused by flowers and other agents (Zhao et al. 2018). The protective role of melatonin in cut flowers of *Anthurium andraeanum* had been illustrated all through storage at low temperatures (Aghdam et al. 2019).

One research measured melatonin levels in field-grown rice plants at three distinct reproductive phases: pre-flowering, flowering, and post-flowering. Melatonin biosynthesis was found to be six times higher in the panicle (flower) than in the flag leaf. Such an induction occurred concurrently with melatonin biosynthesis enzyme induction including tryptophan decarboxylase, tryptamine 5-hydroxylase, and caffeic acid O-methyl transferase, implying that melatonin could play a role in the development of flowers (Park et al. 2013). Melatonin could influence the flowering time modulation in *Arabidopsis* like vitamin C, where an *Arabidopsis* mutant with deficient vitamin C expressed a phenotype with early flowering. It is important to note that flowering was delayed in the mutant with high vitamin C concentrations. In field-grown rice plants, genetically modified plants with high levels of melatonin overexpress serotonin N-acetyltransferase and had three times the internal melatonin concentration compared to the wild-type plants and a twofold increment in the plant biomass. The genetically modified plants had a seven-day delay in flowering. The grains produced from the genetically modified crops rich in melatonin showed an average decrease of 33 percent (Byeon and Back 2014).

The external temperature, gibberellic acid, photoperiodism, autonomy, vernalization, and aging mechanism are among some of the flowering mechanisms found in *Arabidopsis thaliana*. FLC codes for a transcriptional factor within the MADS-box that binds to flowering locus T as well as the Suppressor of overexpression of Constans 1 (SOC1) promoters to control the negative floral transition. Both Flowering locus C as well as Squamosa promoter-binding protein-like produce proteins linked to DELLA which function as inhibitors of the gibberellic acid pathway (Wigge et al. 2005; Li et al. 2016; Yu et al. 2012; Searle et al. 2006).

Research undertaken by Shi and colleagues established a direct connection in both melatonin and floral transition (Shi et al. 2016). This study demonstrated the new function of DELLAs and

Flowering locus C in *Arabidopsis* flowering mediated by melatonin. Thus, plants treated with 500 μM melatonin had a five-day delay in flowering along with having extra rosette leaves as compared to the control with no treatment. According to these findings, the external application of melatonin slows the flowering process. The stabilization of two different DELLAs was enhanced by melatonin while not affecting DELLA transcripts. Furthermore, there was no significant effect of exogenous melatonin on endogenous gibberellic acid levels, implying that melatonin impacts DELLA proteins without affecting gibberellic acid accumulation.

Melatonin increases internal nitric oxide content and could be associated with signaling pathways that are mediated by melatonin (Shi et al. 2015; Mukherjee 2019). Furthermore, nitric oxide induces concentrations of DELLAs, resulting in nitric oxide-mediated root development and photoperiodism (Lozano-Juste and León 2011; Bai et al. 2014). As a result, nitric oxide was found to have an involvement in DELLA stabilization mediated by melatonin, following delayed flowering (Shi et al. 2016). Melatonin also controls the timing of flowering via endorsing Flowering locus C transcription, which inhibits FT transcription. The hypothesis is that FLC is required for melatonin-mediated flowering, but not FT. Exogenous melatonin also has an inhibitory effect on SOC1 by inhibiting DELLA through sphingosine 1-phosphate lyase.

The role of melatonin in flowering has recently been demonstrated in an *Arabidopsis* mutant (Figure 4.1). SNAT2 was found in abundance in flowering plants. The mutant SNAT2 induced lower melatonin content as compared to the wild-type flowers, with no effect on melatonin concentration in the foliage, indicating gene expression in selected tissues in collaboration with SNAT1 (which is more prominent in mature leaves). SNAT2 plants exhibited slightly delayed flowering; however, they had a considerable difference in genes associated with gibberellic acid such as ATMYB33 (a gibberellic acid response inhibitor), ent-kaurene synthase, and FT gene activity compared to exogenous melatonin treatments. According to the authors, the discrepancies in the results between SNAT2 plants and external melatonin treatment in wild-type plants can be attributed to the high dose (0.5 mM) of melatonin which has an inhibitory effect on Arabidopsis seedlings and the likelihood that external melatonin treatment does not have the same flowering effects as internal melatonin (Lee et al. 2019). More research is needed in this area.

Strigolactones are derived from carotenoids that play a role in flowering, germination, shoot branching, and stem secondary growth, lateral root growth, and leaf senescence (Waters et al. 2017; Banerjee and Roychoudhury 2018a,b). Fascinating research on *Arabidopsis thaliana* flowering showed that mutated plants lacking strigolactones had flowered sooner, while externally applied melatonin caused a delay in flowering in all wild types as well as strigolactone mutant plants, although with a threshold of internal melatonin levels of 7 to 8 nanograms/gram FW, whereas flowering accelerated when melatonin concentration in tissues decreased lower than 0.9 to 1.0 nanogram/gram. These findings suggest that strigolactones inhibit SPL gene activity, leading to early blooming if adequate melatonin is not present. Nevertheless, the external application of melatonin in tissues caused a delay in flowering owing to the enhanced gene activity of flowering locus C. As a result, the presence of strigolactones suppresses melatonin signaling and synthesis, hence associating melatonin with strigolactones in the flowering response. Melatonin delays blooming via inducing Flowering locus C expression, which is activated when internal melatonin levels exceed the threshold. It is still unknown how SL regulates melatonin biosynthesis (Zhang et al. 2019).

Melatonin increased the expression of gibberellic acid production genes including gibberellin 20 oxidase 1-like and gibberellin 3-beta-dioxygenase 1-like in cucumbers grown in salt water, conferring increased levels of gibberellins including GA4, and stimulating salt-inhibited development (Zhang et al. 2014). Externally applied melatonin increased GA content in cotton seedlings (Xiao et al. 2019), whereas pear parthenocarpy was promoted by melatonin (Liu et al. 2018b) The discrepant effects of melatonin on few reproductive aspects in relation to vegetative responses including stress conditions remain a fascinating area that needs further investigations.

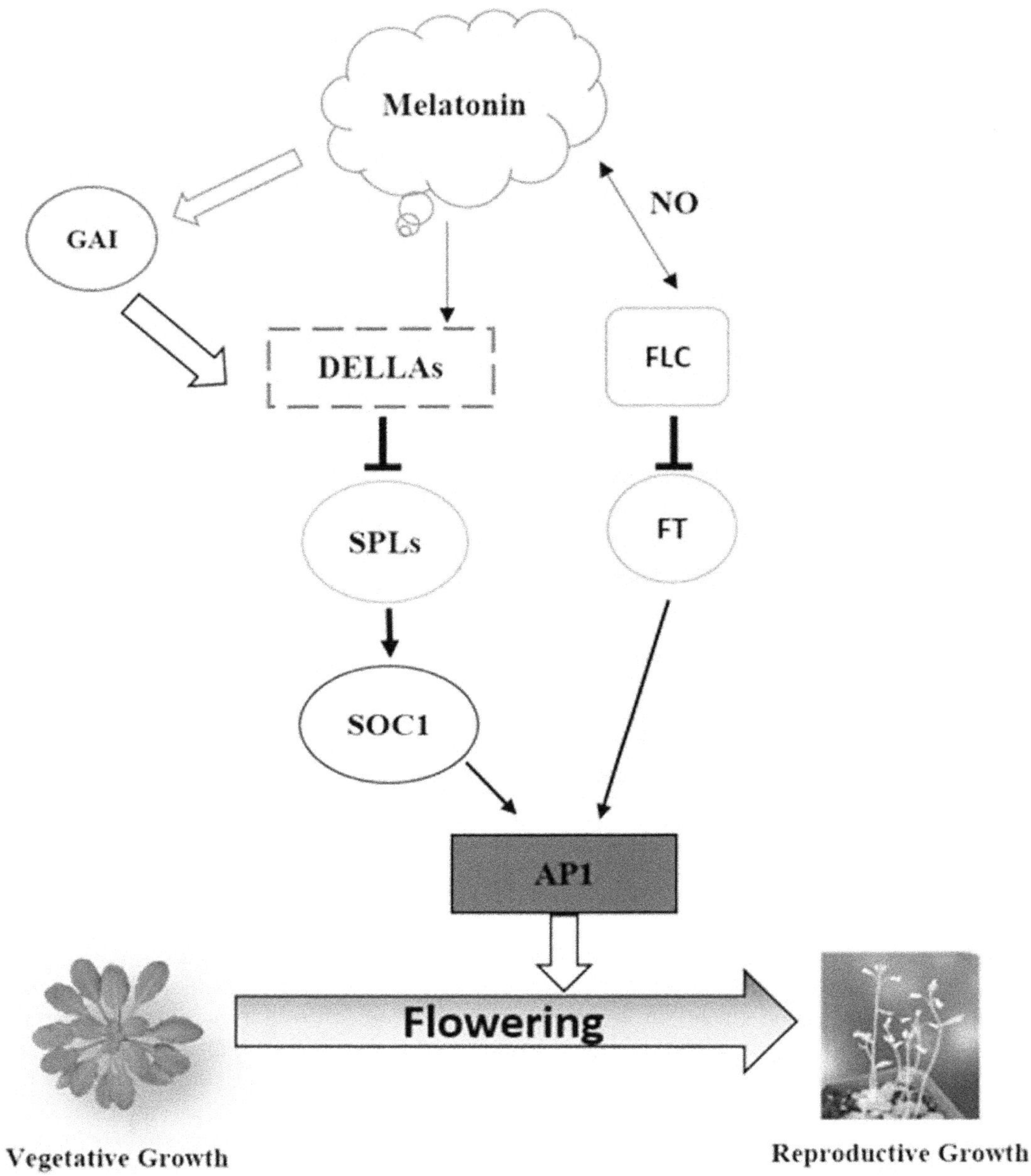

FIGURE 4.1 Proposed role of melatonin in flowering.

4.3 MELATONIN IN FRUIT SET AND DEVELOPMENT

4.3.1 Parthenocarpy

The function of melatonin in fruit growth and setting has been the subject of numerous studies. In preliminary studies, it was found that *Datura metal* fruit bodies contained significant levels of melatonin throughout the first ten days following anthesis, with the fruits having the least quantity and the ovule having the maximum quantity (Murch et al. 2009). Throughout stage II, embryonic development, and endocarp lignification, which have previously been linked to reactive oxygen species buildup in watery seeds, *Prunus avium* fruit had the highest concentrations of melatonin. This indicates that oxidative stress promotes faster melatonin production (Zhao et al. 2013). Melatonin was also accumulated in Micro-Tom tomato fruits all throughout fruit growth stages, with a peak in its final stage where the fruit turned into a reddish-pink color (Okazaki and Ezura 2009). The maximum quantity of melatonin was found in tomato seeds, implying that, as initially suggested, melatonin may have a protective role in seeds (Manchester et al. 2000). The highest level of melatonin was found in green fruit, according to research on the fruit growth (véraison) in Merlot wine grapes (Murch et al. 2010). These maximum levels have also been found in *Hypericum perforatum* throughout the formation of uninucleate formation.

The concentration of melatonin was also linked to an increased potential to regenerate isolated anthers (Murch and Saxena 2002). Melatonin levels were found to be high in the young and ripened red fruit, flowers, as well as seeds of two *Capsicum annuum* cultivars. Melatonin concentration was enhanced in the seeds and fruits considerably as the fruits and seeds matured. Elevated melatonin concentration was identified in cotyledon stage seedlings and reproductive parts and fully grown fruit of both cultivars. This upholds the inference that elevated melatonin concentrations protect these organs from ecological damage (Korkmaz et al. 2014). In an extensive study of Red Fuji apples (*Malus domestica*), the genes responsible for melatonin production were found to be completely interconnected with melatonin synthesis during fruit growth. The fruit melatonin content was found to be inversely related to malondialdehyde content, which is a by-product of lipid peroxidation.

Two major melatonin peaks appeared during fruit development, which usually coincide with rapid fruit growth and increased respiration and reactive oxygen species (Lei et al. 2013). Serotonin, which is needed to produce melatonin, increased in concentration throughout fruit growth across all fruit tissues like seed, mesocarp, jelly, exocarp, columella, as well as embryo; in the *Solanum lycopersicum* study, fully ripe fruit had the highest amount (Hano et al. 2017). Melatonin content of mature seeds was four to ten times greater in various varieties of hemp (*Cannabis sativa*) plants as compared to foliage, stems, or flowers (Allegrone et al. 2019). In any circumstance, it is currently unclear if peaks of melatonin serve as a fruit growth signal or rather as a response to maintain the proper oxidation-reduction status of fruits and seeds throughout the developmental stages. Without a doubt, research on melatonin and fruit development that exhibit the largest impact is those of Zhai et al. (2018); Liu et al. (2018b, 2019a, 2019b). Parthenocarpic fruits develop without ovules, either naturally or artificially.

Phytohormones like gibberellins and auxins have been shown to induce parthenocarpy in a variety of horticulture plants (Ozga and Reinecke 2003; Srivastava and Handa 2005). Melatonin induces parthenocarpy in pears through enhancing Gibberellin 3 beta-hydroxylase enzymes and GA3 synthesis, resulting in fruit comparable to that produced via mechanical pollination by cellular proliferation and fruit mesocarp expansion. An improvement in the levels of Gibberellin 3 beta-hydroxylase enzymes and GA3 levels were reported in ovaries, fruit form index, as well as both fruit longitudinal and transversal diameter, which were usually indicative of gibberellic acid treatment. Melatonin increased the level of active GAs by significantly up regulating GA20ox (which synthesizes active gibberellins) and down regulating GA2ox2 (which synthesizes inactive gibberellins). In contrast to the control, following melatonin therapy, the contents of ovary and seed cell layer numbers had no increment, and upon pollination, the content of indole-3-acetic acid was decreased.

These findings contradict previous findings wherein indole-3-acetic acid induced parthenocarpy in *Solanum melongena* (Du et al. 2016). Following melatonin administration, the expression of genes involved in the auxin signal transduction remained unchanged, indicating that indole-3-acetic acid is not a critical contributor in pear parthenocarpy. Melatonin treatment increased the expression of genes related to chlorophyll, photosystems I and II, as well as saccharides. The genome associated with cell proliferation and cell development altered after melatonin administration according to transcriptome data. Plant cyclins and expansions are essential for cellular proliferation and plant growth (Choi et al. 2006). Melatonin increased the expression of cyclins and nine to 12 differentially regulated expansions in pear parthenocarpy. Melatonin promotes cellular proliferation and growth via the gibberellic acid pathway, as evidenced by histological studies (Liu et al. 2018b).

4.3.2 Melatonin in Fruit Ripening and Senescence

Several pieces of research have been conducted for investigating how melatonin ripens and helps in fruit senescence. Exogenous melatonin effect on the ripening process primarily focuses on the interaction of melatonin with the phytohormone ethene. Tomato fruit applied with 50 μM melatonin exhibited significant alterations in the factors affecting fruit ripening. These factors include biosynthesis enzymes, flavor, ethene signaling, and softening of fruit compared to untreated tomatoes in the first study. Exogenous melatonin increased ethene production and the successive timing of the climacteric peak by up regulating ACC gene activity. Melatonin also increased the gene activity of the NR ethylene receptor and ethylene receptor homolog, as well as the transducing elements including ethylene insensitive 3-like 1 protein, ethylene insensitive 3-like 3 protein, and ethylene-responsive transcription factor 2 (Sun et al. 2015).

Melatonin influenced many proteins that have a role in ripening, fatty acid biosynthesis, flavonoid, cell wall, and carbohydrates in the differential proteome analysis of tomato plant. An increase was observed in the proteins involved in anthocyanin production, which suggests that the external use of melatonin ripens fruit and prevents spoilage (Sun et al. 2016). Additionally, tomato seeds primed with melatonin produced far more vitamin C, calcium, lycopene, and overall increased yield while producing lesser amounts of nitrogen, magnesium, copper, zinc, iron, and manganese. In comparison to the control group, which was irrigated with regular nutrient solution, plants that were provided melatonin-rich solution had significantly improved the content of vitamin C, lycopene, soluble solids, phosphorus, and citric acid. After the administration of melatonin, the levels of glucose, as well as sucrose, improved. Soluble galactose concentrations rose as tomato plants matured, causing ethene synthesis to be stimulated and, as a result, ripening to be promoted. Additionally, plants treated with melatonin had considerably increased citric acid content. Fruits taste best when both organic acids and sugars are present in large proportions (Table 4.1).

Melatonin generally improved tomato quality and yield by 13 percent (Liu et al. 2016). As shown by decreased biomass reduction, decaying rate, and respiratory rate in peach fruit, melatonin administration substantially slowed spoilage in the two peach plants examined. Vitamin C, firmness, and net soluble solid levels were preserved. Applying melatonin after harvesting also delayed the spoilage process and preserved the freshness of the peaches, which may be credited to the ability of melatonin to facilitate antioxidant properties, indicating that treating with melatonin can be thought of as a useful strategy for preserving peaches after harvesting (Gao et al. 2016).

TABLE 4.1
Melatonin Treatment for Fruit Ripening in Different Plant Species

S. No	Plant Species	Mode of action	Best Melatonin concentration	Exposure Time	References
1	Banana	Dipping of fruits	10 mM	3 min	Li et al. (2019)
2	Grape Berry	Dipping of fruits	0.1mM	10 sec	Xu et al. (2017)
3	Litchi	Dipping of fruits	0.4 mM	5 min	Zhang et al. (2018)
4	Plum	Dipping of fruits	0.1 mM	40 days	Bal (2019)
5	Pear	Dipping of fruits	0.1 mM	12 h	Liu et al. (2019b)
6	Strawberry	Dipping of fruits	0.1 mM	5 min	El-Mogy et al. (2019)
7	Tomato	In seeds, roots, and fruits	0.1 mM	2 h	Sun et al. (2016)

Strawberries that were treated with melatonin after harvesting had an increased adenosine triphosphate content, antioxidative properties, increased polyphenol content, and improved decay indicators (Aghdam and Fard 2017). The research by Liu et al. (2018a) yielded comparable results. Strawberries with 0.1–1.0 mM melatonin treatment for five minutes and later kept in a 90 percent humid and 4 °C temperature level showed less decaying and biomass loss. The senescence conditions developed at a later stage due to melatonin administration. The overall phenolic content and flavonoids were increased, which resulted in an increase in the antioxidative properties.

Moreover, melatonin biosynthesis gene expression was enhanced by treating fruits with melatonin. These include T5H, TDC, ASMT, SNAT, and COMT genes, which increase the internal production of melatonin. In recent years, melatonin has been researched in relation to the aromatic profile of strawberry fruits. Treating fruits with melatonin caused a delay in fruit degradation and substantially lowered carbon dioxide levels in comparison to the control group. Most volatile organic compounds (VOCs) were relatively abundant, but certain significant VOCs changed, which could have an impact on how consumers perceive quality (El-Mogy et al. 2019). Melatonin administration increased ethene synthesis in grape berries, which in turn helped to boost polyphenol accumulation (anthocyanins) and antioxidative activity (Xu et al. 2017). Melatonin collected promptly in *Băbească neagră*, starting with vếraison, peaked 94 days following blooming and subsequently showed lower concentrations in the latter phases of berry maturation. In comparison to the control treatment, melatonin treatments at concentrations between 10 and 100 μM enhanced the synthesis of abscisic acid, peroxide, and ethane, while also promoting berry maturation. Comparing the impacts of abscisic acid and peroxide, the inhibition of ethene bioproduction exhibited the most potent inhibition effect upon melatonin-induced ripening of the berries.

Melatonin also boosted abscisic acid-mediated ethene synthesis. These researchers noted that melatonin usage throughout the ripening of the berries is likely to accelerate the ripening process (Xu et al. 2018). Melatonin does have the ability to prolong senescence and preserve the economic importance of harvested pears during cold preservation by reducing ethene formation and the active oxygen species outburst (Zhai et al. 2018). Melatonin and nitric oxide delayed aging in pears by inhibiting the activation of ethene bioproduction genes (ACO and ACS), reducing cellulase and pectinase gene expression, reducing respiratory rates, and decreasing the rate of ethene production (Liu et al. 2019a). In another intriguing study by the researchers, the weight of the fruit was improved by 47.85 percent when melatonin was used in comparison to the control in another intriguing study by the aforementioned researchers. Melatonin enhanced the levels of simple sugars, especially sucrose and sorbitol, throughout fruit maturity, likely due to better carbohydrate formation. Invertase enzymatic efficiency was reduced in fruits that had been exposed to melatonin, although sucrose phosphate synthase efficiency was higher due to increased gene activity.

Melatonin treatment thus increased the saccharose concentration in pears. This particular research reported that pear trees treated with melatonin had sweeter, bigger, and more valuable fruit (Liu et al. 2019b). *Prunus domestica* was submerged for 40 days in varying melatonin solutions at 1 °C immediately after harvest. Melatonin solutions of 0.1- and 1-mM concentrations were found to be the best treatment for lowering biomass loss and retaining the firmness of the fruit. Melatonin-treated plums had higher overall phenolic content, vitamin C concentration, and antioxidative properties compared to the control. The researchers proposed that melatonin application could be a helpful tool for extending plum post-harvest life by ensuring appropriate quality of fruit (Bal 2019). In *Litchi chinensis*, fruits that have been collected at economic ripeness and then stored for five minutes at a temperature of 25 °C while being submerged in 0.4 millimolar melatonin had minimal browned pericarps and delayed discoloring.

Melatonin application hindered the production of reactive oxygen species along with mono aldehyde and also lowered the rate of relative membrane leakage. The application of melatonin significantly increased endogenous melatonin accumulation and delayed the decline of flavonoids, overall phenol content, as well as anthocyanins. Melatonin increased the function of antioxidative enzymatic proteins like catalase, ascorbate peroxidase, and glutathione reductase, as well as the gene activity of four enzymatic proteins involved in oxidized protein repair. Enzymes that cause browning, such as peroxidase and polyphenol oxidase enzymes, had their activities decreased. The results suggest that redox balance maintenance through enhanced antioxidative properties and regulation of repair of ROS-degraded proteins explain the potential of melatonin to prevent browned pericarps and withering in *Litchi chinensis* (Zhang et al. 2018).

Exogenous melatonin application slowed post-harvest banana ripening in research containing many varieties of *Musa acuminata*. This impact was dependent on the concentration, with the most efficient treatment at 500 and 200 μM treatment. External application of melatonin increased the internally present melatonin concentration and decreased ethene synthesis by regulating the expression of 1-Aminocyclopropane-1-carboxylic acid synthase and 1-Aminocyclopropane-1-carboxylic acid oxidase. The researchers suggested the use of melatonin as a fruit ripening indicator in various banana species. The suppression of ethene synthesis and post-harvest ripening by melatonin could potentially be utilized to biologically regulate fruit quality and ripening after harvest (Hu et al. 2017). External application of melatonin caused a considerable reduction of anthracnose in ripe yellow bananas and deferred fruit senescence in another banana study. Melatonin treatment, on the other hand, did not affect *Colletotrichum musae* growth in vitro. Ethene, mitogen-activated protein kinase, and auxin signaling mechanisms were also improved after melatonin treatment, which may improve fruit defense mechanisms by the regulation of physical properties, proteins involved in resistance, catabolites, as well as anabolites (Li et al. 2019).

Several models for melatonin activity towards phytopathogens in fruit bodies have been analyzed for both fungal and viral infections. Melatonin had potent activity against different fungal species, including *Diplocarpon*, *Fusarium*, *Botrytis*, *Alternaria*, *Rhizopus*, *Penicillium*, *Phytophthora*, and *Podosphaera*, usually evoking salicylic acid and jasmonic acid immune responses, thereby extending the storage and overall value of the fruit (Arnao and Hernández-Ruiz 2015b, 2018a, 2018b; Hernández-Ruiz and Arnao 2018; Sharif et al. 2018; Moustafa-Farag et al. 2020).

To summarize, melatonin application could variously impact ethene synthesizing enzymatic proteins, as it is likely to induce 1-Aminocyclopropane-1-carboxylic acid synthase and 1-Aminocyclopropane-1-carboxylic acid oxidase in tomatoes and *Băbească neagră*, encouraging ripening. On the other hand, some enzymes are down regulated by melatonin in pears, strawberries, plums, litchis, and bananas, thereby lowering ethene production and enhancing the storage life by delaying senescence. Understanding the effects produced generally relies on the physical condition of the fruits, the melatonin dosage applied, and the period of melatonin exposure. While senescence is an oxidative mechanism that involves degradation and diseases, ripening is a genetically regulated process that is characterized by the presence of ethene.

Since neither process can be separated at any given period, these processes are controlled in distinct manners. These results imply that melatonin administration might be an effective method for enhancing the quality of the fruit and extending its storage life after harvest (Kumar et al. 2014; Mukherjee 2019; Xu et al. 2019; Arnao and Hernández-Ruiz 2019a, 2019b, 2019c)

4.4 CONCLUSIONS AND FUTURE PERSPECTIVES

Future prospects call for further research into the potential use of melatonin to delay flowering. The application of this information to horticulture plants is essential because there is currently limited information available on *Arabidopsis*. Furthermore, the exact involvement of strigolactones in inhibiting melatonin in flowering should be thoroughly investigated. The underlying effects of melatonin on flowering need further research through a comparative analysis of short-, long-, and neutral-day plant species before arriving at conclusions about flowering. More information on the interactions among melatonin and phytohormones such as cytokinin, auxins, and gibberellic acid is needed to understand the fruit setting process. To fully demonstrate viability of melatonin as a fruit set booster, it is proposed that its ability to induce parthenocarpy in different kinds of crops and fruits be tested.

REFERENCES

Aghdam MS, Fard JR (2017) Melatonin treatment attenuates postharvest decay and maintains nutritional quality of strawberry fruits (*Fragaria*× *anannasa* cv. Selva) by enhancing GABA shunt activity. Food Chem 221:1650–1657

Aghdam MS, Jannatizadeh A, Nojadeh MS, Ebrahimzadeh A (2019) Exogenous melatonin ameliorates chilling injury in cut anthuriumflowers during low temperature storage. Postharvest Biol Technol 148:184–191

Allegrone G, Razzano F, Pollastro F, Grassi G (2019) Determination of melatonin content of different varieties of hemp (*Cannabis sativa* L.) by liquid chromatography tandem mass spectrometry. SN Appl Sci 1:720

Arnao MB, Hernández-Ruiz J (2014) Melatonin: Plant growth regulator and/or biostimulator during stress? Trends Plant Sci 19:789–797

Arnao MB, Hernández-Ruiz J (2015a) Melatonin: Synthesis from tryptophan and its role in higher plants. In: Mello JD (ed) Amino acids in higher plants. CAB Intern, Boston, pp. 390–435

Arnao MB, Hernández-Ruiz J (2015b) Functions of melatonin in plants: A review. J Pineal Res 59:133–150

Arnao MB, Hernández-Ruiz J (2018a) The multi-regulatory properties of melatonin in plants. In: Roshchina VV, Ramakrishna A (eds) Neurotransmitters in plants. Taylor and Francis-CRC, New York, pp. 71–101

Arnao MB, Hernández-Ruiz J (2018b) Melatonin in its relationship to plant hormones. Ann Bot 121:195–207

Arnao MB, Hernández-Ruiz J (2019a) Melatonin and reactive oxygen and nitrogen species: A model for the plant redox network. Melatonin Res 2:152–168

Arnao MB, Hernández-Ruiz J (2019b) Melatonin as a chemical substance or as phytomelatonin rich-extracts for use as plant protector and/or biostimulant in accordance with EC legislation. Agronomy 9:570

Arnao MB, Hernández-Ruiz J (2019c) Melatonin: A new plant hormone and/or a plant master regulator? Trends Plant Sci 24:38–48

Arnao MB, Hernández-Ruiz J (2020) Isphytomelatonin a new plant hormone? Agronomy 10:95

Bai S, Yao T, Li M, Guo X, Zhang Y, Zhu S, He Y (2014) PIF3 is involved in the primary root growth inhibition of *Arabidopsis* induced by nitric oxide in the light. Mol Plant 7:616–625

Bal E (2019) Physicochemical changes in *Santa Rosa* plum fruit treated with melatonin during cold storage. Food Measure 13:1713–1720

Banerjee A, Roychoudhury A (2018a) Melatonin as a regulator of abiotic stress tolerance in plants. In: Singh VP, Singh S, Prasad SM (eds) Mechanisms behind phytohormonal signaling and crop abiotic stress tolerance. Nova Science Publishers, New York, pp. 47–60

Banerjee A, Roychoudhury A (2018b) Strigolactones: Multi-level regulation of biosynthesis and diverse responses in plant abiotic stresses. Acta Physiologiae Plantarum 40:86

Byeon Y, Back K (2014) An increase in melatonin in transgenic rice causes pleiotropic phenotypes, including enhanced seedling growth, delayed flowering, and low grain yield. J Pineal Res 56:408–414

Cano A, Alcaraz O, Arnao MB (2003) Free radical-scavenging activity of indolic compounds in aqueous and ethanolic media. Anal Bioanal Chem 376:33–37.

Choi D, Cho HT, Lee Y (2006) Expansins: Expanding importance in plant growth and development. Physiol Plant 126:511–518

Du L, Bao C, Hu T, Zhu Q, Hu H, He Q, Mao W (2016) SmARF8, a transcription factor involved in parthenocarpy in eggplant. MolGenet Genom 291:93–105

El-Mogy M, Ludlow A, Callum R, Müller C, Rogers J (2019) Postharvest exogenous melatonin treatment of strawberry reduces postharvest spoilage but affects components of the aroma profile. J Berry Res 9:297–307

Gao H, Zhang ZK, Chai HK, Cheng N, Yang Y, Wang DN, Yang T, Cao W (2016) Melatonin treatment delays postharvest senescence and regulates reactive oxygen species metabolism in peach fruit. Postharvest Biol Technol 118:103–110

Hano S, Shibuya T, Imoto N, Ito A, Imanishi S, Aso H, Kanayama Y (2017) Serotonin content in fresh and processed tomatoes and its accumulation during fruit development. Sci Hortic 214:107–113

Hernández-Ruiz J, Arnao MB (2018) Relationship of melatonin and salicylic acid in biotic/abiotic plant stress responses. Agronomy 8:33

Hu W, Yang H, Tie W, Yan Y, Ding Z, Liu Y, Wu C, Wang J, Reiter RJ, Tan DX, Shi H, Xu B, Jin Z (2017) Natural variation in banana varieties highlights the role of melatonin in postharvest ripening and quality. J Agric Food Chem 65:9987–9994

Kolar J (2003) Effects of melatonin on circadian rhythms and photoperiodism in higher plants. Ph.D. Thesis, Faculty of Natural Sciences, Charles University, Prague, Czech Republic

Kolar J, Johnson C, Machackova I (2003) Exogenously applied melatonin affects flowering of the short-day plant *Chenopodiumrubrum*. Physiol Plant 118:605–612

Kolar J, Machackova I (2005) Melatonin in higher plants: Occurrence and possible functions. J Pineal Res 39:333–341

Kolar J, Machackova I, Eder J, Prinsen E, van Dongen W, van Onckelen H, Illnerova H (1997) Melatonin: Occurrence and daily rhythm in *Chenopodium rubrum*. Phytochemistry 44:1407–1413

Kolar J, Machackova I, Illnerova H, Prinsen E, van Dongen W, van Onckelen H (1995) Melatonin in higher plant determined by radioimmunoassay and liquid chromatography-mass spectrometry. Biol Rhythm Res 26:406–409

Korkmaz A, Deger O, Cuci Y (2014) Profiling the melatonin content in organs of the pepper plant during different growth stages. Sci Hortic 172:242–247

Kumar R, Khurana A, Sharma AK (2014) Role of plant hormonesand their interplay in development and ripening of fleshy fruits. J Exp Bot 65:4561–4575

Lee YH, Lee K, Back K (2019) Knockout of *Arabidopsis Serotonin* N-acetyltransferase-2 reduces melatonin levels and delays flowering. Biomolecules 9:712

Lei Q, Wang L, Tan DX, Zhao Y, Zheng XD, Chen H, Li QT, Zuo BX, Kong J (2013) Identification of genes for melatonin synthetic enzymes in Red Fuji apple (*Malus domestica* Borkh. cv. Red) and their expression and melatonin production during fruit development. J Pineal Res 55:443–451

Li M, An F, Li W, Ma M, Feng Y, Zhang X, Guo H (2016) DELLA proteins interact with FLC to repress flowering transition. J Integr Plant Biol 58:642–655

Li T, Wu Q, Zhu H, Zhou Y, Jiang Y, Gao H, Yun Z (2019) Comparative transcriptomic and metabolic analysis reveals the effect of melatonin on delaying anthracnose incidence upon postharvest banana fruit peel. BMC Plant Biol 19:289:1–5

Liu C, Zheng H, Sheng K, Liu W, Zheng L (2018a) Effects of melatonin treatment on the postharvest quality of strawberry fruit. Postharvest Biol Technol 139:47–55

Liu J, Yang J, Zhang H, Cong L, Zhai R, Yang C, Wang Z, Ma F, Xu L (2019a) Melatonin inhibits ethylene synthesis via nitric oxide regulation to delay postharvest senescence in pears. J Agric Food Chem 67:2279–2288

Liu J, Yue R, Si M, Wu M, Cong L, Zhai R, Yang C, Wang Z, Ma F, Xu L (2019b) Effects of exogenous application of melatonin on quality and sugar metabolism in *Zaosu* pear fruit. J Plant Growth Regul 38:1161–1169

Liu J, Zhai R, Liu F, Zhao Y, Wang H, Liu L, Yang C, Wang Z, Ma F, Xu L (2018b) Melatonin induces parthenocarpy by regulating genes in gibberellin pathways in 'Starkrimson' pear (*Pyrus communis* L.). Front Plant Sci 9:946

Liu J, Zhang R, Sun Y, Liu Z, Jin W, Sun Y (2016) The beneficial effects of exogenous melatonin on tomato fruit properties. Sci Hortic 207:14–20

Lozano-Juste J, León J (2011) Nitric oxide regulates DELLA content and PIF expression to promote photomorphogenesis in *Arabidopsis*. Plant Physiol 156:1410–1423

Machackova I, Krekule J (2002) Sixty-five years of searching for the signals that trigger flowering. Russ J Plant Physiol 49:451–459

Manchester LC, Tan DX, Reiter RJ, Park W, Monis K, Qi W (2000) High levels of melatonin in the seeds of edible plants: Possible function in germ tissue protection. Life Sci 67:3023–3029

Moustafa-Farag M, AlmoneafyA, Mahmoud A, Elkelish A, Arnao MB, Li L, Ai S (2020) Melatonin and its protective role against biotic stress impacts on plants. Biomolecules 10:54

Mukherjee S (2019) Recent advancements in the mechanism of nitric oxide signaling associated with hydrogen sulfide and melatonin crosstalk during ethylene-induced fruit ripening in plants. Nitric Oxide 82:25–34

Murch SJ, Alan AR, Cao J, Saxena PK (2009) Melatonin and serotonin in flowers and fruits of *Datura metel* L. J Pineal Res 47:277–283

Murch SJ, Hall BA, Le CH, Saxena PK (2010) Changes in the levels of indoleamine phytochemicals during veraison and ripening of wine grapes. J Pineal Res 49:95–100

Murch SJ, Saxena PK (2002) Mammalian neurohormones: Potential significance in reproductive physiology of St. John's wort (*Hypericum perforatum* L.)? Naturwissenschaften 89:555–560

Okazaki M, Ezura H (2009) Profiling of melatonin in the model tomato (*Solanum lycopersicum* L.) cultivar Micro-Tom. J Pineal Res 46:338–343

Ozga JA, Reinecke DM (2003) Hormonal interactions in fruit development. J Plant Growth Regul 22:73–81

Park S, Le TNN, Byeon Y, Kim YS, Back K (2013) Transient induction of melatonin biosynthesis in rice (*Oryza sativa* L.) during the reproductive stage. J Pineal Res 55:40–45

Samanta S, Banerjee A, Roychoudhury A (2021a) Exogenous melatonin regulates endogenous phytohormone homeostasis and thiol-mediated detoxification in two indica rice cultivars under arsenic stress. Plant Cell Reports 40:1585–1602

Samanta S, Banerjee A, Roychoudhury A (2021b) Melatonin application differentially modulates the enzymes associated with antioxidative machinery and ascorbate glutathione cycle during arsenate exposure in indica rice varieties. Plant Biology 23 (S1): 193–201

Searle I, He Y, Turck F, Vincent C, Fornara F, Kröber S, Amasino RA, Coupland G (2006) The transcription factor FLC confers a flowering response to vernalization by repressing meristem competence and systemic signaling in *Arabidopsis*. Genes Dev 20:898–912

Sharif R, Xie C, Zhang H, Arnao MB, Ali M, Ali Q, Muhammad I, Shalmani A, Nawaz M, Chen P, Li Y (2018) Melatonin and its effects on plant systems. Molecules 23:2352

Shi H, Chen Y, Tan DX, Reiter RJ, Chan Z, He C (2015) Melatonin induces nitric oxide and the potential mechanisms relate to innate immunity against bacterial pathogen infection in *Arabidopsis*. J Pineal Res 59:102–108

Shi H, Wei Y, Wang Q, Reiter RJ, He C (2016) Melatonin mediates the stabilization of DELLA proteins to repress the floral transition in *Arabidopsis*. J Pineal Res 60:373–379

Srivastava A, Handa AK (2005) Hormonal regulation of tomato fruit development: A molecular perspective. J Plant Growth Regul 24:67–82

Sun QQ, Zhang N, Wang J, Cao Y, Li X, Zhang H, Zhang L, Tan DX, Guo YD (2016) A label-free differential proteomics analysis reveals the effect of melatonin on promoting fruit ripening andanthocyanin accumulation upon postharvest in tomato. J Pineal Res 61:138–153

Sun QQ, Zhang N, Wang J, Zhang HJ, Li DB, Shi J, Li R, Weeda S, Zhao B, Ren S, Guo YD (2015) Melatonin promotes ripening and improves quality of tomato fruit during postharvest life. J Exp Bot 66:657–668

Tan DX, Hardeland R, Manchester LC, Poeggeler B, Lopez-Burillo S, Mayo JC, Sainz RM, Reiter RJ (2003) Mechanistic and comparative studies of melatonin and classic antioxidants in terms of their interactions with the ABTS cation radical. J Pineal Res 34:249–259.

Waters MT, Gutjahr C, Bennett T, Nelson DC (2017) Strigolactone signaling and evolution. Annu Rev Plant Biol 68:291–322

Wigge PA, Kim MC, Jaeger KE, Busch W, Schmid M, Lohmann JU, Weigel D (2005) Integration of spatial and temporal information during floral induction in *Arabidopsis*. Science 309:1056

Wolf K, Kolar J, Witters E, van Dongen W, van Onckelen H, Machackova I (2001) Daily profile of melatonin levels in *Chenopodiumrubrum* L. depends on photoperiod. J Plant Physiol 158:1491–1493

Xiao S, Liu L, Wang H, Li D, Bai Z, Zhang Y, Sun H, Zhang K, Li C (2019) Exogenous melatonin accelerates seed germination in cotton (*Gossypium hirsutum* L.). PLoS One 14:e0216575

Xu L, Yue Q, Bian FE, Sun H, Zhai H, Yao Y (2017) Melatonin enhances phenolics accumulation partially via ethylene signaling and resulted in high antioxidant capacity in grape berries. Front Plant Sci 8:1426

Xu L, Yue Q, Xiang G, Bian F, Yao Y (2018) Melatonin promotes ripening of grape berry via increasing the levels of ABA, H2O2, and particularly ethylene. Hortic Res 5:41

Xu T, Chen Y, Kang H (2019) Melatonin is a potential target for improving post-harvest preservation of fruits and vegetables. Front Plant Sci 10:1388

Yu S, Galváo VC, Zhang YC, Horrer D, Zhang TQ, Hao YH, Feng YQ, Wang S, Schmid M, Wang JW (2012) Gibberellin regulates the *Arabidopsis* floral transition through miR156-targeted squamosa promoter binding-like transcription factors. Plant Cell 24:3320–3332

Zhai R, Liu J, Liu F, Zhao Y, Liu L, Fang C, Wang H, Li X, Wang Z, Ma F, Xu L (2018) Melatonin limited ethylene production, softening and reduced physiology disorder in pear (*Pyruscommunis* L.) fruit during senescence. Postharvest Biol Technol 139:38–46

Zhang HJ, Zhang N, Yang RC, Wang L, Sun QQ, Li DB, Cao YY, Weeda S, Zhao B, Ren S, Guo YD (2014) Melatonin promotes seed germination under high salinity by regulating antioxidant systems, ABA and GA4 interaction in cucumber (*Cucumis sativus* L.). J Pineal Res 57:269–279

Zhang Y, Huber DJ, Hu M, Jiang G, Gao Z, Xu X, Jiang Y, Zhang Z (2018) Delay of postharvest browning in litchi fruit by melatonin via the enhancing of antioxidative processes and oxidation repair. J Agric Food Chem 66:7475–7484

Zhang Z, Hu Q, Liu Y, Cheng P, Cheng H, Liu W, Xing X, Guan Z, Fang W, Chen S, Jiang J, Chen F (2019) Strigolactone represses the synthesis of melatonin, thereby inducing floral transition in *Arabidopsis thaliana* in an FLC-dependent manner. J Pineal Res 67:e12582

Zhao D, Wang R, Liu D, Wu Y, Sun J, Tao J (2018) Melatonin and expression of tryptophan decarboxylase gene (TDC) in herbaceous peony (*Paeonia lactiflora* Pall.) flowers. Molecules 23:1164

Zhao Y, Tan DX, Lei Q, Chen H, Wang L, Li QT, Gao Y, Kong J (2013) Melatonin and its potential biological functions in the fruits of sweet cherry. J Pineal Res 55:79–88

5 Melatonin as a Vital Metabolite in Medicinal and Food Plants

Phaniendra Alugoju, K.K Kumaree, Anchalee Prasansuklab, and Tewin Tencomnao

5.1 INTRODUCTION

Melatonin (*N-acetyl*-5-methoxytryptamine) (Figure 5.1) is an indole amine compound that was initially discovered in the pineal gland (an endocrine gland) in the year 1958 by Lerner *et al* (Lerner et al. 1958). Indeed, it is highly conserved among living organisms found ubiquitously in both animal and plant kingdoms. It is also present in a wide range of species such as single-celled eukaryotic organisms (e.g., yeast), prokaryotes (e.g., cyanobacteria, alpha proterobacteria), protozoans, algae (e.g., dinoflagellates, rhodophyta, phaeophyta), fungi, invertebrates (e.g., insects, crustaceans, planarians), vertebrates including humans, and higher plants (Reiter et al. 2003).

In addition to its wide existence in the animal kingdom, it is also present in a variety of plants where it is also essentially implicated in a wide variety of physiological roles in plants. In plants, melatonin was initially discovered in 1995. Melatonin has received great attention in the past few decades because of its multiple physiological roles in both plants as well as in animals. Therefore, this book chapter deals with the exogenous sources of phytomelatonin, biosynthesis of melatonin, and beneficial roles of melatonin in animals. Besides, we have also briefed on the role of phytomelatonin in plants and their physiological processes.

5.2 BIOSYNTHESIS OF MELATONIN IN ANIMALS AND PLANTS

Tryptophan is the precursor molecule for the biosynthesis of melatonin in both animals and plants. In animals, the tryptophan is taken up from the bloodstream by the pinealocytes where the tryptophan is hydroxylated initially to 5-hydroxytryptophan. This first step in the synthesis of melatonin is catalyzed by the enzyme tryptophan-5-hydroxylase. The 5-hydroxytryptophan is then decarboxylated to serotonin by the enzyme aromatic amino acid decarboxylase (5-hydroxytryptamine). Later, the serotonin is further converted into N-acetyl serotonin by the rate-limiting enzyme serotonin N-acetyl transferase (also known as aralkylamine N-acetyltransferase). Finally, N-acetyl serotonin is converted into melatonin (N-acetyl-methoxy tryptamine) by the enzyme N-acetyl serotonin-O-methyl transferase (Odaci and Kaplan 2009). The findings of Garcia-Marin et al. (2015) revealed that the thyroid-stimulating hormone can regulate the activities of serotonin N-acetyl transferase and N-acetyl serotonin-O-methyl transferase, two key enzymes of melatonin biosynthesis.

In the case of plants, the precursor tryptophan is first converted by the enzyme tryptophan decarboxylase to tryptamine, which in turn is converted to serotonin by the enzyme tryptamine-5-hydroxylase. Later, the serotonin is converted to either N-acetyl serotonin or 5-methoxytryptamine which will be subsequently converted to melatonin. In plants, Serotonin N-acetyltransferase is located in plastids, but this plant enzyme does not share homology with the animal serotonin N-acetyltransferase (Hardeland 2015; Banerjee and Roychoudhury 2017). The biosynthetic pathway of melatonin from tryptophan is mentioned in Figure 5.2 (blue lines representing the animal biosynthetic pathway and green lines representing the plant biosynthetic pathway). Though tryptophan is the precursor molecule for the biosynthesis of melatonin in both plants and animals, it is not

DOI: 10.1201/9781003343752-5

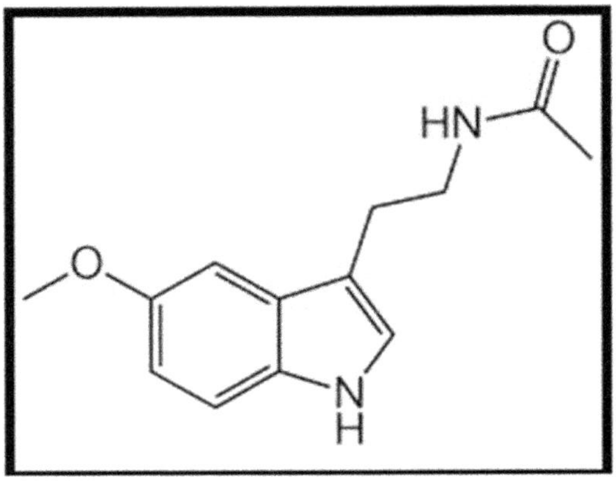

FIGURE 5.1 Melatonin (N-acetyl 5-methoxytryptamine).

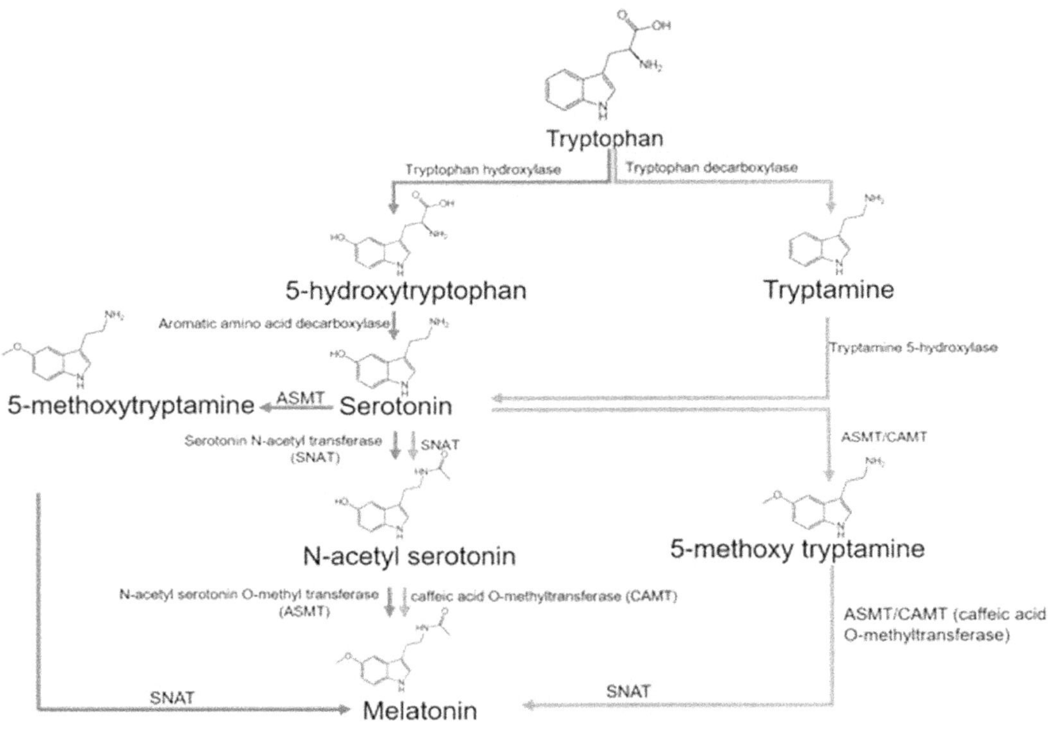

FIGURE 5.2 Biosynthesis of melatonin in animals (blue arrows) and plants (green arrows).

clear whether the same amino acid tryptophan is the precursor for the biosynthesis of melatonin in microorganisms (Que et al. 2020).

5.3 SOURCES OF PHYTOMELATONIN

There is a growing interest in the identification and quantification of melatonin in several organisms from plants to animals. Among animal sourced foods, fish, and eggs constitute the richest sources of melatonin. Nuts constitute the richest sources of melatonin among the plant-derived melatonin (referred to as phytomelatonin) (Meng et al. 2017). In plants, melatonin is usually present in different plant parts including leaves, fruits, vegetables, seeds, and flowers. Therefore, plants serve as the best exogenous source of melatonin for animals and humans (Manchester et al. 2000; Tan et al. 2000). Since melatonin is present in several edible plants as well as medicinal plants, this melatonin enters the human body through the food chain. Owing to its potent antioxidant ability, its entry into human body is beneficial for human health; therefore, it is very essential to keenly identify those edible and medicinal plants that are rich in melatonin. Melatonin has been reported to present in high concentrations in a variety of edible plants, medicinal plants, and fermented foods (Badria 2002; Murch and Saxena 2002). Intake of melatonin-rich plant foods has been reported to have significantly increased the levels of melatonin in blood and 6-sulphatoxymelatonin (melatonin metabolite) in urine (Dericks-Tan et al. 2003). However, several studies have reported different amounts of melatonin in the same plant or plant products which could be due to the usage of different analytical methods as well as the sensitivity of the analytical technique to detect low levels of melatonin.

5.3.1 Edible Plants

Melatonin is present in varying concentrations in a variety of edible plants (listed in Table 5.1) such as almonds, beetroot, broccoli, banana, corn, rice, barley, ginger, *etc.* Thus, melatonin enters our body after consuming such phytomelatonin-rich edible plants or plant products. Higher levels of

TABLE 5.1
Melatonin Levels in Edible Plants

Name of the edible plant	Phytomelatonin Content (ng/g DW or FW)	Reference
Almond	39 DW	(Manchester et al. 2000)
Alfalfa	16 DW	(Manchester et al. 2000)
Barley	82.3 ± 6.0 ng/g FW	(Hernández-Ruiz et al. 2005)
Black mustard	129 DW	(Manchester et al. 2000)
Bell pepper	42 FW	(Korkmaz et al. 2014)
Beetroot	0.002 ng/g FW	(Dubbels et al. 1995)
Broccoli	0.41 ng/g DW	(Aguilera et al. 2015)
Cardamon	15 DW	(Manchester et al. 2000)
Coffee species	5800–6500 DW	(Ramakrishna et al. 2012)
Corn	1.88 ng/g FW	(Badria 2002)
Cauliflower	0.82 ng/g FW	(Badria 2002)
Cucumber	0.59 ng/g FW	(Badria 2002)
Carrot	0.49 ng/g FW	(Badria 2002)
Cabbage	0.11 ng/g FW	(Hattori et al. 1995)
Fenugreek	43 DW	(Manchester et al. 2000)
Fennel	28 DW	(Manchester et al. 2000)
Garlic	0.59 ng/g FW	(Badria 2002)
Ginger	1.42 ng/g FW	(Badria 2002)
Kidney beans	529 DW	(Aguilera et al. 2016)
Oats	90.6 ± 7.7 ng/g FW	(Hernández-Ruiz et al. 2005)
Onion	0.30 ng/g FW	(Badria 2002)
Rice	1.50 ng/g FW	(Badria 2002)
Radish	0.76 ng/g FW	(Badria 2002)
Soybean	0.45 ng/g DW	(Zieliński et al. 2001)

(Continued)

TABLE 5.1 *(Continued)*
Melatonin Levels in Edible Plants

Name of the edible plant	Phytomelatonin Content (ng/g DW or FW)	Reference
Sunflower	29 DW	(Manchester et al. 2000)
Tomato	114 FW	(Stürtz et al. 2011)
Wolfberries	530 DW	(Chen et al. 2003)
White radish	485 DW	(Manchester et al. 2000)
Wheat	124.7±14.9 ng/g FW	(Hernández-Ruiz et al. 2005)

DW – Dry Weight; FW – Fresh Weight

TABLE 5.2
Melatonin Levels in Edible Fruits

Name of the edible plant	Phytomelatonin content (ng/g or pg/g DW or FW)	Reference
Banana	8.9 pg/g	Johns et al. (2013)
Banana	655pg/g	Badria (2002)
Pineapple	302 pg/g	Johns et al. (2013)
Cherry	<17 pg/mL	Burkhardt et al. (2001); Kocadağlı et al. (2014)
Orange	150 pg/g	Johns et al. (2013)
Walnut	137 pg/g	Kocadağlı et al. (2014)
Walnut	3400 pg/g	Reiter et al. (2005)
Strawberry	11.2–1.4 FW	Stürtz et al. (2011)
Grape	1.2 FW	Vitalini et al. (2011b); Iriti and Faoro (2006)
Apple	134 FW	Lei et al. (2013)
Sweet cherry	120–8.0 FW	Zhao et al. (2013)
Pomegranate	0.17 ng/g FW	Badria (2002)
Kiwi	0.02 ng/g FW	Hattori et al. (1995)

DW – Dry Weight; FW – Fresh Weight

melatonin have been reported in almond, alfalfa, bell pepper, barley, wheat, black mustard, cardamon, coffee beans, fenugreek, fennel, kidney beans, tomato, white radish, oats, and wolf berries, whereas low levels of melatonin were reported in beetroot, broccoli, cucumber, carrot, soybean, corn, cauliflower, and rice.

5.3.2 Edible Fruits

A variety of fruits such as banana, pineapple, cherries, oranges, walnuts, strawberries, grapes, apples, sweet cherries, pomegranate, and kiwi fruits were reported to consist of different concentrations of melatonin (Table 5.2). A high concentration of melatonin is reported in cherries (Burkhardt et al. 2001; Kocadağlı et al. 2014), walnuts (Reiter et al. 2005), and banana (Badria 2002; Johns et al. 2013), whereas very low levels were reported in pomegranate and kiwi (Badria 2002). Several studies have demonstrated that the consumption of melatonin-rich fruit juices can enhance serum melatonin levels and total serum antioxidant status. In a crossover study, 12 healthy human volunteers were allowed to drink either juices made from orange or pineapple or banana for one week, and serum melatonin levels, as well as total antioxidant capacity, were observed in the participants. Participants who consumed banana and pineapple had higher melatonin levels in their body, helping in maintaining the circadian rhythms of the body (Sae-Teaw et al. 2013). Similarly, in another study, human volunteers of three different age groups (young – 20 ± 10 years old, middle-aged – 45 ± 10 years old, and elderly – 75 ± 10 years old) were provided to consume red grape juice and the study found a significant increase in total antioxidant levels of serum and urinary 6-sulfatoxymelatonin levels in three different age groups (González-Flores et al. 2012).

5.3.3 Fermented Foods

Very few studies have reported the presence of melatonin in fermented foods and beverages. The amounts of melatonin were found to be present in low levels in a variety of fermented foods and beverages such as fermented dough, bread, kefir, yogurt, wine, beer, and pomegranate wine (Table 5.3). The presence of different levels of melatonin levels in these fermented foods depends on the metabolic activity of different microorganisms and different raw materials used for the production of these fermented foods (Garcia-Moreno et al. 2013; Mena et al. 2012; Vitalini et al. 2011a, 2011b, 2013).

5.3.4 Medicinal Herbs

Consumption of medicinal herbs might increase the blood levels of melatonin and may exert their potent health-promoting effects in humans. Several medicinal plants have been reported to contain different levels of melatonin in a variety of medicinal herbs listed in Table 5.4. Higher levels of melatonin were reported in *Glycyrrhiza uralensis*, *Hypericum perforatum*, *Mentha piperita*, *Salvia officinalis* (Afreen et al. 2006; Chen et al. 2003; Murch and Saxena 2006; Stege et al. 2010), *Piper nigrum* (Padumanonda et al. 2014), whereas low melatonin levels were reported in *Agastache rugosa*, *Aloe vera*, *Achillea millefolium*, *Curcuma aeruginosa*, *Gentiana macrophylla*, *Lonicea japonica*, *Scutellaria amoena*, *Scrophularia ningpoensis*, and *Thymus vulgaris* (Chen et al. 2003).

TABLE 5.3
Concentration of Melatonin in Fermented Foods

Fermented food	ng/mL	Reference
Fermented dough	0.09	Yilmaz et al. (2014)
Bread (crumb and crust)	0.13–0.34	Kocadağlı et al. (2014)
Kefir	-NA-	Kocadağlı et al. (2014)
Yogurt	0.12	Kocadağlı et al. (2014)
Wine	0.1–8	Vitalini et al. (2013) Vitalini et al. (2011a)
Beer	0.06–0.16	Garcia-Moreno et al. (2013)
Pomegranate wines	0.54–5.5	Mena et al. (2012)

NA – not available

TABLE 5.4
Concentration of Melatonin in Medicinal Herbs

Plant name	ng/g FW or DW	Reference
Agastache rugosa	300 DW	Chen et al. (2003)
Aloe vera	516 DW	Chen et al. (2003)
Achillea millefolium	340 DW	Marioni et al. (2008)
Baccaurea ramiflora	43.2 DW	Padumanonda et al. (2014)
Curcuma aeruginosa	120 DW	Chen et al. (2003)
Gentiana macrophylla	180 DW	Chen et al. (2003)
Glycyrrhiza uralensis	34000 DW	Afreen et al. (2006)
Hypericum perforatum	13000 FW	Murch and Saxena (2006)
Lonicea japonica	140 DW	Chen et al. (2003)
Momordica charantia	21.4 DW	Padumanonda et al. (2014)
Morus alba	1510 DW	Chen et al. (2003)
Mentha piperita	19500 DW	Chen et al. (2003)
Piper nigrum	1092.7 DW	Padumanonda et al. (2014)
Salvia officinalis	29000 DW	Stege et al. (2010)
Sesbania glandiflora	26.3 DW	Padumanonda et al. (2014)
Sesbania sesban	8.7 DW	Padumanonda et al. (2014)
Senna tora	10.5 DW	Padumanonda et al. (2014)
Scutellaria amoena	178 DW	Chen et al. (2003)
Scrophularia ningpoensis	342 DW	Chen et al. (2003)
Syzygium aromaticum	446 DW	Chen et al. (2003)
Tanacetum parthenium	1120.0–2086.9 DW	Ansari et al. (2010)
Tripleurospermum disciforme	1305.8–3073.3 DW	Ansari et al. (2010)
Thymus vulgaris	38000 DW	Stege et al. (2010)
Viola philipica	2360 DW	Chen et al. (2003)
Viola odorata	0.8–1.1 DW	Ansari et al. (2010)
Uncaria rhynchophylla	2460 DW	Chen et al. (2003)

DW – Dry Weight; FW – Fresh Weight

5.4 BENEFICIAL ROLES OF MELATONIN IN HUMANS

Melatonin exerts its multiple physiological properties via interaction with G-protein-coupled membrane receptors found in the suprachiasmatic nucleus and in other different sites in the body. Nuclear (ROR/RZR), cytoplasmic (quinone reductase-2, calmodulin, calreticulin) and mitochondrial binding sites, and radical-scavenging properties contribute to the actions of melatonin (Hardeland 2008).

5.4.1 Melatonin and Circadian Rhythms

In vertebrates such as fishes, reptiles, birds, and mammals, melatonin acts as a biological regulator of circadian rhythms such as mood, sleep, sexual behavior, immunological status, etc. (Hardeland et al. 2006; Pandi-Perumal et al. 2006). In humans, these cyclic rhythms are controlled by light via the retina and the melatonin system. Thus, there is an interplay between light and circadian rhythms in humans. Melatonin levels are low in daytime but increased at nighttime. Exposure to light at night can diminish melatonin synthesis, and this could be the possible reason for the increased incidence of breast and colorectal cancer in developed countries. Studies indicate that exposure to constant light can suppress melatonin synthesis, promotes the proliferation of cancer cells, and also increases the uptake of linoleic acid (LA) by cancer cells. In vertebrates, melatonin is usually produced in the pineal gland during early-morning hours of darkness (2–4 AM), even in nocturnal animals, but its production is prevented upon exposure to light at nighttime. Synchronization of human circadian rhythms is very essential for the good health of humans (Pauley 2004). The suprachiasmatic nucleus, the anterior part of the hypothalamus, acts as a pacemaker of the circadian timing system, and it regulates most of the circadian rhythms in humans. This pacemaker is extremely sensitive to ocular light exposure. Lockley et al. conducted a clinical trial and found that humans exposed to 6.5 h of monochromatic blue light (460 nm) induced a twofold greater circadian phase delay than with 555 nm (photopic) light and also suppressed melatonin levels twice as much with 460 nm light than with 555 nm light. Therefore, it can be concluded that blue light regulates the production of melatonin in humans (Lockley et al. 2003). The axons of the retinal ganglion cells (RGCs) in the retina are directly associated with the circadian or biological rhythms. Some of these RGCs contains a photopigment called melanopsin that is sensitive to blue light. It was also reported that melatonin levels were reduced upon exposure to blue light (Pauley 2004).

As discussed above, melatonin biosynthesis during nighttime is regulated by a rate-limiting enzyme serotonin N-acetyl transferase which is expressed in pineal as well as extra pineal sites including retina and other tissues. As this enzyme is essential for the conversion of serotonin to N-acetylserotonin, natural compounds that can modulate the enzymatic activity serotonin N-acetyl transferase may be beneficial for the treatment of patients with altered circadian rhythm disorders. Research studies were conducted to test the efficacy of 267 medicinal plants to inhibit the activity of serotonin N-acetyl transferase. In this study, it was discovered that medicinal herbs such as *Myricae cortex*, *Perillae herba*, and *Eriobotryae folium* were shown to strongly inhibit the serotonin N-acetyl transferase enzyme activity. This suggests that medicinal herbs that can modulate the activity of serotonin N-acetyl transferase can be used for the amelioration of circadian rhythms (Shin et al. 2013).

5.4.2 Antioxidant Role

Melatonin is a potent endogenous antioxidant with an ability to directly detoxify a wide range of free radicals including hydroxyl radical, hydrogen peroxide, nitric oxide, peroxynitrite anion, peroxynitrous acid, and hypochlorous acid (Reiter et al. 2001; Samanta et al. 2021). The indole ring present in the melatonin exhibits high resonance stability, and therefore it is the reactive center of interaction with free radicals. Besides, the amide and methoxy side chain groups of melatonin can also detoxify free radical species; therefore, these groups also contribute for its potent antioxidant ability (Hardeland and Pandi-Perumal 2005). Furthermore, melatonin can exert indirect antioxidant activity by enhancing the activity of other endogenous enzymatic antioxidants and diminishing the prooxidant enzyme levels in the body. Some quinone reductase enzymes catalyze one electron reduction of quinones in presence of molecular oxygen resulting in the formation of semiquinones and subsequent generation of reactive oxygen species (ROS). In contrast to this, NAD(P)H: quinone acceptor oxidoreductase (NQO2), a redox enzyme, catalyzes the two-electron reduction of quinones to hydroquinone but not to semiquinones and therefore prevents the ROS formation. Thus, the indirect antioxidant activity of melatonin is via its ability to enhance the activity of NQO2 and subsequent generation of ROS (Vasiliou et al. 2006). It can also enhance the efficiency of mitochondrial oxidative phosphorylation, thereby preventing electron leakage and subsequent generation of ROS (Reiter et al. 2003).

In contrast to the classical antioxidants, melatonin does not exhibit prooxidative effects. In contrast to the classical antioxidant molecules, melatonin does not undergo redox cycling with the free radicals but can interact with free radicals to form stable end products that will be sent out through the urine. Therefore, melatonin is considered as a suicidal or terminal antioxidant molecule. Melatonin interacts with free radicals and generates other intermediary metabolites such as acetyl methoxykynuramine and hydroxymelatonin which can also function as free radical scavengers

(Schaefer and Hardeland 2009). This phenomenon is defined as the free radical scavenging cascade reaction of the melatonin family. Thus, melatonin initiates a cascade of free radical scavenging process, and because of this cascade of radical scavenging ability, one melatonin molecule can detoxify up to four or more free radicals. This makes melatonin a potent antioxidant. In vivo studies also showed that melatonin administration could efficiently prevent tissue oxidative damage when compared to other classical vitamins, C and E (Tan et al. 2002, 2003).

5.4.3 Anti-Inflammatory Role

Chronic inflammation is critically implicated in the pathogenesis of several human diseases. In vivo studies showed that treatment with melatonin have significant anti-inflammatory effects. Melatonin treatment significantly reduced the edema formation and hyperalgesia in carrageenan-induced paw edema rat model (Padhy and Kumar 2005). Several clinical trials have demonstrated the potent anti-inflammatory effects of melatonin (Cho et al. 2021). Melatonin was shown to inhibit lipopolysaccharide-induced inflammation in human vascular smooth muscles in vitro. Treatment with melatonin inhibited phosphorylation of p38 MAPK, JNK, ERK1/2, IκB-α, and c/EBPβ, and hence multiple signaling and subsequent reduction in the production and expression of proinflammatory markers such as cyclooxygenase-2, tumor necrosis factor-alpha, interleukin-1beta, and nitric oxide. This indicates the potent anti-inflammatory role of melatonin in the treatment of inflammatory diseases (Shi et al. 2012).

5.4.4 Melatonin and Diabetes

Diabetes is a metabolic disease which is associated with an increase in the blood glucose levels. Melatonin has been reported to reduce the levels of lipid peroxides and nitric oxide, thereby ameliorating oxidative stress in streptozotocin (STZ)-induced diabetic rats. In addition, STZ-induced increase of plasma glucose, total lipids, triglyceride, and cholesterol levels were also diminished by the administration of melatonin in the diabetic rat models, indicating the ability of melatonin in effective anti-diabetic therapy to counteract the oxidative stress burden in diabetes mellitus (Anwar and Meki 2003). It was also demonstrated that administration of melatonin prevented gastric oxidative damage and also ameliorated the dyslipidemia in STZ-induced diabetes (Hadjzadeh et al. 2018).

5.4.5 Anti-Cancer Activity

The oncostatic activities have reported in various types of cancers including breast cancer (el-Aziz et al. 2005). Increased uptake of fatty acid linoleic acid and its subsequent metabolism to 13-hydroxyoctadecadienoic acid (a mitogenic signaling molecule) in turn promotes tumor growth. Anti-cancer mechanism of melatonin involves detoxification of free radicals, enhancing the immune function and inducing the apoptosis of cancer cells (Dopfel et al. 2007). Melatonin prevents tumor growth via melatonin receptor and additionally prevents uptake and metabolism of linoleic acid and stimulates tumor melatonin uptake and retention in response to the dietary intake of phytomelatonin (Blask et al. 2004). It is also known that melatonin exhibits both melatonin-receptor and non-receptor mediated actions. Both in vitro and in vivo studies demonstrated that melatonin can inhibit proliferation of melanoma cells, suggesting the possible anti-skin activities of melatonin (Kadekaro et al. 2004). Furthermore, these studies revealed that melatonin exerts anti-skin properties via a melatonin-receptor dependent way (Kadekaro et al. 2004). In another study, researchers conducted in vitro experiments to check whether melatonin can enhance anti-cancer efficacy of fisetin against melanoma cancer cells. The results of this study revealed that combination of melatonin and fisetin markedly decreased cell proliferation, cell migration, and clone formation and induced apoptosis of melanoma cells by increasing PARP cleavage, release of cytochrome C, and inhibiting the expression of pro-inflammatory markers such as COX2 and iNOS. It was also demonstrated that combinatorial treatment was found to be more efficient than the individual treatments (Yi et al. 2014). Similarly, a combination of melatonin and berberine was also shown to exert anti-cancer activity in lung cancer cells by promoting the apoptosis of cancer cells. The anti-cancer mechanism of action of the combinatorial treatment is mainly through the activation of Cyto C dependent apoptotic pathway and inhibition of AP-2β/hTERT, NF-κB/COX-2, and Akt/ERK signaling pathways in lung cancer cells (Lu et al. 2016). Therefore, it can be suggested that melatonin in combination with other anti-cancer drugs may be used as in successful anti-cancer therapeutic strategy (Reiter et al. 2002).

5.5 MELATONIN VS AGING AND NEURODEGENERATION

In humans, serum melatonin levels fluctuate significantly during the aging process. For instance, serum levels are consistently maintained at a young age; however, serum levels decline with aging (Arnao and Hernández-Ruiz 2006). Reduction in melatonin levels were associated with aging (Tan et al. 2010) and subsequent development of neurodegenerative diseases like Alzheimer's disease (Olcese et al. 2009), Parkinson's disease (Videnovic et al. 2014), and stroke (Villapol et al. 2011). Dietary supplementation of melatonin can help in reducing the age-related accumulation of neurotoxic Abeta peptides in the brain (Shukla et al. 2017). Aging is associated with the accumulation of several toxic metals; however, melatonin reduces this metal toxicity seen in aging. Melatonin was also useful for restoring the age-related decline in cholinergic function. Owing to its ability to enhance the potency of other antioxidants, supplementation of melatonin-rich food may help in ameliorating the age-related decline in cognitive function and other neurodegenerative diseases (Hardeland et al. 2006; Lahiri et al. 2005). As the melatonin concentration diminishes with aging, altered biological rhythms are usually seen in senesce animals. However, supplementation of foods rich in melatonin may enhance the active life span (Tan et al. 2010). It has been demonstrated that combined treatment with melatonin and resveratrol synergistically offered neuroprotection via enhancing the expression of redox-regulated enzyme heme oxygenase-1. However, the individual treatments did not influence the expression of heme oxygenase-1, indicating that melatonin exerts its neuroprotective activity by its indirect antioxidant role through stimulating the enzymatic antioxidant systems in the body (Kwon et al. 2011).

The circadian rhythms associated with melatonin levels decrease with aging. Therefore, elderly people exhibit sleep alterations. Food stuff rich in melatonin or tryptophan, the precursor of melatonin, may have beneficial effects in improving sleep disorders in the elderly. In this context, a randomized simple blind study was conducted to investigate the effect of intake of tryptophan rich cereals on the sleep-wake cycles, depression, and anxiety in middle-aged (35 years) and elderly (aged 55–75 years) human subjects. In this study, participants were allowed to consume cereals (at a dose of 22.5 mg tryptophan per 30 g cereals in the control week and 60 mg tryptophan/30 mg cereals in the treatment week) at breakfast and dinner. Following three weeks of dietary intake of cereals, melatonin metabolites were measured in the urine. These study results revealed that intake of a higher dose of tryptophan improved the sleep efficiency, sleep time, immobile time, anxiety, and depression symptoms in the participated elderly. This clearly indicates that tryptophan-enriched cereals may be useful for ameliorating the sleep-wake cycles in the aged people (Bravo et al. 2013). It was also demonstrated that treatment with Novanuit® Triple Action (a combination of melatonin, Vitamin B6, and plant extracts of California poppy, passionflower, and lemon balm) for four weeks improved the sleep efficiency in patients with sleep disorders. This indicates that combinatorial treatment with melatonin may have a significant beneficial impact on insomnia in the elderly (Lemoine et al. 2019).

5.6 MELATONIN AND THE EYE

Melatonin is also synthesized in the retina of the eye. In most of the vertebrates including mammals, melatonin receptors are present in the photoreceptors, horizontal cells of the retina, corneal epithelium, stroma endothelium, and the sclera, indicating that melatonin plays a crucial role in the retina physiology (Head 2001; Lundmark et al. 2006). In a clinical trial, anxious young human subjects were administered with melatonin at a dose of 0.75 mg at night for 10 days. These study results showed that chronic administration of melatonin caused a significant decrease in the thresholds of retinal brightness sensitivity and improved the emotional state in anxious young subjects. This suggests that melatonin supplementation can improve the visual function, thereby can limit anxiety in humans (Ovanesov et al. 2006).

5.7 ROLE OF MELATONIN IN PLANTS

Plants are usually composed of higher levels of melatonin than animals. The discovery of presence of melatonin in plants occurred in 1995. It is present in different parts of plants including leaves, stems, roots, fruits, and seeds. The prime function of melatonin in plants is to serve as a first line of defense against internal and external stress factors (Arnao and Hernández-Ruiz 2015; Kaur et al. 2015; Reiter et al. 2015). In plants, melatonin also plays a pivotal role in a variety of physiological roles such as an antioxidant, growth promoter (Wei et al. 2015), synchronization of photoperiodism, regulation of reproductive physiology, defense against abiotic stress factors such as water stress (Park et al. 2013), drought (Wang et al. 2013), salinity (Wei et al. 2015), and regulation of leaf senescence (Arnao and Hernández-Ruiz 2007; Wang et al. 2013b).

5.7.1 Melatonin as an Antioxidant in Plants

Experimental evidence indicates that phytomelatonin is involved in detoxifying free radicals. Exposure to high levels of ozone leads to the generation of a significant amount of free radicals and subsequent damage in plants. However, exposure to high levels of melatonin reduced ozone- induced damage in plants. Some research studies also showed that melatonin levels increase as an antioxidant response in plants during photosynthesis and/or in plants exposed to UV radiation to detoxify free radicals generated during photosynthesis or upon UV exposure, indicating the possible antioxidant role of melatonin in plants (Manchester et al. 2000; Paredes et al. 2009; Reiter and Tan 2002). It was shown that melatonin offers protection against UV-induced oxidative stress in *Glycyrrhiza uralensis*, a Chinese traditional medicinal plant (Afreen et al. 2006).

5.7.2 Melatonin as a Growth Promoter in Plants

The growth promoting effects of melatonin are similar to that of auxin. For instance, similar to auxin, melatonin promoted growth of coleoptile in monocot plants such as canary grass, wheat, barley, and oat. On the other hand, similar to auxin, melatonin was also shown to inhibit the growth of roots in these plants. This proves the co-participation of melatonin with auxin in plant tissues as auxinic hormones in plants (Hernández-Ruiz et al. 2005). Exogenously, application of melatonin has been shown to induce endogenous indoleacetic acid levels and stimulated root growth in seedlings of *Brassica juncea*. Notably, the low concentration of melatonin (0.1 μM) was reported to exhibit root-growth-promoting effects only in the young seedlings (2 days old) but not in old seedling (4 days old). Conversely, high concentration (100 μM) melatonin showed inhibited root growth in *B. juncea* (Chen et al. 2009). Other studies also reported that root-promoting effects are similar to that of indole-3-acetic acid (Sarropoulou et al. 2012). In another study, authors have conducted in vitro experiments to find the effect of melatonin on the shoot organogenesis in *Mimosa pudica*. In this study, authors have grown the nodal segments of *M. pudica* in MS medium supplemented with melatonin and found that melatonin can promote shoot organogenesis in *M. pudica* (Ramakrishna et al. 2009).

5.7.3 Leaf Senescence

Research findings have confirmed the anti-senescence effects of melatonin in different plants (Arnao and Hernández-Ruiz 2009; Wang et al. 2012). Treatment of barley leaves with melatonin prevented the chlorophyll content, indicating the slowdown of the leaf senescence, suggesting the possible anti-senescence effects of melatonin in plants (Arnao and Hernández-Ruiz 2009). Similarly, exogenous application of melatonin reduced the expression of chlorophyll-degrading genes and subsequent reduction in chlorophyl content, thereby delaying leaf senescence in *Malus domestica* (Wang et al. 2012).

5.7.4 Melatonin in Photoperiodism

It was hypothesized that melatonin functions in plants similar to that in animals. In 1997, work of Kolar et al. using the short-day plant *Chenopodium rubrum* L. revealed that melatonin levels were oscillating between 12 h/12 h light and dark cycles. Melatonin levels were low in light period, whereas significantly elevated levels were noticed in dark phase. However, application of different levels of photoperiods (six, 12, and 18 h of dark period) did not affect the increase in the melatonin levels. Thus, it can be concluded that the synthesis of melatonin is not directly regulated by light, but it shows circadian rhythms similar to that observed in vertebrates (Wolf et al. 2001).

In some phototrophs, melatonin levels are higher at night time. This indicates that melatonin is not generally a mediator of the signal ‘darkness’ in plants. However, in several plants, its formation is stimulated by a variety of environmental factors including visible and/or UV light, high or low temperature, drought stress, salinity stress, and toxic chemicals (Hardeland 2015). Similar to animals, N1-acetyl-N2-formyl-5-methoxykynuramine (AFMK) is also identified as one of the important metabolites in water hyacinth plant. In this plant, melatonin was found to be higher during late light phase. This indicates that, unlike animals, in water hyacinth, melatonin does not act as a signal of darkness but is involved in photosynthesis or photoprotection. Higher levels of these two molecules were noticed when these plants were grown in sunlight. This led to the conclusion that these two molecules act as free radical scavengers and help in coping with soil and water pollutants including toxic heavy metals. Owing to the presence of higher levels of melatonin and AFMK in water hyacinth, the use of this plant for the purpose of phytoremediation can be suggested (Tan et al. 2007a, 2007b).

5.7.5 Protection against Different Stressors

Abiotic factors such as light, temperature, and salinity significantly affect plant growth and productivity. Several studies have reported that melatonin protects plants from a variety of environmental factors including cold stress/low temperature stress (Murch et al. 2009; Posmyk et al. 2009; Zhao et al. 2011). Exposure to cold stress significantly increased the levels of melatonin in the young flower buds of *Datura metel*, indicating that melatonin may play a crucial protective role in the development of reproductive organs in plants (Murch et al. 2009). Application of exogenous melatonin to cucumber (*Cucumis sativus* L.) seeds was shown to improve germination during chilling stress (Posmyk et al. 2009). Melatonin-rich transgenic plants have been demonstrated to possess enhanced resistance to oxidative stress caused by herbicides. Therefore, melatonin-rich transgenic plants may be developed as a future strategy to improve abiotic stress tolerance in plants.

5.8 CONCLUSION

Melatonin is a ubiquitous molecule in nature that has gained widespread popularity. The current advancements in modern science have opened the vast opportunity to utilize this bioactive component as an affordable nutraceutical supplement. Though phytomelatonin has a major application in humans as a supplement for sleeping facilitators, it also has the potential to be therapeutic for several other physiological conditions. Considering its antioxidant, anti-tumor properties, and several other beneficial impacts, it holds to be an excellent supplement for such cases. Henceforth, phytomelatonin-rich species, their identification, and their scientific characterization could be new possibilities open for scientific exploration. Though its application does not have any severe adverse effects, further human clinical trials are very much needed. Moreover, the upcoming generation is eyeing more toward the alternative cost-effective and safer natural therapy. In such a scenario, the production of phytomelatonin as a neutraceutical will be the next goal that can be taken into account.

REFERENCES

Afreen, F., S. M. Zobayed & T. Kozai. 2006. Melatonin in Glycyrrhiza uralensis: Response of plant roots to spectral quality of light and UV-B radiation. *Journal of Pineal Research* 41(2), 108–115.

Aguilera, Y., T. Herrera, V. Benítez, S. M. Arribas, A. L. López de Pablo, R. M. Esteban & M. A. Martín-Cabrejas. 2015. Estimation of scavenging capacity of melatonin and other antioxidants: Contribution and evaluation in germinated seeds. *Food Chemistry* 170, 203–211.

Aguilera, Y., M. Rebollo-Hernanz, T. Herrera, L. T. Cayuelas, P. Rodríguez-Rodríguez, L. de Pablo Á, S. M. Arribas & M. A. Martin-Cabrejas. 2016. Intake of bean sprouts influences melatonin and antioxidant capacity biomarker levels in rats. *Food and Function* 7(3), 1438–1445.

Ansari, M., K. Rafiee, N. Yasa, S. Vardasbi, S. M. Naimi & A. Nowrouzi. 2010. Measurement of melatonin in alcoholic and hot water extracts of *Tanacetum parthenium*, *Tripleurospermum disciforme* and *Viola odorata*. *Daru* 18(3), 173–178.

Anwar, M. M. & A. R. Meki. 2003. Oxidative stress in streptozotocin-induced diabetic rats: Effects of garlic oil and melatonin. *Comparative Biochemistry and Physiology—Part A: Molecular & Integrative Physiology* 135(4), 539–547.

Arnao, M. B. & J. Hernández-Ruiz. 2006. The physiological function of melatonin in plants. *Plant Signaling & Behavior* 1(3), 89–95.

Arnao, M. B. & J. Hernández-Ruiz. 2007. Melatonin promotes adventitious- and lateral root regeneration in etiolated hypocotyls of *Lupinus albus* L. *Journal of Pineal Research* 42(2), 147–152.

Arnao, M. B. & J. Hernández-Ruiz. 2009. Protective effect of melatonin against chlorophyll degradation during the senescence of barley leaves. *Journal of Pineal Research* 46(1), 58–63.

Arnao, M. B. & J. Hernández-Ruiz. 2015. Functions of melatonin in plants: A review. *Journal of Pineal Research* 59(2), 133–150.

Badria, F. A. 2002. Melatonin, serotonin, and tryptamine in some egyptian food and medicinal plants. *Journal of Medicinal Food* 5(3), 153–157.

Banerjee, A. & A. Roychoudhury. 2017. Melatonin as a regulator of abiotic stress tolerance in plants. In: Singh V. P., Singh S., Prasad S. M. (Eds.) *Mechanisms behind phytohormonal signaling and crop abiotic stress tolerance*. Nova Science Publishers, New York, pp. 47–60.

Blask, D. E., R. T. Dauchy, L. A. Sauer & J. A. Krause. 2004. Melatonin uptake and growth prevention in rat hepatoma 7288CTC in response to dietary melatonin: Melatonin receptor-mediated inhibition of tumor linoleic acid metabolism to the growth signaling molecule 13-hydroxyoctadecadienoic acid and the potential role of phytomelatonin. *Carcinogenesis* 25(6), 951–960.

Bravo, R., S. Matito, J. Cubero, S. D. Paredes, L. Franco, M. Rivero, A. B. Rodríguez & C. Barriga. 2013. Tryptophan-enriched cereal intake improves nocturnal sleep, melatonin, serotonin, and total antioxidant capacity levels and mood in elderly humans. *Age (Dordr)* 35(4), 1277–1285.

Burkhardt, S., D. X. Tan, L. C. Manchester, R. Hardeland & R. J. Reiter. 2001. Detection and quantification of the antioxidant melatonin in Montmorency and Balaton tart cherries (*Prunus cerasus*). *Journal of Agricultural and Food Chemistry* 49(10), 4898–4902.

Chen, G., Y. Huo, D. X. Tan, Z. Liang, W. Zhang & Y. Zhang. 2003. Melatonin in Chinese medicinal herbs. *Life Sciences* 73(1), 19–26.

Chen, Q., W. B. Qi, R. J. Reiter, W. Wei & B. M. Wang. 2009. Exogenously applied melatonin stimulates root growth and raises endogenous indoleacetic acid in roots of etiolated seedlings of Brassica juncea. *Journal of Plant Physiology* 166(3), 324–328.

Cho, J. H., S. Bhutani, C. H. Kim & M. R. Irwin. 2021. Anti-inflammatory effects of melatonin: A systematic review and meta-analysis of clinical trials. *Brain, Behavior, and Immunity* 93, 245–253.

Dericks-Tan, J. S., P. Schwinn & C. Hildt. 2003. Dose-dependent stimulation of melatonin secretion after administration of *Agnus castus*. *Experimental and Clinical Endocrinology and Diabetes* 111(1), 44–46.

Dopfel, R. P., K. Schulmeister & E. S. Schernhammer. 2007. Nutritional and lifestyle correlates of the cancer-protective hormone melatonin. *Cancer Detection and Prevention* 31(2), 140–148.

Dubbels, R., R. J. Reiter, E. Klenke, A. Goebel, E. Schnakenberg, C. Ehlers, H. W. Schiwara & W. Schloot. 1995. Melatonin in edible plants identified by radioimmunoassay and by high performance liquid chromatography-mass spectrometry. *Journal of Pineal Research* 18(1), 28–31.

el-Aziz, M. A., H. A. Hassan, M. H. Mohamed, A. R. Meki, S. K. Abdel-Ghaffar & M. R. Hussein. 2005. The biochemical and morphological alterations following administration of melatonin, retinoic acid and Nigella sativa in mammary carcinoma: An animal model. *International Journal of Experimental Pathology* 86(6), 383–396.

Garcia-Marin, R., J. M. Fernandez-Santos, J. Morillo-Bernal, F. Gordillo-Martinez, V. Vazquez-Roman, J. C. Utrilla, A. Carrillo-Vico, J. M. Guerrero & I. Martin-Lacav. 2015. Melatonin in the thyroid gland: Regulation by thyroid-stimulating hormone and role in thyroglobulin gene expression. *Journal of Physiology and Pharmacology* 66(5), 643–652.

Garcia-Moreno, H., J. R. Calvo & M. D. Maldonado. 2013. High levels of melatonin generated during the brewing process. *Journal of Pineal Research* 55.

González-Flores, D., E. Gamero, M. Garrido, R. Ramírez, D. Moreno, J. Delgado, E. Valdés, C. Barriga, A. B. Rodríguez & S. D. Paredes. 2012. Urinary 6-sulfatoxymelatonin and total antioxidant capacity increase after the intake of a grape juice cv: Tempranillo stabilized with HHP. *Food and Function* 3(1), 34–39.

Hadjzadeh, M. A. R., V. Alikhani, S. Hosseinian, B. Zarei & Z. Keshavarzi. 2018. The effect of melatonin against gastric oxidative stress and dyslipidemia in streptozotocin-induced diabetic rats. *Acta Endocrinologica (Buchar)* 14(4), 453–458.

Hardeland, R. 2008. Melatonin, hormone of darkness and more: Occurrence, control mechanisms, actions and bioactive metabolites. *Cellular and Molecular Life Sciences* 65(13), 2001–2018.

Hardeland, R. 2015. Melatonin in plants and other phototrophs: Advances and gaps concerning the diversity of functions. *Journal of Experimental Botany* 66(3), 627–646.

Hardeland, R. & S. R. Pandi-Perumal. 2005. Melatonin, a potent agent in antioxidative defense: Actions as a natural food constituent, gastrointestinal factor, drug and prodrug. *Nutrition & Metabolism (Lond)* 2, 22.

Hardeland, R., S. R. Pandi-Perumal & D. P. Cardinali. 2006. Melatonin. *The International Journal of Biochemistry & Cell Biology* 38(3), 313–316.

Hattori, A., H. Migitaka, M. Iigo, M. Itoh, K. Yamamoto, R. Ohtani-Kaneko, M. Hara, T. Suzuki & R. J. Reiter. 1995. Identification of melatonin in plants and its effects on plasma melatonin levels and binding to melatonin receptors in vertebrates. *Biochemistry and Molecular Biology International* 35(3), 627–634.

Head, K. A. 2001. Natural therapies for ocular disorders, part two: Cataracts and glaucoma. *Alternative Medicine Review* 6(2), 141–166.

Hernández-Ruiz, J., A. Cano & M. B Arnao. 2005. Melatonin acts as a growth-stimulating compound in some monocot species. *Journal of Pineal Research* 39(2), 137–142.

Iriti, M. & F. Faoro. 2006. Grape phytochemicals: A bouquet of old and new nutraceuticals for human health. *Medical Hypotheses* 67(4), 833–838.

Johns, N. P., J. Johns, S. Porasuphatana, P. Plaimee & M. Sae-Teaw. 2013. Dietary intake of melatonin from tropical fruit altered urinary excretion of 6-sulfatoxymelatonin in healthy volunteers. *Journal of Agricultural and Food Chemistry* 61(4), 913–919.

Kadekaro, A. L., L. N. Andrade, L. M. Floeter-Winter, M. D. Rollag, V. Virador, W. Vieira & A. M. Castrucci. 2004. MT-1 melatonin receptor expression increases the antiproliferative effect of melatonin on S-91 murine melanoma cells. *Journal of Pineal Research* 36(3), 204–211.

Kaur, H., S. Mukherjee, F. Baluska & S. C. Bhatla. 2015. Regulatory roles of serotonin and melatonin in abiotic stress tolerance in plants. *Plant Signaling & Behavior* 10(11), e1049788.

Kocadağlı, T., C. Yılmaz & V. Gökmen. 2014. Determination of melatonin and its isomer in foods by liquid chromatography tandem mass spectrometry. *Food Chemistry* 153, 151–156.

Korkmaz, A., O. Değer & Y. Cuci. 2014. Profiling the melatonin content in organs of the pepper plant during different growth stages. *Scientia Horticulturae* 172, 242–247.

Kwon, K. J., J. N. Kim, M. K. Kim, J. Lee, L. J. Ignarro, H. J. Kim, C. Y. Shin & S. H. Han. 2011. Melatonin synergistically increases resveratrol-induced heme oxygenase-1 expression through the inhibition of ubiquitin-dependent proteasome pathway: A possible role in neuroprotection. *Journal of Pineal Research* 50(2), 110–123.

Lahiri, D. K., D. M. Chen, P. Lahiri, S. Bondy & N. H. Greig. 2005. Amyloid, cholinesterase, melatonin, and metals and their roles in aging and neurodegenerative diseases. *Annals of the New York Academy of Sciences* 1056, 430–449.

Lei, Q., L. Wang, D. X. Tan, Y. Zhao, X. D. Zhen, H. Chen, Q. T. Li, B. X. Zuo & J. Kong. 2013. Identification of genes for melatonin synthetic enzymes in 'Red Fuji' apple (Malus domestica Borkh.cv.Red) and their expression and melatonin production during fruit development. *Journal of Pineal Research* 55(4), 443–451.

Lemoine, P., J. C. Bablon & C. Da Silva. 2019. A combination of melatonin, vitamin B6 and medicinal plants in the treatment of mild-to-moderate insomnia: A prospective pilot study. *Complementary Therapies in Medicine* 45, 104–108.

Lerner, A. B., J. D. Case, Y. Takahashi, T. H. Lee & W. Mori. 1958. Isolation of melatonin, the pineal gland factor that lightens melanocyteS1. *Journal of the American Chemical Society* 80(10), 2587–2587.

Lockley, S. W., G. C. Brainard & C. A. Czeisler. 2003. High sensitivity of the human circadian melatonin rhythm to resetting by short wavelength light. *The Journal of Clinical Endocrinology and Metabolism* 88(9), 4502–4505.

Lu, J. J., L. Fu, Z. Tang, C. Zhang, L. Qin, J. Wang, Z. Yu, D. Shi, X. Xiao, F. Xie, W. Huang & W. Deng. 2016. Melatonin inhibits AP-2β/hTERT, NF-κB/COX-2 and Akt/ERK and activates caspase/Cyto C signaling to enhance the antitumor activity of berberine in lung cancer cells. *Oncotarget* 7(3), 2985–3001.

Lundmark, P. O., S. R. Pandi-Perumal, V. Srinivasan & D. P. Cardinali. 2006. Role of melatonin in the eye and ocular dysfunctions. *Visual Neuroscience* 23(6), 853–862.

Manchester, L. C., D. X. Tan, R. J. Reiter, W. Park, K. Monis & W. Qi. 2000. High levels of melatonin in the seeds of edible plants: Possible function in germ tissue protection. *Life Sciences* 67(25), 3023–3029.

Marioni, F., A. Bertoli & L. Pistelli. 2008. A straightforward procedure to biosynthesise melatonin using freshly chopped *Achillea millefolium* L. as reagent. *Phytochemistry Letters* 1(2), 107–110.

Mena, P., A. Gil-Izquierdo, D. A. Moreno, N. Martí & C. García-Viguera. 2012. Assessment of the melatonin production in pomegranate wines. *Lwt—Food Science and Technology* 47, 13–18.

Meng, X., Y. Li, S. Li, Y. Zhou, R. Y. Gan, D. P. Xu & H. B. Li. 2017. Dietary sources and bioactivities of melatonin. *Nutrients* 9(4).

Murch, S. J., A. R. Alan, J. Cao & P. K. Saxena. 2009. Melatonin and serotonin in flowers and fruits of Datura metel L. *Journal of Pineal Research* 47(3), 277–283.

Murch, S. J. & P. K. Saxena. 2002. Mammalian neurohormones: Potential significance in reproductive physiology of St. John's wort (*Hypericum perforatum* L.)? *Naturwissenschaften* 89(12), 555–560.

Murch, S. J., & P. K. Saxena. 2006. A melatonin-rich germplasm line of St John's wort (*Hypericum perforatum* L.). *Journal of Pineal Research* 41(3), 284–287.

Odaci, E., & S. Kaplan. 2009. Chapter 16 melatonin and nerve regeneration. *International Review of Neurobiology* 87, 317–335.

Olcese, J. M., C. Cao, T. Mori, M. B. Mamcarz, A. Maxwell, M. J. Runfeldt, L. Wang, C. Zhang, X. Lin, G. Zhang & G. W. Arendash. 2009. Protection against cognitive deficits and markers of neurodegeneration by long-term oral administration of melatonin in a transgenic model of Alzheimer disease. *Journal of Pineal Research* 47(1), 82–96.

Ovanesov, K. B., I. M. Ovanesova & E. B. Arushanian. 2006. Effects of melatonin and motherwort tincture on the emotional state and visual functions in anxious subjects. *Eksp Klin Farmakol* 69(6), 17–19.

Padhy, B. M. & V. L. Kumar. 2005. Inhibition of Calotropis procera latex-induced inflammatory hyperalgesia by oxytocin and melatonin. *Mediators of Inflammation* 5(6), 360–365.

Padumanonda, T., J. Johns, A. Sangkasat & S. Tiyaworanant. 2014. Determination of melatonin content in traditional Thai herbal remedies used as sleeping aids. *Daru* 22(1), 6.

Pandi-Perumal, S. R., V. Srinivasan, G. J. Maestroni, D. P. Cardinali, B. Poeggeler & R. Hardeland. 2006. Melatonin: Nature's most versatile biological signal? *The FEBS Journal* 273(13), 2813–2838.

Paredes, S. D., A. Korkmaz, L. C. Manchester, D. X. Tan & R. J. Reiter. 2009. Phytomelatonin: A review. *Journal of Experimental Botany* 60(1), 57–69.

Park, S., D. E. Lee, H. Jang, Y. Byeon, Y. S. Kim & K. Back. 2013. Melatonin-rich transgenic rice plants exhibit resistance to herbicide-induced oxidative stress. *Journal of Pineal Research* 54(3), 258–263.

Pauley, S. M. 2004. Lighting for the human circadian clock: Recent research indicates that lighting has become a public health issue. *Medical Hypotheses* 63(4), 588–596.

Posmyk, M. M., M. Bałabusta, M. Wieczorek, E. Sliwinska & K. M. Janas. 2009. Melatonin applied to cucumber (*Cucumis sativus* L.) seeds improves germination during chilling stress. *Journal of Pineal Research* 46(2), 214–223.

Que, Z., T. Ma, Y. Shang, Q. Ge, Q. Zhang, P. Xu, J. Zhang, U. Francoise, X. Liu & X. Sun. 2020. Microorganisms: Producers of melatonin in fermented foods and beverages. *Journal of Agricultural and Food Chemistry* 68(17), 4799–4811.

Ramakrishna, A., P. Giridhar & G. A. Ravishankar. 2009. Indoleamines and calcium channels influence morphogenesis in in vitro cultures of *Mimosa pudica* L. *Plant Signaling & Behavavior* 4(12), 1136–1141.

Ramakrishna, A., P. Giridhar, K. U. Sankar & G. A. Ravishankar. 2012. Melatonin and serotonin profiles in beans of Coffea species. *Journal of Pineal Research* 52(4), 470–476.

Reiter, R. J., L. C. Manchester & D. X. Tan. 2005. Melatonin in walnuts: Influence on levels of melatonin and total antioxidant capacity of blood. *Nutrition* 21(9), 920–924.

Reiter, R. J. & D. X. Tan. 2002. Melatonin: An antioxidant in edible plants. *Annals of the New York Academy of Sciences* 957, 341–344.

Reiter, R. J., D. X. Tan, S. Burkhardt & L. C. Manchester. 2001. Melatonin in plants. *Nutrition Reviews* 59(9), 286–290.

Reiter, R. J., D. X. Tan, J. C. Mayo, R. M. Sainz, J. Leon & Z. Czarnocki. 2003. Melatonin as an antioxidant: Biochemical mechanisms and pathophysiological implications in humans. *Acta Biochimica Polonica* 50(4), 1129–1146.

Reiter, R. J., D. X. Tan, R. M. Sainz, J. C. Mayo & S. Lopez-Burillo. 2002. Melatonin: Reducing the toxicity and increasing the efficacy of drugs. *Journal of Pharmacy and Pharmacology* 54(10), 1299–1321.

Reiter, R. J., D. X. Tan, Z. Zhou, M. H. Cruz, L. Fuentes-Broto & A. Galano. 2015. Phytomelatonin: Assisting plants to survive and thrive. *Molecules* 20(4), 7396–7437.

Sae-Teaw, M., J. Johns, N. P. John & S. Subongkot. 2013. Serum melatonin levels and antioxidant capacities after consumption of pineapple, orange, or banana by healthy male volunteers. *Journal of Pineal Research* 55(1), 58–64.

Samanta, S., A. Banerjee & A. Roychoudhury. 2021. Melatonin application differentially modulates the enzymes associated with antioxidative machinery and ascorbate glutathione cycle during arsenate exposure in indica rice varieties. *Plant Biology* 23(S1), 193–201.

Sarropoulou, V. N., I. N. Therios & K. N. Dimassi-Theriou. 2012. Melatonin promotes adventitious root regeneration in in vitro shoot tip explants of the commercial sweet cherry rootstocks CAB-6P (Prunus cerasus L.), Gisela 6 (P. cerasus × P. canescens), and MxM 60 (P. avium × P. mahaleb). *Journal of Pineal Research* 52(1), 38–46.

Schaefer, M. & R. Hardeland. 2009. The melatonin metabolite N-acetyl-5-methoxykynuramine is a potent singlet oxygen scavenger. *Journal of Pineal Research* 46(1), 49–52.

Shi, D., X. Xiao, J. Wang, L. Liu, W. Chen, L. Fu, F. Xie, W. Huang & W. Deng. 2012. Melatonin suppresses proinflammatory mediators in lipopolysaccharide-stimulated CRL1999 cells via targeting MAPK, NF-κB, c/EBPβ, and p300 signaling. *Journal of Pineal Research* 53(2), 154–165.

Shin, J. C., H. Y. Jung, A. Harikishore, O. D. Kwon, H. S. Yoon, K. T. Kim & B. H. Choi. 2013. The flavonoid myricetin reduces nocturnal melatonin levels in the blood through the inhibition of serotonin N-acetyltransferase. *Biochemical and Biophysical Research Communications* 440(2), 312–316.

Shukla, M., P. Govitrapong, P. Boontem, R. J. Reiter & J. Satayavivad. 2017. Mechanisms of melatonin in alleviating Alzheimer's disease. *Current Neuropharmacology* 15(7), 1010–1031.

Stege, P. W., L. L. Sombra, G. Messina, L. D. Martinez & M. F. Silva. 2010. Determination of melatonin in wine and plant extracts by capillary electrochromatography with immobilized carboxylic multi-walled carbon nanotubes as stationary phase. *Electrophoresis* 31(13).
Stürtz, M., A. B. Cerezo, E. Cantos-Villar & M. C. Garcia-Parrilla. 2011. Determination of the melatonin content of different varieties of tomatoes (*Lycopersicon esculentum*) and strawberries (*Fragaria ananassa*). *Food Chemistry* 127(3), 1329–1334.
Tan, D. X., R. Hardeland, L. C. Manchester, S. D. Paredes, A. Korkmaz, R. M. Sainz, J. C. Mayo, L. Fuentes-Broto & R. J. Reiter. 2010. The changing biological roles of melatonin during evolution: From an antioxidant to signals of darkness, sexual selection and fitness. *Biological Reviews of the Cambridge Philosophical Society* 85(3), 607–623.
Tan, D. X., L. C. Manchester, P. Di Mascio, G. R. Martinez, F. M. Prado & R. J. Reiter. 2007a. Novel rhythms of N1-acetyl-N2-formyl-5-methoxykynuramine and its precursor melatonin in water hyacinth: Importance for phytoremediation. *The FASEB Journal* 21(8), 1724–1729.
Tan, D. X., L. C. Manchester, R. Hardeland, S. Lopez-Burillo, J. C. Mayo, R. M. Sainz & R. J. Reiter. 2003. Melatonin: A hormone, a tissue factor, an autocoid, a paracoid, and an antioxidant vitamin. *Journal of Pineal Research* 34(1), 75–78.
Tan, D. X., L. C. Manchester, P. Helton & R. J. Reiter. 2007b. Phytoremediative capacity of plants enriched with melatonin. *Plant Signaling and Behavior* 2(6), 514–516.
Tan, D. X., L. C. Manchester, R. J. Reiter, W. B. Qi, M. Karbownik & J. R. Calvo. 2000. Significance of melatonin in antioxidative defense system: Reactions and products. *Biological Signals and Receptors* 9(3–4), 137–159.
Tan, D. X., R. J. Reiter, L. C. Manchester, M. T. Yan, M. El-Sawi, R. M. Sainz, J. C. Mayo, R. Kohen, M. Allegra & R. Hardeland. 2002. Chemical and physical properties and potential mechanisms: Melatonin as a broad spectrum antioxidant and free radical scavenger. *Current Topics in Medicinal Chemistry* 2(2), 181–197.
Vasiliou, V., D. Ross & D. W. Nebert. 2006. Update of the NAD(P)H: Quinone oxidoreductase (NQO) gene family. *Human Genomics* 2(5), 329–335.
Videnovic, A., C. Noble, K. J. Reid, J. Peng, F. W. Turek, A. Marconi, A. W. Rademaker, T. Simuni, C. Zadikoff & P. C. Zee. 2014. Circadian melatonin rhythm and excessive daytime sleepiness in Parkinson disease. *JAMA Neurology* 71(4), 463–469.
Villapol, S., S. Fau, S. Renolleau, V. Biran, C. Charriaut-Marlangue & O. Baud. 2011. Melatonin promotes myelination by decreasing white matter inflammation after neonatal stroke. *Pediatric Research* 69(1), 51–55.
Vitalini, S., C. Gardana, P. Simonetti, G. Fico & M. Iriti. 2013. Melatonin, melatonin isomers and stilbenes in Italian traditional grape products and their antiradical capacity. *Journal of Pineal Research* 54.
Vitalini, S., C. Gardana, A. Zanzotto, G. Fico, F. Faoro, P. Simonetti & M. Iriti. 2011a. From vineyard to glass: Agrochemicals enhance the melatonin and total polyphenol contents and antiradical activity of red wines. *Journal of Pineal Research* 51.
Vitalini, S., C. Gardana, A. Zanzotto, P. Simonetti, F. Faoro, G. Fico & M. Iriti. 2011b. The presence of melatonin in grapevine (*Vitis vinifera* L.) berry tissues. *Journal of Pineal Research* 51(3), 331–337.
Wang, P., X. Sun, C. Chang, F. Feng, D. Liang, L. Cheng & F. Ma. 2013a. Delay in leaf senescence of Malus hupehensis by long-term melatonin application is associated with its regulation of metabolic status and protein degradation. *Journal of Pineal Research* 55(4), 424–434.
Wang, P., X. Sun, C. Li, Z. Wei, D. Liang & F. Ma. 2013b. Long-term exogenous application of melatonin delays drought-induced leaf senescence in apple. *Journal of Pineal Research* 54(3), 292–302.
Wang, P., L. Yin, D. Liang, C. Li, F. Ma & Z. Yue. 2012. Delayed senescence of apple leaves by exogenous melatonin treatment: Toward regulating the ascorbate-glutathione cycle. *Journal of Pineal Research* 53(1), 11–20. https://doi.org/10.1111/j.1600-079X.2011.00966.x
Wei, W., Q. T. Li, Y. N. Chu, R. J. Reiter, X. M. Yu, D. H. Zhu, W. K. Zhang, B. Ma, Q. Lin, J. S. Zhang & S. Y. Chen. 2015. Melatonin enhances plant growth and abiotic stress tolerance in soybean plants. *Journal of Experimenatl Botany* 66(3), 695–707.
Wolf, K., J. A. N. Kolář, E. Witters, W. van Dongen, H. van Onckelen & I. Macháčková. 2001. Daily profile of melatonin levels in *Chenopodium rubrum* L. depends on photoperiod. *Journal of Plant Physiology* 158(11), 1491–1493.

Yi, C., Y. Zhang, Z. Yu, Y. Xiao, J. Wang, H. Qiu, W. Yu, R. Tang, Y. Yuan, W. Guo & W. Deng. 2014. Melatonin enhances the anti-tumor effect of fisetin by inhibiting COX-2/iNOS and NF-κB/p300 signaling pathways. *PLoS One* 9(7), e99943.

Yilmaz, C. K., T. Kocadağlı & V. Gökmen. 2014. Formation of melatonin and its isomer during bread dough fermentation and effect of baking. *Journal of Agricultural and Food Chemistry* 62(13), 2900–2905.

Zhao, Y., L. W. Qi, W. M. Wang, P. K. Saxena & C. Z. Liu. 2011. Melatonin improves the survival of cryopreserved callus of *Rhodiola crenulata*. *Journal of Pineal Research* 50(1), 83–88.

Zhao, Y., D. X. Tan, Q. Lei, H. Chen, L. Wang, Q. T. Li, Y. Gao & J. Kong. 2013. Melatonin and its potential biological functions in the fruits of sweet cherry. *Journal of Pineal Research* 55(1), 79–88.

Zieliński, H., B. Lewczuk, B. Przybylska-Gornowicz & H. Kozłowska. 2001. Melatonin in germinated legume seeds as a potentially significant agent for health. In: *Biologically active phytochemicals in food analysis, metabolism, bioavailability and function*. Royal Society of Chemistry, Cambridge, UK, pp. 110–117.

6 Regulation of Photosynthetic and Respiratory Cycles and Nitrogen Assimilation in Higher Plants and Microalgae by Melatonin

Haram Sarfraz and Iffat Zareen Ahmad

6.1 INTRODUCTION

In 1958, *N-acetyl*-5-methoxytryptamine (Melatonin) was discovered for the first time in the pineal gland (Lerner et al., 1958). It was given the name "melatonin" because it can stop melanocyte-stimulating hormones from making skin darker (Rodriguez et al., 2004). N-acetyl-5 hydroxy tryptamine-methyl transferase, tryptophan decarboxylase, 5-hydroxy tryptamine-N-acetyltransferase, and tryptamine-5 hydroxylase catalyze its synthesis. In microalgae, melatonin was first noticed in *Lingulodinium polyedrum* (Vantassel et al., 1995) and has a role in microalgal rhythmic phenomena, cell growth, antioxidant system, and metabolism (Erland & Saxena, 2018; Nawaz et al., 2016). Exogenous melatonin affects microalgal biomass generation and lipid accumulation (Cui et al., 2020; Dong et al., 2019). In the last decade, plant and microalgae physiology research has concentrated on discovering its several functions in higher plants and microalgae. Melatonin is supported by substantial data as a growth stimulant and rooting agent (Sarrou et al., 2014; Zhang et al., 2014).

It is generated in the mitochondria and chloroplasts of plant leaves and roots and then transmitted to the flowers, meristem, and fruits (Wang et al., 2016). Additionally, because of its involvement in higher plants and microalgal growth, melatonin is crucial for plant and microalgal stress response. Plants and microalgae are regularly exposed to environmental stresses. Various melatonin-rich species have shown greater stress tolerance. Ectopic amplification of genes that are involved in melatonin biosynthesis or exogenous therapy may enhance tolerance to a variety of stressors, like severe temperature, salt, drought, chemical responses, and radiation, all of which result in reactive oxygen species (ROS) production (Banerjee & Roychoudhury, 2017). Early research on the extraction of melatonin indicated its function as a free radical scavenger in both microalgae and higher plants (Galano & Reiter, 2018). Later experiments and studies showed its pleiotropic character, which may also have a unique function as an antioxidant in the regulation of reactive nitrogen species (RNS) and reactive oxygen species (ROS) and the suppression of inflammatory responses in plants (Fan et al., 2018; Tan et al., 2007). Melatonin has a greater capacity to scavenge RNS and ROS than vitamin E and glutathione, and it also regulates antioxidant enzymes like catalase, glutathione reductase, superoxide dismutase, and peroxidase (Samanta et al., 2021). It boosts the effectiveness of mitochondrial electron transport chain reactions, reducing electron loss (Reiter et al., 2007; Zhao et al., 2019; Samanta et al., 2020).

The separation and characterization of melatonin in a wide variety of organisms, like *Gonyaulax polyedra*, prompted the investigation of melatonin in further higher plants and autotrophic species

 DOI: 10.1201/9781003343752-6

(Balzer & Hardeland, 1996; Manchester et al., 2000). Melatonin is regarded as a principal regulator since it increases development and growth in plants and microalgae by serving as a signaling molecule linked with a mechanism of defense against various abiotic and biotic challenges, such as drought, cold, salt, as well as nutrient deficits (Arnao & Hernández-Ruiz, 2019; Posmyk & Janas, 2009). It is also implicated in various physiological functions in plants, including photosynthesis and seed germination (Sadak et al., 2020). Abiotic and biotic stressors in higher plants and microalgae result in decreased growth as well as yield, senescence, and even mortality. In response, they generate a variety of defense mechanisms. Melatonin synthesis is among the most essential stress reactions in higher plants, and it has a comparable protective impact as indole-3-acetic acid (IAA) (Li et al., 2019; Martinez et al., 2018). Melatonin study is expanding rapidly, and its roles in several wild and transgenic animals have been identified. In the last ten years, there has been a huge rise in the number of academic articles that talk about how melatonin might affect the physiology and genes of plants and microalgae.

6.2 BIOSYNTHESIS

Isotope tracer tests have proven that plants produce melatonin. It has been hypothesized that higher plants and microalgae produce melatonin through comparable biochemical routes to vertebrates (Figure 6.1). Moreover, the enzymes involved in the manufacture of melatonin have been duplicated. Melatonin production in higher plants involves sequential enzymatic reactions (Kang et al., 2011). Tryptophan decarboxylase (TDC) is the first activated enzyme that converts tryptophan to tryptamine. Numerous species of plants, such as tobacco (Di Fiore et al., 2002), *Catharanthus roseus* (De Luca et al., 1989), rice (Kang et al., 2007b), and pepper (Park et al., 2009), already been cloned by TDC and it has a crucial function in the regulation of serotonin production due to its very low or nonexistent expression. Tryptamine 5-hydroxylase (T5H), an enzyme of the cytochrome P450 family, hydroxylates tryptamine at its C-5 position to produce serotonin. T5H was produced via map-based cloning using rice sekiguchi lesion (*sl*) mutants (Fujiwara et al., 2010), and healthy *Oryza sativa* plants easily express it (Kang et al., 2007b).

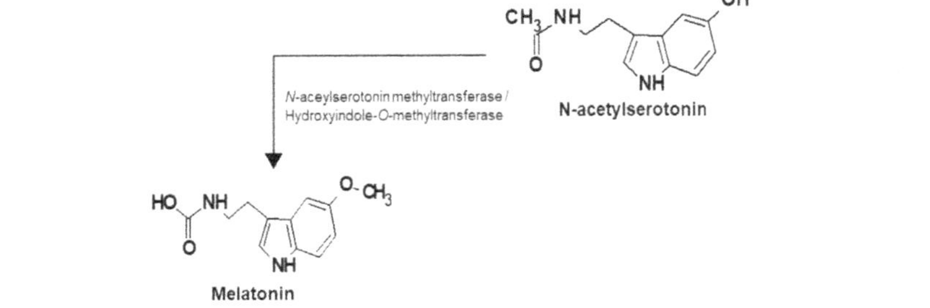

FIGURE 6.1 Biosynthesis pathway of melatonin in plants.

Because of the scarcity of data on associated compounds and the production of melatonin in higher plants and microalgae, the definitive routes have not yet been defined (Byeon et al., 2015). Recently, it was shown that transgenic rice has genes for complete enzymes required in melatonin production (Byeon et al., 2014). On the melatonin biosynthesis route, the last two enzymes are N-aceylserotonin methyltransferase and arylalkylamine N-acetyltransferase. Arylalkylamine N-acetyltransferase has been isolated from the *Chlamydomonas reinhardtii* (Okazaki et al., 2009). Rice serotonin N-acetyltransferase genes have been cloned to encode an enzyme that catalyzes the serotonin to N-acetylserotonin conversion. Using heterologous *Escherichia coli* overexpression, N-aceylserotonin methyltransferase, originally called as hydroxyindole-O-methyltransferase, has been identified from rice (Kang et al., 2013). The enzyme responsible for melatonin production's rate limitation is N-aceylserotonin methyltransferase, which is located in the cytoplasm, whereas rice serotonin N-acetyltransferase protein is located in chloroplasts, revealed by the subcellular localization study (Byeon et al., 2014). In vascular plants, N1-acetyl-N2-formyl-5-methoxykynuramine has been discovered to be a bioactive compound due to its non-enzymatic or enzymatic conversion from melatonin (Tan et al., 2007). Current findings on rice have shown that the melatonin metabolites 2-hydroxymelatonin (99.9%) and 4-hydroxymelatonin (0.05%) are also present (Byeon et al., 2015).

6.3 DETECTION AND QUANTIFICATION METHODS

The accurate extraction of melatonin from plant extracts presented the first difficulty in quantification. Whatever the eventual quantification method utilized, the extraction techniques used for the various matrices may provide a challenge for the detection of melatonin levels. Melatonin levels have been found to range in higher plants from several picograms to micrograms per gramme of material tested. The difficulty of extracting and recovering melatonin from plants is one of the issues with accurate melatonin measurement that has been brought up by a number of researchers (Cao et al., 2006; Pape & Lüning, 2006). There are several ways to extract melatonin at the moment. It is often extracted from plant tissue that has been frozen in liquid nitrogen by means of organic solvents like chloroform, ethyl acetate, or methanol. Nevertheless, inconsistent or poor recovery rates have been found while using aqueous extraction (Burkhardt et al., 2001; Reiter et al., 2005). The melatonin molecule is amphiphilic; therefore, direct sample extraction methods using just organic solvents are also suggested. Hardeland and colleagues noted that extraction effectiveness is a key consideration in the melatonin assessment concentration in photoautotrophic species on a worldwide scale (Hardeland et al., 2007). The relative recovery rates of every extraction method used were seen to have a direct impact on the amount of melatonin in various samples. As a result, the recovery rate will determine how accurate the melatonin level statistics are. Our findings show that, based on the sample, the efficacy (2–20%) of the basic sample extraction method varied when an ultrasonic treatment was added, even though doing so may result in a considerable improvement in melatonin extraction (Arnao & Hernández-Ruiz, 2009a; Cao et al., 2006). The measurement of melatonin in complex matrices has been reported as having additional potential issues, for example, false-positive and false-negative findings, losses via obliteration, overestimations, cross-reactivities in immunological processes, coelution in chromatographic techniques, and so on (Hardeland et al., 2007).

The most common and advised methods for melatonin detection and quantification are liquid chromatography analysis and mass spectrometry identification (LC-MS/MS). In this regard, LC-MS/MS is often employed in conjunction by multiple reaction monitoring and positive electrospray ionization. Melatonin detection has also been carried out using the groundbreaking method of liquid chromatography with time-of-flight/mass spectrometry (LC-TOF/MS). Fewer precise but actual delicate methods like LC by fluorometric or electrochemical monitoring are also useful when used in conjunction with LC-MS/MS identification (Huang & Mazza, 2011; Tan et al., 2012). Because of cross-reactivity with co-extractives, immunological procedures like RIA and ELISA pose significant challenges in plants as opposed to mammals (Hardeland et al., 2007; Van Tassel et al., 2001).

One of the primary issues with any effort to elucidate the importance of melatonin in plants is the lack of comprehensive literature on the melatonin level of edible plants, taking into consideration factors like growing, cultivar, soil, post-harvest conditions, and so on. As melatonin derived from plant sources is ingested, absorbed, and incorporated into the bloodstream, monitoring its levels has generated interest due to potential implications for human food intake. Melatonin is incorporated into the mitochondria and nucleus at the subcellular level after crossing the blood-brain barrier and the placenta. Thus, the potential for modifying the production of blood melatonin in birds and mammals by consuming plant-derived diets with high amounts of melatonin may be interesting (Reiter et al., 2001, 2007).

In general, fruits have the lowest levels of melatonin, while seeds have the greatest levels. It is fascinating to see how much the analytical approach utilized in the investigations affects the quantitative outcomes. The quantitative disparities between white radish and strawberry, which may be up to three times higher, or the tomato fruit, whose levels vary between 0.5 to 114,500 pg/g of tissue, are two examples. Studies conducted in the 1990s tended to discover lower amounts of plant melatonin (Dubbels et al., 1995; Hattori et al., 1995) than might be possible with more recent analyses. The reason for the low recoveries is owing to the absence of control in the extraction procedure of melatonin. In certain situations, it is also because of the employment of immunological detection methods, which may result in wrong negatives and significant interferences. There is very insufficient data on the plant foods' melatonin content, and no correlations are possible, especially in light of the many different factors that can affect growth, including soil, photoperiod, irrigation, harvest, post-harvest preservation techniques, cultivars, and/or species.

Despite the increase in data on wild plants, there is still a dearth of information due to the wide variety of species. *Ipomoea nil* L. shoots contain between 6 pg/g and 1 pg/g of melatonin. In a comprehensive investigation of 108 popular Chinese herbs, similar significant variations in phytomelatonin concentration were also noted. *Gardenia jasminoides* Ellis had melatonin levels 12 ng/g/DW, while 2.3 ng/g/DW was noted in *Viola philippica* Cav. (Chen et al., 2003). The distribution of melatonin often goes as follows: leaves > seeds > roots > flowers > fruits. However, in the majority of instances, information is lacking for at least one organ of the plants being researched. However, some research has shown that it is not possible to consider the melatonin levels in various plant parts to be uniform. In *Lupinus albus* L., for instance, a study demonstrated that melatonin created a gradient of concentration, along the sequence apical > central > basal region including equally hypocotyls as well as roots (Hernández-Ruiz & Arnao, 2008). Strangely, this gradient pattern resembled that of IAA, indicating that both indole chemicals may have physiological effects on plants comparable to one another (Hernandez-Ruiz et al., 2004).

6.4 ROLE IN GROWTH AND DEVELOPMENT

The indole compounds, which are mostly generated from the amino acid tryptophan, are one of the most intriguing classes. These show variations and are abundant in many plants, some of which are thought to have therapeutic value. They are of special significance because they contain IAA, the major prevalent auxin present in plants that, among other physiological activities, promotes plant development. In research comparing the antioxidant capabilities of several indole substances, it was shown that melatonin has a greater antioxidant capacity than common antioxidants, for example, trolox or ascorbic acid, which is the synthetic analogue of vitamin E. The antioxidant activity of melatonin was found to be twice that of trolox or ascorbic acid, in addition to about double that of other indoles, like tryptophan, indole-3-methanol, IAA, indole-3-butyric acid, and indole-3-propionic acid (Cano et al., 2003).

In general, there is a tight link between biological activity and molecular structures. The IAA and melatonin molecular structure reveals that they equally include an indole ring, but their substituents are extremely different. IAA contains an acid assembly at the third position of the indole ring, but melatonin has methoxy and N-acetyl groups in those positions. The indole ring and the acidic group of IAA or the carbonyl group and the indole ring of melatonin, however, are separated by the same amount of space (0.50 nm), which is pretty similar. Thus, it indicates that, under some circumstances, melatonin may mimic the auxin activity. The link among auxinic activity and synthetic chemicals like 2, 4-dichlorophenoxy acetic acid and naphthalene acetic acid is comparable. Melatonin increases the development of faded *Lupinus albus* L. hypocotyls in a comparable manner to that of IAA (Hernandez-Ruiz et al., 2004).

Whenever the aerial meristem was eliminated, exogenous melatonin may have replaced auxin stimulation because the growth response was melatonin-sensitive. Several writers have previously postulated the probable involvement of melatonin as a plant regulator, but this was the initial experimental evidence that convincingly confirmed the auxin-like function of melatonin (Murch & Saxena, 2002). A group discovered comparable melatonin effects in *Hordeum vulgare*, *Triticum aestivum*, and *Phalaris canariensis* (Hernández-Ruiz et al., 2005). In such instances, the growth-enhancing potential of melatonin was equal to 55% more than that of IAA, but its growth-inhibiting impact on the roots was concentration-dependent, comparable to that of IAA. Cotyledons of lupin, which grew under the influence of melatonin that was exogenously supplied, also had a growth-promoting impact (Hernández-Ruiz & Arnao, 2008).

Subsequently, a growth-stimulating impact of melatonin on *Brassica* sp. has been similarly established. Similar to auxin, melatonin has been shown to be capable of modulating the morphogenesis of plants. Melatonin was able to incite rhizogenesis in the pericycle cells of *Lupinus albus* L., and the production of lateral roots from main roots or adventitious roots through stems is a physiological process that is mostly attributed to IAA. Melatonin was also shown to promote this physiological function at a dose comparable to that of IAA (Arnao & Hernández-Ruiz, 2007). Later, it was shown that melatonin had this organogenetic activity in the plants *Oryza sativa*, *Cucumis sativus*, and *Prunus cerasus* (Sarropoulou et al., 2012; Zhang et al., 2013). According to a recent study, melatonin does not duplicate the auxin signaling response (Pelagio-Flores et al., 2012).

Exogenous melatonin inhibits the dark-induced senescence in *Hordeum vulgare* L. leaves (Arnao & Hernández-Ruiz, 2009a). The ability of melatonin to act as an antioxidant, as suggested by us and other researchers, may contribute to its ability to slow the rate of chlorophyll breakdown in ageing leaves, but it might also be linked to the potential modulation of some senescence indicators, as confirmed in apple leaves (*Malus domestica* L.). The overexpression of certain genes related to senescence was reduced, and specific reactive oxygen species scavenging enzyme activities were boosted in melatonin-treated leaves (Wang et al., 2012). This suggests that melatonin is a key factor in causing leaves to fall off. *Prunus* species (Sarropoulou et al., 2012), *Cucumis sativus* L. seedlings (Zhang et al., 2013), and *Malus domestica* (Wang et al., 2013) have all demonstrated this effect. Additionally, cells of the freshwater *Chara australis* treated with melatonin showed an increase in photosystem II efficiency (Lazar et al., 2013). The potential role of melatonin in plants as a regulator of the light-dark cycle has been correspondingly considered. The length and timing of the melatonin signal highlighted the importance of melatonin in photoperiodic control in mammals and birds, where this function has been successfully demonstrated (Reiter, 1993).

In addition to higher plants, the impact of exogenously administered melatonin has additionally been investigated in algae, because they are a rich source of pharmacological substances, possess significant nutritional value, and serve as a key feedstock because of their rapid development and high sugar content (Bruhn et al., 2011). For instance, study using *C. reinhardtii* showed that the exogenous administration of melatonin suppressed the enzyme antioxidants FeSOD and MnSOD in the presence of photo-oxidative stress (Tal et al., 2011). In an additional recent work, melatonin treatment in *H. pluvialis* enhanced tolerance to low nitrogen and high-light situations and boosted astaxanthin production (Ding et al., 2022). Because of its antioxidant property, accumulation of astaxanthin, a red ketocarotenoid, employed extensively in the food and drug industries (Guerin et al., 2003), was associated with a rise in salicylic acid and nitric oxide levels brought about by melatonin. *Spirulina* supplementation is a current trend in human nutrition. The study goals might be broadened to include such feed systems since *Spirulina* is a blue-green algae that supplements vitamins, proteins, and minerals as well as having many therapeutic benefits on humans (Belay et al., 1993).

Endogenous melatonin levels have been seen to oscillate in the following plants: *Chenopodium rubrum* L. (Kolávr et al., 1997; Wolf et al., 2001), *Eichhornia crassipes* (hyacinth) (Tan et al., 2007), and *Vitis vinifera* (Boccalandro et al., 2011), suggesting its possible role as a light-dark regulator. On long photoperiod days, a night-time melatonin peak has been seen in the green macroalga *Ulva* sp. (Tal et al., 2011). Furthermore, another fascinating aspect is the putative defensive potential of melatonin against biotic stress in plants. Many studies have looked into the unique qualities of melatonin as a natural antioxidant for ROS/RNS, including its lack of prooxidant effects. Arnao and Hernández-Ruiz discovered that chemical compounds like zinc sulfate, sodium chloride, and hydrogen peroxide increase the endogenous level of melatonin in a concentration- and time-dependent manner in the roots of *Hordeum vulgare*. The antioxidative defense against chemically induced stress is probably greatly aided by this increase in melatonin levels (Arnao & Hernández-Ruiz, 2009a).

Moreover, zinc sulphate and other chemical stresses raised endogenous melatonin levels in *Lupinus albus* plants by a factor of twelve, most likely by inducing some enzyme(s) in its biosynthesis pathway (Arnao & Hernández-Ruiz, 2009b).

Melatonin may mitigate the damaging effects of stressors on plants, including copper in the soil (Tan et al., 2007), cold (Kang et al., 2007a; Zhao et al., 2011), herbicides (Park et al., 2013), drought, light/dark (Arnao & Hernández-Ruiz, 2013), and UV radiation (Afreen et al., 2006; Conti et al., 2002). Since these stressors tend to raise endogenous melatonin levels, melatonin may support the cellular redox-regulated network and perform a significant physiological role in a number of stress-related situations (Sarropoulou et al., 2012).

Plant melatonin rhythms may serve as a night-time indication, coordinating actions on diurnal and photoperiodic ecological signals, like flowering. Indirect evidence for this theory came from the discovery that melatonin exhibits a strong circadian rhythm in the short-day species *Chenopodium rubrum* L., with the apex coming at night (Kolávr et al., 1997; Wolf et al., 2001), followed by the discovery that exogenously administered melatonin replaced darkness in bioluminescence response of *Lingulodinium polyedrum* Stein (Balzer & Hardeland, 1991) and growth response of *Pterygophora californica* Rupley (Fuhrberg et al., 1996).

The protracted plant *Arabidopsis thaliana* as well as *Chenopodium rubrum* L. have both proven that melatonin limits flowering. Exogenous melatonin can regulate transition to blooming in *Chenopodium rubrum* L. Similar results were seen in *Pharbitis nil* L., *Chenopodium murale* L., *Lemna minor* L., *Spirodela polyrhiza* L., and *Lemna trisulca* L. (Kolávr et al., 1997).

Since the greatest indole concentrations are seen in uninucleate microsporogenesis, and isolated anthers exhibit greater regeneration capability at the period of heightened melatonin content, it has also been hypothesized that melatonin acts as a control in the propagative cycle and in flower maturation of *Hypericum perforatum* L. (Murch & Saxena, 2002). On the other hand, in root cell suspensions of *Daucus carota* L., melatonin administration has been shown to lessen cold-induced cytotoxicity in a way unrelated to the generation of reactive oxygen species. Pretreatment using melatonin for five days dramatically prevented cold-induced apoptosis, as seen by variations in the TUNEL-positive cells, a lower amount of DNA fragmentation, and less morphological abnormalities connected to programmed cell death. Additionally, the levels of the polyamines, spermidine, and putrescine significantly increased in tandem with these modifications (Lei et al., 2004). The authors hypothesized that certain photoperiodic occurrences could be connected to how melatonin affects polyamine levels. However, since traditional plant-development-stimulating hormones like auxin, gibberellins, and cytokinins also raise the number of polyamines, it has been hypothesized that auxin, melatonin, and polyamines may interact. This connection cannot be ruled out since melatonin has been shown to induce auxin-like effects (Hernández-Ruiz et al., 2005).

Melatonin may correspondingly help keep bacteria in a dormant condition or in a differentiated state in fruit tissue. *Hordeum vulgare* L. treatment with melatonin solutions significantly reduced the ageing process, as indicated by the quantity of chlorophyll loss in leaves; this beneficial effect, compared to ageing, was dependent on melatonin concentration (Arnao & Hernández-Ruiz, 2009b). According to these scientists, melatonin might specifically affect the enzymes that break down chlorophyll, viz., red-chlorophyll catabolite reductase, pheophorbide oxygenase, and chlorophyllase.

6.5 CONCLUSION

Since no studies have been published to clarify the relationships between melatonin and the intake of nutrients as well as their transport, it is still unknown how melatonin affects root development or how nutrients are absorbed. The effects of foliar melatonin treatments on the uptake, development, and growth of plants and microalgae are also the subject of relatively little research. Although melatonin is found in all higher plants and microalgae, it has not yet been determined if this indoleamine is produced by all organs. It is necessary to learn how it is transported throughout the whole plant and in macroalgae. Numerous researchers have identified the pleotropic properties of melatonin as well as its auxin-like action. Because it is essential for humans, animals, and plants alike, melatonin may also provide its consumers with a number of financial advantages. High levels of melatonin in plants and microalgae act as antioxidants that help both animals and humans who consume them.

A number of important agricultural products like maize, rice, and wheat have levels of melatonin that are high enough to increase blood levels of the hormone after intake. The potential benefits of melatonin consumption in these items for both health and the economy are clear, given that billions of people throughout the globe rely on them as a significant source of food. Based on existing information and early findings, it is believed that using melatonin as a biostimulator might be a valuable, workable, and affordable technique for agriculture. Additionally, this compound may enhance human health and raise food quality. Numerous studies on melatonin have produced a body of evidence showing the outstanding positive impacts that this substance has on plants and microalgae, particularly under stressful circumstances. More research should be done to find out what role melatonin might play in the growth and development of microalgae and higher plants in the laboratory, in the cutting, vascular reunion, and grafting of plants, in the growth and development of flowers, and in related physiological responses.

REFERENCES

Afreen, F., Zobayed, S. M. A., & Kozai, T. (2006). Melatonin in *Glycyrrhiza uralensis*: Response of plant roots to spectral quality of light and UV-B radiation. *Journal of Pineal Research*, *41*(2), 108–115.

Arnao, M. B., & Hernández-Ruiz, J. (2007). Melatonin promotes adventitious-and lateral root regeneration in etiolated hypocotyls of *Lupinus albus* L. *Journal of Pineal Research*, *42*(2), 147–152.

Arnao, M. B., & Hernández-Ruiz, J. (2009a). Assessment of different sample processing procedures applied to the determination of melatonin in plants. *Phytochemical Analysis*, *20*(1), 14–18.

Arnao, M. B., & Hernández-Ruiz, J. (2009b). Protective effect of melatonin against chlorophyll degradation during the senescence of barley leaves. *Journal of Pineal Research*, *46*(1), 58–63.

Arnao, M. B., & Hernández-Ruiz, J. (2013). Growth conditions determine different melatonin levels in *Lupinus albus* L. *Journal of Pineal Research*, *55*(2), 149–155.

Arnao, M. B., & Hernández-Ruiz, J. (2019). Melatonin: A new plant hormone and/or a plant master regulator? *Trends in Plant Science*, *24*(1), 38–48.

Balzer, I., & Hardeland, R. (1991). Photoperiodism and effects of indoleamines in a unicellular alga, *Gonyaulax polyedra*. *Science*, *253*(5021), 795–797.

Balzer, I., & Hardeland, R. (1996). Melatonin in algae and higher plants-possible new roles as a phytohormone and antioxidant. *Botanica Acta*, *109*(3), 180–183.

Banerjee, A., & Roychoudhury, A. (2017). Melatonin as a regulator of abiotic stress tolerance in plants. In V. P. Singh, S. Singh, & S. M. Prasad (Eds.), *Mechanisms behind phytohormonal signaling and crop abiotic stress tolerance*. Nova Science Publishers, New York, pp. 47–60.

Belay, A., Ota, Y., Miyakawa, K., & Shimamatsu, H. (1993). Current knowledge on potential health benefits of Spirulina. *Journal of Applied Phycology*, *5*(2), 235–241.

Boccalandro, H. E., González, C. V., Wunderlin, D. A., & Silva, M. F. (2011). Melatonin levels, determined by LC-ESI-MS/MS, fluctuate during the day/night cycle in Vitis vinifera cv Malbec: Evidence of its antioxidant role in fruits. *Journal of Pineal Research*, *51*(2), 226–232.

Bruhn, A., Dahl, J., Nielsen Henrik Bangsøand Nikolaisen, L., Rasmussen, M. B., Markager, S., Olesen, B., Arias, C., & Jensen, P. D. (2011). Bioenergy potential of *Ulva lactuca*: Biomass yield, methane production and combustion. *Bioresource Technology*, *102*(3), 2595–2604.

Burkhardt, S., Tan, D. X., Manchester, L. C., Hardeland, R., & Reiter, R. J. (2001). Detection and quantification of the antioxidant melatonin in Montmorency and Balaton tart cherries (*Prunus cerasus*). *Journal of Agricultural and Food Chemistry*, *49*(10), 4898–4902.

Byeon, Y., Lee, H. Y., Lee, K., Park, S., & Back, K. (2014). Cellular localization and kinetics of the rice melatonin biosynthetic enzymes SNAT and ASMT. *Journal of Pineal Research*, *56*(1), 107–114.

Byeon, Y., Tan, D.-X., Reiter, R. J., & Back, K. (2015). Predominance of 2-hydroxymelatonin over melatonin in plants. *Journal of Pineal Research*, *59*(4), 448–454.

Cano, A., Alcaraz, O., & Arnao, M. B. (2003). Free radical-scavenging activity of indolic compounds in aqueous and ethanolic media. *Analytical and Bioanalytical Chemistry*, *376*(1), 33–37.

Cao, J., Murch, S. J., O'Brien, R., & Saxena, P. K. (2006). Rapid method for accurate analysis of melatonin, serotonin and auxin in plant samples using liquid chromatography: Tandem mass spectrometry. *Journal of Chromatography A*, *1134*(1–2), 333–337.

Chen, G., Huo, Y., Tan, D.-X., Liang, Z., Zhang, W., & Zhang, Y. (2003). Melatonin in Chinese medicinal herbs. *Life Sciences*, *73*(1), 19–26.

Conti, A., Tettamanti, C., Singaravel, M., Haldar, C., Pandi-Perumal, R. S., & Maestroni, G. J. M. (2002). Melatonin: An ubiquitous and evolutionary hormone. *Treatise on Pineal Gland and Melatonin*, 105–143.

Cui, J., Yu, C., Zhong, D.-B., Zhao, Y., & Yu, X. (2020). Melatonin and calcium act synergistically to enhance the coproduction of astaxanthin and lipids in *Haematococcus pluvialis* under nitrogen deficiency and high light conditions. *Bioresource Technology*, *305*, 123069.

De Luca, V., Marineau, C., & Brisson, N. (1989). Molecular cloning and analysis of cDNA encoding a plant tryptophan decarboxylase: Comparison with animal dopa decarboxylases. *Proceedings of the National Academy of Sciences*, *86*(8), 2582–2586.

Di Fiore, S., Li, Q., Leech, M. J., Schuster, F., Emans, N., Fischer, R., & Schillberg, S. (2002). Targeting tryptophan decarboxylase to selected subcellular compartments of tobacco plants affects enzyme stability and in vivo function and leads to a lesion-mimic phenotype. *Plant Physiology*, *129*(3), 1160–1169.

Ding, W., Zhao, P., Peng, J., Zhao, Y., Xu, J. W., Li, T., Reiter, R. J., Ma, H., & Yu, X. (2022). Corrigendum: "Melatonin enhances astaxanthin accumulation in the green microalga Haematococcus pluvialis by mechanisms possibly related to abiotic stress tolerance" (Algal Research (2018) 33 (256–265), (S2211926417310640), (10.1016/j.algal.2018.05.021). *Algal Research*, *62*, 102645.

Dong, X., Zhao, Y., Li, T., Huang, L., Zhao, P., Xu, J.-W., Ma, H., & Yu, X. (2019). Enhancement of lipid production and nutrient removal of Monoraphidium sp. FXY-10 by combined melatonin and molasses wastewater treatment. *Journal of the Taiwan Institute of Chemical Engineers*, *99*, 123–131.

Dubbels, R., Reiter, R. J., Klenke, E., Goebel, A., Schnakenberg, E., Ehlers, C., Schiwara, H. W., & Schloot, W. (1995). Melatonin in edible plants identified by radioimmunoassay and by high performance liquid chromatography-mass spectrometry. *Journal of Pineal Research*, *18*(1), 28–31.

Erland, L. A. E., & Saxena, P. K. (2018). Melatonin in plant morphogenesis. *In Vitro Cellular & Developmental Biology-Plant*, *54*(1), 3–24.

Fan, J., Xie, Y., Zhang, Z., & Chen, L. (2018). Melatonin: A multifunctional factor in plants. *International Journal of Molecular Sciences*, *19*(5), 1528.

Fuhrberg, B., Balzer, I., Hardeland, R., Werner, A., & Lüning, K. (1996). The vertebrate pineal hormone melatonin is produced by the brown alga *Pterygophora californica* and mimics dark effects on growth rate in the light. *Planta*, *200*(1), 125–131.

Fujiwara, T., Maisonneuve, S., Isshiki, M., Mizutani, M., Chen, L., Wong, H. L., Kawasaki, T., & Shimamoto, K. (2010). Sekiguchi lesion gene encodes a cytochrome P450 monooxygenase that catalyzes conversion of tryptamine to serotonin in rice. *Journal of Biological Chemistry*, *285*(15), 11308–11313.

Galano, A., & Reiter, R. J. (2018). Melatonin and its metabolites vs oxidative stress: From individual actions to collective protection. *Journal of Pineal Research*, *65*(1), e12514.

Guerin, M., Huntley, M. E., & Olaizola, M. (2003). Haematococcus astaxanthin: Applications for human health and nutrition. *TRENDS in Biotechnology*, *21*(5), 210–216.

Hardeland, R., Pandi-Perumal, S. R., & Poeggeler, B. (2007). Melatonin in plants: Focus on a vertebrate night hormone with cytoprotective properties. *Functional Plant Science and Biotechnology*, *1*(1), 32–45.

Hattori, A., Migitaka, H., Iigo, M., Itoh, M., Yamamoto, K., Ohtani-Kaneko, R., Hara, M., Suzuki, T., & Reiter, R. J. (1995). Identification of melatonin in plants and its effects on plasma melatonin levels and binding to melatonin receptors in vertebrates. *Biochemistry and Molecular Biology International*, *35*(3), 627–634.

Hernández-Ruiz, J., & Arnao, M. B. (2008). Melatonin stimulates the expansion of etiolated lupin cotyledons. *Plant Growth Regulation*, *55*(1), 29–34.

Hernandez-Ruiz, J., Cano, A., & Arnao, M. B. (2004). Melatonin: A growth-stimulating compound present in lupin tissues. *Planta*, *220*(1), 140–144.

Hernández-Ruiz, J., Cano, A., & Arnao, M. B. (2005). Melatonin acts as a growth-stimulating compound in some monocot species. *Journal of Pineal Research*, *39*(2), 137–142.

Huang, X., & Mazza, G. (2011). Application of LC and LC-MS to the analysis of melatonin and serotonin in edible plants. *Critical Reviews in Food Science and Nutrition*, *51*(4), 269–284.

Kang, K., Kong, K., Park, S., Natsagdorj, U., Kim, Y. S., & Back, K. (2011). Molecular cloning of a plant N-acetylserotonin methyltransferase and its expression characteristics in rice. *Journal of Pineal Research*, *50*(3), 304–309.

Kang, K., Lee, K., Park, S., Byeon, Y., & Back, K. (2013). Molecular cloning of rice serotonin N-acetyltransferase, the penultimate gene in plant melatonin biosynthesis. *Journal of Pineal Research*, *55*(1), 7–13.

Kang, S., Kang, K., Lee, K., & Back, K. (2007a). Characterization of rice tryptophan decarboxylases and their direct involvement in serotonin biosynthesis in transgenic rice. *Planta*, *227*(1), 263–272.

Kang, S., Kang, K., Lee, K., & Back, K. (2007b). Characterization of tryptamine 5-hydroxylase and serotonin synthesis in rice plants. *Plant Cell Reports*, *26*(11), 2009–2015.

Kolávr, J., Macháčková, I., Eder, J., Prinsen, E., Van Dongen, W., Van Onckelen, H., & Illnerová, H. (1997). Melatonin: Occurrence and daily rhythm in *Chenopodium rubrum*. *Phytochemistry*, *44*(8), 1407–1413.

Lazar, D., Murch, S. J., Beilby, M. J., & Al Khazaaly, S. (2013). Exogenous melatonin affects photosynthesis in characeae *Chara australis*. *Plant Signaling & Behavior*, *8*(3), e23279.

Lei, X.-Y., Zhu, R.-Y., Zhang, G.-Y., & Dai, Y.-R. (2004). Attenuation of cold-induced apoptosis by exogenous melatonin in carrot suspension cells: The possible involvement of polyamines. *Journal of Pineal Research*, *36*(2), 126–131.

Lerner, A. B., Case, J. D., Takahashi, Y., Lee, T. H., & Mori, W. (1958). Isolation of melatonin, the pineal gland factor that lightens melanocyteS1. *Journal of the American Chemical Society*, *80*(10), 2587.

Li, J., Liu, J., Zhu, T., Zhao, C., Li, L., & Chen, M. (2019). The role of melatonin in salt stress responses. *International Journal of Molecular Sciences*, *20*(7), 1735.

Manchester, L. C., Tan, D.-X., Reiter, R. J., Park, W., Monis, K., & Qi, W. (2000). High levels of melatonin in the seeds of edible plants: Possible function in germ tissue protection. *Life Sciences*, *67*(25), 3023–3029.

Martinez, V., Nieves-Cordones, M., Lopez-Delacalle, M., Rodenas, R., Mestre, T. C., Garcia-Sanchez, F., Rubio, F., Nortes, P. A., Mittler, R., & Rivero, R. M. (2018). Tolerance to stress combination in tomato plants: New insights in the protective role of melatonin. *Molecules*, *23*(3), 535.

Murch, S. J., & Saxena, P. K. (2002). Mammalian neurohormones: Potential significance in reproductive physiology of St. John's wort (*Hypericum perforatum* L.)? *Naturwissenschaften*, *89*(12), 555–560.

Nawaz, M. A., Huang, Y., Bie, Z., Ahmed, W., Reiter, R. J., Niu, M., & Hameed, S. (2016). Melatonin: Current status and future perspectives in plant science. *Frontiers in Plant Science*, *6*, 1230.

Okazaki, M., Higuchi, K., Hanawa, Y., Shiraiwa, Y., & Ezura, H. (2009). Cloning and characterization of a *Chlamydomonas reinhardtii* cDNA arylalkylamine N-acetyltransferase and its use in the genetic engineering of melatonin content in the Micro-Tom tomato. *Journal of Pineal Research*, *46*(4), 373–382.

Pape, C., & Lüning, K. (2006). Quantification of melatonin in phototrophic organisms. *Journal of Pineal Research*, *41*(2), 157–165.

Park, S., Kang, K., Lee, K., Choi, D., Kim, Y.-S., & Back, K. (2009). Induction of serotonin biosynthesis is uncoupled from the coordinated induction of tryptophan biosynthesis in pepper fruits (*Capsicum annuum*) upon pathogen infection. *Planta*, *230*(6), 1197–1206.

Park, S., Lee, D.-E., Jang, H., Byeon, Y., Kim, Y.-S., & Back, K. (2013). Melatonin-rich transgenic rice plants exhibit resistance to herbicide-induced oxidative stress. *Journal of Pineal Research*, *54*(3), 258–263.

Pelagio-Flores, R., Muñoz-Parra, E., Ortiz-Castro, R., & López-Bucio, J. (2012). Melatonin regulates Arabidopsis root system architecture likely acting independently of auxin signaling. *Journal of Pineal Research*, *53*(3), 279–288.

Posmyk, M. M., & Janas, K. M. (2009). Melatonin in plants. *Acta Physiologiae Plantarum*, *31*(1), 1–11.

Reiter, R. J. (1993). The melatonin rhythm: Both a clock and a calendar. *Experientia*, *49*(8), 654–664.

Reiter, R. J., Manchester, L. C., & Tan, D. (2005). Melatonin in walnuts: Influence on levels of melatonin and total antioxidant capacity of blood. *Nutrition*, *21*(9), 920–924.

Reiter, R. J., Tan, D. X., Burkhardt, S., & Manchester, L. C. (2001). Melatonin in plants. *Nutrition Reviews*, *59*(9), 286–290.

Reiter, R. J., Tan, D. X., Manchester, L. C., Simopoulos, A. P., Maldonado, M. D., Flores, L. J., Terron, M. P., & others. (2007). Melatonin in edible plants (phytomelatonin): Identification, concentrations, bioavailability and proposed functions. *World Review of Nutrition and Dietetics*, *97*, 211–230.

Rodriguez, C., Mayo, J. C., Sainz, R. M., Antolín, I., Herrera, F., Martín, V., & Reiter, R. J. (2004). Regulation of antioxidant enzymes: A significant role for melatonin. *Journal of Pineal Research*, *36*(1), 1–9.

Sadak, M. S., Abdalla, A. M., Abd Elhamid, E. M., & Ezzo, M. I. (2020). Role of melatonin in improving growth, yield quantity and quality of *Moringa oleifera* L. plant under drought stress. *Bulletin of the National Research Centre*, *44*(1), 1–13.

Samanta, S., Banerjee, A., & Roychoudhury, A. (2021). Melatonin application differentially modulates the enzymes associated with antioxidative machinery and ascorbate glutathione cycle during arsenate exposure in indica rice varieties. *Plant Biology*, *23*(S1), 193–201.

Samanta, S., Singh, A., Banerjee, A., & Roychoudhury, A. (2020). Exogenous supplementation of melatonin alters representative organic acids and enzymes of respiratory cycle as well as sugar metabolism during arsenic stress in two contrasting indica rice cultivars. *Journal of Biotechnology*, *324*, 220–232.

Sarropoulou, V., Dimassi-Theriou, K., Therios, I., & Koukourikou-Petridou, M. (2012). Melatonin enhances root regeneration, photosynthetic pigments, biomass, total carbohydrates and proline content in the cherry rootstock PHL-C (Prunus avium× Prunus cerasus). *Plant Physiology and Biochemistry*, *61*, 162–168.

Sarrou, E., Therios, I., & Dimassi-Theriou, K. (2014). Melatonin and other factors that promote rooting and sprouting of shoot cuttings in *Punica granatum* cv: Wonderful. *Turkish Journal of Botany*, *38*(2), 293–301.

Tal, O., Haim, A., Harel, O., & Gerchman, Y. (2011). Melatonin as an antioxidant and its semi-lunar rhythm in green macroalga Ulva sp. *Journal of Experimental Botany*, *62*(6), 1903–1910.

Tan, D.-X., Hardeland, R., Manchester, L. C., Korkmaz, A., Ma, S., Rosales-Corral, S., & Reiter, R. J. (2012). Functional roles of melatonin in plants, and perspectives in nutritional and agricultural science. *Journal of Experimental Botany*, *63*(2), 577–597.

Tan, D.-X., Manchester, L. C., Helton, P., & Reiter, R. J. (2007). Phytoremediative capacity of plants enriched with melatonin. *Plant Signaling & Behavior*, *2*(6), 514–516.

Van Tassel, D. L., Roberts, N., Lewy, A., & O'Neill, S. D. (2001). Melatonin in plant organs. *Journal of Pineal Research*, *31*(1), 8–15.

Vantassel, D. L., Roberts, N. J., & Oenill, S. D. (1995). Melatonin from higher-plants-isolation and identification of N-acetyl 5-methoxytryptamine. *Plant Physiology*, *108*(2), 101.

Wang, P., Sun, X., Li, C., Wei, Z., Liang, D., & Ma, F. (2013). Long-term exogenous application of melatonin delays drought-induced leaf senescence in apple. *Journal of Pineal Research*, *54*(3), 292–302.

Wang, P., Yin, L., Liang, D., Li, C., Ma, F., & Yue, Z. (2012). Delayed senescence of apple leaves by exogenous melatonin treatment: Toward regulating the ascorbate: Glutathione cycle. *Journal of Pineal Research*, *53*(1), 11–20.

Wang, R., Yang, X., Xu, H., & Li, T. (2016). Research progress of melatonin biosynthesis and metabolism in higher plants. *Plant Physiology Journal*, *52*, 615–627.

Wolf, K., Kolá\vr, J., Witters, E., van Dongen, W., van Onckelen, H., & Macháčková, I. (2001). Daily profile of melatonin levels in *Chenopodium rubrum* L. depends on photoperiod. *Journal of Plant Physiology*, *158*(11), 1491–1493.

Zhang, N., Zhang, H.-J., Zhao, B., Sun, Q.-Q., Cao, Y.-Y., Li, R., Wu, X.-X., Weeda, S., Li, L., Ren, S., & others. (2014). The RNA-seq approach to discriminate gene expression profiles in response to melatonin on cucumber lateral root formation. *Journal of Pineal Research*, *56*(1), 39–50.

Zhang, N., Zhao, B., Zhang, H.-J., Weeda, S., Yang, C., Yang, Z.-C., Ren, S., & Guo, Y.-D. (2013). Melatonin promotes water-stress tolerance, lateral root formation, and seed germination in cucumber (*Cucumis sativus* L.). *Journal of Pineal Research*, *54*(1), 15–23.

Zhao, D., Yu, Y., Shen, Y., Liu, Q., Zhao, Z., Sharma, R., & Reiter, R. J. (2019). Melatonin synthesis and function: Evolutionary history in animals and plants. *Frontiers in Endocrinology*, *10*, 249.

Zhao, Y., Qi, L.-W., Wang, W.-M., Saxena, P. K., & Liu, C.-Z. (2011). Melatonin improves the survival of cryopreserved callus of *Rhodiola crenulata*. *Journal of Pineal Research*, *50*(1), 83–88.

7 Advancement in Melatonin Research in Plants

Multifaceted Role in Regulating Development and Stress Protection

Faisal Nadeem, Muhammad Saqib, Muhammad Saad, Hafiz Muhammad Jhanzab, Husnain Jawad, Muhammad Zubair and Abdul Qayyum

7.1 INTRODUCTION

Melatonin (*N-acetyl*-5-methoxytryptamine) is a nontoxic biological molecule and important secondary metabolite synthesized in different plant tissues (Sharif et al., 2018). It is an indole molecule (biogenic indoleamine) and has structural resemblance with other substances such as serotonin, tryptophan and indole-3-acetic acid (IAA) (Arnao and Hernández-Ruiz, 2018). The melatonin synthetic pathways vary somewhat between plants and animals, and the amino acid tryptophan is the necessary precursor of melatonin in all taxa (Zhao et al., 2019). In most of the plant species, roots, seeds, leaves, bulbs and flowers are rich sources of melatonin (Reiter et al., 2001). Most of the plant species which are rich in melatonin belong to the families Poaceae, Brassicaceae, Apiaceae, Rosaceae and Vitaceae. However, plants belong to some other families also shown to have a large amount of melatonin (Reiter et al., 2001; Sharif et al., 2018).

Environmental factors (temperature, photoperiod), genotype, developmental stage and method of determination affect the melatonin concentration in plants (Zhao et al., 2012; Byeon and Back, 2014a; Feng et al., 2014). Phytomelatonin has been identified and quantified in more than 140 different plants ranging from picograms to micrograms per gram of tissue (Murch et al., 2009b; Garcia-Parrilla et al., 2009). Hattori et al. (1995) reported that plants including oat, sweet corn, rice, barley and tall fescue exhibited high content of this indole amine. Melatonin was also discovered in several wild perennials and medicinal plants (e.g. *Hypericum perforatum* L.) as well as in flower buds of *Datura metel* L., but its quantity decreased at the floral development stage (Murch, 1997; Murch et al., 2009a; Zohar et al., 2011). High concentration of melatonin was also discovered in edible seeds of alfalfa, flax, fennel, white and black mustard and sunflower (Manchester et al., 2000; Banerjee and Roychoudhury, 2017).

Some studies (Manchester et al., 2000; Murch et al., 2009a) suggested that this hormone in seeds provide antioxidative defense in a relatively dry system which cannot be up-regulated. Therefore, this compound may be an essential compound in seeds protecting germ and reproductive organs of plants from adverse environmental conditions. Melatonin is an important antioxidant and plays a beneficial role in the protection of plants against stressful environments (Sharif et al., 2018). In rice, tryptophan is converted to melatonin via tryptamine but not hydroxytryptophan (Kang et al., 2009; Park, 2011). However, tryptophan (melatonin precursor) availability in plants is theoretically limited as they are able to produce aromatic amino acid via shikimic acid pathway in contrast to animals (Yao et al., 2011). However, it acts as a biostimulator during abiotic stresses and regulates key elements expressed against stressors such as salinity, drought, cold, heat and heavy metals (Zhang

DOI: 10.1201/9781003343752-7

et al., 2017). Plant melatonin appears to be a multi-regulatory molecule with specific functions in plant physiology. Arnao and Hernández-Ruiz (2018) described melatonin as an important modulator of gene expression related to plant hormones (e.g., auxin, carrier proteins) as well as in metabolism of indole-3-acetic acid (IAA), abscisic acid, gibberellins, ethylene and cytokinins. Moreover, in plant immune system, melatonin acts as a key molecule together with other well-known molecules such as hormones and nitric acid (Arnao and Hernández-Ruiz, 2018).

7.2 ROLE OF PHYTOMELATONIN IN PLANT GROWTH AND DEVELOPMENT

7.2.1 Vegetative Development

Hernández-Ruiz et al. (2005) firstly reported melatonin function in enhancing growth of plants which was supported by 10–55% increase in coleoptile growth of wheat, oat, barley and canary grass. Some other studies (Murch et al., 2001; Hernández-Ruiz et al., 2005) explained that melatonin improves root and shoot growth of plants.

As melatonin is chemically similar to IAA (a rooting hormone), it has potential effects on root growth and development. In the last few years, various scientists examined the role of melatonin in root organogenesis and documented that melatonin is a pivotal molecule in root growth and development (Arnao and Hernández-Ruiz, 2015; Liang et al., 2017). The application of melatonin enhanced lateral and adventitious root development in different plants, including rice, cucumber, Arabidopsis and Lupinus (Arnao and Hernández-Ruiz, 2007; Chen et al., 2009; Sun et al., 2021). Melatonin in association with auxins regulates root morphology.

Melatonin has the ability to alter different plant growth stages such as germination, seedling growth, days to flowering, senescence and grain yield (Wang et al., 2013; Byeon and Back, 2014b). In transgenic rice, melatonin enhanced the early growth of seedling but extended the time to flowering and decreased grain yield (Byeon and Back, 2014b). Another study by Park and Back (2012) found that melatonin application (0.5–1 μM) enhanced the growth and root biomass of transgenic rice. A recent study (Wei et al., 2015) provided evidence that coating the seed with melatonin significantly increased the plant height and leaf area of soybean.

In plants, melatonin acts as a night signal and coordinates response to diurnal and photo-periodic environmental cues. Moreover, this compound was also identified in hypocotyls, leaves and fruits and accelerated the vegetative growth in monocot and dicot plants grown in darkness in a similar fashion as IAA (Reiter et al., 2001; Hernández-Ruiz et al., 2004; Arnao and Hernández-Ruiz, 2007). Another study reported that melatonin usage improved the germination and growth of a young seedling by protecting germinating seed and seedling against toxic concentration of heavy metals (Posmyk et al., 2008). Erdal (2019) observed significant increase in plant height and leaf surface area with melatonin application at different doses (10, 100, 1000 μmol L^{-1}) by increasing chlorophyll content and activities of enzymes and gene expression. Melatonin has a pivotal role in improving germination rate by promoting coleoptile elongation and lateral leaflet generation (Park, 2011). Another report indicated that melatonin enhances the vegetative growth of rice, soybean and lupin by improving nutrient uptake efficiency due to better root growth and development (Hernández-Ruiz et al., 2004; Wei et al., 2014; Zhang et al., 2017; Liang et al., 2017).

7.2.2 Reproductive Development

Melatonin is involved in circadian-cycle-related physiological steps of plants. Flowering is one of the most sensitive responses of plants which is tightly regulated by photoperiodism. Interestingly, melatonin effect becomes prominent if applied before a dark period or in the first phase of a dark period (Park, 2011), for example, high dose (100 μM) of melatonin application on a short-day plant reduced flowering (Kolar et al., 2003). Reports (Machackova and Krekule, 2002; Kolar et al., 2003) explained that exogenous application of melatonin at different concentrations (100 and 500 μM) had a weak inhibitory effect on flowering; however, it does not affect the duration and phase of this process. Moreover, melatonin applied to apex and cotyledons of young plants practically remain unmetabolized and undergo only little transport towards other regions. Thus, confusion still exists about melatonin's function in affecting flowering.

Wei et al. (2014) reported that melatonin application in soybean improved the reproductive growth. Furthermore, melatonin affects floral development at different stages (Kolar et al., 2003; Park and Back, 2012) and its level is also altered during fruit development (Okazaki and Ezura, 2009). Phytomelatonin may act as a growth regulatory signal like auxin or as a regulator of reproductive development in plants (Kim et al., 2007). At flowering stage, melatonin and serotonin concentration are high. Therefore, increase in concentration of melatonin and serotonin at a specific stage strongly indicates its vital role in plant reproductive development (Murch and Saxena, 2002).

Further, an increase in melatonin level at a specific stage acts as stimulus for certain other developmental cascades. During phase transition of tetrad to uninucleate microspore which occur between second and third phase, low melatonin and high serotonin concentration are interpretated as triggering substances (Murch and Saxena, 2002). Another research (Murch et al., 2009a) found the protective role of melatonin in the development of flower buds rather than functioning as a stimulus for development. In the fruit development of *D. metel*, a high level of melatonin (two weeks after anthesis) was observed in ovule where earlier stages of seed development were in progress (Murch et al., 2009a).

7.2.3 Potential Roles in Postharvest Fruit Ripening

A lot of studies indicate the effect of melatonin in postharvest development of fruits, especially in climacteric fruits such as banana, apple, peaches, tomato and pear (Sun et al., 2021; Arnao and Hernández-Ruiz, 2019). It helps to hasten fruit ripening and fruit nutritional quality. In-depth analysis revealed that melatonin may trigger ethylene biosynthesis and amplify ethylene signal transduction. However, melatonin and ethylene may work antagonistically to each other. For example, Hu et al. (2017) depicted that ripening in banana was delayed when treated with melatonin.

7.2.4 Anti-Oxidative Role in Plants

Phytomelatonin is a ubiquitous molecule acting as an antioxidant and providing protection against stressful environments (Park, 2011). Antioxidative activities of melatonin against various toxic nitrogen and oxygen species are well reported (Galano et al., 2011) as it is both water and lipid soluble and acts as universal hydrophilic and hydrophobic antioxidant. Some studies (Poeggeler et al., 2002) found melatonin as a powerful antioxidant and highly effective as compared to previously identified substances such as vitamin C, E and K. Moreover, melatonin is of low molecular weight, and due to this, it moves easily between cell compartments and shields reactive oxygen species (ROS) (Bonnefont-Rousselot and Collin, 2010; Samanta et al., 2021). Melatonin application through roots significantly lowers the leaf senescence and enhances photosynthetic efficiency by activating antioxidative enzymes (Wang et al., 2013). Melatonin protects plant biological membrane from oxidative damage through its strong antioxidant potential in several ways: (i) accelerating the activities of antioxidant enzymes, (ii) direct scavenging of free radicals, (iii) protecting and activating antioxidant enzymes, (iv) decreasing free radical production and (v) enhancing mitochondrial transport chain (Leon et al., 2005; Zhao et al., 2019). In another study, melatonin treatment in apple delays the senescence in leaves by enhancing enzymatic activities to scavenge free radicals and enhancing the concentration of ascorbic acid and glutathione (Wang et al., 2012). Exogenous melatonin pretreatment (20 and 100 μM) prevent the stress-induced obstruction of photosystem II and photosynthetic process by alleviating the accumulation of ROS and malondialdehyde and increasing superoxide dismutase (SOD), catalase (CAT) and peroxidase (POD) (Han et al., 2017; Samanta et al., 2022). Melatonin controls ROS generation in two ways: (i) through chemical interaction of melatonin with ROS which causes its inactivation and (ii) by the melatonin-mediated induction of the redox enzymes (i.e., SOD, CAT, POD) that detoxify ROS (Arnao and Hernández-Ruiz, 2019; Khan et al., 2020). Moreover, melatonin induces the accumulation of non-enzymatic antioxidant compounds such as ascorbic acid (AsA) and glutathione (GSH), phenolic compounds, carotenoids and flavonoids (Kim et al., 2007; Zhao et al., 2016; Sharma et al., 2020).

7.2.5 Protective Agent for Crop Improvement and Protection

In plants, melatonin appears to play crucial protective roles against ozone damage, photooxidation of the photosynthetic machinery and ultraviolet damage (Dubbels et al., 1995; Tan et al., 2007a; Paredes et al., 2009). All of these important functions appear through receptor-independent free radical detoxification (Tan et al., 2007b). Melatonin treatment in maize alleviated the oxidative damage and improved the photosynthetic efficiency under drought stress (Huang et al., 2019).

7.3 MELATONIN ROLE IN ABIOTIC STRESS TOLERANCE

7.3.1 Drought Stress

Melatonin is an imperative growth regulator which is known for its crucial roles in plant growth and development (Arnao and Hernández-Ruiz, 2014; Kanwar et al., 2018). Melatonin regulates plant development processes starting from germination, and it is considered to have the same effects as auxin (Hernández-Ruiz et al., 2004). Melatonin and auxin have co-regulatory impacts on plants, and at low concentration, they improve plant growth; however, they have inhibitory effects at higher concentration (Chen et al., 2009). Drought has devastating impacts on growth, physiological functioning and final yield (Hassan et al., 2020a); however, melatonin application favors plant growth under drought stress. Melatonin brings many anatomical changes in plants which favors their survival in drought stress. Likewise, it reduces membrane damage (Yan et al., 2016), forms intact grana lamella of chloroplast (Cui et al., 2017), alleviates the chloroplast damage (Meng et al., 2014; Zhang et al., 2013), safeguards mitochondria (Shi et al., 2015), maintains cell expansion (Khan et al., 2019), leaf thickness (Meng et al., 2014), cuticle formation (Ding et al., 2018) and wax accumulation (Li et al., 2019), thereby improving plant survival under drought stress.

Melatonin protects the plant photosynthetic apparatus from adverse impacts of drought, improving the photosynthetic efficiency (Meng et al., 2014; Liang et al., 2018). Melatonin application prevents chlorophyll degradation and improves stomatal conductance, photosynthesis and transpiration during drought stress (Liang et al., 2018; Karaca and Cekic, 2019). The degradation of chlorophyll is catalyzed by different enzymes including chlorophyllase (Chlase), pheophytinase (PPH) and chlorophyll-degrading peroxidase (Chl-PRX) (Ma et al., 2018; Sharma et al., 2019). The melatonin application down-regulates Chlase, PPH and Chl-PRX (Ma et al., 2018), thereby preventing chlorophyll degradation in drought stress. Pheophorbide-a-oxygenase (PAO) is another enzyme that is involved in the chlorophyll metabolism, and melatonin application down-regulates the transcription level of PAO, resulting in reduced chlorophyll degradation under drought stress (Wang et al., 2013). Additionally, melatonin also improves the efficiency of photosystem-II and photosynthetic electron transportation rate (ETR) (Liang et al., 2019). The non-photochemical quenching substantially increased under drought, which has negative impacts on photosynthetic efficiency. Nonetheless, melatonin application helps the plants facing drought stress in recovering the photosynthetic performance (Wang et al., 2013). The increase in plant leaf area with melatonin application provides another reason for the improved photosynthetic efficiency under drought stress (Campos et al., 2019).

Melatonin protects the plants from adverse impacts of oxidative stress by increasing the scavenging of ROS. The increase in scavenging of ROS by melatonin under drought stress is due to an increase in the efficiency of the anti-oxidant defense system (Meng et al., 2014; Campos et al., 2019; Cui et al., 2017). The stress induced the production of super-oxide anions; however, melatonin controls the production of these anions either by scavenging or reducing their production (Gao et al., 2018). Moreover, melatonin application also increased the scavenging of H_2O_2 in plants in drought conditions (Meng et al., 2014; Zhang et al., 2013), and this is followed by increased detoxification of hydroxyl radicals and other aldehydes involved in inducing oxidative stress in plants (Zhang et al., 2013; Kabiri et al., 2018). The application of melatonin regulates the ascorbate-glutathione cycle and triggers the direct scavenging of ROS (Wang et al., 2013). The increase in ROS scavenging mediated by melatonin protects the plant cell wall and reduces the electrolyte leakage and MDA accumulation in drought stress (Kabiri et al., 2018; Liang et al., 2019).

Melatonin applications improve the activities of ABA-degrading enzymes and ROS-scavenging enzymes including APX, CAT and POD (Li et al., 2015). The increase in activities of the aforementioned enzymes decreases H_2O_2 accumulation, which indicates the direct role of melatonin in H_2O_2 scavenging (Galano et al., 2013; Li et al., 2015). Melatonin also increases the activities of other antioxidant enzymes like SOD, GPX, GR, DHAR and MDHAR (Kabiri et al., 2018) which also protect the plants from the damaging effects of drought stress. Melatonin-induced ROS scavenging in plants is controlled by the regulation of the enzymatic cycle named Asada-Halliwell pathway (Liu et al., 2015a). Moreover, melatonin also regulates the AsA-GSH cycle, which plays a vital role in the detoxification of ROS. This AsA-GSH cycle is regulated by different enzymes such as APX, DHAR, GR and MDHAR (Cui et al., 2017; Wang et al., 2013). GR is considered to be responsible for AsA homeostasis (Foyer and Halliwell, 1976), and melatonin up-regulates the activity of GR under drought stress (Cui et al., 2017). Moreover, GPX is another antioxidant that has the appreciable potential to scavenge the ROS, and melatonin application up-regulates its activity under drought stress and therefore increases ROS scavenging (Bela et al., 2015; Cui et al., 2017). Mitogen-activated protein kinase (MAPK) cascade pathway plays a crucial role in regulating the plant biological processes in drought stress (Zhang et al., 2011). Transcription factors (TFs) including NAC, WRKY, MYB and DREB are the main components of the MAPK signaling pathway in plants under stress conditions (Wang et al., 2018). These TFs regulate the diverse genes responsible for tolerance against different stresses (Hoang et al., 2017). Melatonin regulates the MAPK pathway by regulating these TFs which increase drought tolerance (Gao et al., 2018).

7.3.2 Heat Stress and Cold Stress

Plants require an optimized temperature for their normal growth and development. The below- and above-optimum level temperatures negatively affect the plant growth and productivity. Low and high temperature slow down the germination, limit the nutrient and water uptake and negatively affect the flowering, fruiting and seed production and finally cause plant death. Oxidative stress is the marker of heat stress (HS), and induction of oxidative stress favors the accumulation of ROS and MDA and increases the permeability of membranes (Hassan et al., 2020b). Prolonged heat stress induces the accumulation of excessive H_2O_2, which hinders growth and development. The melatonin application under HS reduces H_2O_2 accumulation and favors the plant survival and ensures better growth and productivity The treatment with MT of wheat seedlings under HS resulted in a sharp decrease in H_2O_2 (Zhao et al., 2012; Liang et al., 2018; Buttar et al., 2020).

The increased activities of antioxidants due to melatonin application improves plant tolerance against heat and cold stress (Janas and Posmyk, 2013; Arnao and Hernandez, 2018). The application of melatonin induces the synthesis and causes up-regulation of genes under cold stress, which protects plants from the damaging effects of oxidative stress (Bajwa et al., 2014). Melatonin application alleviates adverse impacts of stress by up or down-regulating the genes and proteins associated with heat/cold stress, scavenging ROS, modulating polyamine (PA) accumulation, increasing the synthesis of heat shock proteins and chlorophyll contents and affecting the ABA and cytokinin pathways (Shi and Chan, 2014; Turk et al., 2014). Melatonin application also increased plant survival by increasing the synthesis of putrescine and spermidine under cold stress (Lei et al., 2004). In another study, Balabusta et al. (2016) showed that cucumber seeds treated with melatonin had lower ROS and higher SOD activity and ROS detoxification under chilling stress. The exogenously applied melatonin reduced the photo-inhibition by increasing the non-photochemical quenching via induction of violaxanthin de-epoxidase activity in tomato plants under cold stress (Ding et al., 2017). Additionally, melatonin application under HS reduced the electrolyte leakage, MDA and ROS and increased the chlorophyll contents, total proteins and antioxidant activities (Alam et al., 2018).

Seed priming with melatonin regulates the melatonin-associated proteins in seeds exposed to a lower temperature and increased plant tolerance against chilling stress (Kołodziejczyk et al., 2016). The foliar application of melatonin increased the chlorophyll content, photosynthetic efficiency, endogenous melatonin and reduced the ABA content in plants grown under HS conditions (Zhang et al., 2017). Melatonin application increased SOD and CAT activities under chilling stress, reduced the MDA and H_2O_2 accumulation and increased the germination and emergence performance of pepper seedlings (Korkmaz et al., 2017).

The exogenous melatonin application caused a substantial increase in the activities of antioxidant enzymes like SOD, POX, CAT, APX and ascorbic acid, resulting in substantial reduction ROS accumulation and lipid peroxidation under HS (Xu et al., 2010). Melatonin application also modulated ABA metabolism and PA accumulation under chilling stress, which improved cold tolerance and subsequent plant growth (Zhang et al., 2017). Melatonin is remarkably effective in a wide range of conditions (Sharif et al., 2018; Moustafa et al., 2020), including the improvement of antioxidant activities (POD, SOD, CAT, GR and APX) to control the cellular redox homeostasis (Das and Roychoudhury, 2014). Therefore, modulation of the antioxidant defense system by detoxification of ROS may help to reduce impacts of abiotic stresses including the HS (Arnao and Hernández, 2019).

7.3.3 Salt Stress

Photosynthesis is an important process responsible for energy production which is indirectly affected by salinity stress (Li et al., 2017). Melatonin improves photosynthetic efficiency by alleviating stomatal limitation and protecting the photosynthetic apparatus (Meloni et al., 2003; Li et al., 2017). The improvement in photosynthesis is linked with the increased photosynthetic rate (Pn), the efficiency of PS-II and chlorophyll contents (Wang et al., 2016). Melatonin application improves photosynthesis under salinity stress by increasing chlorophyll content (Jiang et al., 2017), photosynthetic rate (Jiang et al., 2016) and reducing the chlorophyll degradation (Liang et al., 2015). Additionally, melatonin also protects the photosynthetic apparatus from damaging impacts of salt stress and therefore improves salt tolerance in plants (Wei et al., 2015; Zhou et al., 2016).

Salinity stress induces the production of ROS, which causes oxidative stress in plants and significantly reduces plant growth and yield (Roychoudhury et al., 2008). The exogenously applied melatonin reduces the salt-induced ROS accumulation (Wang et al., 2016; Zeng et al., 2018) and protects the plants from the damaging effects of salt stress. Melatonin works as a broad spectrum antioxidant which interacts with ROS and directly scavenges ROS (Campos et al., 2012). A single molecule of melatonin scavenges 10 ROS molecules, which indicates its appreciable potential to scavenge ROS (Campos et al., 2012). The application of melatonin decreases the concentration of H_2O_2 and O_2^- by activating different enzymes which also reduces H_2O_2 accumulation (Zeng et al., 2018). The application of melatonin increases the activities of SOD, CAT, POD and APX which, in turn, increases scavenging of ROS under salt stress (Wang et al., 2016). Moreover, melatonin application also improves the activities of APX, CAT, SOD, POD, GR and GPX under salt stress compared to control, which, in turn, reduces the ROS accumulation and favors the plant growth under salt stress (Chen et al., 2018; Jiang et al., 2016). Additionally, melatonin interacts with ROS by improving the concentration of AsA and GSH which, in turn, reduces the ROS accumulation under salt stress (Wang et al., 2016; Chen et al., 2018).

Ionic homeostasis refers to the ability of organisms to maintain the stable concentration of ions in a defined space (Amtmann and Leigh, 2010; Roychoudhury et al., 2013). The application of melatonin improves plant homeostasis under salt stress. Likewise, melatonin considerably increased K^+ accumulation and decreased Na^+ accumulation and maintained higher K^+/Na^+ ratio under salt stress (Jiang et al., 2016). The improved ionic homeostasis under salt stress can be due to the up regulation of different genes like NHX, SOS and AKT. The melatonin application in salt stress up-regulates the transcription levels of MdNHX1 and MdAKT1, which is consistent with higher accumulation of K^+ and maintenance of higher K^+/Na^+ ratio under salt stress (Li et al., 2012). Similarly, melatonin application increased the expression of NHX1 and SOS2, which was correlated with a lower Na^+/K^+ ratio (Zhao et al., 2018). Ca^{2+} signaling plays a crucial role in plants response towards biotic and abiotic stress; however, no evidence is available related to the involvement of Ca^{2+} signaling in melatonin-induced salinity tolerance mechanism. The melatonin, through its interaction with NO, plays an important role in plant response to stress conditions. For instance, melatonin works as a downstream signal for alleviation of sodic alkaline stress in tomatoes (Liu et al., 2015b). These findings indicate that NO is involved in maintaining the ionic homeostasis and improvement in the antioxidant system in plants facing salt stress (Zhao et al., 2018) which triggers salt tolerance. The application of melatonin improves the PA metabolism under salt stress by increasing the conversion of arginine and methionine to PAs, resulting in improved salt tolerance (Ke et al., 2018). Moreover, melatonin also suppressed the activities of two enzymes, viz., PAO (polyamine oxidase) and DAO (diamine oxidase) involved in PA metabolism, which decreased melatonin-induced PA degradation and therefore improved salt tolerance and plant growth (Ke et al., 2018).

7.4 CONCLUSION AND FUTURE PERSPECTIVES

Drought, salinity, high temperature and chilling stress are the main abiotic stresses affecting growth and productivity worldwide. Plant melatonin appears to be a multi-regulatory molecule with important physiological functions in stressful environments. In addition, it is an important modulator of gene expression associated with plant hormones and has a growth stimulatory effect on the crop plant. Most importantly, phytomelatonin appears to have a clear role as an antioxidant which helps in detoxifying free radicals, alleviating the oxidative damage and improving the photosynthetic efficiency of crop plants facing abiotic stresses. Under abiotic stresses, melatonin treatment in crop plants improves the growth and development by protecting the plant cell walls, inhibits electrolyte leakage and causes activation of enzymatic antioxidants. Despite recent progress in highlighting the importance of melatonin in plant growth and development under stressful environments, many regulatory aspects related to melatonin remain to be clarified. There is also a need to find out whether melatonin biosynthesis in cellular compartments varies according to plant species, type of stress and its concentration in plant. Moreover, detection of melatonin in many plants is still unexplored.

REFERENCES

Alam MN, Zhang L, Yang L, Islam R, Liu Y, Luo H, et al. Transcriptomic profiling of tall fescue in response to heat stress and improved thermotolerance by melatonin and 24-epibrassinolide. BMC Genomics. 2018;19:224.

Amtmann A, Leigh R. Ion homeostasis. In Pareek A, Sopory SK, Bohnert HJ (Eds) Abiotic stress adaptation in plants: Physiological, molecular and genomic foundation. Dordrecht, The Netherlands, Springer, 2010.

Arnao MB, Hernández-Ruiz J. Melatonin and its relationship to plant hormones. Annals of Botany. 2018;121:195–207.

Arnao MB, Hernández-Ruiz J. Melatonin and reactive oxygen and nitrogen species: A model for the plant redox network. Melatonin Res. 2019;2:152–168.

Arnao MB, Hernández-Ruiz J. Functions of melatonin in plants: A review. J Pineal Res. 2015;59:133–150.

Arnao MB, Hernández-Ruiz J. Melatonin: Plant growth regulator and/or biostimulator during stress? Trends Plant Sci. 2014;19:789–797.

Arnao MB, Hernández-Ruiz J. Melatonin promotes adventitious- and lateral root regeneration in etiolated hypocotyls of *Lupinus albus* L. J Pineal Res. 2007;42:147–152.

Bajwa VS, Shukla MR, Sherif SM, Murch SJ, Saxena P. Role of melatonin in alleviating cold stress in Arabidopsis thaliana. Journal of Pineal Research. 2014;56:238–245.

Banerjee A, Roychoudhury A. Melatonin as a regulator of abiotic stress tolerance in plants. In Singh VP, Singh S, Prasad SM (Eds) Mechanisms behind phytohormonal signaling and crop abiotic stress tolerance. Nova Science Publishers, New York, 2017, pp. 47–60.

Bela K, Horvath E, Galle A, Szabados L, Tari I, Csiszar J. Plant glutathione peroxidases: Emerging role of the antioxidant enzymes in plant development and stress responses. J Plant Physiol. 2015;176:192–201.

Bonnefont-Rousselot D, Collin F. Melatonin: Action as antioxidant and potential applications in human. Toxicology. 2010;278:55–67.

Buttar ZA, Wu SN, Arnao MB, Wang C, Ullah I, Wang C. Melatonin suppressed the heat stress-induced damage in wheat seedlings by modulating the antioxidant machinery. Plants. 2020;9(7):809.

Byeon Y, Back K. Melatonin synthesis in rice seedlings in vivo is enhanced at high temperatures and under dark conditions due to increased serotonin N-acetyltransferase and N-acetyl serotonin methyl transferase activities. J Pineal Res. 2014a;56:189–195.

Byeon Y, Back K. An increase in melatonin in transgenic rice causes pleiotropic phenotypes, including enhanced seedling growth, delayed flowering, and low grain yield. J Pineal Res. 2014b;56:408–414.

Campos CN, Ávila RG, de Souza KRD, Azevedo LM, Alves JD. Melatonin reduces oxidative stress and promotes drought tolerance in young Coffea arabica L. plants. Agric Water Manag. 2019;211:37–47.

Campos LMO, Hsie SB, Granja AJA, Correia MR, Almeida-Cortez J, Pompelli MF. Photosynthesis and antioxidant activity in Jatropha curcas L. under salt stress. Braz J Plant Physiol. 2012;24:55–67.

Chen Q, Qi WB, Reiter RJ, Wei W, Wang BM. Exogenously applied melatonin stimulates root growth and raises endogenous indoleacetic acid in roots of etiolated seedlings of Brassica juncea. J Plant Physiol. 2009;166:324–328.

Chen Y-E, Mao J-J, Sun L-Q, Huang B, Ding C-B, Gu Y, Liao J-Q, Hu C, Zhang Z-W, Yuan S, et al. Exogenous melatonin enhances salt stress tolerance in maize seedlings by improving antioxidant and photosynthetic capacity. Physiol Plant. 2018;164:349–363.

Cui G, Zhao X, Liu S, Sun F, Zhang C, Xi Y. Beneficial effects of melatonin in overcoming drought stress in wheat seedlings. Plant Physiology and Biochemistry: PPB/Societe francaise de physiologie vegetale. 2017;118:138–149.

Das K, Roychoudhury A. Reactive oxygen species (ROS) and response of antioxidants as ROS-scavengers during environmental stress in plants. Front Environ Sci. 2014;2:53.

Ding F, Wang G, Wang M, Zhang S. Exogenous melatonin improves tolerance to water deficit by promoting cuticle formation in tomato plants. Molecules. 2018;23.

Ding F, Wang M, Liu B, Zhang S. Exogenous melatonin mitigates photoinhibition by accelerating nonphotochemical quenching in tomato seedlings exposed to moderate light during chilling. Frontiers in Plant Science. 2017;8:244.

Dubbels R, Reiter RJ, Klenke E, Goebel A, Schnakenberg E, Ehlers C, Schiwara HW, Schloot W. Melatonin in edible plants identified by radioimmunoassay and by high performance liquid chromatography-mass spectrometry. J Pineal Res. 1995;18:28–31.

Erdal S. Melatonin promotes plant growth by maintaining integration and coordination between carbon and nitrogen metabolisms. Plant Cell Reports. 2019;38(8):1001–1012.

Feng X, Wang M, Zhao Y, Han P, Dai Y. Melatonin from different fruit sources, functional roles, and analytical methods. Trends in Food Science & Technology. 2014;37(1):21–31.

Foyer CH, Halliwell B. The presence of glutathione and glutathione reductase in chloroplasts: A proposed role in ascorbic acid metabolism. Planta. 1976;133:21–25.

Galano A, Tan DX, Reiter RJ. Melatonin as a natural ally against oxidative stress: A physicochemical examination. J Pineal Res. 2011;51:1–16.

Galano A, Tan DX, Reiter RJ. On the free radical scavenging activities of melatonin's metabolites, AFMK and AMK. J Pineal Res. 2013;54:245–257.

Gao W, Zhang Y, Feng Z, Bai Q, He J, Wang Y. Effects of melatonin on antioxidant capacity in naked oat seedlings under drought stress. Molecules. 2018;23:1580.

Garcia-Parrilla MC, Cantos E, Troncoso AM. Analysis of melatonin in foods. J Food Comp Anal. 2009;22:177–183.

Han Q-H, Huang B, Ding C-B, Zhang Z-W, Chen Y-E, Hu C, Zhou L-J, Huang Y, Liao J-Q, Yuan S, Yuan M. Effects of melatonin on anti-oxidative systems and photosystem II in cold-stressed rice seedlings. Frontiers in Plant Science. 2017;8:785.

Hassan MUM, Aamer M, Umer Chattha M, Haiying T, Shahzad B, Barbanti L, Nawaz M, Rasheed A, Afzal A, Liu Y, Guoqin H. The critical role of zinc in plants facing the drought stress. Agriculture. 2020a;10(9):396.

Hassan MUM, Chattha MU, Khan I, Chattha MB, Barbanti L, Aamer M, Iqbal MM, et al. Heat stress in cultivated plants: Nature, impact, mechanisms, and mitigation strategies: A review. Plant Biosystems: An International Journal Dealing with All Aspects of Plant Biology. 2020b;1–24.

Hattori A, Migitaka H, Iigo M, Itoh M, Yamamoto K, Ohtani-Kaneko R, Hara M, Suzuki T, Reiter RJ. Identification of melatonin in plants and its effects on plasma melatonin levels and binding to melatonin receptors in vertebrates. Biochem Mol Biol Int. 1995;35:627–634.

Hernández-Ruiz J, Cano A, Arnao MB. Melatonin: A growth-stimulating compound present in lupin tissues. Planta. 2004;220:140–144.

Hernández-Ruiz J, Cano A, Arnao MB. Melatonin acts as a growth-stimulating compound in some monocot species. Journal of Pineal Research. 2005;39(2):137–142.

Hoang XLT, Nhi DNH, Thu NBA, Thao NP, Tran LP. Transcription Factors and their roles in signal transduction in plants under abiotic stresses. Curr Genom. 2017;18:483–497.

Hu W, Yang H, Tie W, Yan Y, Ding Z, Liu Y, et al. Natural variation in banana varieties highlights the role of melatonin in postharvest ripening and quality. J Agric Food Chem. 2017;65:9987–9994.

Huang B, Chen Y-E, Zhao Y-Q, Ding C-B, Liao J-Q, Hu C, Zhou L-J, Zhang Z-W, Yuan S, Yuan M. Exogenous melatonin alleviates oxidative damages and protects photosystem II in maize seedlings under drought stress. Front Plant Sci. 2019;10:677.

Janas K, Posmyk M. Melatonin, an underestimated natural substance with great potential for agricultural application. Acta Physiologiae Plantarum. 2013;35:3285–3292.

Jiang C, Cui Q, Feng K, Xu D, Li C, Zheng Q. Melatonin improves antioxidant capacity and ion homeostasis and enhances salt tolerance in maize seedlings. Acta Physiol Plant. 2016:38:82.1.

Jiang Y, Liang D, Liao MA, Lin L. Effects of melatonin on the growth of radish Seedlings under salt stress. In Proceedings of the 3rd international conference on renewable energy and environmental technology (ICERE 2017), Hanoi, Vietnam, 25–27 February 2017.

Kabiri R, Hatami A, Oloumi H, Naghizadeh M, Nasibi F, Tahmasebi Z. Foliar application of melatonin induces tolerance to drought stress in Moldavian balm plants (Dracocephalum moldavica) through regulating the antioxidant system. Folia Hortic. 2018;30:155–167.

Kang K, Kim S, Park S, Back K. Senescence-induced serotonin biosynthesis and its role in delaying senescence in tice leaves. Plant Physiol. 2009;150:1380–1393.

Kanwar MK, Yu J, Zhou J. Phytomelatonin: Recent advances and future prospects. J Pineal Res. 2018;65:e12526.

Karaca P, Cekic FÖ. Exogenous melatonin-stimulated defense responses in tomato plants treated with polyethylene glycol. Int J Veg Sci. 2019;1–9.

Ke Q, Ye J, Wang B, Ren J, Yin L, Deng X, Wang S. Melatonin mitigates salt stress in wheat seedlings by modulating polyamine metabolism. Front Plant Sci. 2018;9:914.

Khan A, Numan M, Khan A, Lee IJ, Imran M, Asaf S, Al-Harrasi A. Melatonin: Awakening the defense mechanisms during plant oxidative stress. Plants. 2020;9:407.

Khan MN, Zhang J, Luo T, Liu J, Rizwan M, Fahad S, Xu Z, Hu L. Seed priming with melatonin coping drought stress in rapeseed by regulating reactive oxygen species detoxification: Antioxidant defense system, osmotic adjustment, stomatal traits and chloroplast ultrastructure perseveration. Industrial Crops and Products. 2019;140:111597.

Kim Y-J, Lee YN, Oh YJ, I-y H, Park WJ. What is the role for melatonin in plants? Review on the current status of phytomelatonin research. J Nano Bio Tech. 2007;4:9–14.

Kolar J, Johnson CH, Machackova I. Exogenously applied melatonin affects flowering of the short-day plant Chenopodium rubrum. Physiol Plant. 2003;118:605–612.

Kołodziejczyk I, Dzitkob K, Szewczyk R, Posmyka MM. Exogenous melatonin improves corn (Zea mays L.) embryo proteome in seeds subjected to chilling stress. Journal of Plant Physiology. 2016;193:47–56.

Korkmaz A, Karaca A, Kocacinar F, Cuci Y. The effects of seed treatment with melatonin on germination and emergence performance of pepper seeds under chilling stress. Tarım Bilimleri Dergisi. 2017;23:167–176.

Lei XY, Zhu RY, Zhang GY, Dai YR. Attenuation of cold-induced apoptosis by exogenous melatonin in carrot suspension cells: The possible involvement of polyamines. Journal of Pineal Research. 2004;36:126–131.

Leon J, Acuna-Castroviejo D, Escames G, Tan DX, Reiter RJ. Melatonin mitigates mitochondrial malfunction. J Pineal Res. 2005;38:1–9.

Li C, Tan DX, Liang D, Chang C, Jia D, Ma F. Melatonin mediates the regulation of ABA metabolism, free-radical scavenging, and stomatal behaviour in two Malus species under drought stress. J Exp Bot. 2015;66:669–680.

Li C, Wang P, Wei Z, Liang D, Liu C, Yin L, Jia D, Fu M, Ma F. The mitigation effects of exogenous melatonin on salinity-induced stress in *Malus hupehensis*. J Pineal Res. 2012;53:298–306.

Li H, Chang J, Chen H, Wang Z, Gu X, Wei C, Zhang Y, Ma J, Yang J, Zhang X. Exogenous melatonin confers salt stress tolerance to watermelon by improving photosynthesis and redox homeostasis. Front Plant Sci. 2017;8:295.

Li H, Mo Y, Cui Q, Yang X, Guo Y, Wei C, Yang J, Zhang Y, Ma J, Zhang X. Transcriptomic and physiological analyses reveal drought adaptation strategies in droughttolerant and -susceptible watermelon genotypes. Plant Science. 2019;278:32–43.

Liang B, Ma C, Zhang Z, Wei Z, Gao T, Zhao Q, Ma F, Li C. Long-term exogenous application of melatonin improves nutrient uptake fluxes in apple plants under moderate drought stress. Environ Exp Bot. 2018;155:650–661.

Liang C, Li A, Yu H, Li W, Liang C, Guo S, et al. Melatonin regulates root architecture by modulating auxin response in rice. Front Plant Sci. 2017;8:134.

Liang C, Zheng G, Li W, Wang Y, Hu B, Wang H, Wu H, Qian Y, Zhu X-G, Tan D-X, et al. Melatonin delays leaf senescence and enhances salt stress tolerance in rice. J Pineal Res. 2015;59:91–101.

Liang D, Ni Z, Xia H, Xie Y, Lv X, Wang J, Lin L, Deng Q, Luo X. Exogenous melatonin promotes biomass accumulation and photosynthesis of kiwifruit seedlings under drought stress. Sci Hortic. 2019;246:34–43.

Liu J, Wang W, Wang L, Sun Y. Exogenous melatonin improves seedling health index and drought tolerance in tomato. Plant Growth Regul. 2015a;77:317–326.

Liu N, Gong B, Jin Z, Wang X, Wei M, Yang F, Li Y, Shi Q. Sodic alkaline stress mitigation by exogenous melatonin in tomato needs nitric oxide as a downstream signal. J Plant Physiol. 2015b;186–187:68–77.

Ma X, Zhang J, Burgess P, Rossi S, Huang B. Interactive effects of melatonin and cytokinin on alleviating drought-induced leaf senescence in creeping bentgrass (Agrostis stolonifera). Environ Exp Bot. 2018;145:1–11.

Machackova I, Krekule J. Sixty-five years of searching for the signals that trigger flowering. Russ J Plant Physiol. 2002;49:451–459.

Manchester LC, Tan DX, Reiter RJ, Park W, Monis K, Qi W. High levels of melatonin in the seeds of edible plants: Possible function in germ tissue protection. Life Sci. 2000;67:3023–3029.

Meloni DA, Oliva MA, Martinez CA, Cambraia J. Photosynthesis and activity of superoxide dismutase, peroxidase and glutathione reductase in cotton under salt stress. Environ Exp Bot. 2003;49:69–76.

Meng JF, Xu TF, Wang ZZ, Fang YL, Xi ZM, Zhang ZW. The ameliorative effects of exogenous melatonin on grape cuttings under water-deficient stress: Antioxidant metabolites, leaf anatomy, and chloroplast morphology. Journal of Pineal Research. 2014;57:200–212.

Moustafa FM, Almoneafy A, Mahmoud A, Elkelish A, Arnao MB, Li L, Ai S. Melatonin and its protective role against biotic stress impacts on plants. Biomolecules. 2020;10:54.

Murch SJ. Melatonin in feverfew and other medicinal plants. Lancet. 1997;350:1598–1599.

Murch SJ, Alan AR, Cao J, Saxena PK. Melatonin and serotonin in flowers and fruits of *Datura metel* L. J Pineal Res. 2009a;47:277–283.

Murch SJ, Campbell SS, Saxena PK. The role of serotonin and melatonin in plant morphogenesis: Regulation of auxin-induced root organogenesis in in vitro-cultured explants of St. John's wort (*Hypericum perforatum* L.). In Vitro Cellular & Developmental Biology-Plant. 2001;37(6):786–793.

Murch SJ, Hall BA, Le CH, Saxena PK. Changes in the levels of indoleamine phytochemicals during véraison and ripening of wine grapes. J Pineal Res. 2009b;49:95–100.

Murch SJ, Saxena PK. Mammalian neurohormones: Potential significance in reproductive physiology of St. John's wort (*Hypericum perforatum* L.)? Naturwissenschaften. 2002;89:555–560.

Okazaki M, Ezura H. Profiling of melatonin in the model tomato (*Solanum lycopersicum* L.) cultivar Micro-Tom. J Pineal Res. 2009;46:338–343.
Paredes SD, Korkmaz A, Manchester LC, Tan D-X, Reiter RJ. Phytomelatonin: A review. J Exp Bot. 2009;60:57–69.
Park S, Back K. Melatonin promotes seminal root elongation and root growth in transgenic rice after germination. Journal of Pineal Research. 2012;53(4):385–389.
Park WJ. Melatonin as an endogenous plant regulatory signal: Debates and perspectives. Journal of Plant Biology. 2011;54(3):143–149.
Poeggeler B, Thuermann S, Dose A, Schoenke M, Burkhardt S, Hardeland R. Melatonin's unique radical scavenging properties: Roles of its functional substituents as revealed by a comparison with its structural analogs. J Pineal Res. 2002;33:20–30.
Posmyk MM, Kuran H, Marciniak K, Janas KM. Presowing seed treatment with melatonin protects red cabbage seedlings against toxic copper ion concentrations. Journal of Pineal Research. 2008;45(1):24–31.
Reiter RJ, Tan DX, Burkhardt S, Manchester LC. Melatonin in plants. Nutr Rev. 2001;59:266–290.
Roychoudhury A, Basu S, Sarkar SN, Sengupta DN. Comparative physiological and molecular responses of a common aromatic indica rice cultivar to high salinity with non-aromatic indica rice cultivars. Plant Cell Reports. 2008;27(8):1395–1410.
Roychoudhury A, Paul S, Basu S. Cross-talk between abscisic acid-dependent and abscisic acid-independent pathways during abiotic stress. Plant Cell Reports. 2013;32(7):985–1006.
Samanta S, Banerjee A, Roychoudhury A. Arsenic toxicity is counteracted by exogenous application of melatonin to different extents in arsenic-susceptible and arsenic-tolerant rice cultivars. Journal of Plant Growth Regulation. 2022;41:2210–2231.
Samanta S, Banerjee A, Roychoudhury A. Melatonin application differentially modulates the enzymes associated with antioxidative machinery and ascorbate glutathione cycle during arsenate exposure in indica rice varieties. Plant Biology. 2021;23(S1):193–201.
Sharif R, Xie C, Zhang H, Arnao MB, Ali M, Ali Q, Muhammad I, Shalmani A, Nawaz MA, Chen P, Li Y. Melatonin and its effects on plant systems. Molecules. 2018;23(9):2352.
Sharma A, Wang J, Xu D, Tao S, Chong S, Yan D, Li Z, Yuan H, Zheng B. Melatonin regulates the functional components of photosynthesis, antioxidant system, gene expression, and metabolic pathways to induce drought resistance in grafted Carya cathayensis plants. Sci Total Environ. 2020;713:136675.
Sharma A, Yuan H, Kumar V, Ramakrishnan M, Kohli SK, Kaur R, Thukral AK, Bhardwaj R, Zheng B. Castasterone attenuates insecticide induced phytotoxicity in mustard. Ecotoxicol Environ Saf. 2019;179:50–61.
Shi H, Chan Z. The cysteine2/histidine2-type transcription factor zinc finger of Arabidopsis thaliana 6-activated C-repeat-binding factor pathway is essential for melatonin mediated freezing stress resistance in Arabidopsis. Journal of Pineal Research. 2014;57:185–191.
Shi H, Jiang C, Ye T, Tan D-X, Reiter RJ, Zhang H, Liu R, Chan Z. Comparative physiological, metabolomic, and transcriptomic analyses reveal mechanisms of improved abiotic stress resistance in bermudagrass [Cynodon dactylon (L). Pers.] by exogenous melatonin. Journal of Experimental Botany. 2015;66:681–694.
Sun C, Liu L, Wang L, Li B, Jin C, Lin X. Melatonin: A master regulator of plant development and stress responses. Journal of Integrative Plant Biology. 2021;63(1):126–145.
Tan D-X, Manchester LC, Mascio PD, Martinez GR, Prado FM, Reiter RJ. Novel rhythms of N1-acetyl-N2-formyl-5-methoxykynuramine and its precursor melatonin in water hyacinth: Importance for phytoremediation. FASEB J. 2007a;21:1724–1729.
Tan DX, Manchester LC, Terron MP, Flores LJ, Reiter RJ. One molecule, many derivatives: A never-ending interaction of melatonin with reactive oxygen and nitrogen species? J Pineal Res. 2007b;42:28–42.
Turk H, Erdal S, Genisel M, Atici O, Demir Y, Yanmis D. The regulatory effect of melatonin on physiological, biochemical and molecular parameters in cold-stressed wheat seedlings. Plant Growth Regulation. 2014;74:139–152.
Wang J, Zhang L, Cao Y, Qi C, Li S, Liu L, Wang G, Mao A, Ren S, Guo YD. CsATAF1 positively regulates drought stress tolerance by an ABA-dependent pathway and by promoting ROS scavenging in cucumber. Plant Cell Physiol. 2018;59:930–945.
Wang LY, Liu JL, Wang WX, Sun Y. Exogenous melatonin improves growth and photosynthetic capacity of cucumber under salinity-induced stress. Photosynthetica. 2016;54:19–27.

Wang P, et al. Delayed senescence of apple leaves by exogenous melatonin treatment: Toward regulating the ascorbate-glutathione cycle. J Pineal Res. 2012;53:11–20.

Wang P, Sun X, Chang C, Feng F, Liang D, Cheng L, Ma F. Delay in leaf senescence of Malus hupehensis by long-term melatonin application is associated with its regulation of metabolic status and protein degradation. Journal of Pineal Research. 2013;55(4):424–434.

Wei W, Li Q-T, Chu Y-N, Reiter RJ, Yu X-M, Zhu D-H, ZhangW-K, Ma B, Lin Q, Zhang J-S. Melatonin enhances plant growth and abiotic stress tolerance in soybean plants. J Exp Bot. 2014;66:695–707.

Wei W, Li Q-T, Chu Y-N, Reiter RJ, Yu X-M, Zhu D-H, Zhang W-K, Ma B, Lin Q, Zhang J-S, et al. Melatonin enhances plant growth and abiotic stress tolerance in soybean plants. J Exp Bot. 2015;66:695–707.

Xu XD, Sun Y, Sun B, Zhang J, Guo XQ. Effects of exogenous melatonin on active oxygen metabolism of cucumber seedlings under high temperature stress. Ying Yong Sheng Tai Xue Bao. 2010;21:1295–1300.

Yan W, Hongyan L, Xuejiao M, Xuejuan W, Yuanbing Z. Effect of foliar spraying exogenous melatonin on physiological and biochemical characteristics of dendranthema morifolium. Acta Botanica Boreali-Occidentalia Sinica. 2016;36:2241–2246.

Yao K, Fang J, Yin YL, Feng ZM, Tang ZR, Wu G. Tryptophan metabolism in animals: Important roles in nutrition and health. Front Biosci (Schol Ed). 2011;3:286–297.

Zeng L, Cai JS, Li JJ, Lu GY, Li CS, Fu GP, Zhang XK, Ma HQ, Liu QY, Zou XL, et al. Exogenous application of a low concentration of melatonin enhances salt tolerance in rapeseed (Brassica napus L.) seedlings. J Integ Agr. 2018;17:328–335.

Zhang J, et al. Melatonin alleviates aluminium toxicity through modulating antioxidative enzymes and enhancing organic acid anion exudation in soybean. Funct Plant Biol. 2017;44:961–968.

Zhang J, Shi Y, Zhang X, Du H, Xu B, Huang B. Melatonin suppression of heat-induced leaf senescence involves changes in abscisic acid and cytokinin biosynthesis and signaling pathways in perennial ryegrass (*Lolium perenne* L.). Environmental and Experimental Botany. 2017;138:36–45.

Zhang L, Xi D, Li S, Gao Z, Zhao S, Shi J, Wu C, Guo X. A cotton group C MAP kinase gene, GhMPK2, positively regulates salt and drought tolerance in tobacco. Plant Mol Biol. 2011;77:17–31.

Zhang N, Zhao B, Zhang HJ, Weeda S, Yang C, Yang ZC, Ren S, Guo YD. Melatonin promotes water-stress tolerance, lateral root formation, and seed germination in cucumber (Cucumis sativus L.). J Pineal Res. 2013;54:15–23.

Zhao D, Yu Y, Shen Y, Liu Q, Zhao Z, Sharma R, Reiter R J. Melatonin synthesis and function: Evolutionary history in animals and plants. Frontiers in Endocrinology. 2019;10:249.

Zhao G, Zhao Y, Yu X, Kiprotich F, Han H, Guan R, Wang R, Shen W. Nitric oxide is required for melatonin-enhanced tolerance against salinity stress in rapeseed (*Brassica napus* L.) seedlings. Int J Mol Sci. 2018;19:1912.

Zhao H, Ye L, Wang Y, Zhou X, Yang J, Wang J, Cao K, Zou Z. Melatonin increases the chilling tolerance of chloroplast in cucumber seedlings by regulating photosynthetic electron flux and the ascorbate-glutathione cycle. Front Plant Sci. 2016;7:1814.

Zhao N, Sun Y, Wang DY, Zheng JX. Effects of exogenous melatonin on nitrogen metabolism in cucumber seedlings under high temperature stress. Zhiwu Shengli Xuebao/Plant. Physiol J. 2012;48:557–564.

Zhou X, Zhao H, Cao K, Hu L, Du T, Baluška F, Zou Z. Beneficial roles of melatonin on redox regulation of photosynthetic electron transport and synthesis of D1 protein in tomato seedlings under salt stress. Front Plant Sci. 2016;7:1823.

Zohar R, Izhaki I, Koplovich A, Ben-Shlomo R. Phytomelatonin in the leaves and fruits of wild perennial plants. Phytochem Lett. 2011;4:222–226.

8 Regulation of Diverse Cell Signaling Molecules via Melatonin for Plant Growth and Overcoming Stress

Noor ul huda, Alvina Gul and Haider Ali

8.1 INTRODUCTION

Melatonin is a recently described molecule in the plant kingdom; it is even well-known in animals. The aim of this chapter is to compile the latest studies related to the role of melatonin in the different stages of plants, enhancing plant tolerance against abiotic stresses like high salinity or temperature, drought, toxic molecules, etc. or biotic stresses produced by viruses, bacteria or fungus. These functions must be due to its special structure that provides an antioxidant activity and allows it to scavenge toxic molecules. Melatonin acts as a biostimulator, bio protector and bioregulator in processes such as germination, plant growth and development, flowering, fruit development or in leaf senescence, regulating biosynthesis, transport and accumulation of phytohormones like auxin, gibberellin (GA), abscisic acid (ABA), cytokinin (CK), salicylic acid (SA), jasmonic acid (JA), etc (Samanta et al., 2021a). Moreover, it can regulate other antioxidants, molecules, primary and secondary energy metabolism or cellular homeostasis. However, the mechanisms by which it carries out that process are not studied yet (Pardo-Hernández et al., 2020).

For a long time, melatonin was considered as a substance with hormonal activity exclusive to the animal kingdom, but 30 years later, it was also discovered in the plant kingdom. In 1991, an article was published in which the presence of melatonin in the dinoflagellate Gonyaulax (*Polyedra stein*) was demonstrated, and in 1993, this molecule was described in *Pharbitis nil Choisy* (Japanese morning glory), although it was not published until 1995, later verifying its presence in many other plant species. Currently, the number of studies in plants related to this molecule has grown considerably, describing diverse and varied functions that affect different stages of plant growth and development as well as different organs and tissues (Rehaman et al., 2021).

Some of these functions are the regulation of biosynthesis, perception and signaling of the ethylene (ET), GA, auxins, etc., the induction of fruit ripening, the decrease in the biosynthesis and action of ABA, etc. It has also been associated with routes related to the metabolism of nitrogen, carbohydrates, proteins, lipids, transformation of tricarboxylic acids, in the elimination of Reactive Oxygen Species (ROS), either directly (where its particular structure allows it to donate an electron or ahydrogen atom depending on the type of radical) or indirectly (improving the enzymatic antioxidant), or by reducing the effect of different abiotic stresses, increasing the tolerance of plants to damage caused by high salinity, cold, drought and high temperature (Nawaz et al., 2020).

The current chapter focuses on the review of studies of the different effects of melatonin in each of the processes in which it acts, such as in the mobilization of reserves during germination, the genes whose expression is regulated by this hormone, mechanisms of control against abiotic stress or in its exogenous application in the protection, improvement and yield of crops. This chapter aims to gather the latest knowledge about melatonin in the different processes of plant growth and

DOI: 10.1201/9781003343752-8

development, as well as to expose the different unknowns that have not been resolved about its physiology, its interaction with other molecules or its role as a phytohormone (Tiwari et al., 2021).

8.2 ROLE OF MELATONIN IN PLANT GROWTH

8.2.1 Biosynthesis and Subcellular Localization

The precursor of melatonin biosynthesis is the amino acid tryptophan, in a pathway that requires four steps. The first step is the transformation of tryptophan through decarboxylation to tryptamine, which, in a second reaction and through hydroxylation, generates serotonin. These two steps make up the main route of serotonin formation in plants, although in some species tryptophan previously undergoes hydroxylation and subsequently decarboxylation, in a process similar to that observed in animal cells. Serotonin is converted into N-acetyl-serotonin, a reaction that, in plants, can be catalyzed by two different enzymes (Back, 2021). The last step of biosynthesis involves the formation of melatonin from N-acetyl-serotonin, transformation that can be carried out again by two different enzymes (Banerjee & Roychoudhury, 2017).

There is a variant in melatonin biosynthesis in plants, which uses the 5-methoxy substrate tryptamine, generated from serotonin, as well as a reversible reaction between N-acetyl-serotonin and serotonin. Regarding the subcellular localization, the biosynthesis of melatonin takes place in the cytoplasm, chloroplasts and endoplasmic reticulum. This distribution is due to the presence in these compartments of both the enzymes and the intermediates involved in their biosynthesis. Depending on the species studied, the intermediates and the enzymes, some authors have described four major pathways in the biosynthesis of melatonin in plants, being the chloroplasts and the cytoplasm—the compartments where most reactions occur. Recently, authors also implicate the mitochondria in this process, since it has been detected that SNAT (Serotonin N-acetyltransferase) enzyme catalyzes the transformation of serotonin into N acetyl-serotonin, the last intermediate in its biosynthesis (Tiwari, Lal, Kumar, Mangal, et al., 2021).

8.2.2 Melatonin and Flowering

Flowering is a process in which the plant grows from the vegetative growth stage to the reproductive stage. It is a critical step in the life cycle of those species that reproduce by floral structures and important from the economic point of view. Normally, flowering is induced by different stimuli related to the environment (photoperiod and abiotic factors) or with internal processes (GAs, age, vernalization). These stimuli, which can act synergistically or independently, trigger signaling pathways in which numerous genes are involved, directly or indirectly promoting or inhibiting flowering.

The role of melatonin in flowering and seasonal reproduction has been studied very little. However, in recent years, new data have appeared in studies with a reduced number of species, including *Arabidopsis thaliana*. The role of GAs in flowering induction is given by the regulation exerted by these on the stability of the expression product of the DELLA genes. The absence of DELLA proteins favors flowering by inducing the activity of the FT (flowering locus T) gene, which promotes the activation of genes involved in floral evocation such as AP-1. The exogenous application of melatonin has been shown to be capable of inducing the expression of genes associated with biosynthesis of GAs, in addition to stabilizing the expression products of the DELLA genes, in a pathway independent of GAs. This fact is manifested in a delay in flowering.

On the other hand, phytohormones such as strigolactones (SLs), which participate in multiple processes in plant growth and development, seem to play a role in inducing flowering. In fact, the presence of SLs has been shown to suppress the signaling and synthesis of endogenous melatonin, through a mechanism still unknown, triggering an antagonistic effect to it. It involves the inhibition of the transcription of the SPL (Squamosa-promoter binding protein-like) gene, triggering early flowering. Melatonin also regulates the expression of the FLC (Flowering locus C) gene, which acts as a repressor in the transcription of genes such as FT (Prakash et al., 2019).

The endogenous levels of melatonin in plants are not constant throughout the seasons. The highest levels were found in the fall, during floral senescence, while in spring, the lower levels of melatonin trigger the floral evocation at the level of meristems. Regarding the spectral quality of light, wavelengths in the blue far red rapidly induce an increase in melatonin biosynthesis, while red decreases it. On the other hand, within a given photoperiod, in the species *Paeonia lactiflora Pall*, melatonin peaks are observed during the light-dark cycle of 12 hours, the first at 2 am, during the period of darkness, and another peak at 11 am, during the light period. The latter may correspond to the increase in ROS due to the high irradiance and to the ability of melatonin to eliminate these toxic molecules.

Furthermore, in another study, it was observed how the plants exposed to light contained higher amounts of melatonin than those that grow in the shade, although this difference is not significant, and that those white flowers contained higher melatonin content, so that they could be used with medical uses. The color of the flowers is related to the content of flavonoids, which are secondary metabolites with antioxidant capacities and that can be induced by melatonin (Fatima et al., 2022).

8.2.3 Role of Melatonin in Fruit Development and Ripening

Fruiting is a term that mainly describes three stages of development: fruit set (transformation of the ovary walls in the future), development or filling (characterized due to high cell expansion and division, as well as accumulation of water and reserve substances) and maturation (characterized by external and internal changes that trigger the release of seeds).

Melatonin can be found in flowers, fruits and seeds, although depending on the species, concentration varies greatly from structure to structure. In fact, they have been found in high concentrations in ovules and endocarp of some fruits, while the mesocarp usually presents low concentrations. In tomato, specifically the Micro-Tom variety, melatonin tends to accumulate during the fruit-filling process, showing the maximum concentrations at the beginning of the maturation phase, characterized by a change in its color. These oscillations in the internal concentration of melatonin seem to be regulated by various factors such as species, temperature to which the fruit is subjected during its development as well as the irradiance received by the plant during the fruiting. These changes in melatonin concentration are associated with changes in the activity of enzymes (TDH and T5H) related to their biosynthesis (Franzoni et al., 2019).

Parthenocarpy, which is the formation of fruits without fertilization, seems to be promoted by melatonin. The mechanism involves the induction of the synthesis of GAs, specifically of GA3 and GA4, favorably regulating the GA20ox gene and negatively GA2ox, a fact that has been verified using paclobutrazol (PAC), an inhibitor of the synthesis of this phytohormone. One study showed that the fruit developed by parthenocarpy under the influence of melatonin has an elongated morphology, associated with greater cell division and expansion. Nevertheless, it is not capable of causing normal ovarian development, although it increases the IAA content in the ovary itself and decreases ABA, thus allowing fruit growth (Singh et al., 2020).

The exogenous application of melatonin in the species *Pyrus communis L.*, in the Zaosu variety, produced an increase in weight in the fruit, as well as a change in its morphology during the filling stage, in addition to an increase in soluble sugars sorbitol and sucrose. However, no differences were found in the number of seeds per fruit, the content of organic acids or the firmness of fruit. Fruits treated with melatonin usually present an accumulation of substances related to fruit ripening (simple sugars, anthocyanins, flavonoids, phenols). These exogenous treatments with melatonin also promote ET production, which partially contributes to the accumulation of phenols and promotes their antioxidant activity.

Once the fruit is separated from the plant, the postharvest begins, and melatonin has been shown to extend the quality of the fruit in the conservation stage, as well as during its commercialization.

In a recent study, it was studied that factors such as pH, temperature and light could influence the stability of melatonin in aqueous media, since the treatment by immersion method is frequently used in postharvest. One study result showed that melatonin is more stable in acidic pH, while prolonged exposure to light and high temperatures triggered its degradation. Its application in non-climacteric fruits induces a delay in its senescence, while application in climacteric fruits, such as banana, delays its maturation.

It is necessary to highlight that in September 2019, a cold drop occurred, which created an unstable climate and produced many thunderstorms with torrential rains and flash floods, which affected the production of the trees. Treatments with GABA (Gamma-aminobutyric acid) in lemon trees were very effective in improving crop yield, determined both by production (1 kg/tree) as well as the number of fruits/tree for the two seasons and the two harvest dates, the effect being greater after the application of GABA at 100 mM. This result confirmed that the exogenous application of GABA improves endogenous GABA levels, increasing photosynthetic efficiency in stressed trees and improving stress tolerance through the positive regulation of the GABA shunt (process which conserve and produce GABA supply) towards the growth cycle, tri-carboxylic acids cycle (for energy) and cell membrane stabilization. Soluble solid content, acidity and firmness were not affected by GABA treatments. In conclusion, the efficacy of GABA as an elicitor to increase production and maintain the quality of lemon was demonstrated (Singh et al., 2020).

8.2.4 Role of Melatonin in Germination

Germination is a critical process in the life cycle of a plant, since at this stage, profound biochemical, physiological and morphological changes are experienced. Before this phase is carried out, the seed is generally dormant, induced by cellular dehydration and the presence of ABA. Dormancy is a stage in which the seed is not capable of germinating, either because the conditions are not met optimal outdoors or because the seed does not have the necessary components to initiate said process (Zhang et al., 2017).

Germination consists of three stages. First, imbibition takes place, which is the intake of water from the outside until the seed is fully hydrated. By rehydrating the seed, the biosynthesis of GAs is promoted, and the synthesis of ABA is inhibited, a fact that usually triggers loss of dormancy. In the second phase, the activation of the metabolism takes place in the cell, mainly induced by the biosynthesis of GAs, characterized by the mobilization of reserve substances, the biosynthesis of new molecules, the production of energy as well as the accumulation of molecules of an organic and inorganic nature (Weeda et al., 2014).

In the last stage, the pressure of the combination of both internal and external factors exerted on the seed ends up breaking said structure, allowing the emission of the radicle, a fact that defines the end of germination. During germination, the seeds are extremely sensitive to environmental factors such as salinity, temperature, pH, drought, etc., and that is why the environmental characteristics to which they are exposed must be controlled in crops (Moustafa-Farag et al., 2020b). In addition, depending on the species or even subspecies, different conditions are required for their optimal development (Zhang et al., 2013).

There are several studies which indicate the role of melatonin in the increase of GA levels while decreasing ABA production. One study conducted in cucumber seeds under salt stress showed that melatonin can increase gibberellic acid along with the gene expression regulation of genes GA20ox and GA3ox. Similarly, it promotes ABA catabolism and inhibits the activity of enzymes involved in its synthesis, such as 9-cis epoxy carotenoid dioxygenase (NCED). Furthermore, the antioxidant capacity of melatonin allows it to protect the seeds against reactive oxygen species (ROS) as well as the Reactive Nitrogen Species (RNS), produced by cellular respiration that is carried out to obtain energy or by the effect of both biotic and abiotic stresses. Melatonin induces the expression of genes encoding antioxidant enzymes like SOD, catalase and peroxidase as well as non-enzymatic antioxidant compounds like carotenoids, polyphenols and glutathione, etc., responsible for the detoxification of these substances at the cellular level (Talaat, 2021).

In one study, proteomic analysis showed how melatonin promotes root elongation, increasing the levels of β-Tubulin and Actin, which are associated with cell division. Similarly, the relationship has been described between melatonin and protein synthesis, demonstrating an increase in ribosomal subunits during imbibition where they combine with mRNAs forming polysomes.

On the other hand, at the metabolic level, melatonin regulates the synthesis of some enzymes such as isocitrate lyase and malate synthase, which are involved in β-oxidation and in the glyoxylate cycle in the lipid catabolism, as well as other enzymes involved in glycolysis, glucogenesis and the Calvin cycle. In relation to the mobilization of reserve substances, melatonin has been shown to regulate the activity of enzymes that degrades starch from reserve materials within the endosperm. In addition to intervening in the metabolism of macromolecules (lipids, proteins, sugars, etc.), melatonin is involved in the acquisition and assimilation of sulfur, improving the concentration of chlorophylls, the photosynthesis and biomass production. It has also been found that melatonin increases the biosynthesis of polyamines via improvement of the precursor flux of methionine and arginine, a fact that counteracts its degradation under saline stress conditions.

The effects described earlier associated with seed germination have been obtained using different species. However, there does not seem to be a clear correlation between the concentration used and effects observed. For example, in wheat seeds, the concentration of melatonin used as exogenous treatment was 500 μM, in kiwi seeds 100 μM, while in cotton seeds only 20 μM to obtain an increase in germination. On the other hand, the exogenous application of melatonin was detrimental, even preventing germination, under a certain concentration. Therefore, melatonin treatments must be customized in seeds to improve germination in optimal conditions or under both biotic and abiotic stresses (Tan et al., 2012).

8.2.5 Melatonin and Its Role in Plant Growth and Development

Plant growth and development is a process in which cells undergo, on one hand, cell expansion and division, as well as cell specialization or differentiation to form different plant tissues and organs. The plant growth is usually described in two stages, primary growth in which the plant increases its size longitudinally and secondary growth in which cells allow the thickening of the different structures of the plant. Development not only involves growth but also cell differentiation and transformation at the level of tissues and vegetative organs (Asif et al., 2019).

In all these processes of plant growth and development, the role of auxins has been widely studied. The structural similarity and the fact that they share the same precursor in their biosynthesis (tryptophan) has allowed some studies to establish functional similarities between auxins and melatonin. In this sense, some authors establish a relationship between melatonin and formation of lateral roots, the growth of primary roots and shoots, the size and weight of the plant, contents of primary metabolites, stomatal opening and photosynthesis. In fact, the exogenous application of melatonin has shown that it can promote growth in different plant species (tomato, cherries, cucumber, etc.).

Some authors indicate that this increase is related to the increase in the levels of auxins in plant tissues, although in other studies using transgenic plants from another species, decreased IAA content was noted due to changes in auxin transporters. On the other hand, in some plants (*Zea mays L.*), exogenous application of melatonin does not trigger growth processes associated with auxins, such as coleoptile elongation, root growth inhibition and gene expression of ET biosynthesis. These results have allowed some authors to establish that despite the structural similarities between melatonin and auxins, in corn, the absence of an unbranched side chain with a carboxyl group in its structure disables its ability to act like auxins (Kanwar et al., 2018).

At the root level, melatonin regulates the growth of primary and secondary roots through the control of auxin synthesis and its transport. However, other authors establish that the growth of primary and secondary roots promoted by melatonin is independent of auxin signaling pathways, since melatonin is not capable of activating reporter genes such as DR5, which are normally activated by IAA. The application of high concentrations of melatonin in *Arabidopsis thaliana* L., trigger drastic development of lateral roots by inhibiting primary root growth, while physiological concentrations do not show any negative effect (Arnao and Hernández-Ruiz, 2019). The higher concentrations reduce the expression of the auxin transporter genes PIN 1, 3 and 7 (Khan et al., 2020).

They play an important role in the development of the root meristem. In addition, the polar transport of auxin is necessary for the regulation of its size, since the inhibition of transport provided by the high concentrations of melatonin does not intensify when the plant is treated with 5-Triiodobenzoic acid or TIBA (auxin transport inhibitor), so that a certain number of auxins is necessary. In other studies, at the root level, a relationship between the application of melatonin and ROS levels has been shown.

In fact, in alfalfa plants, the presence of hydrogen peroxide is essential for the induction and formation of lateral roots in transgenic plants with an overexpression of the SNAT enzyme, belonging to the melatonin biosynthesis pathway. From a metabolic point of view, melatonin is also capable of promoting the growth of a plant, maintaining a balance between nitrogen and carbon metabolism. Nitrogen is important as it forms part of the structure of simple molecules such as amino acids or more complex molecules such as proteins, nucleic acids, chlorophyll and phytohormones, while carbon metabolism is provided primarily by photosynthesis and respiration (Turk et al., 2014)

Mitochondria provide carbon skeletons, energy and reducing agents, etc. Melatonin can increase both the activity and the expression of enzymes related to the nitrogen metabolism (nitrate and nitrite reductase, glutamine synthase, etc.), as well as carbon metabolism (RuBisCo, citrate synthase and cytochrome oxidase). In the same way, it was observed how melatonin increases photosynthetic efficiency (specifically the efficiency of the Photosystem II) and positively regulates the expression of genes related to this process.

It has been described that it increases the CO_2 assimilation capacity, stomatal conductance, reduces the chlorophyll degradation and increases leaf size. Regarding leaf growth, high concentrations (approximately 500–1000 μM) of melatonin suppresses cell proliferation and expansion, reducing the rate of cell division. During growth, endoreduplication (DNA replication without cell division) occurs, which causes a change in ploidy, in addition to the formation of a larger nucleus and an increase in the size of the cell.

Melatonin inhibits this process by stabilizing DELLA proteins, whose inhibitory activity reduces plant growth and increases the expression of CYCD genes (family of cyclins type D) that regulate the balance between cell division and expansion; that is, they reduce the endoreduplication. In addition, melatonin can induce an increase in number of leaves, promote bud growth and increase the fresh weight of the plant to physiological concentrations, while at high concentrations, it acts as an inhibitory element. Finally, there are few studies that analyze the combined effects of melatonin with other substances or plant hormones. In this regard, the joint application of melatonin and serotonins induce phytohormones, stress factors and photosynthetic efficiency that work together to direct plant growth. However, the results showed that they can be used independently and cause virtually the same effects, such as radicle and caudal growth, increase in the number and length of the nodes, etc. (Kanwar et al., 2018).

By the combination of melatonin with benzyladenine (BA), which is a synthetic cytokinin responsible for stimulating cell division in plants, the formation of tissue has been observed from hypocotyl callus type and cotyledons with different degrees of success, depending on the species, while its individual application in culture medium does not produce any change in the growth of the explants (Agathokleous et al., 2019).

8.3 ROLE OF MELATONIN IN OVERCOMING STRESS

8.3.1 Role of Melatonin in Abiotic Stress

The adaptation and control of environmental conditions capable of modifying patterns of plant growth and development is crucial from an ecological and an economic point of view. Any type of abiotic stress is capable of negatively modifying growth and development of a plant, depending on its duration and degree of intensity (Shafi et al., 2021). Most of the abiotic stresses such as salinity, drought, heavy metal toxicity etc. negatively modify growth, photosynthesis, ROS production and accumulation, RNS, the flow of the water intake, the availability of metabolites as well as the contents of hormones such as auxins, ABA, CK, ET, brassinosteroids (BRs), etc. (Roychoudhury et al., 2008; Wang et al., 2021).

The presence of melatonin in different cellular compartments and its capacity in the ROS removal has recently been demonstrated. In fact, melatonin activates the ROS detoxification enzymatic mechanisms (SOD, CAT, POX, GR, APX, etc.) and non-enzymatic ones (ascorbic acid, carotenoids, glutathione, etc.). At the cellular level, melatonin biosynthesis in plants is mainly associated with chloroplasts and mitochondria, observing an increase in the expression of enzymes related to their biosynthesis under stress conditions. This translates into less degradation of these cellular compartments associated with the production and accumulation of ROS (Samanta et al., 2021b; Tiwari, Lal, Kumar, Chourasia, et al., 2021).

On the other hand, the increase in melatonin levels has been shown to regulate the osmoprotectants by controlling proline homeostasis at the enzymatic and genetic level, in addition to the content of soluble sugars, Total Soluble Solids (TSS) and soluble proteins (Banerjee & Roychoudhury, 2017). At the cell membrane level, melatonin stabilizes membrane integrity and fluidity by a decrease in the levels of lipid peroxidation. Regarding the efficiency of Photosystem II (PSII) and stomatal conductance, melatonin reverses the effects of stress, improving photosynthetic electron transport, with increases in chlorophyll levels, causing less foliar necrosis and improved transpiration. In addition, it maintains and improves cell expansion and cell wall integrity, avoiding changes in the stomata and in the ultrastructure of the seed (Shafi et al., 2021).

Plants subjected to high temperature stress activate heat shock proteins (HSPs) and chaperones, which prevent the aggregation of denatured proteins by binding to them, as well as the autophagic protein degradation system (Sharma et al., 2020). Melatonin induces the expression of different HSP genes to increase the folding of denatured proteins and related genes for autophagy, improving the thermotolerance of plants. In addition, the system of methylglyoxal detoxification, regulating the enzymatic activity of Glyoxylates I and II (Gly I and Gly II), is activated.

Under stress conditions like heavy metal stress (for example: Cadmium [Cd]), effects on seed elongation, root elongation and fresh weight of root tissues are reversed by melatonin via increase in its biosynthesis (Franzoni et al., 2019). In addition, it eliminates the accumulation of cadmium in tissues via changes in membrane transporters (ABC transporter, PCE2, Nramp6, etc.) and restores redox homeostasis via microRNA398 transcription. The exogenous application of melatonin reduces the content of Cd in leaves, but not in roots, which suggests that melatonin intervenes in the transport of Cd. In addition, it improves the activity H-ATPase and antioxidant enzymes and increases the content of glutathione and phytochelatin, mitigating cadmium stress (Shafi et al., 2021).

Regarding the external application of melatonin in the protection against abiotic stresses, this seems to be dependent on the concentration and the species studied. In several studies, protection at relatively low concentrations has been demonstrated, but when increasing the said optimum concentration, it appears harmful and even, in some cases, fatal for the plant; for example, in cherry or mustard at low concentrations, melatonin relieves the effect of abiotic stresses, but at high concentrations, it prevents the germination of seeds (Moniruzzaman et al., 2018).

8.3.2 Melatonin and Its Role in Biotic Stress

Biotic stress in plants is caused by viroids, viruses, bacteria and fungi, which are one of the main causes of crop loss worldwide due to pests with high spread and infection capacity. Pesticides, fungicides, etc. are used for containment and eradication. However, they are harmful to the environment as they release toxic particles into the air. To avoid the use of these chemical substances, treatments with natural molecules and substances are used. The role of melatonin against biotic stress has recently begun to be studied, finding that it is an efficient molecule for treating the plant and increasing its tolerance against pathogens and that it can reduce the resistance of pathogens to substances and other stresses (Bose & Howlader, 2020).

Exogenous application of melatonin increases the expression of enzymes related to the plant defense, such as chitinase (CHI) and glucanase (GLU), which are related to the degradation of fungal cell wall; Phenylalanine ammonia lyase (PAL), a crucial enzyme for the phenylpropanoid pathway, which plays a key role in disease resistance; and Polyphenol Oxidase (PPO), which catalyzes the phenol oxidation; and those related to apoptosis in pathogenesis. In addition, it increases the transcription of antioxidant enzymes, such as POD, SOD and APX, thus allowing the elimination of ROS. In addition, fungal infection, such as *Botrytis cinerea*, promotes melatonin biosynthesis via an increase in SNAT levels (Zhao et al., 2021).

At the cellular level, the application of melatonin can reduce cell size both in length and thickness, the motility of pathogenic organisms, their division, expansion and cell multiplication (possibly through SA and ET signaling). In addition, it is demonstrated that it can modify the ultrastructure, morphology and cellular components in plant cells. Examples include cellular lipid vesicles (important in the motility and infectious capacity of pathogens), the fragmentation of the cytoplasm or the thickening of the cell wall via the accumulation of cellulose, galactose, xylose and callose. In addition, it reduces the damaged area in the plant, even at high concentrations, avoiding any toxicity (Tan et al., 2012).

At the level of gene expression, melatonin regulates genes related to the metabolism of carbohydrates and amino acids. Increasing the biosynthesis and signaling of jasmonic acid (JA) increases the expression of enzymes such as allene oxide cyclase (AOC) or lipoxygenase and negatively regulates the activity of others such as proteinase inhibitor II (PI II) and MYC2, the negative response regulator to pathogens in JA signaling, and JASMONATE ZIM DOMAIN 1(JAZ1), a gene that regulates JA signaling in an inhibitory manner. It also increases the content of nitric oxide and SA, two molecules that intervene in the defense of the plant against pathogenic organisms (Jahan et al., 2021).

Heat shock proteins or HSPs have an important role in the resistance of pathogens to other stresses. Melatonin inhibits the expression of these proteins in pathogens and induces it in plants, so that plants resist high temperatures, but infectious organisms do not. Through glutamate decarboxylase (GAD), which convert glutamate to carbon dioxide and GABA, HSP90 inhibitor or Proteins of Thermal Shock of 90 kDa, it has been verified how the chaperones are essential so that the melatonin can act. The RAV1 and RAV2 genes, which are involved in the formation of the flower in cassava, and which belong to the apetala2/ethylene response factor family, are essential for disease resistance of the plants. It has been shown via virus-induced gene silencing (VIGS), which prevents the transcription of RAVs and does not allow the effects of melatonin in the response to biotic stresses. They also induce three genes related to the biosynthesis of melatonin, which are TDC, T5H and ASMT (Moustafa-Farag et al., 2019).

The studies show that the application of melatonin offers better results in resistance if the plants are treated twice with this molecule, separated in time, as well as if melatonin is combined with other substances, such as fungicides, pesticides, etc., since they have a greater effect, improving the effectiveness of said substance against different microorganisms. It has been tested with the fungicide Infinito® against *Phytophthora infestans*, with which very little melatonin is needed (Liu et al., 2021).

The inhibition of MFS makes the organism more susceptible to the chemical, and thus damage is minimized in plant. This has also been verified with the combination of melatonin with selenium molecule. Separately, these molecules produce the same effects, such as suppression of cell viability or virulence of *Phytophthora nicotiana*, but their combination enhances these effects, obtaining more beneficial results. Applying a combined treatment, the use of fungicides could be reduced, providing better results in the crops.

8.4 ROLE OF MELATONIN IN REGULATION OF SIGNALING PATHWAYS

8.4.1 Melatonin as a Biostimulant

Many beneficial effects of melatonin on plants are known from inducing germination or the growth and development of roots, stems or shoots, until leaf senescence is delayed, in addition to increasing the tolerance of the plant against abiotic and biotic stresses. These characteristics allow it to act as a biostimulant, bioregulator and bio protector. The levels described in plants are greater than in animals but are not toxic until used in very high concentrations (Mandal et al., 2018).

It meets the conditions to be a biostimulant since it is a molecule of natural origin that increases plant productivity. Biostimulants are not nutrients, but they stimulate the processes of nutrition in addition to improving tolerance to abiotic stress or increasing the quality traits of the plant and increasing the availability of nutrients present in the soil or in the rhizosphere. Also, they optimize the activity of fertilizers (Arnao & Hernández-Ruiz, 2015).

Melatonin has a natural origin, although it can be synthesized easily; it is not toxic, not expensive, dissolves in different solvents such as water, alcohol and even lipids. It is also a small molecule with easy cellular assimilation and compartmentalization and has important antioxidant activities and improves stress tolerance (Talaat, 2021).

As a biostimulant, it regulates the genes related to the different redox pathways of the plant, thus allowing the induction of genes that protect the integrity of the chloroplasts, as well as the pigments it contains and PSII, increasing photosynthesis and thus increasing the metabolism of carbon and nitrogen, important for the increase of vegetal biomass inducing cell division and expansion during germination and during the development and growth of roots, stems, leaves, etc.

8.4.2 Melatonin's Role in Regulating Plant Defense

In addition, it induces hormonal responses and responses to abiotic stresses, such as salinity, drought, high or low temperatures, etc., and biotic stresses, such as fungal or bacterial pests. However, it is capable of inhibiting cell death by inactivating genes related to this, with markers of senescence and with the degradation of chlorophyll, etc.

Its structure allows it to act as a major antioxidant, stabilizing the membrane and maintaining its fluidity, eliminating ROS and RNS, etc. Its role in regulating the growth of a plant is given by its ability to activate signaling genes or biosynthesis of hormones such as ABA, GA, ET, SA or JA, which intervene in the different responses to stress, and to induce the growth of roots, stem, cotyledons, etc. On the other hand, it favors the development of primary, lateral and adventitious roots. All these processes are related to the ability of melatonin to regulate IAA levels through changes in their transport and signaling (Yang et al., 2022).

As a plant protector, melatonin can improve the tolerance of the plant against bacteria, fungi and viruses, activating the immune response and primary and secondary energy metabolism. However, to use it as such, more studies are needed on its impact on the soil, in animals such as arthropods and in aquatic environments. However, in a recent study, the first receptor for melatonin CAND2/PMTR1 in *Arabidopsis thaliana* L., related to the regulation of stomatal closure, has been reported (Bychkov et al., 2022). This process depends on the signaling and transduction cascade of this receptor, which is mediated by hydrogen peroxide and Ca^{2+}, which activate the α subunit of a G protein of the membrane. These are important components in signaling related to growth and plant development, with environmental responses and with stomatal opening. CAND2/PMTR1 is found in the plasma membrane, in various tissues, and its expression is induced by melatonin. Mutants that do not contain this receptor are insensitive to melatonin (Lee & Back, 2016).

This presents melatonin as a phytohormone, and not as a biostimulant, bioregulator or bio protector, although it may have activities dependent and independent of this receptor, as described above. More studies need to be done to verify that melatonin is a phytohormone and, if so, to discover other receptors that mediate other responses of this molecule and its physiology in plants (Kanwar et al., 2018).

8.5 CONCLUSIONS

Melatonin is a molecule widely distributed in different kingdoms. Its functions are well known in animals, where it acts as a hormone, and differ from those found in plants, playing different roles from the germination of the seed to the maturation of the fruit. Of many of these functions, there is very little documented information, and it is necessary to expand studies to solve all the unknowns.

Regarding the use of exogenous melatonin as a treatment, it is known that it increases the endogenous levels of melatonin, promoting the expression of the different genes involved in its biosynthesis. However, when applying melatonin as an exogenous treatment in crops, it should not be generalized, since the optimal levels of this molecule in each species vary enormously, and if applied more than needed, it could be toxic or even deadly to a plant (Lee et al., 2014).

Very few melatonin interactions with other natural substances are known, such as serotonin and phytohormones, or artificial substances, such as fungicides. A comprehensive study is necessary of the synergy between melatonin and different molecules and substances to use it as a treatment against abiotic and biotic stresses and to improve the effect of said substances, thus being able to use less quantity and be more responsible of the environment.

The answer to the question whether it works as a phytohormone in all these processes or can act independently of specific receptors is that, currently, only one receptor is known. For melatonin, more studies would be necessary to unravel its role as a phytohormone in the different plant processes. Finally, it should be noted that although there are still many areas and aspects of melatonin to be studied, especially in the stages of flowering and plant fertilization, it remains to be observed how this molecule is fundamental in plants, both for their proper development and for their survival to various kinds of stresses (Moustafa-Farag et al., 2020a).

REFERENCES

Agathokleous, E., Kitao, M., & Calabrese, E. J. (2019). New insights into the role of melatonin in plants and animals. *Chemico-Biological Interactions*, *299*, 163–167.

Arnao, M. B., & Hernández-Ruiz, J. (2015). Functions of melatonin in plants: A review. *Journal of Pineal Research*, *59*(2), 133–150.

Arnao, M. B., & Hernández-Ruiz, J. (2019). Melatonin: A new plant hormone and/or a plant master regulator? *Trends in Plant Science*, *24*(1), 38–48. https://doi.org/10.1016/j.tplants.2018.10.010

Asif, M., Pervez, A., & Ahmad, R. (2019). Role of melatonin and plant-growth-promoting rhizobacteria in the growth and development of plants. *Clean: Soil, Air, Water*, *47*(6), 1800459.

Back, K. (2021). Melatonin metabolism, signaling and possible roles in plants. *The Plant Journal*, *105*(2), 376–391. https://doi.org/10.1111/tpj.14915

Banerjee, A., & Roychoudhury, A. (2017). Melatonin as a regulator of abiotic stress tolerance in plants. In V. P. Singh, S. Singh, & S. M. Prasad (Eds.), *Mechanisms behind phytohormonal signaling and crop abiotic stress tolerance* (pp. 47–60). New York: Nova Science Publishers.

Bose, S. K., & Howlader, P. (2020). Melatonin plays multifunctional role in horticultural crops against environmental stresses: A review. *Environmental and Experimental Botany*, *176*, 104063.

Bychkov, I. A., Kudryakova, N. V., Shugaev, A. G., Kuznetsov, V. V., & Kusnetsov, V. V. (2022). The melatonin receptor CAND2/PMTR1 is involved in the regulation of mitochondrial gene expression under photooxidative stress. *Doklady Biochemistry and Biophysics*, *502*(1), 15–20. doi:10.1134/S1607672922010021

Fatima, A., Husain, T., Suhel, M., Prasad, S. M., & Singh, V. P. (2022). Implication of nitric oxide under salinity stress: The possible interaction with other signaling molecules. *Journal of Plant Growth Regulation*, *41*(1), 163–177.

Franzoni, G., Trivellini, A., Bulgari, R., Cocetta, G., & Ferrante, A. (2019). Bioactive molecules as regulatory signals in plant responses to abiotic stresses. In *Plant signaling molecules* (pp. 169–182). UK: Woodhead Publishing. https://doi.org/10.1016/B978-0-12-816451-8.00010-1

Jahan, B., Rasheed, F., Sehar, Z., Fatma, M., Iqbal, N., Masood, A., . . . Khan, N. A. (2021). Coordinated role of nitric oxide, ethylene, nitrogen, and sulfur in plant salt stress tolerance. *Stresses*, *1*(3), 181–199.

Kanwar, M. K., Yu, J., & Zhou, J. (2018). Phytomelatonin: Recent advances and future prospects. *Journal of Pineal Research*, *65*(4), e12526.

Khan, A., Numan, M., Khan, A. L., Lee, I.-J., Imran, M., Asaf, S., & Al-Harrasi, A. (2020). Melatonin: Awakening the defense mechanisms during plant oxidative stress. *Plants*, *9*(4), 407.

Lee, H. Y., & Back, K. (2016). Mitogen-activated protein kinase pathways are required for melatonin-mediated defense responses in plants. *Journal of Pineal Research*, *60*(3), 327–335. https://doi.org/10.1111/jpi.12314

Lee, H. Y., Byeon, Y., & Back, K. (2014). Melatonin as a signal molecule triggering defense responses against pathogen attack in Arabidopsis and tobacco. *Journal of Pineal Research*, *57*(3), 262–268.

Liu, Q., Atta, U. R., Wang, R., Liu, K., Ma, X., & Weng, Q. (2021). Defense-related hormone signaling coordinately controls the role of melatonin during Arabidopsis thaliana-Pseudomonas syringae interaction. *European Journal of Plant Pathology*, *160*(3), 707–716. doi:10.1007/s10658-021-02279-8

Mandal, M. K., Suren, H., Ward, B., Boroujerdi, A., & Kousik, C. (2018). Differential roles of melatonin in plant-host resistance and pathogen suppression in cucurbits. *Journal of Pineal Research*, *65*(3), e12505.

Moniruzzaman, M., Ghosal, I., Das, D., & Chakraborty, S. B. (2018). Melatonin ameliorates H2O2-induced oxidative stress through modulation of Erk/Akt/NFkB pathway. *Biological Research*, *51*.

Moustafa-Farag, M., Almoneafy, A., Mahmoud, A., Elkelish, A., Arnao, M. B., Li, L., & Ai, S. (2019). Melatonin and its protective role against biotic stress impacts on plants. *Biomolecules*, *10*(1), 54.

Moustafa-Farag, M., Almoneafy, A., Mahmoud, A., Elkelish, A., Arnao, M. B., Li, L., & Ai, S. (2020a). Melatonin and its protective role against biotic stress impacts on plants. *Biomolecules*, *10*(1), 54. www.mdpi.com/2218-273X/10/1/54

Moustafa-Farag, M., Mahmoud, A., Arnao, M. B., Sheteiwy, M. S., Dafea, M., Soltan, M., . . . Ai, S. (2020b). Melatonin-induced water stress tolerance in plants: Recent advances. *Antioxidants*, *9*(9), 809.

Nawaz, K., Chaudhary, R., Sarwar, A., Ahmad, B., Gul, A., Hano, C., . . . Anjum, S. (2020). Melatonin as master regulator in plant growth, development and stress alleviator for sustainable agricultural production: Current status and future perspectives. *Sustainability*, *13*(1), 294.

Pardo-Hernández, M., López-Delacalle, M., & Rivero, R. M. (2020). ROS and NO regulation by melatonin under abiotic stress in plants. *Antioxidants*, *9*(11), 1078.

Prakash, V., Singh, V. P., Tripathi, D. K., Sharma, S., & Corpas, F. J. (2019). Crosstalk between nitric oxide (NO) and abscisic acid (ABA) signalling molecules in higher plants. *Environmental and Experimental Botany*, *161*, 41–49.

Rehaman, A., Mishra, A. K., Ferdose, A., Per, T. S., Hanief, M., Jan, A. T., & Asgher, M. (2021). Melatonin in plant defense against abiotic stress. *Forests*, *12*(10), 1404.

Roychoudhury, A., Basu, S., Sarkar, S. N., & Sengupta, D. N. (2008). Comparative physiological and molecular responses of a common aromatic indica rice cultivar to high salinity with non-aromatic indica rice cultivars. *Plant Cell Reports*, *27*(8), 1395–1410.

Samanta, S., Banerjee, A., & Roychoudhury, A. (2021a). Exogenous melatonin regulates endogenous phytohormone homeostasis and thiol-mediated detoxification in two indica rice cultivars under arsenic stress. *Plant Cell Reports*, *40*, 1585–1602.

Samanta, S., Banerjee, A., & Roychoudhury, A. (2021b). Melatonin application differentially modulates the enzymes associated with antioxidative machinery and ascorbate glutathione cycle during arsenate exposure in indica rice varieties. *Plant Biology*, *23*(S1), 193–201.

Shafi, A., Singh, A. K., & Zahoor, I. (2021). Melatonin: Role in abiotic stress resistance and tolerance. In *Plant growth regulators* (pp. 239–273). Cham: Springer. https://doi.org/10.1007/978-3-030-61153-8_12

Sharma, L., Priya, M., Kaushal, N., Bhandhari, K., Chaudhary, S., Dhankher, O. P., . . . Nayyar, H. (2020). Plant growth-regulating molecules as thermoprotectants: Functional relevance and prospects for improving heat tolerance in food crops. *Journal of Experimental Botany*, *71*(2), 569–594.

Singh, A. K., Dhanapal, S., & Yadav, B. S. (2020). The dynamic responses of plant physiology and metabolism during environmental stress progression. *Molecular Biology Reports*, *47*(2), 1459–1470.

Talaat, N. B. (2021). Polyamine and nitrogen metabolism regulation by melatonin and salicylic acid combined treatment as a repressor for salt toxicity in wheat (*Triticum aestivum* L.) plants. *Plant Growth Regulation*, *95*(3), 315–329.

Tan, D.-X., Hardeland, R., Manchester, L. C., Korkmaz, A., Ma, S., Rosales-Corral, S., & Reiter, R. J. (2012). Functional roles of melatonin in plants, and perspectives in nutritional and agricultural science. *Journal of Experimental Botany*, *63*(2), 577–597.

Tiwari, R. K., Lal, M. K., Kumar, R., Chourasia, K. N., Naga, K. C., Kumar, D., . . . Zinta, G. (2021). Mechanistic insights on melatonin-mediated drought stress mitigation in plants. *Physiologia Plantarum*, *172*(2), 1212–1226.

Tiwari, R. K., Lal, M. K., Kumar, R., Mangal, V., Altaf, M. A., Sharma, S., . . . Kumar, M. (2021). Insight into melatonin-mediated response and signaling in the regulation of plant defense under biotic stress. *Plant Molecular Biology*, 1–15.

Turk, H., Erdal, S., Genisel, M., Atici, O., Demir, Y., & Yanmis, D. (2014). The regulatory effect of melatonin on physiological, biochemical and molecular parameters in cold-stressed wheat seedlings. *Plant Growth Regulation*, *74*(2), 139–152.

Wang, C., Deng, Y., Liu, Z., & Liao, W. (2021). Hydrogen sulfide in plants: Crosstalk with other signal molecules in response to abiotic stresses. *International Journal of Molecular Sciences*, *22*(21), 12068.

Weeda, S., Zhang, N., Zhao, X., Ndip, G., Guo, Y., Buck, G. A., . . . Ren, S. (2014). Arabidopsis transcriptome analysis reveals key roles of melatonin in plant defense systems. *PloS One*, *9*(3), e93462.

Yang, S., Zhao, Y., Qin, X., Ding, C., Chen, Y., Tang, Z., . . . Yuan, M. (2022). New insights into the role of melatonin in photosynthesis. *Journal of Experimental Botany*. doi:10.1093/jxb/erac230

Zhang, N., Zhang, H.-J., Sun, Q.-Q., Cao, Y.-Y., Li, X., Zhao, B., . . . Guo, Y.-D. (2017). Proteomic analysis reveals a role of melatonin in promoting cucumber seed germination under high salinity by regulating energy production. *Scientific Reports*, *7*(1), 1–15.

Zhang, N., Zhao, B., Zhang, H. J., Weeda, S., Yang, C., Yang, Z. C., . . . Guo, Y. D. (2013). Melatonin promotes water-stress tolerance, lateral root formation, and seed germination in cucumber (*Cucumis sativus* L.). *Journal of Pineal Research*, *54*(1), 15–23.

Zhao, D., Wang, H., Chen, S., Yu, D., & Reiter, R. J. (2021). Phytomelatonin: An emerging regulator of plant biotic stress resistance. *Trends in Plant Science*, *26*(1), 70–82.

9 Role of Melatonin, a Master Regulator in Modulating ROS and NO in Plants during Environmental Stresses

Kakan Ball, Subir Ghosh, Kuntal Bera, Sanjoy Sadhukhan, and Puspendu Dutta

9.1 INTRODUCTION

Plants are continuously subjected to several stresses in nature. Drought, heat, cold, salt, bright light, and pathogen attacks, either individually or in combination, have a detrimental impact on plant growth, reproduction, survival, and limiting crop productivity or agricultural yield (Kul et al. 2019). In the future years, climate change is expected to exacerbate the detrimental impact of environmental stress factors on plant productivity in many parts of the world (Piao et al. 2019), and it raises the obvious question of how humankind will cope with food demand around the world. Plant productivity has been demonstrated to be greatly affected when many stressors are present (Roychoudhury et al. 2008; Zandalinas et al. 2019). According to the UN's Food and Agriculture Organization (FAO), the world population will rise to 9700 billion people by 2050, requiring agricultural systems to generate about 50% more food, fodder, and biofuels than in 2012. Plants respond to environmental stresses through a variety of biochemical and molecular mechanisms, including selective ion absorption and exclusion, ion compartmentalization into the central vacuole, cytoplasmic synthesis and accumulation of organic solutes, changes in membrane composition, and changes in a plant structure (Roychoudhury et al. 2013; Bailey-Serres et al. 2019; Suzuki et al. 2013; Zandalinas et al. 2019). Phytohormones, nitric oxide (NO), hydrogen sulphide, and calcium (Ca^{2+}) are all involved in plant response to its surroundings (Fichman et al. 2019; Nguyen et al. 2018a). Numerous studies have recently revealed that melatonin (MEL), a new and recently discovered plant chemical, plays a significant part in the response of plants to environmental stress and that its exogenous administration has been demonstrated to have a good effect on the reduction of stress in plants (Kul et al. 2019; Martinez et al. 2018). MEL (*N-acetyl*-5-methoxytryptamine), a tryptophan-derived indolic molecule, is a ubiquitous abiotic stress regulator in plants (Wang et al. 2018). MEL was discovered in plants for the first time in 1995 (Dubbels et al. 1995; Hattori et al. 1995) and has since been discovered to play a role in a variety of abiotic stress responses as a pleiotropic signalling molecule (Sun et al. 2021). Furthermore, it is an effective scavenger of both reactive oxygen and nitrogen species (ROS and RNS). MEL is involved in several processes that are important for plant growth and development, as well as many biotic and abiotic stress responses (Arnao & Hernández-Ruiz 2014, 2015, 2019a, 2019b; Byeon & Back 2014; Sun et al. 2015; Van Tassel et al. 2001; Wang et al. 2018; Wei et al. 2015; Zhang et al. 2015). It is also known as a multi-regulatory molecule, and the word "phytomelatonin" was coined to describe it because of its wide range of biological roles (Blask et al. 2004). MEL and phytomelatonin represent identical molecules structurally. The discovery of the

DOI: 10.1201/9781003343752-9

first putative plant MEL receptor in Arabidopsis recently raised the possibility that MEL could be a phytohormone (Arnao & Hernández-Ruiz 2019a, 2020). MEL possesses various phytohormone-like functions which include regulating circadian rhythms, post-harvest fruit ripening, root development and growth, leaf senescence, as well as plant reactions to unfavourable environmental situations, etc. (Arnao & Hernández-Ruiz 2015; Li et al. 2019a, 2012; Shi et al. 2015b, 2016; Sun et al. 2020, 2015; Yin et al. 2013; Zhang et al. 2014a).

9.2 MELATONIN AS A PLANT RESPONSE ELICITOR

MEL, a pleiotropic molecule, is present in all living organisms and has an indole ring structure with low molecular weight. MEL has an amphiphilic or amphipathic molecular property, allowing it to travel into the cytosol, the nucleus, and/or the mitochondria without difficulty through the cell membrane (Debnath et al. 2019; Hardeland et al. 2011). The most ubiquitous MEL derivative in plants is 2-hydroxymelatonin (2-OHM), and its intracellular concentration is almost two orders of magnitude higher than that of MEL. Therefore, it is hypothesised that this MEL derivative may be more effective than MEL in promoting plant stress tolerance (Byeon et al. 2015; Lee & Back 2016a). The amount of MEL in plants varies widely among cultivars, species, growth and developmental stages, tissue types, and even duplications of the same experiment. MEL is present in nearly all plant organs, from leaves to pistils (Debnath et al. 2019; Erland et al. 2015). Although MEL can also be produced in the cytosol, mitochondria and chloroplasts are thought to be the primary sites of MEL production. Reactive oxygen species (ROS) are frequently produced in mitochondria and chloroplasts, and MEL plays a crucial role in free radical defence in these organelles (Back et al. 2016; Erland et al. 2015). It is thought that MEL primarily serves as a regulator of ROS levels since, particularly in plants, MEL is kept at a level that is usually quite constant. However, it can be considered quickly triggered and increased in response to unfavourable circumstances such as cold, heat, salt, thirst, oxidative and nutritional stress, and bacterial infection (Ma et al. 2017; Zhang et al. 2015). On the other hand, it has been demonstrated that high amounts of MEL may be harmful to plants and may hinder growth under some conditions (Posmyk et al. 2008). MEL has been said to play a variety of significant roles in plants. It is a plant hormone that has been implicated in several studies as a potential growth regulator since it affects a variety of physiological functions (Banerjee & Roychoudhury 2017, 2019). MEL controls the development of leaves, shoots, and explants, and contributes to the senescence of leaves. It can govern flowering processes, the development and maturation of fruit and seeds, as well as vegetative growth activities in plants, including roots, leaf ageing, photosynthetic production, and biomass yield (Arnao & Hernández-Ruiz 2014; Kul et al. 2019; Nawaz et al. 2016). There is growing evidence that MEL is critical to phytoremediation (Arnao & Hernández-Ruiz 2014). However, the role of MEL in non-receptor-mediated actions, such as the removal of reactive nitrogen species (RNS) and ROS and enhancement of cellular antioxidant capacity, may be the most significant function that has been identified in living organisms (Debnath et al. 2019). In this manner, when a plant experiences abiotic or biotic stress or detects MEL, MEL controls and rapidly up-regulates the activity of diverse antioxidant enzymes and stress tolerance-related genes, activating downstream signalling transduction pathways (Arnao & Hernández-Ruiz 2015; Wang et al. 2018). By way of MEL signalling elements or secondary messengers like ROS/RNS, the sensing of MEL by receptors may either directly or indirectly regulate the expression of stress-responsive genes. For instance, MEL activates the C_2H_2-type zinc finger transcription factors ZATs in response to cold stress, which acts as upstream regulators of the CBFs/DREBs pathway (Bajwa et al. 2014). Additionally, MEL controls salinity via the SOS pathway (Liu et al. 2015), and MEL modifies the ABA transduction pathway to control the drought-stress response in plants (Sharma et al. 2020). Depending upon the nature of abiotic stress factors, MEL plays various regulating functions in different plant species (Table 9.1).

TABLE 9.1
Roles of MEL in Plant Abiotic Stress Responses

Abiotic Stress	Regulated Element	Plant Species	Reference
High temperature	H_2O_2 and Pro content, AsA levels, the activity of several antioxidant enzymes, GST	Kiwifruit (*Actinidia deliciosa*)	(Liang et al. 2018)
	ROS level, EL, membrane lipid peroxidation, MDA, Chl, total protein, antioxidant enzyme activities	Tall fescue (*Festuca arundinacea*)	(Alam et al. 2018)
	MDA and EL levels, GR, CAT, AsA, GSH, methylglyoxal detoxification, osmoregulation system	Maize (*Zea mays* L.)	(Li et al. 2019b)
	Osmoregulation system activity of APX and CAT	Tomato (*Solanum lycopersicum* L.)	(Ahammed et al. 2019)
	ROS levels, RuBisCo activity	Tomato (*Solanum lycopersicum* L.)	(Ahammed et al. 2019)
	ROS accumulation in the anthers, activity of antioxidant enzymes, heat shock protein	Tomato (*Solanum lycopersicum* L. cv. Micro-Tom)	(Qi et al. 2018)
Low temperature	CBFs/DREBs, COR15a, CAMTA1, ZAT10, ZAT12.	*Arabidopsis thaliana*	(Bajwa et al. 2014)
	Levels of AsA and GSH, SOD, APX, MDHAR, DHAR, and GR in the AsA—GSH cycle	Cucumber (*Cucumis sativus* L.)	(Bałabusta et al. 2016)
	Antioxidant enzymes especially SOD and GSSG-R, synthesis of glutathione, GSH/GSSG ratio	Cucumber (*Cucumis sativus* L.)	(Posmyk et al. 2009)
	ABA, downstream cold-responsive genes such as EnCBF9, EnCBF14, and EnCOR14	*Elymus nutans*	(Fu et al. 2017)
	Photosynthetic capacity, antioxidant potential, redox homeostasis	Tea (*Camellia sinensis* L.)	(Li et al. 2018)
	Accumulation of ROS, lipid peroxidation	Tomato (*Solanum lycopersicum* L. cv. Micro-Tom)	(Wang et al. 2020)
	Arginine pathway activity, PAs, electrolyte, MDA, NO accumulation	Tomato (*Lycopersicon esculentum*)	(Aghdam et al. 2020)
Salinity	ROS levels, PET, D1 protein.	Tomato (*Solanum lycopersicum* cv. Jin Peng Yi Hao)	(Zhou et al. 2016b)
	Stomatal closure, light energy absorption, PET in photosystem II, activities of antioxidant enzymes	Watermelon (*Citrullus lanatus*)	(Li et al. 2017b)
	Scavenging ROS, activities of antioxidant enzymes	Alfalfa (*Medicago sativa*)	(Cen et al. 2020)
	Antioxidant enzymes, ABA	(*Fragaria ananassa*)	(Zahedi et al. 2020)
	NR and NOA1- dependent NO concentration	Rapeseed (*Brassica napus* L.)	(Zhao et al. 2018)
	SOD isoforms (Cu/Zn SOD and Mn SOD), NO	Sunflower (*Helianthus annuus* L.)	(Arora & Bhatla 2017)
	GR (glutathione reductase) activity, GSH content	Sunflower (*H. annuus* L.)	(Kaur & Bhatla 2016)
	Na^+ accumulation, uptake K^+, antioxidant enzyme activity, AsA—GSH	Tomato (*S. lycopersicum* L.)	(Liu et al. 2015)

(Continued)

TABLE 9.1 *(Continued)*
Roles of MEL in Plant Abiotic Stress Responses

Abiotic Stress	Regulated Element	Plant Species	Reference
Drought	Photosynthetic rate, Chl degradation, SOD, POD, CAT	Cucumber (*C. sativus* L.)	(Zhang et al. 2013)
	Levels of H_2O_2 and $O_2^{\cdot-}$, SOD, POD, CAT and APX activities, MAPKs, TFs (WRKY1, DREB2 and MYB)	Naked oat (*Avena nuda* L.)	(Gao et al. 2018)
High light stress	Photosynthetic efficiency, activities of antioxidants enzymes, soluble proteins, Pro	Maize (*Zea mays* L.)	(Ahmad et al. 2019)
	D1 protein, photosynthetic activities, antioxidative defence system	Maize (*Zea mays* L.)	(Huang et al. 2019)
	Photosystem II efficiency, leaf area index, the activity of SOD, POD, CAT, and MDA	Soybean (*Glycine max* L.)	(Zhang et al. 2019a)
	ROS scavenging activity, photosynthetic activity, soluble sugars, Pro	Chinese hickory (*Carya cathayensis*)	(Sharma et al. 2020)
	Photosynthetic pigments phenolic, antioxidant enzyme systems, MDA content	*Moringa oleifera* L.	(Sadak et al. 2020)
	SOD, GR, CAT, APX, NR, NadDe	Alfalfa (*Medicago sativa* L.)	(Antoniou et al. 2017)
	Scavenging ROS, antioxidant enzymes	*Arabidopsis thaliana*	(Lee & Back 2018)
	Photosystem efficiency, oxidative damage	*Arabidopsis thaliana*	(Bychkov et al. 2019)
	Photosynthetic parameters, Chl fluorescence parameters, stomatal apertures, APX, CAT, POD	*Malus hupehensis*	(Wei et al. 2019)
Waterlogging stress	Photosynthesis, oxidative damage	*Malus baccata*	(Zheng et al. 2017)
	Ethylene production, PA content	Alfalfa plants (*Medicago sativa* L.)	(Zhang et al. 2019b)
Fe deficiency	Remobilizing cell wall Fe, chlorosis	*Arabidopsis thaliana*	(Zhou et al. 2016a)
N deficiency	N contents and nitrate levels, NR and GS activities	Winter wheat (*Triticum aestivum* L.)	(Qiao et al. 2019)
S deficiency	ROS content, Chl content, photosynthesis, enzymes involved in S transport and metabolism	Tomato (*Solanum lycopersicum* L. cv. Ailsa Craig)	(Hasan et al. 2018)
Salinity + high temperature stress combination	Antioxidant capacity, photosynthesis parameters, APX, GR, GPX, Ph-GPX	Tomato (*Solanum lycopersicon* cv. Boludo)	(Martinez et al. 2018)
Low temperature + drought stress combination	Several transporter proteins, Pro content, mitochondrial structure	Rice (*Oryza sativa*)	(Lee & Back. 2016a)
High light + N deficiency stress combination	Astaxanthin, cAMP signalling pathway, signalling cascade of NO-mediated MAPK	*Haematococcus pluvialis*	(Ding et al. 2018b)
Fe deficiency + salinity stress Combination	NO, H_2S	Pepper (*Capsicum annuum* L.)	(Kaya et al. 2020)

9.3 A MASTER PLANT REGULATOR: MELATONIN

The significance of MEL in plants as a growth stimulant with both *in vitro* and *in vivo* effects has been demonstrated by an excellent research study. Although MEL content was discovered to vary in different plants, it was proven that MEL is naturally produced endogenously in a variety of plant species, including medicinal herbs, crops, and fruit, as detailed by Fan and his co-workers in their study in 2018 (Fan et al. 2018). There have been numerous regulatory aspects of this molecule as well as the underlying mechanisms described, including its functions in plant growth, the development of aerial organs, root morphology, and the floral transition (Arnao & Hernández-Ruiz 2014, 2019a; Erland et al. 2015; Hernández-Ruiz et al. 2005; Hernández et al. 2015; Pelagio-Flores et al. 2012; Reiter et al. 2010; Shi et al. 2016; Wei et al. 2015).

9.3.1 Melatonin as Growth Promoter in Plants

Numerous phytohormones, particularly auxin, have a role in the growth and development of plants. MEL, a kind of indoleamine, and indole-3-acetic acid (IAA) have the same precursor molecule, tryptophan; hence, MEL ought to be involved in the control of plant growth and development (Fan et al. 2018). The biosynthesis of numerous different natural and synthetic plant growth regulators, including auxin, abscisic acid, gibberellins, cytokinins, ethylene, polyamines, jasmonic acid, and salicylic acid, can all be controlled by MEL, an independent plant growth regulator. MEL has growth-promoting effects on plants which are comparable to those of other plant growth regulators (Arnao & Hernández-Ruiz 2017, 2018). Additionally, MEL can postpone senescence while safeguarding photosynthetic systems and associated subcellular components and functions like cytokinins. Many researchers have also reported its function as a gene regulator of ethylene-related factors in fruit ripening and post-harvest processes (Abd El-Naby et al. 2019; Koyama et al. 2013). To assess the precise impact of MEL on plant growth, we must first have a thorough understanding of all the changes in gene expression that MEL modulates. In reality, there may be many interactions between MEL and many other plant growth regulators. Later, it was discovered that IAA enhanced root organogenesis and that different concentrations of endogenous MEL and repressors of the transport of serotonin and MEL suppressed cytokinin-induced shoot organogenesis (Murch et al. 2001). Hernández-Ruiz et al. (2005) revealed that MEL administration lengthened the coleoptiles of canary grass (*Phalaris canariensis*), wheat (*Triticum aestivum*), barley (*Hordeum vulgare*), and oat (*Avena sativa*), establishing the first evidence for the direct involvement of MEL in promoting plant growth. MEL used in agriculture is an uncharted territory that needs more research.

9.3.1.1 Root Organogenesis

MEL and IAA share a chemical resemblance, which has drawn attention to its potential role in root growth and development. The root-promoting impact of MEL seems to be particularly pronounced in reports of improved lateral and adventitious rooting, which are the most prevalent rooting structures other than primary roots. It also stands out that this occurs in a dose-dependent manner; low levels of MEL first stimulate plant root growth with increasing concentrations, and after a maximum response is attained, the highest levels of MEL become inhibitory. While this maximum concentration varies significantly between plant species, 10 M has an inhibitory effect in cherry rootstocks (*Prunus avium* x *Prunus cerasus*), and 500 M still has a stimulating effect in other species (Sarropoulou et al. 2012a, 2012b). In a variety of plants, including *Lupinus albus*, rice, cucumber, and Arabidopsis, exogenous melatonin treatment increased root organogenesis, such as lateral root and adventitious root formation (Arnao & Hernández-Ruiz 2007; Chen et al. 2019; Pelagio-Flores et al. 2012; Zhang et al. 2014b). MEL increased the amount of adventitious and lateral roots in Arabidopsis by up to two and three times, respectively (Pelagio-Flores et al. 2012). MEL administration induced the growth of adventitious roots in tomato (*Solanum lycopersicum*) by increasing NO production, which affected the expression of several auxin-related genes, including those involved in auxin transport, signal transduction, and auxin accumulation (Wen et al. 2016). We can use MEL to alter root architecture for effective water and nutrient uptake by better understanding the intricate connections between MEL and auxin during root growth.

9.3.1.2 Transition of Flowering

Recent research has shown a new function for MEL in the regulation of growth: the timing of flowering (Arnao & Hernández-Ruiz 2020). The first plant in which a circadian-regulating function for MEL was discovered was *Chenopodiumrubrum* (Wolf et al. 2001). MEL regulates a few crucial stages of the floral transition but does not affect the flowering stage. Shi et al. (2016) identified a mechanism for this impact, and the significance of MEL in the transition from vegetative to reproductive development in Arabidopsis was further described. In the context of the floral transition in Arabidopsis, Zhang et al. (2019c) discovered a clear connection between MEL and strigolactones, a class of phytohormones and signalling molecules produced from carotenoid pigments. Plant MEL levels seem to change according to their circadian rhythm. For instance, under a 12-h-day: 12-h-night regime, Kolář et al. (1997) observed an oscillation in endogenous melatonin levels in *Chenopodium rubrum*, with low levels during the light period and a significant increase in darkness. It is fair to believe that MEL also synchronises a variety of biological activities to diurnal and photoperiodic environmental cues, such as flowering, a short-day response if phytomelatonin serves as a night signal.

9.3.1.3 Photosynthetic Activity

Maintaining chlorophyll activity is essential for avoiding leaf senescence, which eventually has an impact on yields. Therefore, MEL function in controlling the photosynthetic process merits special consideration. MEL has been discovered to be effective in plants for increasing photosynthetic activity while maintaining redox homeostasis, in addition to aiding in the preservation of chlorophyll (Sharif et al. 2018). Through the increased levels of MEL, it safeguards a variety of agricultural or horticultural crops, including wheat, barley, delicious cherries, and many others (Sharma & Zheng 2019). For instance, in apple trees, MEL therapy increased photosystem II efficiency, enabling the leaves to maintain higher rates of CO_2 assimilation and stomatal conductance (Wang et al. 2012). Environmental stressors including heat and UV rays cause plants to produce ROS, which, in turn, causes damage to reaction centre D1 protein binding and chlorophyll (Yamamoto et al. 2008). In plants, MEL functions as an antioxidant to prevent the deterioration of chlorophyll. Under abiotic stressors, MEL delays the senescence condition of leaves in both monocots and dicots (Arnao & Hernández-Ruiz 2019a). There has been a further investigation into the biological roles of MEL in post-harvest senescence and chlorophyll preservation, both of which have significant implications for agricultural science.

9.4 ROS-MEDIATED PLANT STRESS RESPONSE AND ITS RELATION TO MEL

The production of ROS and its presence within cells are an inescapable aspect of oxygen-evolving and oxygen-dependent life forms. The majority of the plant cell's organelles, including the apoplast, chloroplast, mitochondria, peroxisomes, and cytoplasm, are constantly producing ROS. Cells have developed diverse and effective detoxification processes through a network that integrates antioxidant enzymes and non-enzymatic substances in response to the detrimental effects of ROS. Superoxide dismutase (SOD), catalase (CAT), peroxidase (POD), ascorbate peroxidase (APX), and glutathione peroxidase (GPX) are well-known antioxidant enzymes, whereas glutathione (GSH), ascorbic acid (AsA), proline (Pro), carotenoids, alpha tocopherols, etc. are representative non-enzymatic substances (Das & Roychoudhury 2014). In addition, plants often accumulate a variety of metabolites, such as sugars and amino acids, to defend themselves against abiotic stress (Abd Elhamid et al. 2014; Dreyer & Dietz 2018; Wang et al. 2020). Different phenolics (Dawood & Sadak 2014) and flavonoids (Kim et al. 2010) are additional antioxidant molecules that enhance plant tolerance. However, we will only concentrate on the increasingly well-known and significant antioxidant—that is, MEL.

In the regulation of several biological processes, including plant growth, development, and responses to biotic and/or abiotic stress, ROS play a crucial role as signalling molecules. Temporal-spatial coordination between ROS and other signals depends on the generation of stress-specific molecules, compounds, and hormones that mediate stress responses in plants (Baxter et al. 2014). Stressed plants have higher ROS levels than non-stressed plants. The ROS, especially $OH^{\cdot-}$, $O_2^{\cdot-}$, and H_2O_2, function as signalling molecules and activate several systems involved in stress tolerance. It has recently been demonstrated that MEL is one of these ROS-regulated compounds. It is demonstrated that ROS can activate the genes involved in the MEL production pathway, increasing the endogenous levels of MEL in the plant (Zhan et al. 2019). Endogenous MEL can therefore function as an efficient ROS scavenger (Kaur et al. 2015; Tan et al. 2002). Thus, MEL regulates ROS levels in two different ways: either by inactivating ROS chemically through its interaction with them (Baxter et al. 2014) or by inducing the redox enzymes that detoxify ROS, such as SOD, CAT, POD, GPX, and APX (Khan et al. 2020; Samanta et al. 2021) through MEL. A representative non-enzymatic antioxidant substance, such as GSH and AsA, phenolic compounds (Dawood & Sadak 2014), flavonoids via the NO-dependent pathway (Kim et al. 2010), and carotenoids (Lee & Back 2018; Sharma et al. 2020), are also induced by MEL. Similar to how endogenous MEL was shown to interact with its plant receptor (CAND2/PMTR1), which seems to be generated by MEL, responses to stressors were activated (Zhan et al. 2019). In short, MEL converts its actions into a feedback system with many regulated components of the redox network, including ROS and RNS (Arnao & Hernández-Ruiz 2019a). The underlying mechanisms governing MEL production in plants and role of MEL in improving plants responses under stressful conditions have been presented in Figure 9.1.

9.5 RELATIONSHIP BETWEEN THE NO-PLANT MEDIATED STRESS RESPONSE AND MEL

Another important signalling molecule in plant physiology is nitric oxide (NO). The method of producing NO that has been most well described and demonstrated involves plants, which reduce nitrite to NO through a variety of non-enzymatic and/or enzymatic mechanisms (Astier et al. 2018). One of the proteins that have been described as being able to catalyse the synthesis of NO from nitrites in the enzymatic NO-generating pathway is nitrate reductase (NR). The first limiting stage of nitrate assimilation is the reduction of nitrate to nitrite by the multifunctional cytoplasmic enzyme NR using NADH as an electron donor (Campbell 2001). This enzyme also contains nitrite: NO reductase activity (Ni-NR activity), from which NO is generated; however, under normal circumstances, the second reaction only accounts for 1% of the capacity of NR to reduce nitrates. However, certain reactions occur in anoxic environments (Rockel et al. 2002; Yamasaki & Sakihama 2000; Yamasaki et al. 1999). Additionally, it has been demonstrated in *Arabidopsis thaliana* that NR can interact with the NOFNiR (nitric oxide–forming nitrite reductase) partner protein to create NO from nitrite. The NO-producing activity, in contrast to the Ni-NR action, takes place in normoxia and is not hampered by nitrate (Gupta & Igamberdiev 2011; Planchet et al. 2005). Additionally, mitochondrial electron transport chain (mETC), which functions under anaerobic circumstances and is dependent on the availability of nitrite, can convert nitrite into NO. In addition to these reductive pathways, other pieces of evidence show that plants also have an oxidative process for producing NO that is arginine-dependent (Astier et al. 2018). At this time, it is clear that the transcriptomes of embryophytes lack conventional nitric oxide synthases (NOSs). The existence of NOS-like activity

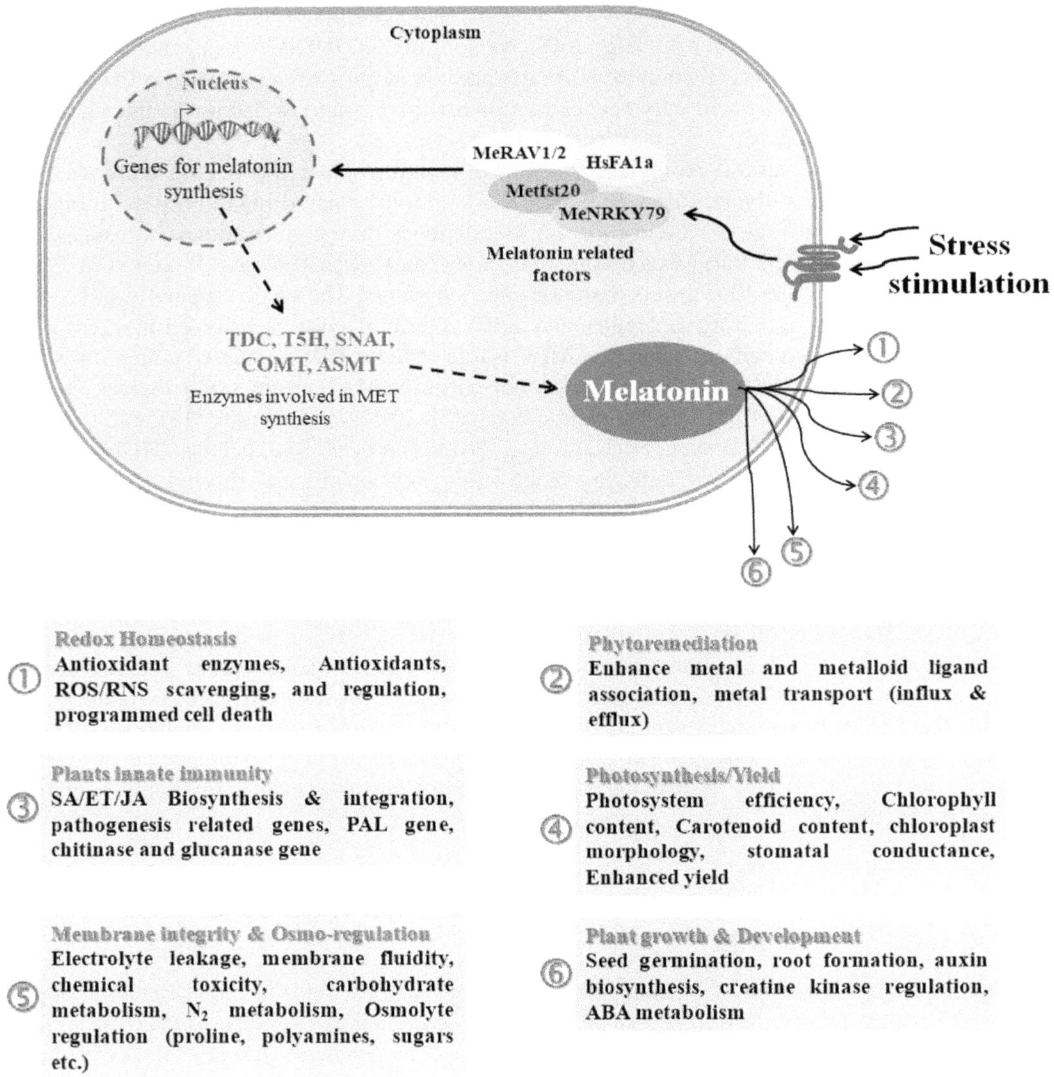

FIGURE 9.1 Mechanisms governing melatonin production in plants and various actions of melatonin in supporting plants under stressful conditions

is, nevertheless, supported by many pieces of data that have been described in numerous studies. Arginine, or at least the arginine metabolic pathways, is necessary for this action. The existence of NOS-like activity is, nevertheless, supported by some pieces of data that have been described in numerous studies. Arginine, or at least the arginine metabolic pathways, is necessary for this action. For a better understanding of NO production in plants, the proteins, substrates, and cofactors involved must be identified and described (Astier et al. 2018; Corpas & Barroso 2017).

Together with NO, MEL promotes plant growth and mediates the ability of plants to withstand abiotic stress by strengthening their antioxidant defence system. It has been proposed that the MEL-induced adventitious root formation in tomato (*Solanum lycopersicum* L.) plants occurred under normal growth conditions as a result of increased NO levels brought about by the up-regulation of NR, the down-regulation of S-nitrosoglutathione reductase (GSNOR), and auxin signalling (Wen et al. 2016). Under external stimuli and developmental cues, NO, similar to MEL, can directly scavenge free radicals and lessen oxidative damage in a receptor-independent manner (Aydogan et al. 2006). Due to the up-regulation of associated genes, MEL can up-regulate NO synthesis, scavenge excess NO, and promote the accumulation of NO by enhancing the activity of NOS-like protein (arginine metabolic pathway) (Aghdam et al. 2019; Liu et al. 2019). When oxygen is present, NO nitrosation can quickly convert MEL to N-Nitrosomelatonin (NOMET) at various pH circumstances (Blanchard et al. 2000). However, NOMET is an effective NO donor in cell cultures when serotonin and its derivatives are present (Berchner-Pfannschmidt et al. 2008; Kopczak et al. 2007). On the other hand, NO increases the expression of TDC, T5H, SNAT, and COMT (genes of the enzymes from the MEL biosynthesis pathway) to raise the levels of MEL through a cGMP-dependent pathway (He & He 2020) Given that they interact separately and via several signalling pathways, the interplay between NO and MEL exhibits some complexity (Zhu et al. 2019).

9.6 PLANT MELATONIN ACTS AS AN ANTIOXIDANT

The role of MEL as an antioxidant is one of the most intriguing concepts regarding the function of MEL in plants. Due to its high antioxidant properties, MEL may alter ROS signalling cascades, interrupting long-distance communications and having a significant impact on cellular contexts and responses. Oxygen species, both radical and non-radical, make up ROS. They could be more complex molecules like hydrogen peroxide (H_2O_2) and hydroxyl radicals ($OH^{\cdot-}$), or simpler molecules like superoxide molecules (Ray et al. 2012). Under typical photosynthetic and respirational conditions, chloroplasts and mitochondria generate the majority of reactive oxygen and reactive nitrogen species (ROS and RNS), respectively. ROS production in chloroplasts is dependent on how light and chlorophyll (Chl) interact (Hasanuzzaman et al. 2020). Another important cellular location for NO synthesis that controls ROS metabolism is the peroxisome. MEL and its metabolites interact directly with ROS and RNS to conduct antioxidant action. One molecule of MEL can scavenge up to ten molecules of ROS/RNS through this non-enzymatic mechanism according to its high antioxidant activity (Back 2021). Additionally, MEL indirectly affects the levels of certain antioxidants like ascorbic acid and glutathione as well as the activity of key antioxidant enzymes like glutathione peroxidase and SOD (Arnao & Hernández-Ruiz 2021).

In plants, oxidant stress, which manifests as a disruption of chemical equilibrium between oxidant and antioxidant chemical species in the cell, alters redox homeostasis, and speeds up ROS build-up, is usually the source of damage brought on by environmental stress (Nguyen et al. 2018b). Under stressful circumstances, ROS can lead to oxidative alteration of proteins in chloroplasts. Furthermore, under environmental stress, many hormones (including JA, SA, and ABA) interact in the crosstalk with ROS to activate ROS detoxification (Lemos et al. 2016; Lv et al. 2019; Postiglione & Muday 2020; Wang et al. 2021a). Trehalose and anthocyanins are reported to control ROS metabolism in addition to hormones (Naing & Kim 2021; Sarkar & Sadhukhan 2022; Yang et al. 2022).

According to Pardo-Hernández et al. (2020), MEL is one of the chemicals that regulates ROS levels. It is well known for its antioxidant properties and for removing ROS from the environment. Free radical species are neutralised by MEL by single-electron and hydrogen transfer (Galano & Reiter 2018). The respiratory burst NADPH oxidase (RBOH) in Arabidopsis was necessary for MEL-induced ROS generation in salt tolerance (Chen et al. 2017), stomatal closure regulation (Li et al. 2020), and lateral root growth (Chen et al. 2019). MEL promotes H_2O_2 build-up and the Ca^{2+} wave to activate the C-Repeat Binding Factor (CBF) pathway in watermelon (*Citrullus lanatus*), which confers plant cold resistance (Chang et al. 2021a). Additionally, MEL application increased the accumulation of soluble sugars, free proline, and other compatible protectants for the maintenance of osmotic pressure and cell integrity under certain abiotic stress conditions, such as low temperature, high salinity, and excessive levels of heavy metals (Chang et al. 2021b; Talaat 2021; Turk et al. 2014). All of these results support the notion that MEL supplementation improved the antioxidant defence system to lessen the effects of various biotic and abiotic stresses (Samanta et al. 2022).

9.7 ROLES OF MELATONIN IN PLANT STRESS RESPONSES

The remarkable antioxidant properties of MEL are complemented by its exceptional ability to scavenge free radicals against harmful and damaging reactive species, such as nitric oxide, hydrogen peroxide, singlet oxygen, hypochlorous acid, and many more. MEL also serves as a protective mechanism and part of the defence against a variety of biotic and abiotic stimuli.

9.7.1 Melatonin and Biotic Stress

Biologic stress is frequently defined in a horticulture context as the loss of fruits and vegetables owing to infectious diseases caused by fungi, bacteria, viruses, nematodes, and/or insect damage. These biotic stress agents are directly linked to plants, from which they obtain nutrients as well as the ability to die and decompose. Pathogenic stress in horticulture is a significant factor in pre- and post-harvest crop losses. Since MEL is a promising eco-friendly chemical, it might become a potential substitute method for initiating plant defence against biotic stress. According to growing data over the past ten years, MEL protects plants from biotic stress, which results in significant yield and minimise financial losses in crops all over the world (Arnao & Hernández-Ruiz 2014, 2015; Moustafa-Farag et al. 2019; Weeda et al. 2014). MEL is highly effective at both directly suppressing the pathogen and eliciting a defence response in host plants, according to numerous recent studies (Moustafa-Farag et al. 2019). Up-regulation of defence genes, thickening of the cell wall, ROS scavenging, and an increase in NO generation are the main mechanisms that enable MEL-induced biotic stress in plants (Sharif et al. 2018; Shi et al. 2015a; Wei et al. 2018; Zhao et al. 2019). In MEL-treated plants, cell wall thickening results from the accumulation of callose, cellulose, galactose, and xylose, which induces the biochemical basis of disease resistance (Qian et al. 2015; Zhao et al. 2015). Additionally, MEL therapy increased resistance of cotton to *Verticillium dahlia* by increasing metabolic flux into the phenylpropanoid, mevalonate, and gossypol pathways, which resulted in the build-up of lignin and gossypol (Li et al. 2019a). Genes involved in SA, ET, and JA signal transduction were highly up-regulated by MEL, indicating a close relationship between MEL and these phytohormones, according to a thorough analysis of genome-wide gene expression in MEL-treated *Arabidopsis* (Lee et al. 2014). In addition to playing a crucial role in MEL-mediated immunity in plants, MAPKs are significant molecular sensors of stress signals in plants (Lee & Back 2016b; Pitzschke et al. 2009). MEL-triggered innate immunity signal transduction is mediated by MAPK cascades and oxidative signal inducible 1 (OXI1) kinases (Arnao & Hernández-Ruiz 2018; Lee & Back 2017). The interactions of MEL with other components should be better understood by further research using genomic and proteomic techniques.

9.7.2 Melatonin against Abiotic Stress

Melatonin plays a crucial part in how plants react to abiotic stress factors. Recent studies on melatonin-mediated reactions to several significant abiotic stress elements will be covered in this part.

9.7.2.1 Drought Stress

Crops used for both agriculture and horticulture are extremely susceptible to drought stress. The physiochemical and molecular levels of these plants experience numerous negative alterations when they are subjected to dry conditions, which impede their growth and yield (Hussain et al. 2018). Drought causes oxidative stress and stimulates stress signalling pathways that are mediated by ROS, Ca^{2+}, and hormones. MEL defends crops by preventing damage to the photosynthetic apparatus, triggering the antioxidative defences, controlling oxidative stress, and activating other defence mechanisms (Sharma & Zheng 2019). Melatonin can improve stomatal performance during drought stress. MEL induction through the roots improves several parameters, including electrolyte leakage, chlorophyll stability and relative water content, photosynthetic efficiency, and stomatal conductance

(Kabiri et al. 2018). Additionally, plants under drought stress which receive MEL treatment show half as much ABA as plants which do not receive MEL treatment. As a result of the decreased H_2O_2 and ABA content, stomata perform better and can handle these stresses (Ding et al. 2018a). Endogenous MEL levels change as a result of the activation of MEL biosynthesis genes (including TDC, ASMT, COMT, and SNAT) during periods of dehydration. There have been reports of drought-induced MEL production in Arabidopsis, barley, bermudagrass, apples, grapevines, and rice (Moustafa-Farag et al. 2020). Additionally, MEL administration during drought can enhance fruit quality, hypocotyl elongation, lateral root formation, seed germination, and vegetative growth (Hosseini et al. 2021; Sun et al. 2021).

9.7.2.2 Salt Stress

Salt stress has emerged as a serious global issue, restricting agricultural output and causing significant economic losses globally (Su et al. 2021). Osmotic stress and ion toxicity are two negative impacts that salt exerts on plants (Van Zelm et al. 2020). Plants also employ a variety of different strategies to deal with these stressors, including ion compartmentalization, selective ion exclusion, altered photosynthetic pathways, compatible solute production, induction of antioxidant enzymes, altered membrane structure, up-regulation of gene expression, and stimulation of phytohormones (Parida & Das 2005). Recently, auxin-independent MEL effects have been seen in a variety of plant growth regulators. The salinity stress-induced restriction of plant development was reported to reduce abruptly by the addition of 0.1 M MEL, and it was found to be important for preserving plant photosynthetic capacity (Arnao & Hernández-Ruiz 2014). To control redox equilibrium, MEL occasionally interacts with other signalling molecules like NO. For instance, in the cotyledons of sunflower seedlings under salt stress, Kaur & Bhatla (2016) discovered an intriguing interplay between MEL and NO in modifying GSH levels and glutathione reductase activity. Salt stress has been shown to raise endogenous MEL levels and alter the expression patterns of crucial MEL biosynthesis enzymes (Arnao & Hernández-Ruiz 2009). For instance, compared to wild-type lines, VvSNAT1 overexpression Arabidopsis lines display greener leaves, improved growth, and greater germination rates (Wu et al. 2021b). Exogenous MEL therapy controls antioxidant enzyme expression and activity, polyamine metabolism, and NO signalling to protect plants against salt stress (Zhan et al. 2019). Additionally, MEL significantly maintains the ionic balances to mitigate the effects of salt stress. For instance, MEL treatment induces salinity tolerance in maize seedlings by increasing KCl accumulation, decreasing NaCl accretion, and maintaining a greater KCl/NaCl ratio (Jiang et al. 2016). *Dracocephalum kotschyi* develops salt stress tolerance as a result of the interaction between Ca and MEL (Vafadar et al. 2020). In conclusion, MEL increases photosynthetic efficiency, KCl influx, and NaCl efflux, lowers ROS formation, boosts antioxidant activities, and accumulates suitable solutes to improve plant development under salt stress.

9.7.2.3 Heat Stress

Heat stress (HS) is regarded as the most significant threat to food security in this century since it severely inhibits plant growth and drastically lowers crop yield (Hassan et al. 2021). MEL plays an important function in boosting the enzymes of nitrogen metabolism, increasing nitrogen content, and lowering ammonia levels at high temperatures during heat stress. In one instance, the production of MEL was reported to increase in rice under heat stress circumstances, demonstrating MEL function in heat stress tolerance (Byeon & Back 2014). Increased MEL levels in green microalgae *Ulva* sp. at high temperature confirm the role of hormone in promoting heat stress tolerance (Tal et al. 2011). In tomato plants, MEL therapy improved not only cell membrane stability but also metabolic gene expression, plant growth, and the reduction of heat stress–related harm to the antioxidant defence system (Jahan et al. 2019). MEL-mediated heat stress responses are primarily mediated by heat shock factors. Shi et al. (2015c) provided evidence that MEL-induced thermotolerance in Arabidopsis is mediated by heat shock factors (HSFA2 and HSA32) and heat shock proteins (HSP90 and HSP101). It was reported that MEL treatment increased photosynthesis in kiwifruit by successfully modulating carbon fixation under heat stress (Liang et al. 2019). Similarly, wheat MEL supplementation reduced the harm caused by HS by triggering the antioxidant system (Buttar et al. 2020). MEL considerably slowed down the HS-induced senescence of leaves as evidenced by decreased leaf yellowing and a higher Fv/Fm ratio, which decreased the formation of ROS (Jahan et al. 2021). Accordingly, it is inferred that MEL minimises the impact of HS via enhancing photosynthetic efficiency, safeguarding the photosynthetic apparatus, lowering ROS and ABA accumulation, and raising antioxidant activities and HSP expression.

9.7.2.4 Cold Stress

Low-temperature stress, which causes chilling injury, is a significant abiotic stress, particularly for tropical or warm-climate plants. The physiological, molecular, and metabolic activity of plants is significantly altered by cold stress, which also affects antioxidant activity and membrane permeability (Bajwa et al. 2014; Hu et al. 2016). According to Mishra et al. (2011), crop development and production are severely restricted by cold stress, which also has disastrous effects on plants. The majority of the detrimental effects are similar to signs of drought stress, including cell membrane damage, disruption of the electron transport chain, excessive ROS production, unbalanced ion homeostasis, etc. (Hussain et al. 2018). Recent research has demonstrated the critical function of melatonin in preventing cold damage to plants by increasing the activity of specific transcription factors such as the C_2H_2 zinc-finger transcription factors ZAT10 and ZAT12 and C-Repeat Binding Factors (Bajwa et al. 2014). It was also reported that plants primed with MEL displayed a high level of tolerance to cold stress (Ding et al. 2017a). MEL has been shown to serve as a long-distance signal that can travel from roots to shoots (Tan et al. 2007). MEL increases cold tolerance in grafted watermelons by enhancing root-to-shoot communication and movement (Li et al. 2021). By enhancing chilled storage tolerance, exogenous MEL therapy helps preserve the quality of fruits, vegetables, and cut flowers. For instance, pre-treating loquat fruit with MEL before storage causes phenolic compounds to accumulate and lignin levels to drop, reducing flavour and nutritional loss brought on by chilling injury during cold storage (Wang et al. 2021b). The amounts of secondary metabolites such as amino acids, organic acids, carbohydrates, and sugars can be increased in Bermuda grass after MEL administration and thereby increases cold tolerance (Hu et al. 2016). MEL supplementation in hulless barley seedlings similarly affects the rhythmic expression patterns of circadian clock genes, controlling their growth and improving cold tolerance (Chang et al. 2021b). Together, these exogenous treatments and endogenous MEL-level augmentations in a variety of plants, including tomato, pea, and rice seedlings, have been reported. These treatments have had a significant impact on the growth, morphological changes, photosynthetic carbon fixation, photosystem II ability, and overall antioxidant enzyme activity of the plants.

9.7.2.5 Heavy Metal Stress

Heavy metals and compounds found in the roots and shoots of plants cause a variety of morphological alterations and growth inhibitions. The production of food worldwide is severely threatened by heavy metals (HMs). Due to human activity, their concentration in agricultural soil is quickly rising (Hassan et al. 2019; Khan et al. 2021; Rehman et al. 2022; Umer Chattha et al. 2021). As a free radical scavenger, MEL can combat dangerous pollutants (Nawaz et al. 2016). Due to overall bio-stimulatory effect of MEL on plants and its propensity to mobilise harmful metals through phytochelatins, transport, and sequestration, plants are highly tolerant to heavy metals. In one experiment, 0.1 M MEL concentrations improved the amount of chlorophyll, plant growth, and photosynthetic assimilation in watermelon seedlings while simultaneously controlling the concentration of vanadium ions in the roots and shoots (Nawaz et al. 2018). Inducing MEL in maize plants reduced the amount of lead contamination that resulted from its accumulation under stressful circumstances (Okant & Kaya 2019). MEL applied to tobacco (*Nicotiana tabacum* L.) leaves increases its resistance to cadmium (Wang et al. 2019). When wheat seedlings were exposed to aluminum stress, exogenous MEL administration boosted antioxidant levels, activated antioxidant enzymes, and reduced ROS generation (Sun et al. 2020). Through NO signalling, MEL also assists plants under heavy metal stress in regaining redox equilibrium. When applied to stressed plants, MEL also prevents membrane leakage, morphological distortion, and cell death caused by lead (Kobylińska et al. 2017; Li et al. 2016). Strawberry seedlings cultivated in Cd-contaminated soil displayed significantly lower growth rates, biomass output, chlorophyll levels, and antioxidant enzyme activities. However, MEL administration (200 mmol) has demonstrated a significant increase in growth biomass output due to improved APX, CAT, POD, and SOD activities as well as decreased MDA formation (Wu et al. 2021a). Together, these results indicate that MEL is involved in the adaptation of plants to heavy metal stress and functions in several ways to facilitate this process.

9.8 EXOGENOUS APPLICATION OF MEL

A new era in plant adaptation processes has been ushered in by MEL. As a result, research into the effects of exogenous MEL on stressed plants has become increasingly important to scientists interested in plant survival and adaptation. Exogenous MEL therapy can enhance a variety of abiotic stress tolerance in plants, including salt, drought, water logging, heat, cold, and heavy metal toxicity. The increase in endogenous MEL production following exogenous MEL therapy is largely responsible for these improvements in stress tolerance. Additionally, these enhancements are commonly dose-dependent. MEL can increase the activity of transcription factors linked to stress, including WRKY, zinc finger transcription factors, and NAC domain-containing proteins (Zhang et al. 2014b). Similar to this, it can modify the expression of genes involved in stress defence in plants and control the calcium signal associated with stress and the expression of stress receptors (Weeda et al. 2014). Exogenous application of MEL can inhibit the expression and up-regulation of key genes for the Chl degradation, viz., pheophorbide-a-oxygenase (PAO) and senescence-associated gene 12 (SAG12), thus preventing the breakdown of chlorophyll. By decreasing ROS and accelerating photosynthetic rates, MEL can preserve the chlorophyll content during senescence (Wang et al. 2013).

Exogenous MEL exposure during salinity stress might increase the expression of ion channel genes (such as NHX1 and AKT1) in leaves, which can maintain ion homeostasis and increase salinity resistance in plants (Li et al. 2012). Exogenous MEL treatments increased the growth, photosynthetic capability, antioxidant activity, and chlorophyll content of cucumbers growing under salinity stress conditions, although they decreased the ROS level and oxidative damage (Wang et al. 2016). Zhou et al. (2016b) have studied the effects of MEL treatments on tomato photosynthetic activity under salt conditions. From another study, it has also been reported that MEL treatments increase K^+ and decrease NaCl concentration, and it finally enhances K^+/Na^+ balance and tolerance of potato to salt stress (Yu et al. 2018).

Exogenous MEL treatments also enhance the plant tolerance to water deficiency stress, according to earlier investigations. Cui et al. (2017) showed that MEL treatments reduced ROS and membrane damage while enhancing antioxidant activity to mitigate the harmful effects of drought stress on wheat. MEL treatments have also been reported to improve the Fv/Fm ratio, which may have a protective effect on plants under water shortage situations and thereby help the plants to recover from drought stress. According to Fleta-Soriano et al. (2017), MEL treatments cause improve photosystem II, which, in turn, results in a conserving factor in maize under drought stress. Exogenous MEL treatments have boosted indole acetic acid (IAA) and zeatin production while decreasing H_2O_2 and aminocyclopropane-1-carboxylic acid (ACC) production, which has improved photosynthetic capacity and water usage efficiency (Li et al. 2017a).

Exogenous MEL therapies have been demonstrated in numerous trials to minimise the toxic effects of various heavy metals, including cadmium, aluminum, copper, vanadium, nickel, etc. According to Tang et al. (2015), foliar MEL treatments increased the ability of eggplant to photosynthesise under cadmium stress. They hypothesised that stressed plants with higher MEL concentrations had higher photosynthetic capacities and that the optimal concentration for reducing cadmium stress was 150 μmol L^{-1}. Exogenous MEL therapies for *Glycine max* were investigated by Zhang et al. (2017) to see if they could reduce aluminum-induced phytotoxicity. They made note of the dose-dependent nature of MEL impact on Al stress. MEL dosages of 0.1 and 1 mM enhanced root development and decreased H_2O_2 concentration, but 100 and 200 mM doses had the opposite effect (Tang et al. 2015). Extreme temperature levels were demonstrated to promote MEL biosynthesis. Additionally, exogenous MEL treatments have reportedly aided in shielding plants from temperature swings (Tan et al. 2000). According to some research, plants exposed to abiotic stressors including cold and heat could be strengthened by the antioxidant capacity of MEL (Arnao & Hernández-Ruiz 2018; Janas & Posmyk 2013). Through the production of polyamines, exogenous MEL therapy can prevent cold-induced apoptosis. Polyamines play a role in the pathways that control programmed cell death, either directly or indirectly (Moschou & Roubelakis-Angelakis 2014). According to Lei et al. (2004), MEL improved carrot cell viability because it promoted polyamines like putrescine and spermidine production in response to cold stress. Similar findings were made by Bałabusta et al. (2016), who discovered that MEL-Osmo-primed cucumber seeds detoxified ROS during chilling stress by having decreased ROS levels and greater superoxide dismutase (SOD) activity. It has been shown that exogenous MEL treatments improve non-photochemical quenching by increasing violaxanthin de-epoxidase activity in tomato plants under chilling stress, which reduces photoinhibition (Ding et al. 2017b). In a different investigation, MEL (50 and 500 μM) was used to prime maize seeds to ascertain the priming-induced alterations under chilling stress. Plant tolerance to chilling was increased by priming with MEL via controlling MEL-associated proteins in seeds exposed to lower temperatures (Kołodziejczyk et al. 2016). To enhance germination under cold stress, Posmyk et al. (2009) studied the effects of osmo- and hydro-priming with MEL application on cucumber (*Cucumis sativus*) seeds.

According to Xiang-dong et al. (2010), external MEL applications significantly increased the levels of nonenzymatic antioxidants like ascorbic acid and vitamin E as well as enzymatic antioxidants like SOD, POX, CAT, and APX in cucumber under high-temperature stress. It has been reported that at 15°C, seed germination increases to 50–60%, and the percentage has been further boosted by the addition of 25–100 μM MEL. Following these findings, it has been revealed that low amounts of MEL produce oxidative alterations in proteins, whereas high levels of MEL protect cell membranes from peroxidation in cucumber seeds under chilling stress.

In plants that had been pre-treated with melatonin, various genes associated with various primary and secondary plant physiological metabolisms (nitrogen metabolism, carbohydrate metabolism, tri-carboxylic acid transformation, transport, hormone metabolism, metal handling, and redox) were characterised by gene ontology enrichment studies. MEL is essential for the control of multiple distinct stress-related genes. For instance, MEL can conserve the amount of chlorophyll, and treating Arabidopsis with MEL greatly reduces the activity of the light-regulating enzyme chlorophyllase (CLH1), which contributes to chlorophyll breakdown (Weeda et al. 2014). To create an effective system that shields plants from harsh environments, MEL works in cooperation with antioxidant enzymes to enhance photosynthesis, postpone leaf senescence, moderate changes in leaf ultra-structure, delay the manufacture of metabolites, and modify stress genes. All of these findings imply that MEL is essential for surviving adverse environmental conditions.

9.9 FUTURE PERSPECTIVES AND CONCLUSION

MEL is crucial for plant functioning, and because MEL has so many beneficial physiological effects on plants, research on plants is becoming more and more popular. MEL modifies various biochemical, molecular, and physiological processes to increase plant stress tolerance. MEL improves photosynthesis efficiency and shields the photosynthetic machinery from oxidative damage caused by environmental stresses. Through the activation of antioxidant enzymes, the accumulation of suitable solutes, and the increased expression of stress-responsive genes, the application of MEL under diverse conditions decreases the formation of ROS. Plant MEL accumulation is influenced by both genetic and environmental factors. Arabidopsis and rice are two model plants that are mostly used in current research on the molecular regulation of MEL. However, the synthesis and metabolic pathways of plant MEL are yet unknown, and its complicated metabolic activities may entail the oxidation of free radicals, enzymes, and pseudoenzymes. It is necessary to research the precise function of MEL in signalling pathways under various stressors. Another essential component of MEL that considerably raises stress tolerance is the regulation of gene expression and interactions with various hormones. However, endogenous MEL is insufficient to handle demanding situations. Thus, more extensive research in diverse crops under signal and in conjunction with various conditions is needed to understand the cellular signalling pathways caused by MEL. Further research should be done on the molecular mechanism by which MEL increases the expression of genes that respond to stress and antioxidants. It would be possible to determine how MEL is affected by UV radiation, ozone, and pH stress through more thorough transcriptome and proteomic research. Finally, it is important to look into the patterns of MEL application in plant responses to separate and combined challenges in the field. Finally, it may be concluded that there is a lot of research opportunity for improving our knowledge on how MEL affects fundamental life processes in all plant kingdoms and for developing novel strategies to advance plant cultivation and agriculture.

REFERENCES

Abd Elhamid, E. M., Sadak, M. S., & Tawfik, M. M. (2014). Alleviation of adverse effects of salt stress in wheat cultivars by foliar treatment with antioxidant 2: Changes in some biochemical aspects, lipid peroxidation, antioxidant enzymes and amino acid contents. *Agricultural Sciences*, *5*(13), 1269.

Abd El-Naby, S. K. M., Mohamed, A. A. A., & El-Naggar, Y. I. M. (2019). Effect of melatonin, GA3 and NAA on vegetative growth, yield and quality of 'canino' apricot fruits. *Acta Scientiarum Polonorum Hortorum Cultus*, *18*, 167–174.

Aghdam, M. S., Luo, Z., Jannatizadeh, A., Sheikh-Assadi, M., Sharafi, Y., Farmani, B., . . . & Razavi, F. (2019). Employing exogenous melatonin applying confers chilling tolerance in tomato fruits by upregulating ZAT2/6/12 giving rise to promoting endogenous polyamines, proline, and nitric oxide accumulation by triggering arginine pathway activity. *Food Chemistry*, *275*, 549–556.

Aghdam, M. S., Luo, Z., Li, L., Jannatizadeh, A., Fard, J. R., & Pirzad, F. (2020). Melatonin treatment maintains nutraceutical properties of pomegranate fruits during cold storage. *Food Chemistry*, *303*, 125385.

Ahammed, G. J., Xu, W., Liu, A., & Chen, S. (2019). Endogenous melatonin deficiency aggravates high temperature-induced oxidative stress in *Solanum lycopersicum* L. *Environmental and Experimental Botany*, *161*, 303–311.

Ahmad, S., Kamran, M., Ding, R., Meng, X., Wang, H., Ahmad, I., . . . & Han, Q. (2019). Exogenous melatonin confers drought stress by promoting plant growth, photosynthetic capacity and antioxidant defense system of maize seedlings. *PeerJ*, *7*, e7793.

Alam, M. N., Zhang, L., Yang, L., Islam, M., Liu, Y., Luo, H., . . . & Chan, Z. (2018). Transcriptomic profiling of tall fescue in response to heat stress and improved thermotolerance by melatonin and 24-epibrassinolide. *BMC Genomics*, *19*(1), 1–14.

Antoniou, C., Chatzimichail, G., Xenofontos, R., Pavlou, J. J., Panagiotou, E., Christou, A., & Fotopoulos, V. (2017). Melatonin systemically ameliorates drought stress-induced damage in *Medicago sativa* plants by modulating nitro-oxidative homeostasis and proline metabolism. *Journal of Pineal Research*, *62*(4), e12401.

Arnao, M. B., & Hernández-Ruiz, J. (2007). Melatonin promotes adventitious-and lateral root regeneration in etiolated hypocotyls of *Lupinus albus* L. *Journal of Pineal Research*, *42*(2), 147–152.

Arnao, M. B., & Hernández-Ruiz, J. (2009). Chemical stress by different agents affects the melatonin content of barley roots. *Journal of Pineal Research*, *46*(3), 295–299.

Arnao, M. B., & Hernández-Ruiz, J. (2014). Melatonin: Plant growth regulator and/or biostimulator during stress? *Trends in Plant Science*, *19*(12), 789–797.

Arnao, M. B., & Hernández-Ruiz, J. (2015). Functions of melatonin in plants: A review. *Journal of Pineal Research*, *59*(2), 133–150.

Arnao, M. B., & Hernández-Ruiz, J. (2017). Growth activity, rooting capacity, and tropism: Three auxinic precepts fulfilled by melatonin. *Acta Physiologiae Plantarum*, *39*(6), 1–9.
Arnao, M. B., & Hernández-Ruiz, J. (2018). Melatonin and its relationship to plant hormones. *Annals of Botany*, *121*(2), 195–207.
Arnao, M. B., & Hernández-Ruiz, J. (2019a). Melatonin: A new plant hormone and/or a plant master regulator? *Trends in Plant Science*, *24*(1), 38–48.
Arnao, M. B., & Hernández-Ruiz, J. (2019b). Role of melatonin to enhance phytoremediation capacity. *Applied Sciences*, *9*(24), 5293.
Arnao, M. B., & Hernández-Ruiz, J. (2020). Melatonin in flowering, fruit set and fruit ripening. *Plant Reproduction*, *33*(2), 77–87.
Arnao, M. B., & Hernández-Ruiz, J. (2021). Melatonin as a plant biostimulant in crops and during post-harvest: A new approach is needed. *Journal of the Science of Food and Agriculture*, *101*(13), 5297–5304.
Arora, D., & Bhatla, S. C. (2017). Melatonin and nitric oxide regulate sunflower seedling growth under salt stress accompanying differential expression of Cu/Zn SOD and Mn SOD. *Free Radical Biology and Medicine*, *106*, 315–328.
Astier, J., Gross, I., & Durner, J. (2018). Nitric oxide production in plants: An update. *Journal of Experimental Botany*, *69*(14), 3401–3411.
Aydogan, S., Yerer, M. B., & Goktas, A. (2006). Melatonin and nitric oxide. *Journal of Endocrinological Investigation*, *29*(3), 281–287.
Back, K. (2021). Melatonin metabolism, signaling and possible roles in plants. *The Plant Journal*, *105*(2), 376–391.
Back, K., Tan, D. X., & Reiter, R. J. (2016). Melatonin biosynthesis in plants: Multiple pathways catalyze tryptophan to melatonin in the cytoplasm or chloroplasts. *Journal of Pineal Research*, *61*(4), 426–437.
Bailey-Serres, J., Parker, J. E., Ainsworth, E. A., Oldroyd, G. E., & Schroeder, J. I. (2019). Genetic strategies for improving crop yields. *Nature*, *575*(7781), 109–118.
Bajwa, V. S., Shukla, M. R., Sherif, S. M., Murch, S. J., & Saxena, P. K. (2014). Role of melatonin in alleviating cold stress in *Arabidopsis thaliana*. *Journal of Pineal Research*, *56*(3), 238–245.
Bałabusta, M., Szafrańska, K., & Posmyk, M. M. (2016). Exogenous melatonin improves antioxidant defense in cucumber seeds (*Cucumis sativus* L.) germinated under chilling stress. *Frontiers in Plant Science*, *7*, 575.
Banerjee, A., & Roychoudhury, A. (2017). Melatonin as a regulator of abiotic stress tolerance in plants. In V. P. Singh, S. Singh, & S. M. Prasad (Eds.), *Mechanisms behind phytohormonal signaling and crop abiotic stress tolerance* (pp. 47–60). New York: Nova Science Publishers.
Banerjee, A., & Roychoudhury, A. (2019). Melatonin application reduces fluoride uptake and toxicity in rice seedlings by altering abscisic acid, gibberellin, auxin and antioxidant homeostasis. *Plant Physiology and Biochemistry*, *145*, 164–173.
Baxter, A., Mittler, R., & Suzuki, N. (2014). ROS as key players in plant stress signalling. *Journal of Experimental Botany*, *65*(5), 1229–1240.
Berchner-Pfannschmidt, U., Tug, S., Trinidad, B., Becker, M., Oehme, F., Flamme, I., . . . & Kirsch, M. (2008). The impact of N-nitrosomelatonin as nitric oxide donor in cell culture experiments. *Journal of Pineal Research*, *45*(4), 489–496.
Blanchard, B., Pompon, D., & Ducrocq, C. (2000). Nitrosation of melatonin by nitric oxide and peroxynitrite. *Journal of Pineal Research*, *29*(3), 184–192.
Blask, D. E., Dauchy, R. T., Sauer, L. A., & Krause, J. A. (2004). Melatonin uptake and growth prevention in rat hepatoma 7288CTC in response to dietary melatonin: Melatonin receptor-mediated inhibition of tumor linoleic acid metabolism to the growth signaling molecule 13-hydroxyoctadecadienoic acid and the potential role of phytomelatonin. *Carcinogenesis*, *25*(6), 951–960.
Buttar, Z. A., Wu, S. N., Arnao, M. B., Wang, C., Ullah, I., & Wang, C. (2020). Melatonin suppressed the heat stress-induced damage in wheat seedlings by modulating the antioxidant machinery. *Plants*, *9*(7), 809.
Bychkov, I., Kudryakova, N., Andreeva, A., Pojidaeva, E., & Kusnetsov, V. (2019). Melatonin modifies the expression of the genes for nuclear-and plastid-encoded chloroplast proteins in detached Arabidopsis leaves exposed to photooxidative stress. *Plant Physiology and Biochemistry*, *144*, 404–412.
Byeon, Y., & Back, K. (2014). Melatonin synthesis in rice seedlings in vivo is enhanced at high temperatures and under dark conditions due to increased serotonin N-acetyltransferase and N-acetylserotonin methyltransferase activities. *Journal of Pineal Research*, *56*(2), 189–195.

Byeon, Y., Tan, D. X., Reiter, R. J., & Back, K. (2015). Predominance of 2-hydroxymelatonin over melatonin in plants. *Journal of Pineal Research*, *59*(4), 448–454.

Campbell, W. H. (2001). Structure and function of eukaryotic NAD (P) H: Nitrate reductase. *Cellular and Molecular Life Sciences CMLS*, *58*(2), 194–204.

Cen, H., Wang, T., Liu, H., Tian, D., & Zhang, Y. (2020). Melatonin application improves salt tolerance of alfalfa (*Medicago sativa* L.) by enhancing antioxidant capacity. *Plants*, *9*(2), 220.

Chang, J., Guo, Y., Li, J., Su, Z., Wang, C., Zhang, R., . . . & Li, H. (2021b). Positive interaction between H2O2 and Ca2+ mediates melatonin-induced CBF pathway and cold tolerance in watermelon (*Citrullus lanatus* L.). *Antioxidants*, *10*(9), 1457.

Chang, T., Zhao, Y., He, H., Xi, Q., Fu, J., & Zhao, Y. (2021a). Exogenous melatonin improves growth in hulless barley seedlings under cold stress by influencing the expression rhythms of circadian clock genes. *PeerJ*, *9*, e10740.

Chen, J., Li, H., Yang, K., Wang, Y., Yang, L., Hu, L., . . . & Shi, Z. (2019). Melatonin facilitates lateral root development by coordinating PAO-derived hydrogen peroxide and Rboh-derived superoxide radical. *Free Radical Biology and Medicine*, *143*, 534–544.

Chen, Z., Xie, Y., Gu, Q., Zhao, G., Zhang, Y., Cui, W., . . . & Shen, W. (2017). The AtrbohF-dependent regulation of ROS signaling is required for melatonin-induced salinity tolerance in Arabidopsis. *Free Radical Biology and Medicine*, *108*, 465–477.

Corpas, F. J., & Barroso, J. B. (2017). Nitric oxide synthase-like activity in higher plants. *Nitric Oxide: Biology and Chemistry*, *68*, 5–6.

Cui, G., Zhao, X., Liu, S., Sun, F., Zhang, C., & Xi, Y. (2017). Beneficial effects of melatonin in overcoming drought stress in wheat seedlings. *Plant Physiology and Biochemistry*, *118*, 138–149.

Das, K., & Roychoudhury, A. (2014). Reactive oxygen species (ROS) and response of antioxidants as ROS-scavengers during environmental stress in plants. *Frontiers in Environmental Science*, *2*, 53.

Dawood, M. G., & Sadak, M. S. (2014). Physiological role of glycinebetaine in alleviating the deleterious effects of drought stress on canola plants (*Brassica napus* L.). *Middle East J. Agric. Res.*, *3*(4), 943–954.

Debnath, B., Islam, W., Li, M., Sun, Y., Lu, X., Mitra, S., . . . & Qiu, D. (2019). Melatonin mediates enhancement of stress tolerance in plants. *International Journal of Molecular Sciences*, *20*(5), 1040.

Ding, F., Liu, B., & Zhang, S. (2017a). Exogenous melatonin ameliorates cold-induced damage in tomato plants. *Scientia Horticulturae*, *219*, 264–271.

Ding, F., Wang, G., Wang, M., & Zhang, S. (2018a). Exogenous melatonin improves tolerance to water deficit by promoting cuticle formation in tomato plants. *Molecules*, *23*(7), 1605.

Ding, F., Wang, M., Liu, B., & Zhang, S. (2017b). Exogenous melatonin mitigates photoinhibition by accelerating non-photochemical quenching in tomato seedlings exposed to moderate light during chilling. *Frontiers in Plant Science*, *8*, 244.

Ding, W., Zhao, Y., Xu, J. W., Zhao, P., Li, T., Ma, H., . . . & Yu, X. (2018b). Melatonin: A multifunctional molecule that triggers defense responses against high light and nitrogen starvation stress in Haematococcus pluvialis. *Journal of Agricultural and Food Chemistry*, *66*(29), 7701–7711.

Dreyer, A., & Dietz, K. J. (2018). Reactive oxygen species and the redox-regulatory network in cold stress acclimation. *Antioxidants*, *7*(11), 169.

Dubbels, R., Reiter, R. J., Klenke, E., Goebel, A., Schnakenberg, E., Ehlers, C., . . . & Schloot, W. (1995). Melatonin in edible plants identified by radioimmunoassay and by high performance liquid chromatography-mass spectrometry. *Journal of Pineal Research*, *18*(1), 28–31.

Erland, L. A., Murch, S. J., Reiter, R. J., & Saxena, P. K. (2015). A new balancing act: The many roles of melatonin and serotonin in plant growth and development. *Plant Signaling & Behavior*, *10*(11), e1096469.

Fan, J., Xie, Y., Zhang, Z., & Chen, L. (2018). Melatonin: A multifunctional factor in plants. *International Journal of Molecular Sciences*, *19*(5), 1528.

Fichman, Y., Miller, G., & Mittler, R. (2019). Whole-plant live imaging of reactive oxygen species. *Molecular Plant*, *12*(9), 1203–1210.

Fleta-Soriano, E., Díaz, L., Bonet, E., & Munné-Bosch, S. (2017). Melatonin may exert a protective role against drought stress in maize. *Journal of Agronomy and Crop Science*, *203*(4), 286–294.

Fu, J., Wu, Y., Miao, Y., Xu, Y., Zhao, E., Wang, J., . . . & Hu, T. (2017). Improved cold tolerance in Elymus nutans by exogenous application of melatonin may involve ABA-dependent and ABA-independent pathways. *Scientific Reports*, *7*(1), 1–11.

Galano, A., & Reiter, R. J. (2018). Melatonin and its metabolites vs oxidative stress: From individual actions to collective protection. *Journal of Pineal Research*, *65*(1), e12514.

Gao, W., Zhang, Y., Feng, Z., Bai, Q., He, J., & Wang, Y. (2018). Effects of melatonin on antioxidant capacity in naked oat seedlings under drought stress. *Molecules*, *23*(7), 1580.

Gupta, K. J., & Igamberdiev, A. U. (2011). The anoxic plant mitochondrion as a nitrite: NO reductase. *Mitochondrion*, *11*(4), 537–543.

Hardeland, R., Cardinali, D. P., Srinivasan, V., Spence, D. W., Brown, G. M., & Pandi-Perumal, S. R. (2011). Melatonin: A pleiotropic, orchestrating regulator molecule. *Progress in Neurobiology*, *93*(3), 350–384.

Hasan, M., Liu, C. X., Pan, Y. T., Ahammed, G. J., Qi, Z. Y., & Zhou, J. (2018). Melatonin alleviates low-sulfur stress by promoting sulfur homeostasis in tomato plants. *Scientific Reports*, *8*(1), 1–12.

Hasanuzzaman, M., Bhuyan, M. B., Parvin, K., Bhuiyan, T. F., Anee, T. I., Nahar, K., . . . & Fujita, M. (2020). Regulation of ROS metabolism in plants under environmental stress: A review of recent experimental evidence. *International Journal of Molecular Sciences*, *21*(22), 8695.

Hassan, M. U., Chattha, M. U., Khan, I., Chattha, M. B., Aamer, M., Nawaz, M., . . . & Khan, T. A. (2019). Nickel toxicity in plants: Reasons, toxic effects, tolerance mechanisms, and remediation possibilities: A review. *Environmental Science and Pollution Research*, *26*(13), 12673–12688.

Hassan, M. U., Chattha, M. U., Khan, I., Chattha, M. B., Barbanti, L., Aamer, M., . . . & Aslam, M. T. (2021). Heat stress in cultivated plants: Nature, impact, mechanisms, and mitigation strategies: A review. *Plant Biosystems: An International Journal Dealing with All Aspects of Plant Biology*, *155*(2), 211–234.

Hattori, A., Migitaka, H., Iigo, M., Itoh, M., Yamamoto, K., Ohtani-Kaneko, R., . . . & Reiter, R. J. (1995). Identification of melatonin in plants and its effects on plasma melatonin levels and binding to melatonin receptors in vertebrates. *Biochemistry and Molecular Biology International*, *35*(3), 627–634.

He, H., & He, L. F. (2020). Crosstalk between melatonin and nitric oxide in plant development and stress responses. *Physiologia Plantarum*, *170*(2), 218–226.

Hernández, I. G., Gomez, F. J. V., Cerutti, S., Arana, M. V., & Silva, M. F. (2015). Melatonin in Arabidopsis thaliana acts as plant growth regulator at low concentrations and preserves seed viability at high concentrations. *Plant Physiology and Biochemistry*, *94*, 191–196.

Hernández-Ruiz, J., Cano, A., & Arnao, M. B. (2005). Melatonin acts as a growth-stimulating compound in some monocot species. *Journal of Pineal Research*, *39*(2), 137–142.

Hosseini, M. S., Samsampour, D., Zahedi, S. M., Zamanian, K., Rahman, M. M., Mostofa, M. G., & Tran, L. S. P. (2021). Melatonin alleviates drought impact on growth and essential oil yield of lemon verbena by enhancing antioxidant responses, mineral balance, and abscisic acid content. *Physiologia Plantarum*, *172*(2), 1363–1375.

Hu, Z., Fan, J., Xie, Y., Amombo, E., Liu, A., Gitau, M. M., . . . & Fu, J. (2016). Comparative photosynthetic and metabolic analyses reveal mechanism of improved cold stress tolerance in bermudagrass by exogenous melatonin. *Plant Physiology and Biochemistry*, *100*, 94–104.

Huang, B., Chen, Y. E., Zhao, Y. Q., Ding, C. B., Liao, J. Q., Hu, C., . . . & Yuan, M. (2019). Exogenous melatonin alleviates oxidative damages and protects photosystem II in maize seedlings under drought stress. *Frontiers in Plant Science*, *10*, 677.

Hussain, H. A., Hussain, S., Khaliq, A., Ashraf, U., Anjum, S. A., Men, S., & Wang, L. (2018). Chilling and drought stresses in crop plants: Implications, cross talk, and potential management opportunities. *Frontiers in Plant Science*, *9*, 393.

Jahan, M. S., Guo, S., Sun, J., Shu, S., Wang, Y., Abou El-Yazied, A., . . . & Hasan, M. M. (2021). Melatonin-mediated photosynthetic performance of tomato seedlings under high-temperature stress. *Plant Physiology and Biochemistry*, *167*, 309–320.

Jahan, M. S., Shu, S., Wang, Y., Chen, Z., He, M., Tao, M., . . . & Guo, S. (2019). Melatonin alleviates heat-induced damage of tomato seedlings by balancing redox homeostasis and modulating polyamine and nitric oxide biosynthesis. *BMC Plant Biology*, *19*(1), 1–16.

Janas, K. M., & Posmyk, M. M. (2013). Melatonin, an underestimated natural substance with great potential for agricultural application. *Acta Physiologiae Plantarum*, *35*(12), 3285–3292.

Jiang, C., Cui, Q., Feng, K., Xu, D., Li, C., & Zheng, Q. (2016). Melatonin improves antioxidant capacity and ion homeostasis and enhances salt tolerance in maize seedlings. *Acta Physiologiae Plantarum*, *38*(4), 1–9.

Kabiri, R., Hatami, A., Oloumi, H., Naghizadeh, M., Nasibi, F., & Tahmasebi, Z. (2018). Foliar application of melatonin induces tolerance to drought stress in Moldavian balm plants (*Dracocephalum moldavica*) through regulating the antioxidant system. *Folia Horticulturae*, *30*(1), 155.

Kaur, H., & Bhatla, S. C. (2016). Melatonin and nitric oxide modulate glutathione content and glutathione reductase activity in sunflower seedling cotyledons accompanying salt stress. *Nitric Oxide*, *59*, 42–53.

Kaur, H., Mukherjee, S., Baluska, F., & Bhatla, S. C. (2015). Regulatory roles of serotonin and melatonin in abiotic stress tolerance in plants. *Plant Signaling & Behavior*, *10*(11), e1049788.

Kaya, C., Higgs, D., Ashraf, M., Alyemeni, M. N., & Ahmad, P. (2020). Integrative roles of nitric oxide and hydrogen sulfide in melatonin-induced tolerance of pepper (*Capsicum annuum* L.) plants to iron deficiency and salt stress alone or in combination. *Physiologia Plantarum*, *168*(2), 256–277.

Khan, A., Numan, M., Khan, A. L., Lee, I. J., Imran, M., Asaf, S., & Al-Harrasi, A. (2020). Melatonin: Awakening the defense mechanisms during plant oxidative stress. *Plants*, *9*(4), 407.

Khan, I., Awan, S. A., Rizwan, M., Ali, S., Hassan, M. J., Brestic, M., Zhang, X., & Huang, L. (2021). Effects of silicon on heavy metal uptake at the soil-plant interphase: A review. *Ecotoxicology and Environmental Safety*, *222*, 112510.

Kim, T. Y., Jo, M. H., & Hong, J. H. (2010). Protective effect of nitric oxide against oxidative stress under UV-B radiation in maize leaves. *Journal of Environmental Science International*, *19*(12), 1323–1334.

Kobylińska, A., Reiter, R. J., & Posmyk, M. M. (2017). Melatonin protects cultured tobacco cells against lead-induced cell death via inhibition of cytochrome c translocation. *Frontiers in Plant Science*, *8*, 1560.

Kolář, J., Macháčková, I., Eder, J., Prinsen, E., Van Dongen, W., Van Onckelen, H., & Illnerová, H. (1997). Melatonin: Occurrence and daily rhythm in *Chenopodium rubrum*. *Phytochemistry*, *44*(8), 1407–1413.

Kołodziejczyk, I., Dzitko, K., Szewczyk, R., & Posmyk, M. M. (2016). Exogenous melatonin improves corn (*Zea mays* L.) embryo proteome in seeds subjected to chilling stress. *Journal of Plant Physiology*, *193*, 47–56.

Kopczak, A., Korth, H. G., Groot, H. D., & Kirsch, M. (2007). N-nitroso-melatonin releases nitric oxide in the presence of serotonin and its derivatives. *Journal of Pineal Research*, *43*(4), 343–350.

Koyama, F. C., Carvalho, T. L., Alves, E., Da Silva, H. B., De Azevedo, M. F., Hemerly, A. S., & Garcia, C. R. (2013). The structurally related auxin and melatonin tryptophan-derivatives and their roles in *Arabidopsis thaliana* and in the human malaria parasite *Plasmodium falciparum*. *Journal of Eukaryotic Microbiology*, *60*(6), 646–651.

Kul, R., Esringü, A., Dadasoglu, E., Sahin, Ü., Turan, M., Örs, S., . . . & Yildirim, E. (2019). Melatonin: Role in increasing plant tolerance in abiotic stress conditions. *Abiotic and Biotic Stress in Plants*, *1*, 19.

Lee, H. J., & Back, K. (2016a). 2-Hydroxymelatonin promotes the resistance of rice plant to multiple simultaneous abiotic stresses (combined cold and drought). *Journal of Pineal Research*, *61*(3), 303–316.

Lee, H. Y., & Back, K. (2016b). Mitogen-activated protein kinase pathways are required for melatonin-mediated defense responses in plants. *Journal of Pineal Research*, *60*, 327–335.

Lee, H. Y., & Back, K. (2017). Melatonin is required for H 2 O 2-and NO-mediated defense signaling through MAPKKK 3 and OXI 1 in *Arabidopsis thaliana*. *Journal of Pineal Research*, *62*(2), e12379.

Lee, H. Y., & Back, K. (2018). Melatonin induction and its role in high light stress tolerance in *Arabidopsis thaliana*. *Journal of Pineal Research*, *65*(3), e12504.

Lee, H. Y., Byeon, Y., & Back, K. (2014). Melatonin as a signal molecule triggering defense responses against pathogen attack in Arabidopsis and tobacco. *Journal of Pineal Research*, *57*(3), 262–268.

Lei, X. Y., Zhu, R. Y., Zhang, G. Y., & Dai, Y. R. (2004). Attenuation of cold-induced apoptosis by exogenous melatonin in carrot suspension cells: The possible involvement of polyamines. *Journal of Pineal Research*, *36*(2), 126–131.

Lemos, M., Xiao, Y., Bjornson, M., Wang, J. Z., Hicks, D., Souza, A. D., . . . & Dehesh, K. (2016). The plastidial retrograde signal methyl erythritol cyclopyrophosphate is a regulator of salicylic acid and jasmonic acid crosstalk. *Journal of Experimental Botany*, *67*(5), 1557–1566.

Li, C., He, Q., Zhang, F., Yu, J., Li, C., Zhao, T., . . . & Chen, J. (2019a). Melatonin enhances cotton immunity to Verticillium wilt via manipulating lignin and gossypol biosynthesis. *The Plant Journal*, *100*(4), 784–800.

Li, C., Wang, P., Wei, Z., Liang, D., Liu, C., Yin, L., . . . & Ma, F. (2012). The mitigation effects of exogenous melatonin on salinity-induced stress in *Malus hupehensis*. *Journal of Pineal Research*, *53*(3), 298–306.

Li, D., Wei, J., Peng, Z., Ma, W., Yang, Q., Song, Z., . . . & Chen, Q. (2020). Daily rhythms of phytomelatonin signaling modulate diurnal stomatal closure via regulating reactive oxygen species dynamics in Arabidopsis. *Journal of Pineal Research*, *68*(3), e12640.

Li, D., Zhang, D., Hongguang, W. A. N. G., Yanming, L. I., & Ruiqi, L. I. (2017a). Physiological response of plants to polyethylene glycol (PEG-6000) by exogenous melatonin application in wheat. *Zemdirbyste-Agriculture*, *104*(3).

Li, H., Chang, J., Chen, H., Wang, Z., Gu, X., Wei, C., . . . & Zhang, X. (2017b). Exogenous melatonin confers salt stress tolerance to watermelon by improving photosynthesis and redox homeostasis. *Frontiers in Plant Science*, *8*, 295.

Li, H., Guo, Y., Lan, Z., Xu, K., Chang, J., Ahammed, G. J., . . . & Zhang, X. (2021). Methyl jasmonate mediates melatonin-induced cold tolerance of grafted watermelon plants. *Horticulture Research*, *8*.

Li, M. Q., Hasan, M. K., Li, C. X., Ahammed, G. J., Xia, X. J., Shi, K., . . . & Zhou, J. (2016). Melatonin mediates selenium-induced tolerance to cadmium stress in tomato plants. *Journal of Pineal Research*, *61*(3), 291–302.

Li, X., Wei, J. P., Scott, E. R., Liu, J. W., Guo, S., Li, Y., . . . & Han, W. Y. (2018). Exogenous melatonin alleviates cold stress by promoting antioxidant defense and redox homeostasis in *Camellia sinensis* L. *Molecules*, *23*(1), 165.

Li, Z. G., Xu, Y., Bai, L. K., Zhang, S. Y., & Wang, Y. (2019b). Melatonin enhances thermotolerance of maize seedlings (*Zea mays* L.) by modulating antioxidant defense, methylglyoxal detoxification, and osmoregulation systems. *Protoplasma*, *256*(2), 471–490.

Liang, D., Gao, F., Ni, Z., Lin, L., Deng, Q., Tang, Y., . . . & Xia, H. (2018). Melatonin improves heat tolerance in kiwifruit seedlings through promoting antioxidant enzymatic activity and glutathione S-transferase transcription. *Molecules*, *23*(3), 584.

Liang, D., Ni, Z., Xia, H., Xie, Y., Lv, X., Wang, J., . . . & Luo, X. (2019). Exogenous melatonin promotes biomass accumulation and photosynthesis of kiwifruit seedlings under drought stress. *Scientia Horticulturae*, *246*, 34–43.

Liu, J., Yang, J., Zhang, H., Cong, L., Zhai, R., Yang, C., . . . & Xu, L. (2019). Melatonin inhibits ethylene synthesis via nitric oxide regulation to delay postharvest senescence in pears. *Journal of Agricultural and Food Chemistry*, *67*(8), 2279–2288.

Liu, N., Gong, B., Jin, Z., Wang, X., Wei, M., Yang, F., . . . & Shi, Q. (2015). Sodic alkaline stress mitigation by exogenous melatonin in tomato needs nitric oxide as a downstream signal. *Journal of Plant Physiology*, *186*, 68–77.

Lv, R., Li, Z., Li, M., Dogra, V., Lv, S., Liu, R., . . . & Kim, C. (2019). Uncoupled expression of nuclear and plastid photosynthesis-associated genes contributes to cell death in a lesion mimic mutant. *The Plant Cell*, *31*(1), 210–230.

Ma, Y., Jiao, J., Fan, X., Sun, H., Zhang, Y., Jiang, J., & Liu, C. (2017). Endophytic bacterium Pseudomonas fluorescens RG11 may transform tryptophan to melatonin and promote endogenous melatonin levels in the roots of four grape cultivars. *Frontiers in Plant Science*, *7*, 2068.

Martinez, V., Nieves-Cordones, M., Lopez-Delacalle, M., Rodenas, R., Mestre, T. C., Garcia-Sanchez, F., . . . & Rivero, R. M. (2018). Tolerance to stress combination in tomato plants: New insights in the protective role of melatonin. *Molecules*, *23*(3), 535.

Mishra, A., Mishra, K. B., Höermiller, I. I., Heyer, A. G., & Nedbal, L. (2011). Chlorophyll fluorescence emission as a reporter on cold tolerance in *Arabidopsis thaliana* accessions. *Plant Signaling & Behavior*, *6*(2), 301–310.

Moschou, P. N., & Roubelakis-Angelakis, K. A. (2014). Polyamines and programmed cell death. *Journal of Experimental Botany*, *65*(5), 1285–1296.

Moustafa-Farag, M., Almoneafy, A., Mahmoud, A., Elkelish, A., Arnao, M. B., Li, L., & Ai, S. (2019). Melatonin and its protective role against biotic stress impacts on plants. *Biomolecules*, *10*(1), 54.

Moustafa-Farag, M., Mahmoud, A., Arnao, M. B., Sheteiwy, M. S., Dafea, M., et al. (2020). Melatonin-induced water stress tolerance in plants: Recent advances. *Antioxidants*, *9*, 809.

Murch, S. J., Campbell, S. S., & Saxena, P. K. (2001). The role of serotonin and melatonin in plant morphogenesis: Regulation of auxin-induced root organogenesis in in vitro-cultured explants of St. John's wort (*Hypericum perforatum* L.). *In Vitro Cellular & Developmental Biology-Plant*, *37*(6), 786–793.

Naing, A. H., & Kim, C. K. (2021). Abiotic stress-induced anthocyanins in plants: Their role in tolerance to abiotic stresses. *Physiologia Plantarum*, *172*(3), 1711–1723.

Nawaz, M. A., Huang, Y., Bie, Z., Ahmed, W., Reiter, R. J., Niu, M., & Hameed, S. (2016). Melatonin: Current status and future perspectives in plant science. *Frontiers in Plant Science*, *6*, 1230.

Nawaz, M. A., Jiao, Y., Chen, C., Shireen, F., Zheng, Z., Imtiaz, M., . . . & Huang, Y. (2018). Melatonin pretreatment improves vanadium stress tolerance of watermelon seedlings by reducing vanadium concentration in the leaves and regulating melatonin biosynthesis and antioxidant-related gene expression. *Journal of Plant Physiology*, *220*, 115–127.

Nguyen, C. T., Kurenda, A., Stolz, S., Chételat, A., & Farmer, E. E. (2018a). Identification of cell populations necessary for leaf-to-leaf electrical signaling in a wounded plant. *Proceedings of the National Academy of Sciences*, *115*(40), 10178–10183.

Nguyen, H. C., Lin, K. H., Ho, S. L., Chiang, C. M., & Yang, C. M. (2018b). Enhancing the abiotic stress tolerance of plants: From chemical treatment to biotechnological approaches. *Physiologia Plantarum*, *164*(4), 452–466.

Okant, M., & Kaya, C. (2019). The role of endogenous nitric oxide in melatonin-improved tolerance to lead toxicity in maize plants. *Environmental Science and Pollution Research*, *26*(12), 11864–11874.

Pardo-Hernández, M., López-Delacalle, M., & Rivero, R. M. (2020). ROS and NO regulation by melatonin under abiotic stress in plants. *Antioxidants*, *9*(11), 1078.

Parida, A. K., & Das, A. B. (2005). Salt tolerance and salinity effects on plants: A review. *Ecotoxicology and Environmental Safety*, *60*(3), 324–349.

Pelagio-Flores, R., Muñoz-Parra, E., Ortiz-Castro, R., & López-Bucio, J. (2012). Melatonin regulates Arabidopsis root system architecture likely acting independently of auxin signaling. *Journal of Pineal Research*, *53*(3), 279–288.

Piao, S., Liu, Q., Chen, A., Janssens, I. A., Fu, Y., Dai, J., . . . & Zhu, X. (2019). Plant phenology and global climate change: Current progresses and challenges. *Global Change Biology*, *25*(6), 1922–1940.

Pitzschke, A., Schikora, A., & Hirt, H. (2009). MAPK cascade signalling networks in plant defence. *Current Opinion in Plant Biology*, *12*(4), 421–426.

Planchet, E., Jagadis Gupta, K., Sonoda, M., & Kaiser, W. M. (2005). Nitric oxide emission from tobacco leaves and cell suspensions: Rate limiting factors and evidence for the involvement of mitochondrial electron transport. *The Plant Journal*, *41*(5), 732–743.

Posmyk, M. M., Bałabusta, M., Wieczorek, M., Sliwinska, E., & Janas, K. M. (2009). Melatonin applied to cucumber (*Cucumis sativus* L.) seeds improves germination during chilling stress. *Journal of Pineal Research*, *46*(2), 214–223.

Posmyk, M. M., Kuran, H., Marciniak, K., & Janas, K. M. (2008). Presowing seed treatment with melatonin protects red cabbage seedlings against toxic copper ion concentrations. *Journal of Pineal Research*, *45*(1), 24–31.

Postiglione, A. E., & Muday, G. K. (2020). The role of ROS homeostasis in ABA-induced guard cell signaling. *Frontiers in Plant Science*, *11*, 968.

Qi, Z. Y., Wang, K. X., Yan, M. Y., Kanwar, M. K., Li, D. Y., Wijaya, L., . . . & Zhou, J. (2018). Melatonin alleviates high temperature-induced pollen abortion in *Solanum lycopersicum*. *Molecules*, *23*(2), 386.

Qian, Y., Tan, D. X., Reiter, R. J., & Shi, H. (2015). Comparative metabolomic analysis highlights the involvement of sugars and glycerol in melatonin-mediated innate immunity against bacterial pathogen in Arabidopsis. *Scientific Reports*, *5*(1), 1–11.

Qiao, Y., Yin, L., Wang, B., Ke, Q., Deng, X., & Wang, S. (2019). Melatonin promotes plant growth by increasing nitrogen uptake and assimilation under nitrogen deficient condition in winter wheat. *Plant Physiology and Biochemistry*, *139*, 342–349.

Ray, P. D., Huang, B. W., & Tsuji, Y. (2012). Reactive oxygen species (ROS) homeostasis and redox regulation in cellular signaling. *Cellular Signalling*, *24*(5), 981–990.

Rehman, S., Chattha, M. U., Khan, I., Mahmood, A., Hassan, M. U., Al-Huqail, A. A., . . . & El-Esawi, M. A. (2022). Exogenously applied trehalose augments cadmium stress tolerance and yield of mung bean (*Vigna radiata* L.) grown in soil and hydroponic systems through reducing cd uptake and enhancing photosynthetic efficiency and antioxidant defense systems. *Plants*, *11*(6), 822.

Reiter, R. J., Tan, D. X., & Fuentes-Broto, L. (2010). Melatonin: A multitasking molecule. *Progress in Brain Research*, *181*, 127–151.

Rockel, P., Strube, F., Rockel, A., Wildt, J., & Kaiser, W. M. (2002). Regulation of nitric oxide (NO) production by plant nitrate reductase in vivo and in vitro. *Journal of Experimental Botany*, *53*(366), 103–110.

Roychoudhury, A., Basu, S., Sarkar, S. N., & Sengupta, D. N. (2008). Comparative physiological and molecular responses of a common aromatic indica rice cultivar to high salinity with non-aromatic indica rice cultivars. *Plant Cell Reports*, *27*(8), 1395–1410.

Roychoudhury, A., Paul, S., & Basu, S. (2013). Cross-talk between abscisic acid-dependent and abscisic acid-independent pathways during abiotic stress. *Plant Cell Reports*, *32*(7), 985–1006.

Sadak, M. S., Abdalla, A. M., Abd Elhamid, E. M., & Ezzo, M. I. (2020). Role of melatonin in improving growth, yield quantity and quality of *Moringa oleifera* L. plant under drought stress. *Bulletin of the National Research Centre*, *44*(1), 1–13.

Samanta, S., Banerjee, A., & Roychoudhury, A. (2021). Melatonin application differentially modulates the enzymes associated with antioxidative machinery and ascorbate glutathione cycle during arsenate exposure in indica rice varieties. *Plant Biology*, *23*(S1), 193–201.
Samanta, S., Banerjee, A., & Roychoudhury, A. (2022). Arsenic toxicity is counteracted by exogenous application of melatonin to different extents in arsenic-susceptible and arsenic-tolerant rice cultivars. *Journal of Plant Growth Regulation*, *41*, 2210–2231.
Sarkar, A. K., & Sadhukhan, S. (2022). Imperative role of trehalose metabolism and trehalose-6-phosphate signaling on salt stress responses in plants. *Physiologia Plantarum*, *174*(1), e13647.
Sarropoulou, V. N., Dimassi-Theriou, K., Therios, I., & Koukourikou-Petridou, M. (2012a). Melatonin enhances root regeneration, photosynthetic pigments, biomass, total carbohydrates and proline content in the cherry rootstock PHL-C (Prunus avium× Prunus cerasus). *Plant Physiology and Biochemistry*, *61*, 162–168.
Sarropoulou, V. N., Therios, I. N., & Dimassi-Theriou, K. N. (2012b). Melatonin promotes adventitious root regeneration in in vitro shoot tip explants of the commercial sweet cherry rootstocks CAB-6P (Prunus cerasus L.), Gisela 6 (P. cerasus× P. canescens), and MxM 60 (P. avium× P. mahaleb). *Journal of Pineal Research*, *52*(1), 38–46.
Sharif, R., Xie, C., Zhang, H., Arnao, M. B., Ali, M., Ali, Q., . . . & Li, Y. (2018). Melatonin and its effects on plant systems. *Molecules*, *23*(9), 2352.
Sharma, A., Wang, J., Xu, D., Tao, S., Chong, S., Yan, D., . . . & Zheng, B. (2020). Melatonin regulates the functional components of photosynthesis, antioxidant system, gene expression, and metabolic pathways to induce drought resistance in grafted Carya cathayensis plants. *Science of the Total Environment*, *713*, 136675.
Sharma, A., & Zheng, B. (2019). Melatonin mediated regulation of drought stress: Physiological and molecular aspects. *Plants*, *8*(7), 190.
Shi, H., Chen, Y., Tan, D. X., Reiter, R. J., Chan, Z., & He, C. (2015a). Melatonin induces nitric oxide and the potential mechanisms relate to innate immunity against bacterial pathogen infection in Arabidopsis. *Journal of Pineal Research*, *59*(1), 102–108.
Shi, H., Reiter, R. J., Tan, D. X., & Chan, Z. (2015b). Indole-3-acetic acid inducible 17 positively modulates natural leaf senescence through melatonin-mediated pathway in Arabidopsis. *Journal of Pineal Research*, *58*(1), 26–33.
Shi, H., Tan, D. X., Reiter, R. J., Ye, T., Yang, F., & Chan, Z. (2015c). Melatonin induces class A1 heat-shock factors (HSFA 1s) and their possible involvement of thermotolerance in Arabidopsis. *Journal of Pineal Research*, *58*(3), 335–342.
Shi, H., Wei, Y., Wang, Q., Reiter, R. J., & He, C. (2016). Melatonin mediates the stabilization of DELLA proteins to repress the floral transition in A rabidopsis. *Journal of Pineal Research*, *60*(3), 373–379.
Su, J., Yang, X., Shao, Y., Chen, Z., & Shen, W. (2021). Molecular hydrogen: Induced salinity tolerance requires melatonin signalling in *Arabidopsis thaliana*. *Plant, Cell & Environment*, *44*(2), 476–490.
Sun, C., Liu, L., Wang, L., Li, B., Jin, C., & Lin, X. (2021). Melatonin: A master regulator of plant development and stress responses. *Journal of Integrative Plant Biology*, *63*(1), 126–145.
Sun, C., Lv, T., Huang, L., Liu, X., Jin, C., & Lin, X. (2020). Melatonin ameliorates aluminum toxicity through enhancing aluminum exclusion and reestablishing redox homeostasis in roots of wheat. *Journal of Pineal Research*, *68*(4), e12642.
Sun, Q., Zhang, N., Wang, J., Zhang, H., Li, D., Shi, J., . . . & Guo, Y. D. (2015). Melatonin promotes ripening and improves quality of tomato fruit during postharvest life. *Journal of Experimental Botany*, *66*(3), 657–668.
Suzuki, N., Miller, G., Salazar, C., Mondal, H. A., Shulaev, E., Cortes, D. F., . . . & Mittler, R. (2013). Temporal-spatial interaction between reactive oxygen species and abscisic acid regulates rapid systemic acclimation in plants. *The Plant Cell*, *25*(9), 3553–3569.
Tal, O., Haim, A., Harel, O., & Gerchman, Y. (2011). Melatonin as an antioxidant and its semi-lunar rhythm in green macroalga Ulva sp. *Journal of Experimental Botany*, *62*(6), 1903–1910.
Talaat, N. B. (2021). Polyamine and nitrogen metabolism regulation by melatonin and salicylic acid combined treatment as a repressor for salt toxicity in wheat (*Triticum aestivum* L.) plants. *Plant Growth Regulation*, *95*(3), 315–329.
Tan, D. X., Manchester, L. C., Di Mascio, P., Martinez, G. R., Prado, F. M., & Reiter, R. J. (2007). Novel rhythms of N1-acetyl-N2-formyl-5-methoxykynuramine and its precursor melatonin in water hyacinth: Importance for phytoremediation. *The FASEB Journal*, *21*(8), 1724–1729.

Tan, D. X., Manchester, L. C., Reiter, R. J., Qi, W. B., Karbownik, M., & Calvo, J. R. (2000). Significance of melatonin in antioxidative defense system: Reactions and products. *Neurosignals*, *9*(3–4), 137–159.
Tan, D. X., Reiter, R. J., Manchester, L. C., Yan, M. T., El-Sawi, M., Sainz, R. M., . . . & Hardeland, R. (2002). Chemical and physical properties and potential mechanisms: Melatonin as a broad spectrum antioxidant and free radical scavenger. *Current Topics in Medicinal Chemistry*, *2*(2), 181–197.
Tang, Y., Li, J., & Li, H. (2015). *6th international conference on manufacturing science and engineering*, 656–659. Guangzhou: Atlantis Press, China.
Turk, H., Erdal, S., Genisel, M., Atici, O., Demir, Y., & Yanmis, D. (2014). The regulatory effect of melatonin on physiological, biochemical and molecular parameters in cold-stressed wheat seedlings. *Plant Growth Regulation*, *74*(2), 139–152.
Umer Chattha, M., Arif, W., Khan, I., Soufan, W., Bilal Chattha, M., Hassan, M. U., . . . & Qari, S. H. (2021). Mitigation of cadmium induced oxidative stress by using organic amendments to improve the growth and yield of mash beans [*Vigna mungo* (L.)]. *Agronomy*, *11*(11), 2152.
Vafadar, F., Amooaghaie, R., Ehsanzadeh, P., Ghanati, F., & Sajedi, R. H. (2020). Crosstalk between melatonin and Ca2+/CaM evokes systemic salt tolerance in *Dracocephalum kotschyi*. *Journal of Plant Physiology*, *252*, 153237.
Van Tassel, D. L., Roberts, N., Lewy, A., & O'Neill, S. D. (2001). Melatonin in plant organs. *Journal of Pineal Research*, *31*(1), 8–15.
Van Zelm, E., Zhang, Y., & Testerink, C. (2020). Salt tolerance mechanisms of plants. *Annual Review of Plant Biology*, *71*, 403–433.
Wang, D., Chen, Q., Chen, W., Guo, Q., Xia, Y., Wu, D., . . . & Liang, G. (2021a). Melatonin treatment maintains quality and delays lignification in loquat fruit during cold storage. *Scientia Horticulturae*, *284*, 110126.
Wang, L. Y., Liu, J. L., Wang, W. X., & Sun, Y. (2016). Exogenous melatonin improves growth and photosynthetic capacity of cucumber under salinity-induced stress. *Photosynthetica*, *54*(1), 19–27.
Wang, M., Duan, S., Zhou, Z., Chen, S., & Wang, D. (2019). Foliar spraying of melatonin confers cadmium tolerance in *Nicotiana tabacum* L. *Ecotoxicology and Environmental Safety*, *170*, 68–76.
Wang, M., Zhang, S., & Ding, F. (2020). Melatonin mitigates chilling-induced oxidative stress and photosynthesis inhibition in tomato plants. *Antioxidants*, *9*(3), 218.
Wang, P., Sun, X., Li, C., Wei, Z., Liang, D., & Ma, F. (2013). Long-term exogenous application of melatonin delays drought-induced leaf senescence in apple. *Journal of Pineal Research*, *54*(3), 292–302.
Wang, P., Yin, L., Liang, D., Li, C., Ma, F., & Yue, Z. (2012). Delayed senescence of apple leaves by exogenous melatonin treatment: Toward regulating the ascorbate: Glutathione cycle. *Journal of Pineal Research*, *53*(1), 11–20.
Wang, Y., Mostafa, S., Zeng, W., & Jin, B. (2021b). Function and mechanism of jasmonic acid in plant responses to abiotic and biotic stresses. *International Journal of Molecular Sciences*, *22*(16), 8568.
Wang, Y., Reiter, R. J., & Chan, Z. (2018). Phytomelatonin: A universal abiotic stress regulator. *Journal of Experimental Botany*, *69*(5), 963–974.
Weeda, S., Zhang, N., Zhao, X., Ndip, G., Guo, Y., Buck, G. A., . . . & Ren, S. (2014). Arabidopsis transcriptome analysis reveals key roles of melatonin in plant defense systems. *PloS One*, *9*(3), e93462.
Wei, W., Li, Q. T., Chu, Y. N., Reiter, R. J., Yu, X. M., Zhu, D. H., . . . & Chen, S. Y. (2015). Melatonin enhances plant growth and abiotic stress tolerance in soybean plants. *Journal of Experimental Botany*, *66*(3), 695–707.
Wei, Y., Chang, Y., Zeng, H., Liu, G., He, C., & Shi, H. (2018). RAV transcription factors are essential for disease resistance against cassava bacterial blight via activation of melatonin biosynthesis genes. *Journal of Pineal Research*, *64*(1), e12454.
Wei, Z., Li, C., Gao, T., Zhang, Z., Liang, B., Lv, Z., . . . & Ma, F. (2019). Melatonin increases the performance of *Malus hupehensis* after UV-B exposure. *Plant Physiology and Biochemistry*, *139*, 630–641.
Wen, D., Gong, B., Sun, S., Liu, S., Wang, X., Wei, M., . . . & Shi, Q. (2016). Promoting roles of melatonin in adventitious root development of *Solanum lycopersicum* L. by regulating auxin and nitric oxide signaling. *Frontiers in Plant Science*, *7*, 718.
Wolf, K., Kolář, J., Witters, E., van Dongen, W., van Onckelen, H., & Macháčková, I. (2001). Daily profile of melatonin levels in *Chenopodium rubrum* L. depends on photoperiod. *Journal of Plant Physiology*, *158*(11), 1491–1493.
Wu, S., Wang, Y., Zhang, J., Gong, X., Zhang, Z., Sun, J., . . . & Wang, Y. (2021a). Exogenous melatonin improves physiological characteristics and promotes growth of strawberry seedlings under cadmium stress. *Horticultural Plant Journal*, *7*(1), 13–22.

Wu, Y., Fan, X., Zhang, Y., Jiang, J., Sun, L., Rahman, F. U., & Liu, C. (2021b). VvSNAT1 overexpression enhances melatonin production and salt tolerance in transgenic Arabidopsis. *Plant Physiology and Biochemistry*, *166*, 485–494.

Xiang-dong, X., Yan, S., Bo, S., Jian, Z., & Xiao-qin, G. (2010). Effects of exogenous melatonin on active oxygen metabolism of cucumber seedlings under high temperature stress. *Yingyong Shengtai Xuebao*, *21*(5).

Yamamoto, Y., Aminaka, R., Yoshioka, M., Khatoon, M., Komayama, K., Takenaka, D., . . . & Yamamoto, Y. (2008). Quality control of photosystem II: Impact of light and heat stresses. *Photosynthesis Research*, *98*(1), 589–608.

Yamasaki, H., & Sakihama, Y. (2000). Simultaneous production of nitric oxide and peroxynitrite by plant nitrate reductase: In vitro evidence for the NR-dependent formation of active nitrogen species. *FEBS Letters*, *468*(1), 89–92.

Yamasaki, H., Sakihama, Y., & Takahashi, S. (1999). An alternative pathway for nitric oxide production in plants: New features of an old enzyme. *Trends in Plant Science*, *4*(4), 128–129.

Yang, Y., Yao, Y., Li, J., Zhang, J., Zhang, X., Hu, L., . . . & Xie, J. (2022). Trehalose alleviated salt stress in tomato by regulating ROS metabolism, photosynthesis, osmolyte synthesis, and trehalose metabolic pathways. *Frontiers in Plant Science*, 623.

Yin, L., Wang, P., Li, M., Ke, X., Li, C., Liang, D., . . . & Ma, F. (2013). Exogenous melatonin improves Malus resistance to Marssonina apple blotch. *Journal of Pineal Research*, *54*(4), 426–434.

Yu, Y., Wang, A., Li, X., Kou, M., Wang, W., Chen, X., . . . & Sun, J. (2018). Melatonin-stimulated triacylglycerol breakdown and energy turnover under salinity stress contributes to the maintenance of plasma membrane H+: ATPase activity and K+/Na+ homeostasis in sweet potato. *Frontiers in Plant Science*, *9*, 256.

Zahedi, S. M., Hosseini, M. S., Abadía, J., & Marjani, M. (2020). Melatonin foliar sprays elicit salinity stress tolerance and enhance fruit yield and quality in strawberry (Fragaria× ananassa Duch.). *Plant Physiology and Biochemistry*, *149*, 313–323.

Zandalinas, S. I., Sengupta, S., Burks, D., Azad, R. K., & Mittler, R. (2019). Identification and characterization of a core set of ROS wave-associated transcripts involved in the systemic acquired acclimation response of Arabidopsis to excess light. *The Plant Journal*, *98*(1), 126–141.

Zhan, H., Nie, X., Zhang, T., Li, S., Wang, X., Du, X., . . . & Song, W. (2019). Melatonin: A small molecule but important for salt stress tolerance in plants. *International Journal of Molecular Sciences*, *20*(3), 709.

Zhang, H. J., Zhang, N. A., Yang, R. C., Wang, L., Sun, Q. Q., Li, D. B., . . . & Guo, Y. D. (2014a). Melatonin promotes seed germination under high salinity by regulating antioxidant systems, ABA and GA 4 interaction in cucumber (*Cucumis sativus* L.). *Journal of Pineal Research*, *57*(3), 269–279.

Zhang, J., Zeng, B., Mao, Y., Kong, X., Wang, X., Yang, Y., . . . & Chen, Q. (2017). Melatonin alleviates aluminium toxicity through modulating antioxidative enzymes and enhancing organic acid anion exudation in soybean. *Functional Plant Biology*, *44*(10), 961–968.

Zhang, M., He, S., Zhan, Y., Qin, B., Jin, X., Wang, M., . . . & Wu, Y. (2019a). Exogenous melatonin reduces the inhibitory effect of osmotic stress on photosynthesis in soybean. *PloS One*, *14*(12), e0226542.

Zhang, N., Sun, Q., Zhang, H., Cao, Y., Weeda, S., Ren, S., & Guo, Y. D. (2015). Roles of melatonin in abiotic stress resistance in plants. *Journal of Experimental Botany*, *66*(3), 647–656.

Zhang, N., Zhang, H. J., Zhao, B., Sun, Q. Q., Cao, Y. Y., Li, R., . . . & Guo, Y. D. (2014b). The RNA-seq approach to discriminate gene expression profiles in response to melatonin on cucumber lateral root formation. *Journal of Pineal Research*, *56*(1), 39–50.

Zhang, N., Zhao, B., Zhang, H. J., Weeda, S., Yang, C., Yang, Z. C., . . . & Guo, Y. D. (2013). Melatonin promotes water-stress tolerance, lateral root formation, and seed germination in cucumber (*Cucumis sativus* L.). *Journal of Pineal Research*, *54*(1), 15–23.

Zhang, Q., Liu, X., Zhang, Z., Liu, N., Li, D., & Hu, L. (2019b). Melatonin improved waterlogging tolerance in alfalfa (*Medicago sativa*) by reprogramming polyamine and ethylene metabolism. *Frontiers in Plant Science*, *10*, 44.

Zhang, Z., Hu, Q., Liu, Y., Cheng, P., Cheng, H., Liu, W., . . . & Chen, F. (2019c). Strigolactone represses the synthesis of melatonin, thereby inducing floral transition in Arabidopsis thaliana in an FLC-dependent manner. *Journal of Pineal Research*, *67*(2), e12582.

Zhao, G., Zhao, Y., Yu, X., Kiprotich, F., Han, H., Guan, R., . . . & Shen, W. (2018). Nitric oxide is required for melatonin-enhanced tolerance against salinity stress in rapeseed (*Brassica napus* L.) seedlings. *International Journal of Molecular Sciences*, *19*(7), 1912.

Zhao, H., Xu, L., Su, T., Jiang, Y., Hu, L., & Ma, F. (2015). Melatonin regulates carbohydrate metabolism and defenses against *Pseudomonas syringae* pv. tomato DC 3000 infection in *Arabidopsis thaliana*. *Journal of Pineal Research*, *59*(1), 109–119.

Zhao, L., Chen, L., Gu, P., Zhan, X., Zhang, Y., Hou, C., . . . & Wang, Q. C. (2019). Exogenous application of melatonin improves plant resistance to virus infection. *Plant Pathology*, *68*(7), 1287–1295.

Zheng, X., Zhou, J., Tan, D. X., Wang, N., Wang, L., Shan, D., & Kong, J. (2017). Melatonin improves waterlogging tolerance of *Malus baccata* (Linn.) Borkh: Seedlings by maintaining aerobic respiration, photosynthesis and ROS migration. *Frontiers in Plant Science*, *8*, 483.

Zhou, C., Liu, Z., Zhu, L., Ma, Z., Wang, J., & Zhu, J. (2016a). Exogenous melatonin improves plant iron deficiency tolerance via increased accumulation of polyamine-mediated nitric oxide. *International Journal of Molecular Sciences*, *17*(11), 1777.

Zhou, X., Zhao, H., Cao, K., Hu, L., Du, T., Baluška, F., & Zou, Z. (2016b). Beneficial roles of melatonin on redox regulation of photosynthetic electron transport and synthesis of D1 protein in tomato seedlings under salt stress. *Frontiers in Plant Science*, *7*, 1823.

Zhu, Y., Gao, H., Lu, M., Hao, C., Pu, Z., Guo, M., . . . & Huang, X. (2019). Melatonin-nitric oxide crosstalk and their roles in the redox network in plants. *International Journal of Molecular Sciences*, *20*(24), 6200.

10 Melatonin-Phytohormone Interaction in Regulating Plant Growth and Stress Tolerance

Ankur Singh and Aryadeep Roychoudhury

10.1 INTRODUCTION

Plants, being sessile organisms, have to encounter multiple abiotic stresses throughout their lives. To cope with this situation, they have evolved various survival strategies that help them to thrive under such harsh conditions. One of the most important biomolecules which has gained popularity in recent decades is melatonin, which acts as an effective protective agent. Melatonin is a ubiquitous molecule which was initially identified in the pineal glands of animals in 1958, followed by its characterization in microorganisms in 1991 and lastly in plants in 1995 (Fan et al. 2018). After its discovery, melatonin research was mostly conducted on animals for around four decades. However, in plants, melatonin has been shown to participate in various abiotic stress tolerance as an important signaling molecule. Along with this, it is also known to be a potent scavenger of both reactive nitrogen species (RNS) and reactive oxygen species (ROS) (Arnao and Hernández-Ruiz 2019a). Melatonin also regulates various processes in plants such as leaf development, fruit maturation, root organogenesis and senescence (Arnao and Hernández-Ruiz 2019b). Recently, Tan and Reiter (2020) reported that melatonin mostly occurs in chloroplast and mitochondria of leaves, roots, stems, flower buds, petals, fruits and seeds of various plants species.

Since its discovery, the interaction between melatonin and other phytohormones has been widely studied. According to Liang et al. (2017), melatonin can mimic the activity of auxin, thus acting as an upstream regulator of various auxin-related transcription factors such as MYB, bHLH, WRKY, NAC and HD-ZIP. The interaction of melatonin with other major phytohormones such as abscisic acid (ABA), cytokinin, gibberellic acid (GA), ethylene, jasmonic acid (JA), indoleacetic acid (IAA) and salicylic acid has also been studied in plants exposed to abiotic stresses such as drought, salinity, heavy metal toxicity and extreme temperature (Moustafa-Farag et al. 2020; Weeda et al. 2014). In addition, exogenous melatonin can regulate the expression of the previously mentioned phytohormones, which eventually modulates the biosynthesis and function of protective metabolites, thus enhancing the tolerance capability of plants exposed to harsh environmental conditions (Sun et al. 2021; Zhang et al. 2015).

Considering its wide role in inducing plant tolerance capability, melatonin has attracted the interest of various researchers and plants scientists. In recent times, the increasing harmful effects of global warming, climate change, rapid population and industrial growth followed by soil salinization have altogether drastically affected the production of crops, which is a major cause of concern. Exogenous application of melatonin has shown to restore the growth and yield of plants. Over the past decade, considerable progress has been made in the general understanding of melatonin action in plants. Therefore, an updated review highlighting recent findings in melatonin-mediated abiotic stress tolerance is needed. In this chapter, we have summarized the biosynthetic and catabolic pathways of melatonin. We have also focused on the defensive roles of melatonin against various abiotic stresses, as well as its interaction with major phytohormones that helps in improving plant stress tolerance.

DOI: 10.1201/9781003343752-10

10.2 METABOLISM OF MELATONIN IN PLANTS

10.2.1 Melatonin Biosynthesis

Melatonin is a stable molecule with low molecular weight (Wu et al. 2021). In animals, melatonin biosynthesis is mostly performed by four major enzymes, that is, tryptophan hydroxylase, aromatic amino acid decarboxylase, serotonin N-acetyltransferase and N-acetylserotonin methyltransferase (Tan et al. 2016). However, as compared to that of animals, melatonin is formed in plants by six biosynthetic enzymes.

Tryptophan is the precursor of melatonin which plants are able to synthesize *de novo* via shikimate pathway. Tryptophan is initially decarboxylated into tryptamine by the enzymatic action of L-tryptophan decarboxylase (TDC) that eventually gets converted into serotonin by tryptamine 5-hydroxylase (T5H). In addition, tryptophan also gets converted into 5-hydroxytryptophan by tryptophan hydroxylase (TPH), followed by its decarboxylation to serotonin by TDC. Due to affinity of TDC to both tryptophan and 5-hydroxytryptophan, both routes are possible; however, decarboxylation of tryptophan followed by its hydroxylation is more frequent in plants (Back et al. 2016). Conversion of serotonin into melatonin involves three enzymes, that is, serotonin N-acetyltransferase (SNAT) (controls the rate limiting step of melatonin synthesis), acetylserotonin O-methyltransferase (ASMT) and caffeic acid 3-O-methyltransferase (COMT) in a two-step reaction. The first enzyme catalyzes an acetylation, whereas the other two enzymes are methyltransferases (Back et al. 2016).

Of all the previously mentioned six enzymes involved in plant melatonin biosynthesis, TDC, ASMT and COMT are localized in cytoplasm (Zhou et al. 2020; Zhao et al. 2021), T5H is found in endoplasmic reticulum (Back 2021) and SNAT is expressed in chloroplast (Back 2021). In addition to the enzyme localization, melatonin biosynthesis also widely varies in plants depending on the plant growth and stressed conditions. Under normal conditions, the level of serotonin is low in plants, and, thus, formation of melatonin occurs via tryptamine/serotonin/N-acetylserotonin and, finally, to melatonin, where tryptophan serves as the precursor molecule (Ye et al. 2019). However, in contrast to this, in a stressed environment or during senescence, the level of melatonin intermediates such as serotonin, tryptophan and tryptamine is higher in plants (Back 2021). In this situation, plants mostly prefer formation of melatonin via tryptamine/serotonin/5-methoxytrptamine (Back et al. 2016).

According to Hardeland (2019), chloroplast and mitochondria are able to synthesise melatonin in plants due to their evolutionary histories. Chloroplast evolved in plants from photosynthetic bacteria, whereas mitochondria from ingested α-proteobacteria during evolution. Both the organelles retained their ability to synthesise melatonin (Zhao et al. 2019). Recently, Tan and Reiter (2020) reported that during normal condition, chloroplast is the major site of melatonin synthesis, whereas during stressed condition when synthesis is blocked in chloroplast, mitochondrial synthesis of melatonin occurs.

10.2.2 Catabolic Pathway

As compared to that of animals, the understanding about catabolic pathway of melatonin is still in its early phase. Melatonin can be degraded via both non-enzymatic and enzymatic transformation routes (Back 2021). Non-enzymatic degradation of melatonin with singlet oxygen or free radicals is very much important with regard to tolerance capability of plants. Along with this, a non-enzymatic dioxygenation and hydroxylation mode of degradation of melatonin has also been reported earlier by Hardeland (2015) and Tan et al. (2002), respectively. According to Hardeland et al. (2003), dioxygenation of melatonin is mostly caused by the sequential action of superoxide anion and abstraction of electrons or hydrogen ions from free radicals. Non-enzymatic hydroxylation in different positions of the indole moiety are caused by interactions with free radicals of higher reactivity, usually hydroxyl radicals or another hydrogen-abstracting radical followed by combination with hydroxyl radicals (Tan et al. 2014).

Enzymatic degradation of melatonin occurs via hydroxylation, demethylation, deacetylation and dioxygenation; however, due to the presence of various isoforms of cytochrome P_{450}, demethylation and hydroxylations do occur mostly (Zhong et al. 2002). Of all the possibilities, only the hydroxylation reaction has been identified in a plant that is carried by the enzymatic action of melatonin 2-hydroxylase rather than cytochrome P_{450} enzymes (Byeon et al. 2015a). Melatonin 2-hydroxylase converts melatonin into 2-hydroxymelatonin which have tautomeric equilibrium with its stable form, that is, 3-acetamidoethyl-5-methoxyindolin-2-one (Hardeland 2010). Apart from this, only one hydroxylated product of melatonin, namely 4-hydroxymelatonin, is only detected in plants (Byeon et al. 2015b).

10.3 ROLE OF MELATONIN AND ITS INTERACTION WITH OTHER PHYTOHORMONES TO REGULATE ABIOTIC STRESS TOLERANCE LEVEL OF PLANTS

Plants, being sessile organisms, are subjected to a variety of environmental pressures. To survive under these conditions, they have developed a wide range of protective metabolites. Recent studies have established melatonin as a universal stress regulator in plants (Wang et al. 2018). Various reports have shown the promising role of melatonin against abiotic stresses such as drought, salinity, extreme temperature and heavy metals.

10.3.1 Drought Stress

Drought stress inhibits the growth and development of plants and also causes a severe threat to crop quality and productivity by altering the molecular, biological and physiological features of plants. Drought stress induces the formation of ROS, which eventually causes oxidative stress in plants. Drought stress induces the formation of melatonin in plants by up regulating the expression of melatonin biosynthetic genes (Shi et al. 2015a). Recent reports have also shown that exogenous application of melatonin was effective in maintaining its endogenous levels, which eventually restored the cell turgor pressure, chlorophyll and photosynthetic machineries and cell stability under drought stress (Ding et al. 2018). Recently, Sun et al. (2021) and Hosseini et al. (2021) also showed that melatonin treatment maintains seed germination, fruit quality and vegetative growth of plants during drought stress.

Exogenous application of melatonin can also efficiently trigger the enzymatic antioxidants in plants. Recently, Sadak and Bakry (2020) reported that exogenous application of melatonin enhanced the activity of catalase (CAT), superoxide dismutase (SOD) and peroxidase (POD), which suppressed the accumulation of ROS during drought. Tiwari et al. (2021) showed that melatonin treatment enhances the ascorbate-glutathione cycle that enhances the tolerance capability of plants. Exogenous application of melatonin also regulates the drought-related transcription factors and MAPKs. Gao et al. (2018) demonstrated that melatonin application up regulated the expression levels of transcription factors (*DREB2*, *WRKY1* and *MYB*) and MAPKs (*Asmap1* and *Aspk11*).

Crosstalk between melatonin and other phytohormones also plays a pivotal role in regulating the drought stress tolerance mechanism in plants. Tiwari et al. (2020) reported that exogenous application of melatonin was effective in repressing the drought-induced suppression of flowering in plants by enhancing the endogenous GA content of the plants. Earlier, Li et al. (2015) reported that melatonin application down regulated and up regulated the expression of ABA anabolic genes and catabolic genes, respectively, which eventually resulted in lower ABA content during drought stress, which in turn helped in regulating the stomatal function of the plants. Similar reports were also recently published by Chen et al. (2020) and Qi et al. (2021) where they showed that melatonin application reduced the endogenous ABA level, which helps in the regulation of stomatal closure.

Based on these reports, it can be concluded that negative and positive interactions between melatonin and ABA and GA, respectively, are responsible for regulating the drought tolerance capability of plants.

10.3.2 Salt Stress

Salt stress is considered as one the major limiting factors for growth and productivity of the crops. Salt stress possesses toxicity effects on plants by ion poisoning and osmotic stress (Van Zelm et al. 2020). Salt stress alters gene expression, mRNA stability and translational regulation, thus altering the abundance of proteins in plants (Sharma and Dietz 2009; Van Zelm et al. 2020).

The promising role of melatonin in inducing the tolerance level of plants such as maize, rice, wheat, cotton and cucumber is widely proven (Liang et al. 2015; Zhang et al. 2021; Zhou et al. 2016; Ke et al. 2018; Chen et al. 2018). Zhan et al. (2019) reported that exogenous application of melatonin helps to protect plants by enhancing the expression and activity of enzymatic antioxidants, nitrous oxide signaling and polyamine metabolism. Salt stress also hampers the ionic balance of the cells. To maintain the tolerance level of the plants, the level of Na^+ and K^+ in the cells needs to be maintained. Li et al. (2012) reported that melatonin-mediated ion homeostasis is associated with the expression of *NHX* and *AKT*, which encode two important ion channels. Along with this, Zhang et al. (2017a) showed that exogenous melatonin treatment enhanced the energy formation in the cells that maintained the activity of plasma membrane H^+-ATPase which regulated K^+/Na^+ homeostasis in cucumber seedlings exposed to salt stress. Similarly, Li et al. (2017) also reported that exogenous melatonin application inhibited the accumulation of Na^+ and Cl^- ions in the rice seedlings by enhancing the transcription of *Chloride channel 1* and *Chloride channel 2* in the roots and leaves and *Salt overly sensitive 1* in the roots. Salt stress also enhances the endogenous level of melatonin by enhancing the expression of the biosynthetic enzymes (Arnao and Hernández-Ruiz 2009). Recently, Wu et al. (2021) showed that overexpression of *SNAT1* in *Arabidopsis* enhanced the tolerance capability of plants by enhancing the endogenous melatonin levels. Thus, the role of melatonin in regulating salt tolerance in plants is quite evident.

Melatonin also interacts with other phytohormones to regulate the salt tolerance level of the plants. Xu et al. (2019) reported that exogenous application of melatonin enhanced the endogenous level of ethylene in salt-stressed grapevine. They further reported that melatonin application regulated the expression of *MYB108A*, which eventually triggered the expression of ethylene biosynthetic gene *ACS1*. In another study, Zhang et al. (2014) reported that NaCl treatment for 16 h up regulated the expression of ABA catabolic genes (*CsCYP707A1* and *CsCYP707A2*) and down regulated the expression of ABA anabolic genes (*CsNED2*) during the early stage of germination, which suggest that melatonin regulated salt stress in cucumber by lowering the endogenous ABA content. At the same time, they also reported that melatonin positively up regulated the expression of GA biosynthesis genes (*GA20ox* and *GA3ox*) during NaCl treatment that eventually led to a significant increase in GA levels (especially GA4). Similar observation was also reported by Park et al. (2021) where they showed that exogenous melatonin application reduced the level of ABA and enhanced the endogenous content of salicylic acid in salt-stressed *Brassica juncea* that eventually restored the growth and development of the plants. Thus, based on these studies, it can be concluded that interaction between melatonin and other phytohormones plays a major role in regulating salt tolerance level of plants.

10.3.3 Heavy Metal Toxicity

Due to the ever-growing world population and rapid industrialization, the release of toxic metals has significantly increased in the environment. Excess accumulation of heavy metals in the soil leads to their uptake and deposition in plant tissues, which promote the formation of ROS and also represses the growth and development of plants. Along with this, excess deposition of heavy metals in the plant tissues also hampers the Kreb's cycle, carbohydrate metabolism and photosynthesis in plants (Ni et al. 2018).

Exogenous application of melatonin enhances the endogenous content of melatonin in plants that strengthen the heavy metal tolerance capability of plants. According to Ni et al. (2018), exogenous application of melatonin promoted the level of endogenous melatonin in cadmium-stressed wheat seedling that eventually maintained the H_2O_2 homeostasis and activity of enzymatic antioxidants. Similarly, Sun et al. (2020) reported that exogenous melatonin level restored the antioxidative potential and lipid peroxidation in aluminum-stressed plants. A similar report was also published by Jahan et al. (2020) where they showed that melatonin application alleviated nickel toxicity by enhancing mineral and nutrient uptake and formation of secondary metabolites in tomato seedlings. Earlier, Hasan et al. (2015) also showed that melatonin effectively ameliorated cadmium-induced toxicity by inducing the H^+-ATPase activity and glutathione and phytochelatin content that sequester the deposited cadmium in the cells of tomato seedlings.

Several reports have also showed that the interaction of melatonin with other phytohormones is crucial for maintaining the tolerance level of crops. Samanta et al. (2021) reported that exogenous application of melatonin ameliorated the arsenic-induced perturbations by triggering the endogenous level of melatonin (by up regulating the expression of melatonin biosynthetic genes, that is, *TDC*, *SNAT* and *ASMT*) and GA (by up regulating the expression of GA anabolic genes *GA3ox*). They also showed that melatonin application further enhanced the endogenous ABA content of the rice seedling by up regulating and down regulating the expression of anabolic (*NCED3*) and catabolic (*ABA8ox*) genes involved in ABA metabolism, which eventually restored the level of protective metabolites in the seedlings. However, a contrasting report was recently published by Li et al. (2022) where they showed that pre-treatment of rice seedlings with melatonin reduced the endogenous ABA content and restored the GA content in the seedlings, which was highly effective in maintaining the growth of the seedlings by reducing the level of ROS and also regulating the level of protective metabolites. Thus, based on these works, it can be concluded that the interaction between melatonin and other phytohormones plays a pivotal role in controlling the tolerance capability of the plants; however, these reports are still in the very early stage, and more works still need to be done in the coming time.

10.3.4 Extreme Temperature

10.3.4.1 Heat Stress

Heat stress is becoming a huge concern for the scientific community worldwide due to global warming. Frequent heat waves affect plants at physiological, transcriptional, post-transcriptional and epigenetic levels (Zhao et al. 2020). To combat the negative effects of heat stress, plants induce the endogenous melatonin level that regulates the thermotolerance of the plants. Ahammed et al. (2019) reported that down regulating the expression of *TDC* and *COMT1* further aggravated the temperature-induced oxidative stress in plants due to lower content of endogenous melatonin. Similarly, overexpression of melatonin biosynthetic genes, that is, *ASMT* and *SNAT*, up regulated the formation of melatonin, which eventually led to higher thermotolerance (Xu et al. 2016; Wang et al. 2020). Exogenous melatonin application can also modulate the activity of heat shock factors in plants that are the master mediators of heat responses. According to Shi et al. (2015b), in *Arabidopsis*, melatonin-mediated thermotolerance was associated with the up regulation of HSFA2 and HSA32, as well as HSP90 and HSP101. Recently, Wang et al. (2020) further added that heat shock protein 40 interacted with *SlSNAT* to regulate melatonin biosynthesis and promoted thermotolerance by maintaining the RuBisCo enzyme stability under heat stress.

Similar to that of other abiotic stresses, crosstalk between melatonin and other phytohormones is crucial for regulating the heat tolerance level of the plants. Zhang et al. (2017b) demonstrated that exogenous application of melatonin further aggravated the endogenous melatonin and cytokinin level and reduced the ABA level, which eventually suppressed the heat-induced senescence in perennial ryegrass. Exogenous melatonin application also reduced the level of ABA by up regulating the expression of ABA catabolic genes (*CYP707A1* and *CYP707A2*) and down regulating the expression of ABA anabolic genes (*NCED3*) in heat-stressed *Glycine max* (Imran et al. 2021). They also demonstrated that enhanced melatonin level restored the level of salicylic acid and expression of heat shock factors that helped in lowering the heat-induced oxidative damage.

10.3.4.2 Cold Stress

As compared to that of heat stress, cold stress is not considered to be that harmful for the plants. However, extremely cold environment can also lead to redox imbalance and induce excessive ROS accumulation in plant tissues. In response to cold stress, plants generally enhance the level of melatonin that protects them from fatal injuries. Kang et al. (2010) reported that overexpression of *SNAT* in rice transgenics improved the cold tolerance level of the seedlings. Similar to that of other stresses, exogenous melatonin application alleviated the negative effects of cold stress by modulating the activity of antioxidants that reduced the accumulation of ROS and lipid peroxidation (Hu et al. 2016). Aghdam et al. (2019) also reported that cut flowers treated with melatonin led to lower accumulation of H_2O_2 due to higher 2,2-diphenyl-1-picrylhydrazyl scavenging capacity under cold stress than did untreated flowers.

Additionally, the content of phytohormones in plants during cold stress is also regulated by melatonin. For example, the expression level of a key gene (*ClAOC1*) in JA biosynthesis and a key gene (*ClAMI1*) in IAA biosynthesis were up regulated in watermelon leaves after melatonin application under cold stress (Chang et al. 2020; Li et al. 2021). Consequently, the JA and IAA content increased, thus improving photosynthesis and redox homeostasis (Chang et al. 2021).

10.4 CONCLUSION AND FUTURE PERSPECTIVE

Melatonin is an important metabolite that plays a pivotal role in abrogating the negative effects of abiotic stresses in plants. In this chapter, we tried to summarize the regulatory mechanisms that drive the melatonin-mediated abiotic stress tolerance in plants. Melatonin can directly scavenge the ROS formed in the cells, thus improving the cellular redox status. It can also regulate the expression and activity of enzymes involved in stress responses, phytohormone pathway and antioxidant production.

Exogenous melatonin application can enhance the biochemical and physiological processes to enhance the tolerance capability of the plants. In spite of such a promising role of melatonin against abiotic stress, the signaling, transcriptional and epigenetic regulation of melatonin still remains unclear. Crosstalk between melatonin and other phytohormones also occurs in plants; however, the detailed mechanism of this crosstalk is not completely understood, and further studies need to be done in the near future. Additionally, the beneficial role of melatonin is tested under laboratory conditions and on model plants; however, large-scale agricultural and commercial applications of melatonin still need to be done. The effects of melatonin other than on model plants also need to be investigated. Additionally, due to such a promising role of melatonin, transgenic technologies might be considered as a potential tool for enhancing the endogenous melatonin content which might come in handy to enhance the tolerance level of the plants. Therefore, it can be concluded that melatonin treatment can help in restoring the growth and yield of the plants that can fulfill the food requirements of the ever-growing population of the world in the future.

REFERENCES

Aghdam MS, Jannatizadeh A, Nojadeh MS, Ebrahimzadeh A (2019) Exogenous melatonin ameliorates chilling injury in cut anthurium flowers during low temperature storage. Postharvest Biol Technol 148: 184–191.

Ahammed GJ, Xu W, Liu A, Chen S (2019) Endogenous melatonin deficiency aggravates high temperature-induced oxidative stress in *Solanum lycopersicum* L. Environ Exp Bot 161: 303–311.

Arnao MB, Hernández-Ruiz J (2009) Chemical stress by different agents affects the melatonin content of barley roots. J Pineal Res 46: 295–299.
Arnao MB, Hernández-Ruiz J (2019a) Melatonin and reactive oxygen and nitrogen species: A model for the plant redox network. Melatonin Res 2: 152–168.
Arnao MB, Hernández-Ruiz J (2019b) Melatonin: A new plant hormone and/or a plant master regulator? Trends Plant Sci 24: 38–48.
Back K (2021) Melatonin metabolism, signaling and possible roles in plants. Plant J 105: 376–391.
Back K, Tan DX, Reiter RJ (2016) Melatonin biosynthesis in plants: Multiple pathways catalyze tryptophan to melatonin in the cytoplasm or chloroplasts. J Pineal Res 61: 426–437.
Byeon Y, Lee HY, Hwang OJ, Lee HJ, Lee K, Back K (2015a) Coordinated regulation of melatonin synthesis and degradation genes in rice leaves in response to cadmium treatment. J Pineal Res 58: 470–478.
Byeon Y, Tan D-X, Reiter RJ, Back K (2015b) Predominance of 2-hydroxymelatonin over melatonin in plants. J Pineal Res 59: 448–454.
Chang J, Guo Y, Zhang Z, Wei C, Zhang Y, Ma, J, Yang J, Zhang X, Li H (2020) CBF-responsive pathway and phytohormones are involved in melatonin-improved photosynthesis and redox homeostasis under aerial cold stress in watermelon. Acta Physiol Plant 42: 159.
Chang T, Zhao Y, He H, Xi Q, Fu J, Zhao Y (2021) Exogenous melatonin improves growth in hulless barley seedlings under cold stress by influencing the expression rhythms of circadian clock genes. Peer J 9: e10740.
Chen K, Li GJ, Bressan RA, Song CP, Zhu JK, Zhao Y (2020) Abscisic acid dynamics, signaling, and functions in plants. J Integr Plant Biol 62: 25–54.
Chen Z, Gu Q, Yu X, Huang L, Xu S, Wang R, Shen W, Shen W (2018) Hydrogen peroxide acts downstream of melatonin to induce lateral root formation. Ann Bot 121: 1127–1136.
Ding F, Wang G, Wang M, Zhang S (2018) Exogenous melatonin improves tolerance to water deficit by promoting cuticle formation in tomato plants. Molecules 23: 1605.
Fan J, Xie Y, Zhang Z, Chen L (2018) Melatonin: A multifunctional factor in plants. Int J Mol Sci 19: 1528.
Gao W, Zhang Y, Feng Z, Bai Q, He J, Wang YJ (2018) Effects of melatonin on antioxidant capacity in naked oat seedlings under drought stress. Molecules 23: 1580.
Hardeland R (2010) Melatonin metabolism in the central nervous system. Curr Neuropharmacol 8: 168–181.
Hardeland R (2015) Melatonin in plants and other phototrophs: Advances and gaps concerning the diversity of functions. J Exp Bot 66: 627–646.
Hardeland R (2019) Melatonin in the evolution of plants and other phototrophs. Melatonin Res 2: 10–36.
Hardeland R, Poeggeler B, Niebergall R, Zelosko V (2003) Oxidation of melatonin by carbonate radicals and chemiluminescence emitted during pyrrole ring cleavage. J Pineal Res 34: 17–25.
Hasan MK, Ahammed GJ, Yin L, Shi K, Xia X, Zhou Y, Yu J, Zhou J (2015) Melatonin mitigates cadmium phytotoxicity through modulation of phytochelatins biosynthesis, vacuolar sequestration, and antioxidant potential in *Solanum lycopersicum* L. Front Plant Sci 6: 601.
Hosseini MS, Samsampour D, Zahedi SM, Zamanian K, Rahman MM, Mostofa MG, Tran LP (2021) Melatonin alleviates drought impact on growth and essential oil yield of lemon verbena by enhancing antioxidant responses, mineral balance, and abscisic acid content. Physiol Plant 172: 1363–1375.
Hu Z, Fan J, Xie Y, Amombo E, Liu A, Gitau MM, Khaldun ABM, Chen L, Fu J (2016) Comparative photosynthetic and metabolic analyses reveal mechanism of improved cold stress tolerance in bermudagrass by exogenous melatonin. Plant Physiol Biochem 100: 94–104.
Imran M, Aaqil Khan M, Shahzad R, Bilal S, Khan M, Yun BW, Khan AL, Lee IJ (2021) Melatonin ameliorates thermotolerance in soybean seedling through balancing redox homeostasis and modulating antioxidant defense, phytohormones and polyamines biosynthesis. Molecules 26: 5116.
Jahan MS, Guo S, Baloch AR, Sun J, Shu S, Wang Y, Ahammed GJ, Kabir K, Roy R (2020) Melatonin alleviates nickel phytotoxicity by improving photosynthesis, secondary metabolism and oxidative stress tolerance in tomato seedlings. Ecotoxicol Environ Saf 197: 110593.
Kang K, Lee K, Park S, Kim YS, Back K (2010) Enhanced production of melatonin by ectopic overexpression of human serotonin N-acetyltransferase plays a role in cold resistance in transgenic rice seedlings. J Pineal Res 49: 176–182.
Ke Q, Ye J, Wang B, Ren J, Yin L, Deng X, Wang S (2018) Melatonin mitigates salt stress in wheat seedlings by modulating polyamine metabolism. Front Plant Sci 9: 914.

Li C, Tan DX, Liang D, Chang C, Jia D, Ma F (2015) Melatonin mediates the regulation of ABA metabolism, free-radical scavenging, and stomatal behaviour in two Malus species under drought stress. J Exp Bot 66: 669–680.
Li C, Wang P, Wei Z, Liang D, Liu C, Yin L, Jia D, Fu M, Ma F (2012) The mitigation effects of exogenous melatonin on salinity-induced stress in *Malus hupehensis*. J Pineal Res 53: 298–306.
Li H, Guo Y, Lan Z, Xu K, Chang J, Ahammed GJ, Ma J, Wei C, Zhang X (2021) Methyl jasmonate mediates melatonin-induced cold tolerance of grafted watermelon plants. Hortic Res 8: 57.
Li R, Wu L, Shao Y, Hu Q, Zhang H (2022) Melatonin alleviates copper stress to promote rice seed germination and seedling growth via crosstalk among various defensive response pathways. Plant Physiol Biochem 179: 65–77.
Li X, Yu B, Cui Y, Yin Y (2017) Melatonin application confers enhanced salt tolerance by regulating Na^+ and Cl^- accumulation in rice. Plant Growth Regul 83: 441–454.
Liang C, Li A, Yu H, Li W, Liang C, Guo S, Zhang R, Chu C (2017) Melatonin regulates root architecture by modulating auxin response in rice. Front Plant Sci 8: 134.
Liang C, Zheng G, Li W, Wang Y, Hu B, Wang H, Wu H, Qian Y, Zhu XG, Tan DX, Chen SY, Chu C (2015) Melatonin delays leaf senescence and enhances salt stress tolerance in rice. J Pineal Res 59: 91–101.
Moustafa-Farag M, Mahmoud A, Arnao MB, Sheteiwy MS, Dafea M, Soltan M, Elkelish A, Hasanuzzaman M, Ai S (2020) Melatonin-induced water stress tolerance in plants: Recent advances. Antioxidants (Basel) 9: 809.
Ni J, Wang Q, Shah FA, Liu W, Wang D, Huang S, Fu S, Wu L (2018) Exogenous melatonin confers cadmium tolerance by counter-balancing the hydrogen peroxide homeostasis in wheat seedlings. Molecules 23: 799.
Park HS, Kazerooni EA, Kang SM, Al-Sadi AM, Lee IJ (2021) Melatonin enhances the tolerance and recovery mechanisms in *Brassica juncea* (L.) Czern. under saline conditions. Front Plant Sci 12: 593717.
Qi H, Xia FN, Xiao S (2021) Autophagy in plants: Physiological roles and post-translational regulation. J Integr Plant Biol 63: 161–179.
Sadak MS, Bakry BA (2020) Alleviation of drought stress by melatonin foliar treatment on two flax varieties under sandy soil. Physiol Mol Biol Plants 26: 907–919.
Samanta S, Banerjee A, Roychoudhury A (2021) Exogenous melatonin regulates endogenous phytohormone homeostasis and thiol-mediated detoxification in two indica rice cultivars under arsenic stress. Plant Cell Rep 40: 1585–1602.
Sharma SS, Dietz KJ (2009) The relationship between metal toxicity and cellular redox imbalance. Trends Plant Sci 14: 43–50.
Shi H, Jiang C, Ye T, Tan DX, Reiter RJ, Zhang H, Liu R, Chan Z (2015a) Comparative physiological, metabolomic, and transcriptomic analyses reveal mechanisms of improved abiotic stress resistance in bermudagrass [*Cynodon dactylon* (L). Pers.] by exogenous melatonin. J Exp Bot 66: 681–694.
Shi H, Tan DX, Reiter RJ, Ye T, Yang F, Chan ZJ (2015b) Melatonin induces class A1 heat-shock factors (HSFA1s) and their possible involvement of thermotolerance in *Arabidopsis*. J. Pineal Res 58: 335–342.
Sun C, Liu L, Wang L, Li B, Jin C, Lin X (2021) Melatonin: A master regulator of plant development and stress responses. J Integr Plant Biol 63: 126–145.
Sun C, Lv T, Huang L, Liu X, Jin C, Lin X (2020) Melatonin ameliorates aluminum toxicity through enhancing aluminum exclusion and reestablishing redox homeostasis in roots of wheat. J Pineal Res 68: e12642.
Tan DX, Hardeland R, Back K, Manchester LC, Alatorre-Jimenez MA, Reiter RJ (2016) On the significance of an alternate pathway of melatonin synthesis via 5-methoxytryptamine: Comparisons across species. J Pineal Res 61: 27–40.
Tan DX, Hardeland R, Manchester LC, Galano A, Reiter RJ (2014) Cyclic-3-hydroxymelatonin (C3HOM), a potent antioxidant, scavenges free radicals and suppresses oxidative reactions. Curr Med Chem 21: 1557–1565.
Tan DX, Reiter RJ (2020) An evolutionary view of melatonin synthesis and metabolism related to its biological functions in plants. J Exp Bot 71: 4677–4689.
Tan DX, Reiter RJ, Manchester LC, Yan MT, El-Sawi M, Sainz RM, Mayo JC, Kohen R, Allegra M, Hardeland R (2002) Chemical and physical properties and potential mechanisms: Melatonin as a broad spectrum antioxidant and free radical scavenger. Curr Top Med Chem 2: 181–197.
Tiwari RK, Lal MK, Kumar R, Chourasia KN, Naga KC, Kumar D, Das SK, Zinta G (2021) Mechanistic insights on melatonin-mediated drought stress mitigation in plants. Physiol Plant 172: 1212–1226.

Tiwari RK, Lal MK, Naga KC, Kumar R, Chourasia KN, Subhash S, Kumar D, Sharma S (2020) Emerging roles of melatonin in mitigating abiotic and biotic stresses of horticultural crops. Sci Hortic 272: 109592.

Van Zelm E, Zhang Y, Testerink C (2020) Salt tolerance mechanisms of plants. Annu Rev Plant Biol 71: 403–433.

Wang X, Zhang H, Xie Q, Liu Y, Lv H, Bai R, Ma R, Li X, Zhang X, Guo YD, Zhang N (2020) *SlSNAT* interacts with HSP40, a molecular chaperone, to regulate melatonin biosynthesis and promote thermotolerance in tomato. Plant Cell Physiol 61: 909–921.

Wang Y, Reiter RJ, Chan Z (2018) Phytomelatonin: A universal abiotic stress regulator. J Exp Bot 69: 963–969.

Weeda S, Zhang N, Zhao X, Ndip G, Guo Y, Buck GA, Fu C, Ren S (2014) Arabidopsis transcriptome analysis reveals key roles of melatonin in plant defense systems. PLoS One. 9: e93462.

Wu Y, Fan X, Zhang Y, Jiang J, Sun L, Rahman FU, Liu C (2021) *VvSNAT1* overexpression enhances melatonin production and salt tolerance in transgenic *Arabidopsis*. Plant Physiol Biochem 166: 485–494.

Xu L, Xiang G, Sun Q, Ni Y, Jin Z, Gao S, Yao Y (2019) Melatonin enhances salt tolerance by promoting MYB108A-mediated ethylene biosynthesis in grapevines. Hortic Res 6: 114.

Xu W, Cai SY, Zhang Y, Wang Y, Ahammed GJ, Xia XJ, Shi K, Zhou YH, Yu JQ, Reiter RJ, Zhou J (2016) Melatonin enhances thermotolerance by promoting cellular protein protection in tomato plants. J Pineal Res 61: 457–469.

Ye T, Yin X, Yu L, Zheng SJ, Cai WJ, Wu Y, Feng YQ (2019) Metabolic analysis of the melatonin biosynthesis pathway using chemical labeling coupled with liquid chromatography-mass spectrometry. J Pineal Res 66: e12531.

Zhan H, Nie X, Zhang T, Li S, Wang X, Du X, Tong W, Song W (2019) Melatonin: A small molecule but important for salt stress tolerance in plants. Int J Mol Sci 20: 709.

Zhang HJ, Zhang N, Yang RC, Wang L, Sun QQ, Li DB, Cao YY, Weeda S, Zhao B, Ren S, Guo YD (2014) Melatonin promotes seed germination under high salinity by regulating antioxidant systems, ABA and GA_4 interaction in cucumber (*Cucumis sativus* L.). J Pineal Res 57: 269–279.

Zhang J, Shi Y, Zhang X, Du H, Xu B, Huang B (2017b) Melatonin suppression of heat-induced leaf senescence involves changes in abscisic acid and cytokinin biosynthesis and signaling pathways in perennial ryegrass (*Lolium perenne* L.). Environ Exp Bot 138: 36–45.

Zhang N, Sun Q, Zhang H, Cao Y, Weeda S, Ren S, Guo YD (2015) Roles of melatonin in abiotic stress resistance in plants. J Exp Bot 66: 647–656.

Zhang N, Zhang HJ, Sun QQ, Cao YY, Li X, Zhao B, Wu P, Guo YD (2017a) Proteomic analysis reveals a role of melatonin in promoting cucumber seed germination under high salinity by regulating energy production. Sci Rep 7: 503.

Zhang Y, Fan Y, Rui C, Zhang H, Xu N, Dai M, Chen X, Lu X, Wang D, Wang J, Wang J, Wang Q, Wang S, Chen C, Guo L, Zhao L, Ye W (2021) Melatonin improves cotton salt tolerance by regulating ROS scavenging system and Ca^{2+} signal transduction. Front Plant Sci 12: 693690.

Zhao D, Yao Z, Zhang J, Zhang R, Mou Z, Zhang X, Li Z, Feng X, Chen S, Reiter RJ (2021) Melatonin synthesis genes N-acetylserotonin methyltransferases evolved into caffeic acid O-methyltransferases and both assisted in plant terrestrialization. J Pineal Res 71: e12737.

Zhao D, Yu Y, Shen Y, Liu Q, Zhao Z, Sharma R, Reiter RJ (2019) Melatonin synthesis and function: Evolutionary history in animals and plants. Front Endocrinol (Lausanne) 10: 249.

Zhao J, Lu Z, Wang L, Jin B (2020) Plant responses to heat stress: Physiology, transcription, noncoding RNAs, and epigenetics. Int J Mol Sci 22: 117.

Zhong L, Wang K, Tan J, Li W, Li S (2002) Putative cytochromeP450 genes in rice genome (*Oryza sativa* L. ssp. indica) and their EST evidence. Sci China C Life Sci 45: 512–517.

Zhou X, Zhao H, Cao K, Hu L, Du T, Baluška F, Zou Z (2016) Beneficial roles of melatonin on redox regulation of photosynthetic electron transport and synthesis of D1 protein in tomato seedlings under salt stress. Front Plant Sci 7: 1823.

Zhou Y, Liao L, Liu X, Liu B, Chen X, Guo Y, Huang C, Zhao Y, Zeng Z (2020) Crystal structure of *Oryza sativa* TDC reveals the substrate specificity for TDC-mediated melatonin biosynthesis. J Adv Res 24: 501–511.

11 Crosstalk between Melatonin and Gasotransmitters in Plant Adaptation against Abiotic and Biotic Stress

Pankajini Samal, Meera Kumari Kar and Arup Kumar Mukherjee

11.1 INTRODUCTION

Melatonin (*N-acetyl*-5-methoxytryptamine) was first identified in the bovine pineal gland of mammals in 1958 (Lerner et al. 1958). Melatonin is found in all organisms including plants, animals and microbes, especially bacteria. However, the function of melatonin varies from species to species. In animals, melatonin affects many physiological activities like food processing, locomotion, circadian rhythms, sleeping behaviour, body temperature, retinal physiology, immunological regulation, sexual behaviour and seasonal reproduction. In 1995, melatonin was first discovered in higher plants by Dubbles group (Dubbels et al. 1995) and Hattori groups (Hattori et al. 1995) individually. Their report says that in the case of plants, melatonin regulates the photoperiodic behaviour, plant growth and development and immune system for stress management. Evidence from a number of studies has established that the exogenous treatment of melatonin never affects plants as the enzymatic biosynthesis of melatonin is similar to that in animals (Samanta et al. 2020; Debnath et al. 2019).

In plants, melatonin plays a role of multi-regulator biomolecule which can tolerate both biotic and abiotic stress, can act as a growth regulator of both vegetative and reproductive development as well as photosynthetic efficiency and biomass yield (Arnao and Hernández-Ruiz 2018).

11.2 BIOSYNTHESIS AND FUNCTION OF MELATONIN

In plants, melatonin is distributed in various parts like root, stem and leaves. The biosynthesis of melatonin is controlled by multiple factors such as some environmental factors like sunlight, temperature and plant development processes including leaf development, fruit ripening, senescence, etc. (Byeon et al. 2012; Dl et al. 2016; Okazaki and Ezura 2009; Shi, Reiter, et al. 2015). In addition to these factors, melatonin synthesis is stimulated due to many environmental stresses such as drought, cold, heat stress and ultraviolet radiations which affect the plant growth and development (Arnao and Hernández-Ruiz 2013; Byeon and Back 2014; Afreen, Zobayed, and Kozai 2006). Generally, the synthesis of melatonin in plants starts from tryptophan with the help of various enzymes (Fan et al. 2018). The first step is the decarboxylation of tryptophan with the help of the enzyme tryptophan decarboxylase (TDH), followed by hydroxylation of amine by tryptamine-5-hydroxylase (T5H). Then T5H converts tryptamine to serotonin which ultimately gives melatonin, followed by two steps (Posmyk and Janas 2009). In the first step, serotonin is converted to N-acetylserotonin via N-acetyltransferase. Then the end product melatonin is produced upon catalysis of N-acetylserotonin via N-acetylserotonin methyltransferase (ASMT). In the second step, conversion of serotonin to 5-methoxytryptamine takes place by the help of ASMT and then the synthesis of melatonin via serotonin N-acetyltransferase (SNAT) (Raza et al.

DOI: 10.1201/9781003343752-11

2022). Though the site of melatonin biosynthesis has not been explored properly, some evidences from the previous studies recommend that chloroplast and mitochondria are the major sites for melatonin biosynthesis in plants (Tan and Reiter 2020; Tan et al. 2013; Byeon et al. 2014).

In one of the earliest studies, Murch and Saxena (2002) demonstrated that indole-3-acetic acid (IAA) is also synthesized from tryptamine in an alternative step by the catalyst tryptamine deaminase (TDA), and it is one of the intermediates of serotonin as well as melatonin biosynthesis (Figure 11.1). Hence, auxin and phytomelatonin have some similar functions as they are

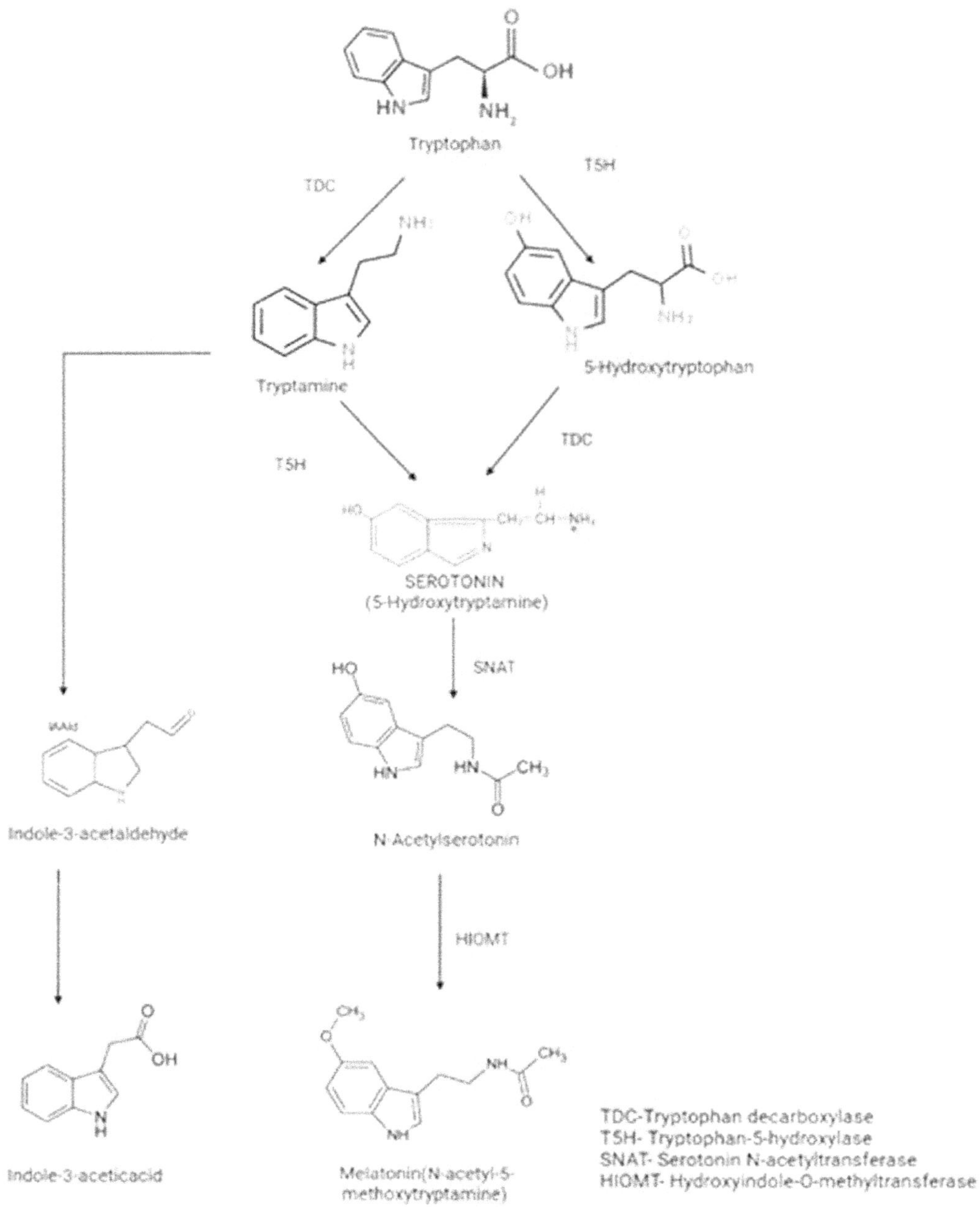

FIGURE 11.1 Biosynthetic pathway of melatonin in plants.

sharing the same biosynthetic pathways. It can be otherwise said that phytomelatonin controls the development and growth of plants by regulating auxin production and transport (Banerjee and Roychoudhury 2017).

Generally, phytomelatonin is considered to be an antioxidant which plays a key role in controlling reactive oxygen (ROS) and reactive nitrogen species (RNS) and other harmful free radicals and oxidative molecules under stress conditions. In addition, it also shows an involvement in various physiological processes such as seed germination, photosynthesis, growth, development of plant organs and increased tolerance against biotic and abiotic stresses (Arnao and Hernández-Ruiz 2019).

11.3 ROLE OF MELATONIN IN ASSOCIATION WITH GASOTRANSMITTERS FOR ABIOTIC STRESS MANAGEMENT

Plant hormones are involved in various developmental processes such as seed germination, flowering, plant growth, seed dormancy and senescence. So the phytohormones play an influential role in the adaptive behaviour of plants against any harmful external stimuli, whether they are biotic or abiotic stresses. Plants are vulnerable to various negative factors like environmental stress (low/high temperature, salinity, less/excess water), ultraviolet radiation and heavy metal, leading to a great yield loss to crops. To tackle these stresses, plants activate their immune system, and many endogenous molecules get activated (Figure 11.2). Among them, some endogenously synthesized signal molecules are gasotransmitters, which include nitric oxide (NO), hydrogen sulfide (H_2S), carbon monoxide (CO), methane (CH_4) and reactive oxygen species (ROS). Recent studies revealed a correlation between these gasotransmitters (NO, H_2S, CO, CH_4) and melatonin for abiotic stress tolerance in plants (Wang et al. 2022; Tan et al. 2007; Arnao, Cano, and Hernández-Ruiz 2022). The increase in concentration to these reactive signaling molecules triggers specific cellular and antioxidant activities. Phytomelatonin, being a master regulator of plant abiotic stress management, balances the ROS level by controlling the activities of redox enzymes such as super oxide dismutase, peroxidase, catalase, glutathione reductase, etc. (Arnao and Hernández-Ruiz 2019; Gu et al. 2022). Previous studies have established that melatonin also plays an important role in metabolic processes by regulating primary and secondary metabolism (Gu et al. 2022). Iqbal et al. (2021) reported that the crosstalk between melatonin and H_2S increased carbohydrate metabolism in plants in response to heat stress. Phytomelatonin shows significant involvement in cold temperature stress as it was reportedly induced during cold stress in cucumber seedlings (Feng et al. 2021). There are also evidences suggesting that exogenous application of melatonin improved low temperature stress-induced loss of seed germination in rice. Melatonin, in association with ABA-INSENSITIVE 5 (OsABI5) signals, reduced the low temperature stress during seed germination in rice (Li et al. 2021).

It has been shown that exogenous application of melatonin enhanced nitrogen metabolism and also established oxidative equilibrium during aluminum stress condition in maize (Ren et al. 2022). Furthermore, in Chinese cabbage seedlings, melatonin increased tolerance against cadmium stress by inhibiting NO synthesis (Wang et al. 2021).

Salinity stress is a common abiotic stress which causes tremendous yield losses to various agriculturally important crops. Generally, it affects seed germination, delayed growth of plants and promotes senescence, resulting in cell death (Roychoudhury et al. 2013). Plants develop various ways for tolerating these stresses both at the molecular level and biochemical level. Among them, phytomelatonin plays a key role in managing these stresses by interacting with some gasotransmitters (Figure 11.2). During salt stress, the interplay between phytomelatonin and NO controls the seedling growth and maintains the oxidative homeostasis (Arora and Bhatla 2017) in sunflower seedlings. A recent study by Zhao et al. (2018) established that application of melatonin and NO resisted the salt stress–induced seedling growth inhibition as well as balanced the redox homeostasis in *Brassica napus*.

A recent investigation showed the crosstalk between melatonin and NO, which counteracted drought-induced loss in soya bean. They studied the detailed mechanism in the physiological,

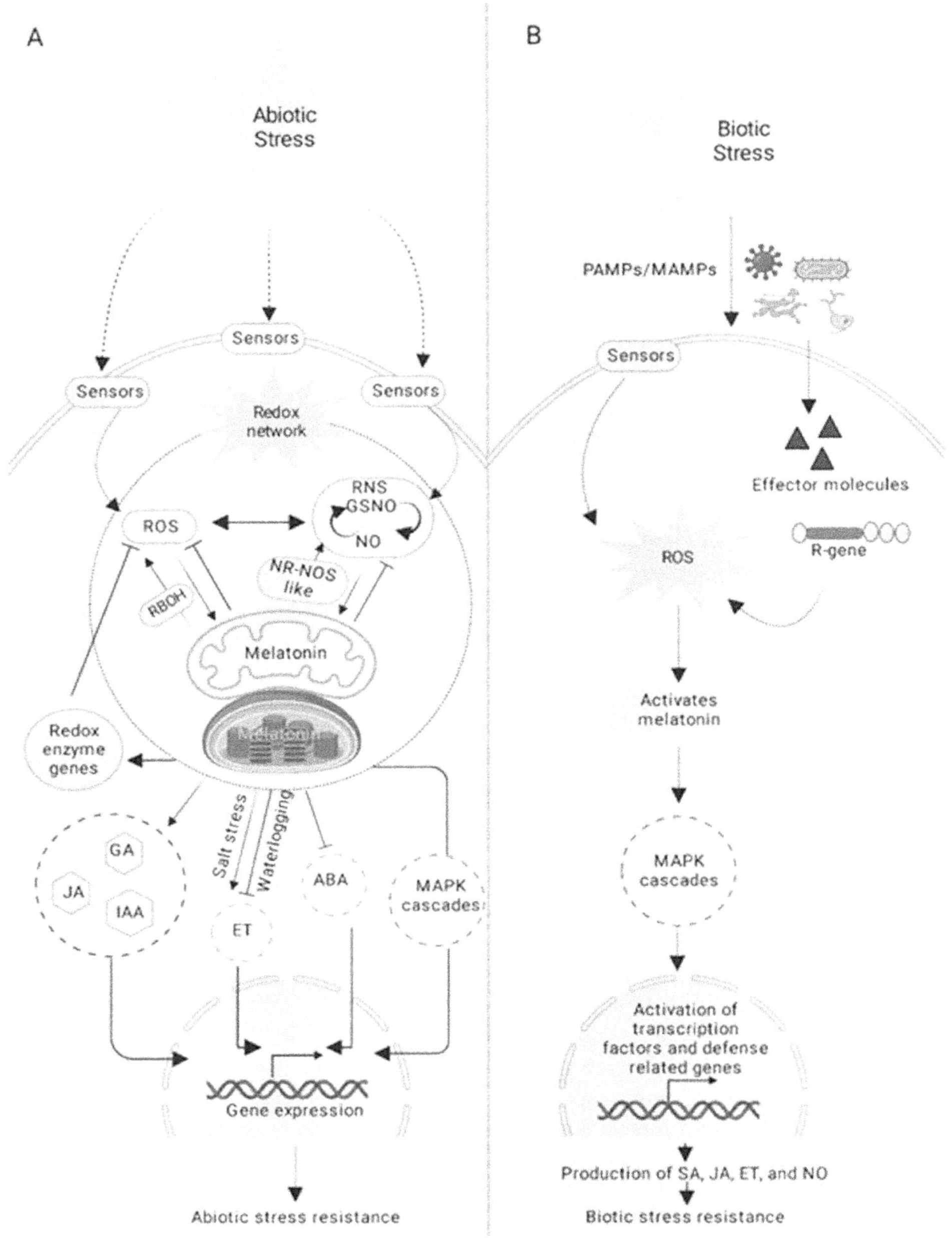

FIGURE 11.2 A model showing melatonin-mediated plant stress management. A. Role of melatonin in abiotic stress tolerance. Various abiotic stresses induce the accumulation of ROS (reactive oxygen species) and RNS (reactive nitrogen species), which promote endogenous melatonin in plants. Melatonin-mediated feedback mechanism can control the levels of ROS and RNS by inducing different redox enzymes. MAPK (mitogen-activated protein kinase) and other phytohormone signals are directly activated as melatonin mediated feedback mechanism under abiotic stresses in plants. RBOH, respiratory burst oxidase homolog, GSNO, S-nitroso glutathione, NOS, nitric oxide synthase. B. A schematic showing the role of melatonin in the regulation of biotic stress tolerance in plants.

molecular and antioxidant level and concluded that melatonin enhanced tolerance against drought stress by triggering down regulated accumulation of NO (Imran et al. 2021). Another evidence of melatonin as an anti-stress molecule is 2-hydroxy-melatonin, which showed its involvement in tolerance mechanism against combined stresses such as cold and drought in tobacco, tomato and cucumber (Lee and Back 2019).

Overall, these studies highlighted the potential usefulness of melatonin and its coordinate mode of action along with gasotransmitters and other signaling molecules against various abiotic stresses in plants (Samanta et al. 2021).

11.4 MELATONIN IN ASSOCIATION WITH GASOTRANSMITTERS AGAINST BIOTIC STRESS

Like abiotic stresses, plants are also vulnerable to various pathogens and microorganisms causing many diseases to plants (Figure 11.2). Previous studies suggested that melatonin plays a significant role in plant biotic stress management (Aftab and Roychoudhury 2021). There are also evidences that melatonin treatment enhanced the disease resistance against pathogens in many plants (He and He 2020). Yin et al. (2013) reported that pre-treatment of apple with melatonin improved resistance against Marssonina apple blotch, a fungal disease. They also observed stable maintenance of H_2O_2 concentration at steady state level as well as enhancement in activities of antioxidant enzymes and pathogenesis-related proteins during plant-pathogen interaction. The exogenous application of melatonin on *Arabidopsis* and tobacco leaves induced a series of defense related genes, salicylic acid and ethylene responsive genes, which are the key players of plant defense mechanism against the bacterial pathogen *Pseudomonas syringae* DC3000 (Lee, Byeon, and Back 2014). In addition, Shi, Chen, et al. (2015) further analysed that during bacterial infection, the levels of endogenous nitric oxide (NO) increased due to the exogenous application of melatonin. Both melatonin and NO play a significant role in the enhancement of disease resistance against *Pseudomonas syringae* in *Arabidopsis*. This result indicates the coordinate action of melatonin and NO, one of the major gasotransmitters of plants for tolerating the pathogen infestation (Figure 11.2). Similarly, another study showed that enhanced innate immunity induced by melatonin was due to the accumulation of sugars and glycerol against *Pseudomonas syringae* in tobacco (Qian et al. 2015).

In continuation to their previous research, Lee and Back (2016) again investigated for MAPK (mitogen-activated protein kinase) activity of *Arabidopsis* and tobacco after treatment with melatonin to examine whether the enhanced pathogen resistance is associated with MAPK cascade (Figure 11.2). They found two melatonin-induced MAPK kinase, MAPK6 and MAPK3, in *Arabidopsis* and tobacco triggered by MAPK signaling through MAPK kinase 4/5/7/9/. Melatonin functions as a first line of defense in plants and acts via H_2O_2 and NO dependent pathways by interacting with the MAPK cascades and oxidative burst pathways (Lee and Back 2017). One earlier study revealed that, in the case of banana, the involvement of melatonin in the regulation of heat shock protein increased disease resistance against *Fusarium* wilt (Wei et al. 2017). Potato late blight caused by *Phytophthora infestans* is a serious threat to potato cultivation and its yield. A previous research investigated the direct effects of melatonin in controlling infection caused by *P. infestans* in potato (Zhang et al. 2017). Melatonin application helped to inhibit the mycelial growth of *P. infestans* resulting in devitalization of potato late blight. Furthermore, the efficiency of fungicide and melatonin was also studied by applying both against *P. infestans* where significant increase in resistance against potato late blight was noted. To better understand the mechanism of resistance, transcriptome profiling was carried out where some differentially expressed genes were affected due to melatonin. These genes are responsive for fungal stress tolerance and plant defense strategy. So this study provides an intensive understanding about the role of melatonin in plant disease management. Recent study has evidenced exogenous application of 100 μM melatonin increased resistance against tobacco mosaic virus (TMV) in *Nicotiana glutinosa* and increased relative expression of pathogenesis-related (PR) genes 1 and 5 in *Solanum lycopersicum*. The melatonin-mediated viral resistance was due to the accumulation of salicylic acid (SA) and nitric oxide (NO) (Zhao et al. 2019).

Taken together, these studies support the notion that the crosstalk and interplay between melatonin and the gasotransmitters help plants to tolerate various stresses caused by fungi, bacteria and viruses.

11.5 CONCLUDING REMARKS AND FUTURE PROSPECTIVE

This chapter critically discusses some of the prominent functions of melatonin and its association with gasotransmitters to combat various biotic and abiotic stresses in plants. Melatonin, along with other plant hormones, shows a clear involvement in plant stress signaling. The interplay with ROS and RNS establish the fact that melatonin plays a significant role in maintaining oxidative homeostasis in plants during adverse situations. However, further research on these aspects might explore more closely the links between melatonin and signaling molecules of plants as well as animals. Melatonin is also a key factor in circadian rhythms of some plant species which needs a detailed and extensive investigation in the future. Overall, the present report enlightens the important role of the natural plant hormone melatonin as a master regulator of biotic and abiotic stress management (Figure 11.2).

REFERENCES

Afreen, F., S. M. A. Zobayed, and T. Kozai. 2006. "Melatonin in Glycyrrhiza Uralensis: Response of Plant Roots to Spectral Quality of Light and UV-B Radiation." *Journal of Pineal Research* 41 (2): 108–115. https://doi.org/10.1111/j.1600-079X.2006.00337.x.

Aftab, Tariq, and Aryadeep Roychoudhury. 2021. "Crosstalk Among Plant Growth Regulators and Signaling Molecules during Biotic and Abiotic Stresses : Molecular Responses and Signaling Pathways." *Plant Cell Reports* 40 (11): 2017–2019. https://doi.org/10.1007/s00299-021-02791-5.

Arnao, Marino B., Antonio Cano, and Josefa Hernández-Ruiz. 2022. "Phytomelatonin: An Unexpected Molecule with Amazing Performances in Plants." *Journal of Experimental Botany* 73 (17): 5779–5800. https://doi.org/10.1093/jxb/erac009.

Arnao, Marino B., and Josefa Hernández-Ruiz. 2013. "Growth Conditions Determine Different Melatonin Levels in *Lupinus Albus* L." *Journal of Pineal Research* 55 (2): 149–155. https://doi.org/10.1111/jpi.12055.

Arnao, Marino B., and Josefa Hernández-Ruiz. 2018. "Melatonin and Its Relationship to Plant Hormones." *Annals of Botany* 121 (2): 195–207. https://doi.org/10.1093/aob/mcx114.

Arnao, Marino B., and Josefa Hernández-Ruiz. 2019. "Melatonin: A New Plant Hormone and/or a Plant Master Regulator?" *Trends in Plant Science* 24 (1): 38–48. https://doi.org/10.1016/j.tplants.2018.10.010.

Arora, Dhara, and Satish C. Bhatla. 2017. "Author's Accepted Manuscript." *Free Radical Biology and Medicine*. https://doi.org/10.1016/j.freeradbiomed.2017.02.042.

Banerjee A., and A Roychoudhury. 2017" Melatonin as a Regulator of Abiotic Stress Tolerance in Plants." In: Singh V. P., Singh S., Prasad S. M. (Eds.) *Mechanisms Behind Phytohormonal Signaling and Crop Abiotic Stress Tolerance*, New York, Nova Science Publishers, pp. 47–60.

Byeon, Yeong, and Kyoungwhan Back. 2014. "Melatonin Synthesis in Rice Seedlings in Vivo Is Enhanced at High Temperatures and Under Dark Conditions Due to Increased Serotonin N-Acetyltransferase and N-Acetylserotonin Methyltransferase Activities." *Journal of Pineal Research* 56 (2): 189–195. https://doi.org/10.1111/jpi.12111.

Byeon, Yeong, Hyoung Yool Lee, Kyungjin Lee, Sangkyu Park, and Kyoungwhan Back. 2014. "Cellular Localization and Kinetics of the Rice Melatonin Biosynthetic Enzymes SNAT and ASMT." *Journal of Pineal Research* 56 (1): 107–114. https://doi.org/10.1111/jpi.12103.

Byeon, Yeong, Sangkyu Park, Young Soon Kim, Don Hee Park, Sungbeom Lee, and Kyoungwhan Back. 2012. "Light-Regulated Melatonin Biosynthesis in Rice during the Senescence Process in Detached Leaves." *Journal of Pineal Research* 53 (1): 107–111. https://doi.org/10.1111/j.1600-079X.2012.00976.x.

Debnath, Biswojit, Waqar Islam, Min Li, Yueting Sun, Xiaocao Lu, Sangeeta Mitra, Mubasher Hussain, Shuang Liu, and Dongliang Qiu. 2019. "Melatonin Mediates Enhancement of Stress Tolerance in Plants." *International Journal of Molecular Sciences* 20 (5). https://doi.org/10.3390/ijms20051040.

Dl, Van Tassel, N. Roberts, A. Lewy, O. Neill Sd, David L. Van Tassel, Nicholas Roberts, Alfred Lewy, and Sharman D. O. Neill. 2016. "Melatonin in Plants." *Serotonin and Melatonin*, 25–36. https://doi.org/10.1201/9781315369334-10.

Dubbels, R., R. J. Reiter, E. Klenke, A. Goebel, E. Schnakenberg, C. Ehlers, H. W. Schiwara, and W. Schloot. 1995. "Melatonin in Edible Plants Identified by Radioimmunoassay and by High Performance Liquid Chromatography-Mass Spectrometry." *Journal of Pineal Research* 18 (1): 28–31. https://doi.org/10.1111/j.1600-079X.1995.tb00136.x.

Fan, Jibiao, Yan Xie, Zaichao Zhang, and Liang Chen. 2018. "Melatonin: A Multifunctional Factor in Plants." *International Journal of Molecular Sciences* 19 (5): 1–14. https://doi.org/10.3390/ijms19051528.

Feng, Yiqing, Xin Fu, Lujie Han, Chenxiao Xu, Chaoyue Liu, Huangai Bi, and Xizhen Ai. 2021. "Nitric Oxide Functions as a Downstream Signal for Melatonin-Induced Cold Tolerance in Cucumber Seedlings." 12 (July): 1–20. https://doi.org/10.3389/fpls.2021.686545.

Gu, Quan, Qingqing Xiao, Ziping Chen, and Yi Han. 2022. "Crosstalk between Melatonin and Reactive Oxygen Species in Plant Abiotic Stress Responses: An Update." *International Journal of Molecular Sciences* 23 (10): 5666. https://doi.org/10.3390/ijms23105666.

Hattori, A., H. Migitaka, M. Iigo, M. Itoh, K. Yamamoto, R. Ohtani-Kaneko, M. Hara, T. Suzuki, and R. J. Reiter. 1995. "Identification of Melatonin in Plants and Its Effects on Plasma Melatonin Levels and Binding to Melatonin Receptors in Vertebrates." *Biochemistry and Molecular Biology International* 35 (3): 627–634.

He, Huyi, and Long Fei He. 2020. "Crosstalk between Melatonin and Nitric Oxide in Plant Development and Stress Responses." *Physiologia Plantarum* 170 (2): 218–226. https://doi.org/10.1111/ppl.13143.

Imran, Muhammad, Raheem Shazad, Saqib Bilal, Qari Muhammad Imran, Murtaza Khan, Sang-Mo Kang, Abdul Latif Khan, Byung-Wook Yun, and In-Jung Lee. 2021. "Exogenous Melatonin Mediates the Regulation of Endogenous Nitric Oxide in Glycine Max L. to Reduce Effects of Drought Stress." *Environmental and Experimental Botany* 188: 104511. https://doi.org/10.1016/j.envexpbot.2021.104511.

Iqbal, Noushina, Mehar Fatma, Harsha Gautam, Shahid Umar, Adriano Sofo, Ilaria D'ippolito, and Nafees A. Khan. 2021. "The Crosstalk of Melatonin and Hydrogen Sulfide Determines Photosynthetic Performance by Regulation of Carbohydrate Metabolism in Wheat under Heat Stress." *Plants* 10 (9). https://doi.org/10.3390/plants10091778.

Lee, Hye-Jung, and Kyoungwhan Back. 2019. "2-Hydroxymelatonin Confers Tolerance against Combined Cold and Drought Stress in Tobacco, Tomato, and Cucumber as a Potent Anti-Stress Compound in the Evolution of Land Plants." *Melatonin Research* 2 (2): 35–46. https://doi.org/10.32794/mr11250020.

Lee, Hyoung Yool, and Kyoungwhan Back. 2016. "Mitogen-Activated Protein Kinase Pathways Are Required for Melatonin-Mediated Defense Responses in Plants." *Journal of Pineal Research* 60 (3): 327–335. https://doi.org/10.1111/jpi.12314.

Lee, Hyoung Yool, and Kyoungwhan Back. 2017. "Melatonin Is Required for H2O2- and NO-Mediated Defense Signaling through MAPKKK3 and OXI1 in Arabidopsis Thaliana." *Journal of Pineal Research* 62 (2). https://doi.org/10.1111/jpi.12379.

Lee, Hyoung Yool, Yeong Byeon, and Kyoungwhan Back. 2014. "Melatonin as a Signal Molecule Triggering Defense Responses against Pathogen Attack in Arabidopsis and Tobacco." *Journal of Pineal Research* 57 (3): 262–268. https://doi.org/10.1111/jpi.12165.

Lerner, Aaron B., James D. Case, Yoshiyata Takahashi, Teh H. Lee, and Wataru Mori. 1958. "Isolation of Melatonin, the Pineal Gland Factor That Lightens Melanocytes1." *Journal of the American Chemical Society* 80 (10): 2587. https://doi.org/10.1021/ja01543a060.

Li, Ruiqing, Meng Jiang, Yue Song, and Huali Zhang. 2021. "Melatonin Alleviates Low-Temperature Stress via ABI5-Mediated Signals During Seed Germination in Rice (Oryza Sativa L.)." *Frontiers in Plant Science* 12 (September): 1–16. https://doi.org/10.3389/fpls.2021.727596.

Murch, Susan J., and Praveen K. Saxena. 2002. "Melatonin: A Potential Regulator of Plant Growth and Development?" *In Vitro Cellular and Developmental Biology: Plant* 38 (6): 531–536. https://doi.org/10.1079/IVP2002333.

Okazaki, Masateru, and Hiroshi Ezura. 2009. "Profiling of Melatonin in the Model Tomato (*Solanum Lycopersicum* L.) Cultivar Micro-Tom." *Journal of Pineal Research* 46 (3): 338–343. https://doi.org/10.1111/j.1600-079X.2009.00668.x.

Posmyk, Małgorzata M., and Krystyna M. Janas. 2009. "Melatonin in Plants." *Acta Physiologiae Plantarum* 31 (1): 1–11. https://doi.org/10.1007/s11738-008-0213-z.

Qian, Yongqiang, Dun Xian Tan, Russel J. Reiter, and Haitao Shi. 2015. "Comparative Metabolomic Analysis Highlights the Involvement of Sugars and Glycerol in Melatonin-Mediated Innate Immunity against Bacterial Pathogen in Arabidopsis." *Scientific Reports* 5 (June): 1–11. https://doi.org/10.1038/srep15815.

Raza, Ali, Sidra Charagh, Pedro García-caparrós, Atikur Rahman, H. Ogwugwa, Faisal Saeed, Wanmei Jin, et al. 2022. "Biotechnology in Agriculture and the Food Chain Melatonin-Mediated Temperature Stress Tolerance in Plants." *GM Crops & Food* 13 (1): 196–217. https://doi.org/10.1080/21645698.2022.2106111.

Ren, Jianhong, Xiaoxiao Yang, Ning Zhang, Lu Feng, Chunying Ma, Yuling Wang, Zhenping Yang, and Juan Zhao. 2022. "Melatonin Alleviates Aluminum-Induced Growth Inhibition by Modulating Carbon and Nitrogen Metabolism, and Reestablishing Redox Homeostasis in Zea Mays L." *Journal of Hazardous Materials* 423 (PB): 127159. https://doi.org/10.1016/j.jhazmat.2021.127159.

Roychoudhury A., S. Paul, and S. Basu. (2013). "Cross-Talk between Abscisic Acid-Dependent and Abscisic Acid-Independent Pathways during Abiotic Stress." *Plant Cell Reports* 32 (7): 985–1006.

Samanta S., A. Banerjee, and A. Roychoudhury. (2021). "Exogenous Melatonin Regulates Endogenous Phytohormone Homeostasis and Thiol-Mediated Detoxification in Two Indica Rice Cultivars Under Arsenic Stress." *Plant Cell Reports* 40: 1585–1602.

Samanta, Santanu, Ankur Singh, Aditya Banerjee, and Aryadeep Roychoudhury. 2020. "Exogenous Supplementation of Melatonin Alters Representative Organic Acids and Enzymes of Respiratory Cycle as Well as Sugar Metabolism during Arsenic Stress in Two Contrasting Indica Rice Cultivars." *Journal of Biotechnology* 324: 220–232. https://doi.org/10.1016/j.jbiotec.2020.10.013.

Shi, Haitao, Yinhua Chen, Dun Xian Tan, Russel J. Reiter, Zhulong Chan, and Chaozu He. 2015. "Melatonin Induces Nitric Oxide and the Potential Mechanisms Relate to Innate Immunity against Bacterial Pathogen Infection in Arabidopsis." *Journal of Pineal Research* 59 (1): 102–108. https://doi.org/10.1111/jpi.12244.

Shi, Haitao, Russel J. Reiter, Dun Xian Tan, and Zhulong Chan. 2015. "Indole-3-Acetic Acid Inducible 17 Positively Modulates Natural Leaf Senescence through Melatonin-Mediated Pathway in Arabidopsis." *Journal of Pineal Research* 58 (1): 26–33. https://doi.org/10.1111/jpi.12188.

Tan, Dun Xian, Lucien C. Manchester, Xiaoyan Liu, Sergio A. Rosales-Corral, Dario Acuna-Castroviejo, and Russel J. Reiter. 2013. "Mitochondria and Chloroplasts as the Original Sites of Melatonin Synthesis: A Hypothesis Related to Melatonin's Primary Function and Evolution in Eukaryotes." *Journal of Pineal Research* 54 (2): 127–138. https://doi.org/10.1111/jpi.12026.

Tan, Dun Xian, Lucien C. Manchester, Maria P. Terron, Luis J. Flores, and Russel J. Reiter. 2007. "One Molecule, Many Derivatives: A Never-Ending Interaction of Melatonin with Reactive Oxygen and Nitrogen Species?" *Journal of Pineal Research* 42 (1): 28–42. https://doi.org/10.1111/j.1600-079X.2006.00407.x.

Tan, Dun Xian, and Russel J. Reiter. 2020. "An Evolutionary View of Melatonin Synthesis and Metabolism Related to Its Biological Functions in Plants." *Journal of Experimental Botany* 71 (16): 4677–4689. https://doi.org/10.1093/jxb/eraa235.

Wang, Tao, Jinxue Song, Ze Liu, Zili Liu, and Jin Cui. 2021. "Melatonin Alleviates Cadmium Toxicity by Reducing Nitric Oxide Accumulation and IRT1 Expression in Chinese Cabbage Seedlings." *Environmental Science and Pollution Research* 28 (12): 15394–15405. https://doi.org/10.1007/s11356-020-11689-w.

Wang, Yueqiao, Pengfei Cheng, Gan Zhao, Longna Li, and Wenbiao Shen. 2022. "Phytomelatonin and Gasotransmitters: A Crucial Combination for Plant Physiological Functions." *Journal of Experimental Botany* 73 (17): 5851–5862. https://doi.org/10.1093/jxb/erac159.

Wei, Yunxie, Wei Hu, Qiannan Wang, Hongqiu Zeng, Xiaolin Li, Yu Yan, Russel J. Reiter, Chaozu He, and Haitao Shi. 2017. "Identification, Transcriptional and Functional Analysis of Heat-Shock Protein 90s in Banana (Musa Acuminata L.) Highlight Their Novel Role in Melatonin-Mediated Plant Response to Fusarium Wilt." *Journal of Pineal Research* 62 (1). https://doi.org/10.1111/jpi.12367.

Yin, Lihua, Ping Wang, Mingjun Li, Xiwang Ke, Cuiying Li, Dong Liang, Shan Wu, et al. 2013. "Exogenous Melatonin Improves Malus Resistance to Marssonina Apple Blotch." *Journal of Pineal Research* 54 (4): 426–434. https://doi.org/10.1111/jpi.12038.

Zhang, Shumin, Xianzhe Zheng, Russel J. Reiter, Shun Feng, Ying Wang, Sen Liu, Liang Jin, Zhengguo Li, Raju Datla, and Maozhi Ren. 2017. "Melatonin Attenuates Potato Late Blight by Disrupting Cell Growth,

Stress Tolerance, Fungicide Susceptibility and Homeostasis of Gene Expression in Phytophthora Infestans." *Frontiers in Plant Science* 8. https://doi.org/10.3389/fpls.2017.01993.

Zhao, Gan, Yingying Zhao, Xiuli Yu, Felix Kiprotich, Han Han, Rongzhan Guan, Ren Wang, and Wenbiao Shen. 2018. "Nitric Oxide Is Required for Melatonin-Enhanced Tolerance against Salinity Stress in Rapeseed (Brassica Napus L.) Seedlings." *International Journal of Molecular Sciences* 19 (7): 1–22. https://doi.org/10.3390/ijms19071912.

Zhao, L., L. Chen, P. Gu, X. Zhan, Y. Zhang, C. Hou, Z. Wu, Y. F. Wu, and Q. C. Wang. 2019. "Exogenous Application of Melatonin Improves Plant Resistance to Virus Infection." *Plant Pathology* 68 (7): 1287–1295. https://doi.org/10.1111/ppa.13057.

12 Phytomelatonin
A Versatile Abiotic Stress Regulator in Plants

Gouranga Upadhyaya, Vishmita Sethi and Sreeramaiah N. Gangappa

12.1 INTRODUCTION TO MELATONIN PHYTOHORMONE: THE MANAGER OF ALL ENVIRONMENTAL STRESSORS

Melatonin, chemically termed *N-acetyl*-5-methoxytryptamine, is an evolutionarily conserved pleiotropic molecule with a ubiquitous presence in all living organisms. Initially discovered as a mammalian neurohormone, melatonin in plant biology is being increasingly revealed for its numerous roles in the growth and development of plants, most importantly its stress management ability. In addition to stress tolerance, melatonin is crucial owing to its direct role in biomass increase, seed germination, fruit ripening, delayed flowering and many more physiological processes (Zhang et al., 2021b). Increasing global warming and other climatic anomalies have resulted in a rise of various types of stresses and also resulted in combined stresses, that is, simultaneous exposure to more than one stress. Several plant species, either rich in melatonin or exogenously treated with melatonin, have shown significant ability to impede the constantly looming threat of environmental stress factors.

Melatonin as a regulatory hormone was first observed in plants in 1995 and later proved to be a constituent of higher plants with varying concentrations ranging from picogram to microgram in almost every plant tissue (Arnao and Hernández-Ruiz, 2009). Melatonin is amphiphilic in nature, that is, it can cross cell membranes easily and move into the subcellular compartments. Initially, experimental evidence proved melatonin to be a growth-enhancing and rooting molecule (Zhang et al., 2015); however, its role in plant stress tolerance has recently grabbed the limelight. Further, melatonin is also being revealed to be involved in stress-triggered developmental switches, including flowering, fruiting and ageing. It is regarded as a multidirectional stress manager owing to its ability to defend against a wide array of abiotic and biotic stress conditions. Multiple defense strategies employed by melatonin, including effective scavenging of reactive oxygen/nitrogen species (ROS and RNS), stimulation of antioxidant enzymes and stress-specific genes and transcription factors, make it a master regulator of plant stress defense mechanism. Moreover, the crosstalk feature of melatonin with other phytohormones makes it an interesting multifunctional molecule with wide research possibilities.

Melatonin is being enormously researched for its application as a biostimulant in agriculture (Arnao and Hernández-Ruiz, 2021). The exogenous application of melatonin and/or ectopic overexpression of melatonin biosynthetic genes can improvise plant defense response to combat several different types of adverse environmental factors like high temperature, cold, drought, waterlogging, salinity, high light, ultraviolet radiation, heavy metals, etc., all of which lead to ROS-mediated oxidative stress (Banerjee and Roychoudhury, 2017; Hassan et al., 2022). Hence, these biotechnological approaches are being studied to deduce the probable potential of melatonin in agricultural sectors. The present chapter emphasizes on the versatile role of melatonin in stress tolerance mechanism

 DOI: 10.1201/9781003343752-12

of plants, especially against abiotic stress, along with insights into how exogenous melatonin and genetically engineered melatonin synthesis is being employed as a biostimulant in agricultural sectors to fortify the crops against environmental stressors.

12.2 ANABOLIC AND CATABOLIC PATHWAYS OF MELATONIN IN PLANTS

Melatonin anabolism or synthesis was first reported by Murch et al. (2000) from tryptophan as the precursor in *Hypericum perforatum* plants. Like melatonin synthesis in animals, the plant melatonin biosynthesis pathway (MBP) follows four enzymatic steps, with tryptophan as the primary precursor, as demonstrated in Figure 12.1. Tryptophan acting as the precursor for MBP is derived from the shikimic acid pathway. The genes encoding enzymes in every step of MBP have been identified and cloned from various plant species like rice, pepper, *Catharanthus roseus*, tobacco, etc. (Zhang et al., 2015). There exists more than one biosynthetic pathway. Interestingly, all the pathways essentially involve serotonin as the intermediate molecule for melatonin synthesis in plants and animals. The enzymes of plant MBP include the following:

- Tryptophan decarboxylase (TDC)
- Tryptamine 5-hydroxylase (T5H)
- Serotonin N-acetyltransferase (SNAT)
- N-acetylserotonin methyltransferase (ASMT)/Caffeic acid O-methyltransferase (COMT)

In normal conditions, while the TDC enzyme catalyzes the first reaction of MBP in plants, T5H takes the initial step in animals. TDC causes decarboxylation of tryptophan in plants, leading to the formation of tryptamine. On the other hand, tryptophan hydroxylation by T5H occurs at the first step in animals, forming 5-hydroxytryptophan. Post-decarboxylation, the next step,

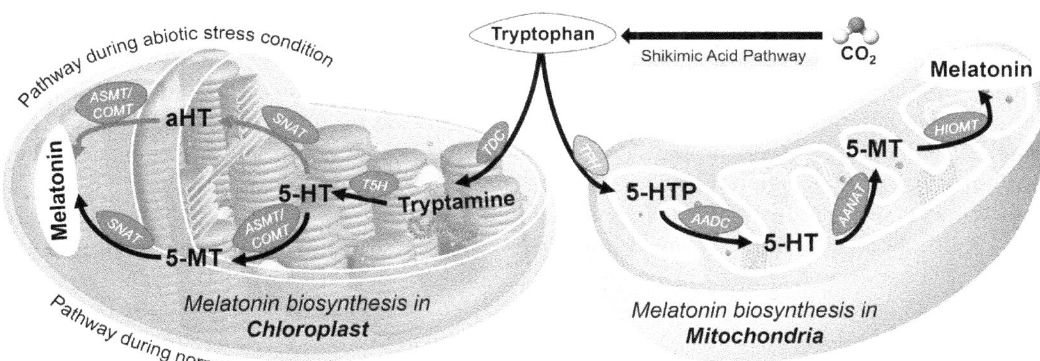

FIGURE 12.1 Melatonin biosynthesis pathway in plants. Schematic representation of melatonin synthesis pathway in the chloroplast (left panel) and mitochondria (right panel). In case of plants, the chloroplast-derived pathway of melatonin synthesis generally occurs, which is also stress-dependent. Sometimes, the chloroplast pathway may be blocked, and then the pathway bypass through the mitochondria-derived synthesis of melatonin. 5-HT, 5-hydroxytryptamine/serotonin; aHT, hydroxytryptamine; 5-MT, 5-methoxytryptamine; 5-HTP, 5-hydroxytryptophan; TDC, tryptophan decarboxylase; T5H, tryptamine 5-hydroxylase; SNAT, serotonin N-acetyltransferase; ASMT, N-acetylserotonin methyltransferase; COMT, caffeic acid O-methyltransferase; TPH, tryptophan hydroxylase; AADC, aromatic amino acid decarboxylase; AANAT, arylalkylamine N-acetyltransferase; HIOMT, hydroxyindole-O-methyltransferase.

involves hydroxylation at the C-5 position of tryptamine by the cytochrome P450 enzyme: T5H forming serotonin (or 5-hydroxytryptamine or 5-HT). Contrastingly, in animals, 5-hydroxytryptophan gets decarboxylated by TDC to form serotonin. The product serotonin acts as the key intermediary product of MBP, be it in plants or animals. Synthesis of serotonin from tryptophan is a highly active metabolic process, suggesting that significantly higher metabolic flow occurs for serotonin production rather than the succeeding end-product melatonin. Thus, TDC can be regarded as a rate-limiting step in the serotonin biosynthetic pathway. Subsequent melatonin production from serotonin requires two enzymatic reactions catalyzed by SNAT, ASMT and COMT, and as evidenced by many studies, this reaction has a comparatively low metabolic capacity. The first step is N-acetylserotonin (or N-acetyl-5-hydroxytryptamine or aHT) generation from serotonin by SNAT, which then gets O-methylated to form melatonin by either ASMT or COMT. SNAT is an essential key rate-limiting enzyme, and plants with silenced SNAT depict altered phenotypes with greater sensitivity to some abiotic stresses (Lee et al., 2019). Under abiotic stress conditions, these two enzymatic reactions are simply reversed, as shown in Figure 12.1.

The melatonin biosynthetic pathways differ in different subcellular compartments (Zeng et al., 2022). Various reports demonstrate that although chloroplast is the main site and the preferred subcellular organelle for melatonin biosynthesis, in some instances, it takes place in the mitochondria as well, particularly when the chloroplast pathway is blocked. Depending on the site of biosynthesis, the quantities of serotonin and melatonin synthesized vary massively owing to the different anabolic and catabolic flows. For example, while tryptophan and serotonin store up profusely in the ageing rice leaves, tryptamine and N-acetylserotonin are not formed sufficiently. This occurs due to the rapid conversion of tryptamine to serotonin by T5H and the slow activity of SNAT for converting serotonin to N-acetylserotonin. Moreover, serotonin does not accumulate much when a different enzyme(s) competing for serotonin as a substrate are present concurrently in the same subcellular organelle. The variable subcellular site for melatonin production with multiple biosynthetic pathways plays crucial roles in the induction of melatonin synthesis, maintenance of steady-state levels of melatonin and the response reaction to various environmental stressors.

On the other hand, melatonin catabolism or degradation is comparatively less probed. Two distinct routes of melatonin degradation have been reported till date: enzymatic and non-enzymatic. Non-enzymatic degradation is quite similar to that in animals and is catalyzed by the generated ROS/free radicals or by photocatalytic reactions by UV light (Zeng et al., 2022). These non-enzymatic degradative processes may be more important in plants, especially under adverse environmental conditions. For example, nitrosation of melatonin to form N-nitrosomelatonin (NOmela) contributes to long-distance redox signaling (Blanchard et al., 2000). Secondly, the enzymatic degradation of melatonin has been documented in many reports (Tan and Reiter, 2020). For example, indoleamine 2, 3-dioxygenase (IDO) constitutively expressed in transgenic tomatoes (*Solanum lycopersicum*) show a marked reduction in melatonin content (Okazaki et al., 2010). Another report demonstrates the enzymatic hydroxylation of melatonin into 2-,4- and 6-hydroxymelatonin in plants (Tan and Reiter, 2020). Among these, the concentration of 2-hydroxymelatonin (2-OHMel) at times exceeds that of melatonin in plants and is the most dominating one with a vital role in many physiological and biochemical cellular activities, especially under challenging conditions (Tan and Reiter, 2020). Interestingly, endogenously elicited and exogenously supplied 2-OHMel also improves plant tolerance to different abiotic stress conditions (Back et al., 2016; Tan and Reiter, 2020). Another identified melatonin hydroxylation metabolite is cyclic 3-hydroxymelatonin (c3-OHMel), formed by melatonin 3-hydroxylase (M3H) (Choi and Back, 2019). Other degradation products of melatonin include its deacetylation to 5-methoxytryptamine (5-MT) by aryl acylamidase (AAA), as observed in dinoflagellates, rhodophyceae and vertebrates (Tan et al., 2012; Hardeland, 2015), but these have not been reported in plants so far.

12.3 MELATONIN AND ITS FIGHT AGAINST VARIOUS ABIOTIC STRESS FACTORS

In natural environmental conditions, plants are continuously exposed to multiple abiotic stressors like high temperature, drought, cold, flooding, salinity, heavy metals, etc. These external factors affect the growth and development of plants by impairing physiological and biochemical processes such as photosynthesis, respiration, solute translocation, etc. All these factors have a negative effect on redox homeostasis through increased ROS accumulation at the cellular level (Das and Roychoudhury, 2014). The primary signature of such oxidative injury is the increase in lipid peroxidation rate, electrolyte leakage, membrane damage, injury of photosynthetic apparatus and consequent reduction of photosynthetic efficiency (Foyer and Harbinson, 2019; Mano et al., 2002; Upadhyaya et al., 2021b). Hence, ROS production damages plant growth, causing a massive reduction in crop yield, quality and biomass (Farooq et al., 2009; Roychoudhury et al. 2008).

In response to these oxidative damages, plants have developed an array of adaptive and defense mechanisms including the synthesis of plant growth regulators; one such phytohormone is melatonin (Sun et al., 2021). The sharp increase in endogenous melatonin concentration inside plant cells in response to environmental fluctuations or stress has been reported by many scientists across the world. A report by Hernández-Ruiz and Arnao (2008) suggests that both barley and lupin roots produced a significant amount of melatonin levels under stress conditions. In lupin plant tissues, exposure to chemical stressors $ZnSO_4$ or NaCl were noted to show the most prominent up regulation of endogenous melatonin levels, followed by chilling temperature and drought stress (Arnao and Hernández-Ruiz, 2013). Even a hypoxic atmosphere around the roots triggered a slight increase in root melatonin levels. Under pH stress, alkaline conditions decreased the root melatonin amount while acidification amplified it (Arnao and Hernández-Ruiz, 2013). The melatonin content changes were evidently dose- and time-dependent. Hence, the authors concluded that melatonin biosynthesis is universally triggered in response to all these environmental stressors. Further, many other stress factors have been studied for their effect on melatonin content in plants. The exogenous application of melatonin has been reported to enhance the synthesis of other growth hormones like auxin, ethylene, gibberellic acid, etc., thus fortifying plants against stress conditions. Interestingly, melatonin can sometimes regulate its own biosynthesis, which strengthens the plant's ability to tolerate stress. Working as an anti-stress agent, melatonin enhances plant growth, seed germination and biomass production by maintaining the rate of photosynthesis, metabolic activity and synthesis of key antioxidant enzymes superoxide dismutase (SOD), peroxidase (POX), catalase (CAT), etc. Therefore, there is a growing interest among researchers in investigating the role of melatonin and, consequently, its function as a biostimulator in agriculture. Hence, phytomelatonin research geared up toward applying exogenous melatonin and melatonin-based priming for agricultural purposes.

12.3.1 Exogenous Melatonin Supply as a Defense Measure

Application of exogenous melatonin in the form of a foliar spray, seedling treatment and/or root dipping during various types of stress conditions has resulted in significantly improved resistance of plants through elicitation of multiple defense pathways. This section focuses on recent studies about melatonin-induced responses to different abiotic stress factors and insights into their mechanistic basis.

12.3.1.1 Melatonin in Drought Stress

Intensifying global warming has led to an increase in frequent drought spells throughout. Drought-mediated injury often results from oxidative stress–triggered increase in cellular ROS, Ca^{2+} and hormones. The transcription factors associated with this drought-induced cellular response include NACs, MYBs, AP2/EREBPs, bZIPs, HDs and bHLHs. Hence, drought stress adaptive mechanisms result from a complex regulatory network of various cellular molecules (Upadhyaya et al., 2021a). Melatonin has been found to play a vital role in drought regulation. Melatonin pretreatment to drought-exposed plants modulates various physiological processes like osmoregulation, seed germination, photosynthesis, primary/secondary metabolism, senescence and hormonal crosstalk. Endogenous melatonin level owing to up regulated expression of MBP genes has been reported to increase in response to drought stress in many plant species like Arabidopsis and other crop plants such as barley, Bermuda grass, apple, grapevine and rice (Moustafa-Farag et al., 2020; Liang et al., 2018a). The application of exogenous melatonin also enhanced endogenous melatonin levels. It thus helps to maintain the turgor pressure, chlorophyll recovery, photosynthetic machinery, seed germination, lateral root formation, hypocotyl elongation and fruit quality (Zeng et al., 2022; Sun et al., 2021).

(Zhang et al., 2013) first reported melatonin-mediated drought suppression under *in vitro* conditions in PEG-stressed cucumber plants (*Cucumis sativus* L.), as it improved seed germination and lateral root development. Subsequently, Wang et al. (2013) observed that melatonin ameliorated drought-induced oxidative damage and photosynthetic abnormalities in drought-stressed apples. The role of melatonin in combating drought stress has been reported by many, as detailed in Table 12.1.

TABLE 12.1
Effect of Exogenous Melatonin in Drought Stress

Plant specimen	Melatonin concentration	Effect of melatonin in drought stress	Reference
Solanum lycopersicum	0.1 mM, 10–40 ppm	Decreased compatible solute contents (e.g., proline and soluble sugars) and increased photosynthesis ability	(Liu et al., 2015; Ibrahim et al., 2020)
Arabidopsis	50 μM	Up regulation of the expression of stress-responsive genes (*COR15A*, *RD22*, etc.)	(Shi et al., 2015b)
Cynodon dactylon	4–100 μM	Modulation of photorespiratory, carbohydrate and nitrogen metabolism level	(Shi et al., 2015a)
Malus	100 μM	Changes stomatal conductance and reduces ROS level	(Li et al., 2015)
Zea mays	100 μM	Helps to recover plants from seven-day-long drought stress by maintaining photosynthetic rate and redox signaling	(Ye et al., 2016)
Triticum aestivum	500 μM	Uplift the water holding ability and expression level of GST, APX, GPX, DHAR, GPX and efficiency of PS II while in drought stress (60% water loss)	(Cui et al., 2017)
Zea mays	100 μM	Improve the photosynthetic characteristics and activities of antioxidants enzymes (APX, CAT, SOD, POX) during drought stress (60% water loss)	(Ahmad et al., 2019)
Rapeseed	500 μM	Improves stomatal and ultrastructural traits along with osmotic adjustment	(Khan et al., 2019)
Moringa oleifera	150 μM	Stimulates auxin synthesis, growth rate, plant yield and reduces ROS accumulation	(Sadak et al., 2020)
Glycine max	50 and 100 μM	Seedling pretreatment with melatonin helps to improve plant growth	(Imran et al., 2021)
Coffea arabica	100 μM	Regulation of photosynthetic efficiency and reduction in oxidative damage	(Cherono et al., 2021)
Zea mays	25–75 μM	Enhances seed filling rate, seed weight and chlorophyll content	(Ahmad et al., 2021)
Lemon verbena	100–200 μM	Improves growth and production of essential oils by regulating mineral homeostasis and osmolyte accumulation	(Hosseini et al., 2021)

The direct (ROS scavenging and heavy metal chelation) or indirect roles (activation of antioxidants, protection to photosynthetic apparatus and regulation of stress-related transcription factors) of melatonin have been studied in many cereals, pulses, oilseeds, vegetables, fruit crops and trees. Melatonin has been found to promote seedling growth and germination under drought-stress conditions, similar to auxin (Huang et al., 2019). Melatonin has also been reported to enhance the carbon assimilation efficiency, stomatal conductance, photochemical efficiency of PSII and RuBisCO accumulation, resulting in delay in senescence by lowering the H_2O_2 levels with high APX activity (Sharma and Zheng, 2019). Another report suggests that melatonin-mediated photosynthesis improvement in drought stress results from the protected chloroplast structures (Cui et al., 2017). In addition, melatonin also regulates fruit ripening by stimulating ethylene biosynthesis, like in tomatoes (Sun et al., 2021). Many other pathways have also been reported to be influenced by melatonin under drought stress, like the MAPK pathway (Zhang et al., 2011), GS-GOGAT enzymes (Liang et al., 2018b), GA, IAA biosynthetic pathways (Yang et al., 2019), CK biosynthetic pathways (Ma et al., 2018). The transcription factors like *WRKY1*, *DREB2* and *MYB* have been reported to be up regulated by melatonin treatment in oat seedlings (Gao et al., 2019).

The primary defense mechanism employed by melatonin is alleviating the stress-induced oxidative damage and rendering protection to the stomatal function to prevent water loss. To prevent oxidative damage, melatonin amplifies the activities of antioxidant enzymes (ascorbate peroxidase [APX], CAT, SOD, peroxidase, dehydroascorbate reductase [DHAR]) along with other antioxidants like ascorbic acid, glutathione (GSH), phenolic compounds, etc. to impede ROS accumulation. On the other hand, melatonin also enhanced drought tolerance by protecting the nitro-oxidative homeostasis in *Medicago sativa* (Antoniou et al., 2017) and by protecting the photosynthetic apparatus from drought-induced damage. Secondly, melatonin can help in stomatal closure by interacting with abscisic acid (ABA) to enable tolerance to drought stress (Chen et al., 2020a). Melatonin also triggered down regulation of an ABA biosynthesis gene, 9-cis-epoxycarotenoid dioxygenase 3 gene MdNCED3 and up regulation of its catabolic genes MdCYP707A1 and A2, which ultimately led to reduced ABA content and preserved stomatal function (Li et al., 2015). Moreover, as a part of the crosstalk of melatonin with other hormones, GA biosynthesis was found to be elevated by melatonin application to alleviate drought-induced suppression of flowering (Tiwari et al., 2020). Water loss can also be prohibited by melatonin by stimulating biosynthesis and deposition of cuticular waxes in plant parts (Ding et al., 2018).

12.3.1.2 Melatonin in High-Temperature Stress

Ambient temperature is an essential prerequisite for the growth and development of plants. Temperature above or below the optimum level is highly detrimental for plants. Notably, high temperature or heat stress is another direct effect of global warming on crop yield and quality. High-temperature stress often leads to oxidative damage by ROS generation and shows peroxidative damage to the cellular membranes, making them lose their stability. Loss of cellular membrane structure leads to extreme damage to almost all cellular constituents; hence, the plant gets damaged beyond repair.

Like drought stress, high-temperature stress also results in a notable increase in endogenous melatonin content, which helps reinforce thermotolerance of plants. For example, a study by Jahan et al. (2019) showed that melatonin alleviated high-temperature-elicited impairment in tomato seedlings by redox balance homeostasis and modulation of polyamine and NO biosynthesis. In continuation, melatonin-mediated repression of high-temperature-induced senescence in tomato was executed through ABA- and GA-signaling pathways (Jahan et al., 2021b). Melatonin treatment in high-temperature-exposed tomato seedlings boosted CO_2 assimilation and enhanced the activities of RuBisCo and Fructose 1,6-bisphosphatase (Jahan et al., 2021a). The addition of melatonin augmented the electron transport efficiency and helped to foster the dissipation of excess excitation energy. Endogenous melatonin deficit intensified high-temperature-induced oxidative stress in

TABLE 12.2
Effect of Exogenous Melatonin in High-Temperature or Heat Stress

Plant specimen	Melatonin concentration	Effect of melatonin in high-temperature stress	Reference
Solanum lycopersicum	10 μM	Proteome protection via HSPs and autophagic refolding or protein degradation	(Xu et al., 2016)
Lolium perenne	100 μM	Alter abscisic acid and cytokinin biosynthesis	(Zhang et al., 2017b)
Actinidia deliciosa	200 μM	Promotes antioxidant enzymes and transcription of glutathione S-transferase	(Liang et al., 2018b)
Solanum lycopersicum	100 μM	Modulates polyamine and nitric oxide biosynthesis and balances cellular ROS levels	(Jahan et al., 2019)
Oryza sativa	200 μM	Reduces stress-mediated photosystem damage and chlorophyll synthesis	(Barman et al., 2019)
Triticum aestivum	100 μM	Lowers the oxidative damage through higher expression of antioxidants and anti-stress responsive genes	(Buttar et al., 2020)
Radish	11–67 mg/L	Increases the biomass, chlorophyll and carotenoid content, soluble protein and soluble solid content, auxin and abscisic acid contents along with increase in the activity of antioxidant enzymes (APX and POX)	(Jia et al., 2020)
Triticum aestivum	20 μM and 100 μM	Crosstalk between melatonin and hydrogen sulfide protects plants from heat stress–induced oxidative damage and photosynthetic injury, lowering of TBARS and H_2O_2 contents	(Iqbal et al., 2021)

tomato plants (Ahammed et al., 2019). Melatonin-mediated high-temperature stress tolerance has been demonstrated in many other plants, as depicted in Table 12.2.

As a part of the defense program induced by melatonin, it causes the activation of the heat stress pathway. Heat shock factors (HSFs) like HSFA2, HSA32 and heat shock proteins (HSP90 and HSP101) contributed to melatonin-induced thermotolerance in Arabidopsis (Shi et al., 2015c). Melatonin is also reported to promote cellular protein protection from high-temperature stress in tomato plants with the help of HSP production and autophagic pathway to enable refolding and/or degradation of denatured or damaged proteins (Xu et al., 2016). Moreover, a recent study reported that HSP40-SlSNAT interaction is directly related to melatonin biosynthesis-mediated thermotolerance by maintaining RuBisCo enzyme stability (Wang et al., 2020). Another study noted that melatonin treatment led to an increase in Ca^{2+} uptake with the help of a non-selective Ca^{2+} permeable cation channel (Çelik and Nazıroğlu, 2012).

12.3.1.3 Melatonin in Cold Stress

Cold stress, that is, temperature below the optimum level, does have a damaging effect on plant growth and physiology. Specifically, metabolic activity, membrane permeability and antioxidant activities are compromised. An increase in endogenous melatonin content and exogenous supply of melatonin both showed similar anti-cold-stress effects by variable mechanisms in different plant species, as noted in Table 12.3. For example, melatonin exerts its antioxidant activity by directly scavenging accumulated ROS, preventing malondialdehyde formation and electrolyte leakage and boosting the levels of polyamines, sucrose and proline along with the increase in the activities of antioxidant enzyme and non-enzymatic antioxidants (Zeng et al., 2022). Melatonin also showed anti-apoptotic effects on carrot suspension cells, possibly with the help of polyamines (Lei et al.,

TABLE 12.3
Effect of Exogenous Melatonin in Cold Stress

Plant specimen	Melatonin concentration	Effect of melatonin in cold stress	Reference
Cucumis sativus	25–50 M	Seed treatment with melatonin improves germination rate and photosynthetic efficiency	(Posmyk et al., 2009)
Triticum aestivum	1 mM	Amelioration of oxidative stress by increase in APX, SOD, CAT, GPX enzyme level	(Turk et al., 2014)
Cynodondactylon	100 μM	Protects photosystem II	(Fan et al., 2015)
Hordeum vulgare	1 μM	Priming with melatonin helps to culminate drought-cold-stress-induced subcellular antioxidant systems and ABA levels	(Li et al., 2016)
Cynodondactylon	100 μM	Maintains carbohydrates, organic acids and amino acid level and balances antioxidant system and metabolic homeostasis	(Hu et al., 2016)
Solanum lycopersicum	100 μmol	Accelerates non-photochemical quenching	(Ding et al., 2017b)
Oryza sativa	20–100 μM	Enhancement of growth rate, photosynthetic efficiency, alleviation of ROS and malondialdehyde formation	(Han et al., 2017)
Solanum lycopersicum	100 μM	Enhance the level of cold-induced genes and collectively increase photosynthetic carbon assimilation, polyamines, sucrose and proline	(Ding et al., 2017a)
Lycopersicon esculentum	100 μM	Protects photosynthetic apparatus and increases electron transfer rate and quantum yield of PSI and PSII	(Yang et al., 2018)
Camellia sinensis	100 μM	Enhances photosynthetic rate, antioxidant potential, chlorophyll level and reduces ROS generation	(Li et al., 2018)
Cucumis sativus	100 μM	NO acts in melatonin-induced stress tolerance	(Feng et al., 2021)
Citrullus lanatus	150 μM	Promotes methyl jasmonate level, which further helps to enhance cold tolerance	(Li et al., 2021)
Cucurbita lunatus	150 μM	Helps to improve MeJA-induced melatonin biosynthesis in leaves	(Li et al., 2021)

2004). Thus, melatonin plays a significant role in maintaining redox homeostasis under low-temperature stress. Consequently, melatonin also shows a protective effect on photosynthetic apparatus by preventing stress-induced inhibitions of photosystems I and II. Moreover, melatonin has been reported to up regulate the expression of cold stress–related genes and transcription factors, like *C-repeat binding factors* (*CBFs*), *COR15a* (transcription factor in freezing tolerance), *CAMTA1* and activators of ROS-related *ZAT10* and *ZAT12* antioxidant genes, which may further aid in the biosynthesis of cold-protecting compounds (Bajwa et al., 2014).

12.3.1.4 Melatonin in Salinity Stress

Salt stress is a major global problem affecting agricultural yield across the world. Salt stress affects plant health by disrupting osmotic imbalance and ion balance (Van Zelm et al., 2020). Plant phospholipid/protein kinase signals and ABA have been reported to dominate salt stress–induced cellular responses. One of the significant impacts of salt stress is noted on the photosynthetic efficiency owing to the closing of stomata and the direct effect on photosystem protein structures (Meloni et al., 2003).

TABLE 12.4
Effect of Exogenous Melatonin in Salinity Stress

Plant specimen	Melatonin concentration	Effect of melatonin on salinity stress	Reference
Triticum aestivum	1 µM	Regulates polyamine biosynthesis, enhance photosynthetic rate and IAA content and reduces H_2O_2 generation in 100 mM salt stress	(Ke et al., 2018)
Brassica napus	1 µM	Enhances the nitric oxide level that helps to scavenge ROS	(Zhao et al., 2018)
Solanum lycopersicum	150 µmol	Maintains the balance of PS II function and reduces oxidative injury under 150 mM salt stress	(Yin et al., 2019)
Avena nuda	100 µM	Regulates the growth level and expression of marker antioxidant genes (APX, CAT, POD, SOD) under 150 mM salt stress	(Gao et al., 2019)
Oryza sativa	200 µM	Enhances polyamine content, K^+/Na^+ pump ratio and activity of nitric oxide synthase and reduces ion leakage at 150 mM salt stress condition	(Yan et al., 2020)
Cucumis sativus	300 µM	Alters the expression of antioxidant genes and reduces ROS level due to 150 mM salt stress	(Zhang et al., 2020)
Gossypium hirsutum	20 µM	Application of exogenous melatonin to cotton seed, enhances seed germination rate and salt tolerance ability	(Chen et al., 2020b)
Triticum aestivum	70 µM	Enhances root and shoot growth rate	(Talaat, 2021)
Beta vulgaris	100 µM	Maintains K^+/Na^+ homeostasis and reduces superoxide anions ($O_2^{\bullet-}$) and hydrogen peroxide (H_2O_2) under 600 µM salt stress	(Zhang et al., 2021a)
Glycine max	0.10 mM	Protects plant from 100 mM salt stress by enhancing pigment synthesis and PSII activity and reducing ROS level	(Alharbi et al., 2021)
Brassica juncea	1 µM	Enhances recovery rate by enhancing plant height, leaf length, leaf width and gas exchange ability	(Park et al., 2021)
Oryza sativa	50–500 µM	Maintains sodium/potassium ion balance and reduces oxidative damage	(Xie et al., 2021)

In response to salt stress, melatonin has been found to render tolerance to crops like maize, wheat, cucumber, tomato, cotton and rice (Zeng et al., 2022; Hassan et al., 2022; Sun et al., 2021). Salt stress has been shown to increase the endogenous melatonin content by the up regulation of crucial melatonin biosynthetic enzymes (Arnao and Hernández-Ruiz, 2009). Similarly, exogenous melatonin application renders plant resistance to salt stress through various mechanisms as listed in Table 12.4. Melatonin can prevent salt stress–induced damage by its antioxidant activities or restoring ion homeostasis. Direct scavenging of free radicals like H_2O_2 or activating the antioxidant enzymes is a commonly employed defense technique (Roychoudhury et al. 2007).

Additionally, melatonin-NO interaction helped to balance GSH and glutathione reductase activity in sunflower seedlings (Kaur and Bhatla, 2016). Along with that, melatonin treatment results in nutrient uptake, polyamine biosynthesis, improvement in chlorophyll biosynthesis, photosynthesis rate, sugar accumulation, etc. Melatonin foliar sprays have also been observed to enrich fruit yield and quality in strawberry (Zahedi et al., 2020). Another strategy melatonin employs is maintaining ion balance to counteract the effects of salt stress. In rice, melatonin-mediated protection from salinity stress occurs through maintenance of K^+ homeostasis by NADPH oxidase–dependent control of plasma membrane K^+ transporters (Liu et al., 2020). Melatonin helps K^+ retention through

OsRBOHF-dependent ROS burst that activates K^+ uptake transporters among a series of stress-responsive genes. Melatonin-led increase in K^+ concentration is related to a concomitant decrease in Na^+ ions; thus, a higher K/Na ratio helps maize seedlings tolerate salinity stress (Jiang et al., 2016). Moreover, melatonin also induces the breakdown of triacylglycerol and assists in K/Na homeostasis associated with energy turnover in salinity-stressed sweet potato (Yu et al., 2018). The Ca^{2+}/CaM (Ca^{2+}/calmodulin) and melatonin interaction are also associated with salt stress tolerance (Vafadar et al., 2020). Lastly, melatonin has also been found to crosstalk with ethylene, gibberellic acid and brassinosteroids, which plays a pivotal role in stress tolerance (Xu et al., 2019; Zhang et al., 2014, 2021b).

12.3.1.5 Melatonin in Heavy Metal Stress

Anthropogenic activities have led to an immense increase in the concentration of heavy metals like cadmium (Cd), arsenic (As), aluminium (Al), copper (Cu), zinc (Zn), chromium (Cr), iron (Fe), lead (Pb) and nickel (Ni) in the agricultural soil, and, as a result, the plants are getting severely affected (Samanta and Roychoudhury, 2021). Heavy metal stress often inhibits plant growth and development by triggering ROS accretion and subsequent damage to cellular constituents and organelles, like chloroplasts, mitochondria, etc. As a result, major cellular processes like photosynthesis and carbohydrate metabolism get hampered by heavy metal toxicity (Roychoudhury et al., 2012; Hasan et al., 2015). An article by Hoque et al. (2021) reviewed that heavy metal stress caused an induction in the endogenous melatonin biosynthesis in the root tissue of *Hordeum vulgare* (barley), *Solanum lycopersicum* (tomato), *Glycine max* (soybean) and *Lupinus albus* (lupin), leaves of *Nicotiana tabacum* (tobacco), *Arabidopsis thaliana*, tomato and the seedlings of *Oryza sativa* (rice). Among the heavy metals, Cd stress severely injures cellular organelles like chloroplasts, mitochondria and endoplasmic reticulum, which leads to cytosolic leakage of endogenous serotonin N-acetyltransferase (SNAT) enzymes, thus resulting in melatonin synthesis in the cytoplasm (Lee and Back, 2017). Another report states that exposure to 50 µM Al^{3+} induced melatonin accumulation in the roots of soybean plant, possibly owing to the disrupted cell membrane, but Al^{3+} stress in rice plants could not hamper the chloroplasts to synthesize melatonin even at 500 µM. Therefore, it was concluded that regardless of the plant species, concentration and exposure duration act as the rate-limiting aspects for endogenous melatonin synthesis. Moreover, exogenous melatonin application has also been demonstrated to mitigate the negative effects of various heavy metals in many plant species either via direct scavenging of ROS and/or via indirect way of increasing antioxidant enzymes activities, photosynthetic efficiency and metabolite content, as demonstrated in **Table 12.5**.

Melatonin contributes to heavy metal tolerance by enhancing antioxidant activity via ROS or NO signaling and expediting the exclusion or sequestration of heavy metals in plant cells (Rodriguez et al., 2004). For example, melatonin inhibited Al stress in *Brassica rapa* (Chinese cabbage) by regulating osmotic balance and protecting cell membrane potential and the photosynthetic system (Tang et al., 2017). In another study in soybean roots, melatonin reduced Al-induced oxidative stress by an upsurge in citrate and malate exudation along with the higher activity of antioxidant enzymes (Zhang et al., 2017a). Exogenous foliar melatonin spray rendered rosemary herbs tolerant to Al stress by increasing the non-protein thiols, phytochelatins (PCs) and other osmoregulatory constituents (Farouk and Al-Amri, 2019). In the case of *Medicago sativa* (alfalfa) seedlings and tobacco plants, melatonin was found to lower Cd accumulation and maintain redox balance (Gu et al., 2017; Wang et al., 2019). An interesting fact reported is that exogenous melatonin could regulate HA7 (PM H^+-ATPases 7), *NRAMP1* (natural-resistance-associated macrophage protein 1), NRAMP3, HMA4 (P-type HM ATPase 4), PCR2 (plant Cd resistance protein 2), NAS1 (nicotianamine synthase 1), MT2 (metallothionein 2), ABCC1 (ATP-binding cassette transporter C1) and MHX (magnesium proton exchanger protein) genes involved in Cd uptake and translocation in apple plants (He et al., 2020). Hence, melatonin can be concluded to directly participate in regulating Cd uptake and translocation. External melatonin has also been noted for its anti-Cd-stress effects on *Cucumis sativus* (cucumber) seedlings, tomato plants, *Spinacia oleracea*, *Cucumis melo*

TABLE 12.5
Effect of Exogenous Melatonin during Toxicity to Several Pollutants

Plant specimen	Type of stress	Melatonin concentration	Effect of melatonin in heavy metal stress	Reference
Brassica oleracea	Copper (0.5 mM)	1–100 μM	Increased the germination rate by 17% by inhibiting membrane peroxidation and protein denaturation	(Posmyk et al., 2008)
Solanum lycopersicum	Cadmium (100 mM)	500 μM	Modulation of phytochelatins biosynthesis, antioxidant ability and vacuolar sequestration	(Hasan et al., 2015)
Nicotiana tabacum	Lead	200 nM	Reduce DNA damage and maintain cellular homeostasis	(Kobylińska et al., 2017)
Malus hupehensis	Alkaline	5 μM	Mitigates stress-mediated ROS generation	(Gong et al., 2017)
Triticum aestivum	Cadmium (200 mM)	50 mM	Reduces H_2O_2 level, MDA content and electrolyte leakage and enhances antioxidant genes expression	(Ni et al., 2018)
Zea mays	Lead	0.05–0.10 mM	Improves NO level and stress tolerance though enhancement of antioxidant enzymes activity and reduction in MDA	(Okant and Kaya, 2019)
Nicotiana tabacum	Cadmium	100 μmol L^{-1}	Modulating the expression of Cd-transport genes.	(Wang et al., 2019)
Triticum aestivum	Cadmium	50 or 100 μM	Reduction in H_2O_2 and MDA level	(Kaya et al., 2019)
Oryza sativa	Fluoride	20 μM	Ameliorates fluoride-induced toxicity	(Banerjee and Roychoudhury, 2019)
Solanum lycopersicum	Nickel (50 μM)	100 μM	Increase in photosynthetic efficacy, accumulation of secondary metabolites and antioxidant defense system	(Jahan et al., 2020)
Triticum aestivum	Chromium (100 mg/kg)	2 mM	Enhancement in chlorophyll and gas exchange attributes and antioxidant enzyme activity	(Seleiman et al., 2020)
Spinacia oleracea	Arsenic and cadmium (150 mg/kg)	100 μM	Prevents photosynthetic apparatus damage and chlorophyll synthesis and enhances the activity of antioxidant enzymes	(Asif et al., 2020)
Cucumis sativus	Iron	100 μM	Enhances internal melatonin level, Chl content, CO_2 assimilation rate, alleviates oxidative damage, activates antioxidant enzymes, promotes secondary metabolites and enzymes, alters Fe acquisition	(Ahammed et al., 2020)
Solanum lycopersicum	Nickel	100 μM	Altering the expression of melatonin-biosynthesis-related genes as well as Chl-biosynthesis-related genes	(Jahan et al., 2020)

(melon) and other plants by various mechanisms (Hoque et al., 2021). Defense against Pb stress was also noted in maize, safflower, tobacco suspension cells and Bermuda grass plants (Okant and Kaya, 2019; Namdjoyan et al., 2020; Kobylińska and Posmyk, 2016; Xie et al., 2018) In *Brassica napus* (canola) and *Origanum majorana* plants, protection from Cr stress has been reported following melatonin application (Ayyaz et al., 2020; Farouk and Al-Amri, 2019).

Interestingly, it was found that water hyacinth plants that can survive in highly polluted environments are rich in melatonin content (Rezania et al., 2015; Tan et al., 2007). Melatonin's ability to boost mobilization, absorption and pollutant sequestration by plants can be used to implement melatonin in phytoremediation. It was reported by Cao et al. (2019) that a melatonin-mediated increase in glutathione and phytochelatins in cucumber promoted chelation, vacuolar sequestration and cell wall trapping of excess Cu^{2+}. Such melatonin-led pollutant sequestration has been noted in the case of cadmium, arsenic, mercury and lead in other plant species. Hence, melatonin might be a successful recovery bioagent in chemically contaminated soils, especially in the case of hyperaccumulating plants.

12.3.1.6 Melatonin in Other Types of Abiotic Stresses

Many other environmental stresses, like waterlogged conditions, ultraviolet (UV) radiation, etc., can also be severely harmful to the growth and development of plants. For example, heavy rainfall and poor drainage system led to waterlogging stress for plants, significantly affecting crop growth and yield. Waterlogging stress creates an anaerobic respiratory condition, which further results in oxidative stress (ROS generation). Mainly, waterlogging affects the root tissue, followed by the leaf tissue wilting. Melatonin has been shown to significantly protect crop growth and productivity during waterlogging. It also plays an inductive role in melatonin biosynthesis genes and melatonin production (Hassan et al., 2022). Plants faced with waterlogging have been reported to up regulate melatonin biosynthesis genes. The same is induced by exogenous melatonin as well. Up regulated endogenous melatonin in waterlogging tolerance has been studied in Alfalfa (Zhang et al., 2019), *Malus* (Zheng et al., 2017), *Prunus* (Gu et al., 2021b), etc. Melatonin plays a crucial role in reducing waterlogging-induced hypoxia stress by optimizing metabolism in roots and regulating hypoxia-related genes (e.g., calcineurin B-like protein-interacting protein kinase and ethylene response factors) (Gu et al., 2021a). Further, it enhances anaerobic respiration in the aerenchyma and suppresses the regulation of metabolic enzymes (Gu et al., 2021b). During waterlogging, exogenous melatonin abates stomatal closure, leaf senescence and photosynthetic loss (Gu et al., 2021b). Exogenous melatonin helps in the post-synthesis regulation of endogenous melatonin (Zheng et al., 2017).

Another key abiotic stress that affects plant growth is UV-mediated stress. UV light is the most widely known free radical generator in plants. Prolonged exposure to UV light results in an imbalance in cellular redox signaling and causes damage to several cellular components (Yao et al., 2021). Melatonin biosynthesis is regulated by light. An increase in light intensity, especially UV-B, has been found to stimulate melatonin production. During UV stress, melatonin application attenuates UV-induced damage to DNA and UVB-induced ROS in plants like *Nicotiana sylvestris* (Zhang et al., 2012), *Ocimum basilicum* (Nazir et al., 2020), *Malus hupehensis* (Wei et al., 2019), etc. Melatonin participates in the UVB signaling network through up regulating key developmental genes such as Elongated Hypocotyl 5 (*HY5*), *Constitutive Photomorphogenic 1* (*COP1*), *HY5 Homolog* (*HYH*), and Repressor of UvB Photomorphogenesis 1(*RUP1*) and *RUP2*, resulting in UV tolerance (Yao et al., 2021). Additionally, melatonin can improve the photosynthetic rate and oxidative injury by maintaining the redox state. The exogenous application of melatonin helps to mitigate light stress by modulating the expression of genes involved in the antioxidant pathway (Zeng et al., 2022; Hassan et al., 2022). Flavonoids are well-known plant secondary metabolites which play a pivotal role in UV stress alleviation. Interestingly, melatonin also participates in flavonoid-mediated UV-light stress tolerance (Wei et al., 2019).

12.3.2 Genetic Engineering of Melatonin Biosynthesis Genes as a Defense Tool

Melatonin has a critical role in plant abiotic defense mechanisms. As exemplified in the previous section, various types of abiotic stress can be mitigated by exogenous melatonin application. To illustrate whether this stress tolerance specifically results from melatonin biosynthesis activation or not, several reports not only showed the enhancement in the endogenous melatonin level but also checked the expression of melatonin biosynthesis pathway genes. Interestingly, scientists found coordinated expression levels of melatonin biosynthesis genes against variable abiotic conditions. For example, during water deprived state, *TDC*, *ASMT*, *COMT* and *SNAT* genes were up regulated significantly to fight against drought stress by melatonin enhancement (Moustafa-Farag et al., 2020). Supporting this, drought stress–mediated melatonin induction and upliftment of melatonin intermediate pathway enzyme activity has been noticed in plants like Arabidopsis, rice, grapevine, barley, etc. (Sun et al., 2021). The application of exogenous melatonin was also reported to trigger the endogenous melatonin level through the activation of candidate genes of melatonin biosynthesis pathway (Zeng et al., 2022).

To date, most of the study about abiotic stress tolerance of melatonin has been undertaken on salt stress. Melatonin plays a crucial role in salt stress mitigation, particularly in mRNA stability, translational regulation of stress-responsive genes and redox homeostasis. (Arnao and Hernández-Ruiz, 2009) reported that salt stress could up regulate MBP gene expression (SNAT, ASMT, COMT, etc.), leading to increased melatonin synthesis (Gao et al., 2019; Chen et al., 2020a; Zeng et al., 2022). Similarly, cold and hot environments typically stimulate melatonin accumulation by up regulating the intermediate biosynthesis genes (Han et al., 2017; Li et al., 2021; Zeng et al., 2022; Hassan et al., 2022). The role of melatonin in various heavy metal stress is well explored. Like salt, drought, cold and heat stress, heavy metals can also trigger the expression of genes involved in melatonin biosynthesis, resulting in a significant increase in plant tolerant ability (Hassan et al., 2022; Asif et al., 2020; Zeng et al., 2022). Melatonin also has a role in the regulation of plant ability to withstand waterlogging. Waterlogging stress leads to the activation of endogenous melatonin synthesis genes (e.g., *TDC*, *T5H*, *ASMT*, *COMT* and *SNAT*), resulting in waterlogging resistance (Moustafa-Farag et al., 2020). Even exogenous melatonin application induces endogenous melatonin levels through activation of biosynthesis genes during waterlogging stress in alfalfa, apple, peach, etc. (Gu et al., 2017; Zheng et al., 2017; Zeng et al., 2022). Pathway intermediate enzymes are either rate-limiting or specifically expressed in certain plant developmental stages for plant melatonin synthesis. Therefore, if their expression can be triggered constitutively, they will be able to activate melatonin biosynthesis constitutively. Previously, it has been stated that higher production of melatonin leads to higher abiotic stress tolerance. Hence, scientists have undertaken the genetic manipulation approach for melatonin biosynthesis genes to enhance its production. Studies related to the MBP genes, their native host and overexpression details have been enlisted in **Table 12.6**. These studies report that overexpression of genes like *SNAT*, *COMT*, etc. increases melatonin formation and the

TABLE 12.6
Overexpression (OE) of Animal/Plant Melatonin Biosynthesis Genes in Various Crops and Their Effect on Stress Tolerance

Organism	Gene	OE in	Type of stress tolerance	Function	Reference
Homo sapiens	*SNA*	*Oryza sativa*	Cold	Significantly enhanced melatonin biosynthesis and elevated photosynthesis	(Kang et al., 2010)
Sheep	*oAANAT and oHIOMT*	*Solanum lycopersicum*	Drought	Enhanced melatonin level, reduced plant dehydration and enhanced growth rate	(Wang et al., 2014)

TABLE 12.6 *(Continued)*
Overexpression (OE) of Animal/Plant Melatonin Biosynthesis Genes in Various Crops and Their Effect on Stress Tolerance

Organism	Gene	OE in	Type of stress tolerance	Function	Reference
Malus zumi	*ASMT1*	*Arabidopsis*	Drought	Increased melatonin content two to fourfold higher than wild-type-mediated melatonin-derived ROS detoxification	(Zuo et al., 2014)
Ovine (Sheep)	*AANAT and HIOMT*	*Panicum virgatum*	Salt	Transgenic plants exhibit higher melatonin level; melatonin positively affected growth, flowering and salt tolerance by modulating the expression of *APL3*, *SL1*, *FT1*, *FLP3*, *MADS6*, *MADS15* and *NHX1* gene	(Huang et al., 2017)
Pseudomonas fluorescens	*TDC*	*Vitis vinifera*	Salt	Enhanced expression of grapevine tryptophan decarboxylase and serotonin N-acetyltransferase genes helps to ameliorate salt stress through melatonin biosynthesis	(Ma et al., 2017)
Oryza sativa	*SNAT*	*Oryza sativa*	Cadmium	Higher melatonin accumulation helps to resist cadmium and senescence and increases grain yield	(Lee and Back, 2017)
Medicago sativa	*SNAT*	*Arabidopsis*	Cadmium	Produces melatonin in root tissue, which helps to modulate heavy metal transporter and redox homeostasis	(Gu et al., 2017)
Solanum lycopersicum	*COMT1*	*Solanum lycopersicum*	Salt (800 mM NaCl)	Increases melatonin content; enhances salt tolerance by limiting superoxide and hydrogen peroxide formation	(Liu et al., 2019)
Triticum aestivum	*COMT*	*Arabidopsis*	Drought	Stimulates melatonin production, higher drought tolerance through higher fresh weight and root system modulation	(Yang et al., 2019)
Solanum lycopersicum	*SlCOMT1*	*Solanum lycopersicum*	Salt	Increases melatonin level; helps to maintain Na^+/K^+ balance and reduce ion damage by modulating salt overly sensitive (SOS) pathway; also elevates the expression of stress-responsive genes	(Sun et al., 2020)
Malus zumi	*ASMT1*	*Solanum lycopersicum*	Salt	Enhanced melatonin formation; plants exhibited healthy phenotype viz., higher plant height, fresh weight; root formation, relative water content (RWC) and Chl content with balanced Fv/Fm ratio	(Zhuang et al., 2020)
Vitis vinifera	*SNAT1*	*Arabidopsis*	Salt	Enhanced melatonin production and salt resistance; plants had higher germination rate, fresh weight, better root system and higher ROS mitigating ability	(Wu et al., 2021)
Vitis vinifera	*ASMT1*	*Nicotiana benthamiana*	Osmotic	Significantly uplifts melatonin production; maintains osmotic pressure, reduces water loss, MDA formation, ROS generation and longer root formation	(Yu et al., 2022)

subsequent stress alleviation ability. On the other hand, few studies rely on the knockdown of the biosynthesis gene to prove their effect on melatonin levels and stressors. In Arabidopsis, *snat1* and *cand2* (a melatonin receptor) mutants could not contribute to osmotic stress–induced melatonin accumulation (Wang et al., 2021). Similarly, in tomato, *COMT1* and *TDC* silencing also magnified high-temperature-induced oxidative damage owing to their defect in endogenous melatonin biosynthesis (Ahammed et al., 2019).

12.4 CONCLUSIONS AND FUTURE AIMS FOR MELATONIN-MEDIATED PROTECTION FROM ABIOTIC STRESS

Melatonin is an emergent molecule with versatile potencies in plant cells, including plant growth stimulation, stress protection, flowering and fruit ripening primarily by regulating the cellular redox network or coordinating with other phytohormones (Sun et al., 2021). Figure 12.2 depicts

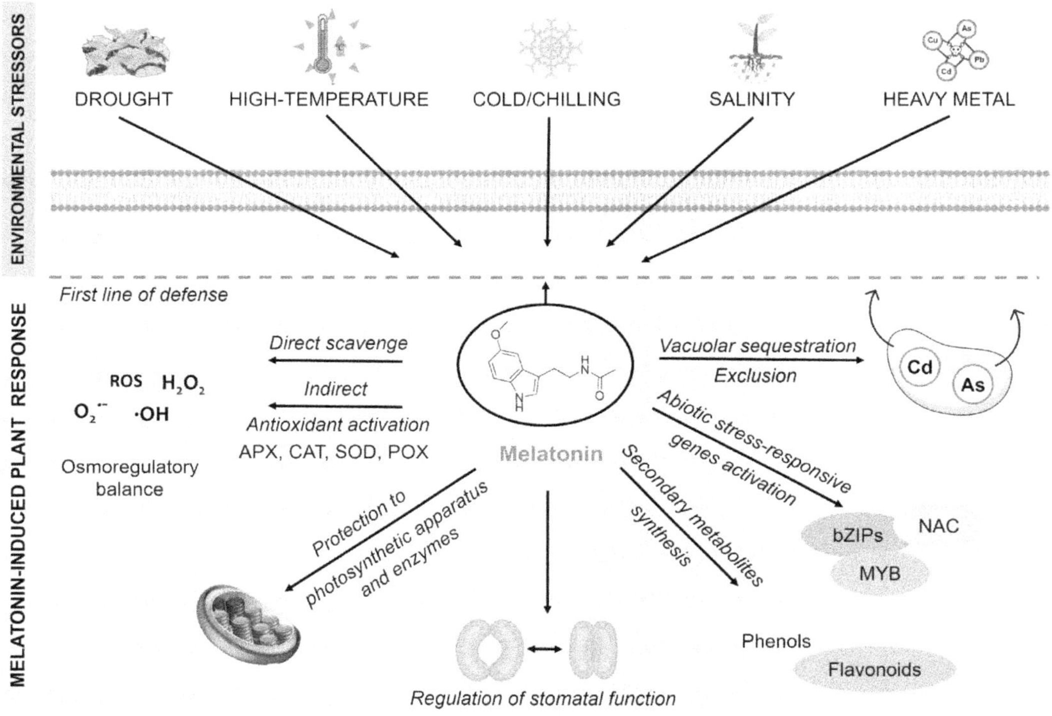

FIGURE 12.2 Melatonin-induced defense responses in plant cells against various environmental stressors. In response to different types of adverse environmental conditions, like drought, high temperature, cold, salinity, heavy metals, etc., melatonin acts as one of the key defense regulatory molecules. Melatonin is a well-known antioxidant which is capable of preventing oxidative damage originating from any type of environmental stress factor. Melatonin can either directly scavenge ROS or activate the antioxidants to maintain the osmoregulatory balance. Melatonin also induces the production of secondary metabolites (like phenols, flavonoids, etc.) and stress-responsive genes and transcription factors (like bZIPs, MYB, NAC, etc.). By this mechanism, melatonin has been reported to protect the photosynthetic apparatus and induce chlorophyll biosynthesis and Calvin cycle enzymes. Particularly, in case of salinity stress, melatonin can interact or act via the abscisic acid (ABA) hormone to control the stomatal function for maintaining the cell turgor. Another mechanism observed is in the case of heavy metal stress, where melatonin directly helps in vacuolar sequestration or exclusion from the cells.

multidirectional mode of action of melatonin against different abiotic stressors. Further probing into the melatonin signaling pathway, including the various melatonin receptors, can open up new directions into the functional roles of melatonin in stress protection and/or plant physiology. Therefore, many scientists prefer referring to melatonin as a master regulator in plant biology instead of a phytohormone unless proven otherwise. A prospect also lies in exploring the triggering mechanism behind melatonin signaling, that is, how melatonin synthesis and metabolism are switched on or off. Since identifying N-nitrosomelatonin (nitrosation product of melatonin by NO), further studies are pending on how melatonin or by-product N-nitrosomelatonin is transduced to longer distances.

Moreover, since an exogenous supply of melatonin has been demonstrated to serve as the first shield line fighting oxidative stress (a secondary product of every environmental stress factor), melatonin could be implemented as a biostimulator to cultivate fortified plants. Thus, the potency of exogenous melatonin supply in the agricultural world should be judged in field trials. The effect of melatonin under combined stress conditions should be experimented to validate the feasibility of melatonin as a biostimulant in real-life agricultural practice. Another imperative subject is how melatonin coordinates and cooperates with other phytohormones in diverse physiological processes. Further research directions associated with melatonin include the epigenetic programming of melatonin-associated genes and the melatonin target genes. Multiple experimental reports have shown that melatonin can regulate the expression of various enzymes, receptors and transcription factors involved in auxin, gibberellic acid, salicylic acid, ethylene, abscisic acid, nitric oxide, strigolactones and brassinosteroids. Phytohormones can be generated to clarify their function. Even melatonin signaling at the soil microbe-root interface relevant to plant growth, ion transport and nutrient allocation can be deeply investigated as a novel research field. As a master regulator in plant systems, melatonin application in agricultural practice should be attempted for commercial translations to contribute to global food security.

REFERENCES

Ahammed, Golam Jalal, Meijuan Wu, Yaqi Wang, Yaru Yan, Qi Mao, Jingjing Ren, Ronghui Ma, Airong Liu, and Shuangchen Chen. "Melatonin alleviates iron stress by improving iron homeostasis, antioxidant defense and secondary metabolism in cucumber." *Scientia Horticulturae* 265 (2020): 109205.

Ahammed, Golam Jalal, Wen Xu, Airong Liu, and Shuangchen Chen. "Endogenous melatonin deficiency aggravates high temperature-induced oxidative stress in *Solanum lycopersicum* L." *Environmental and Experimental Botany* 161 (2019): 303–311.

Ahmad, Shakeel, Muhammad Kamran, Ruixia Ding, Xiangping Meng, Haiqi Wang, Irshad Ahmad, Shah Fahad, and Qingfang Han. "Exogenous melatonin confers drought stress by promoting plant growth, photosynthetic capacity and antioxidant defense system of maize seedlings." *PeerJ* 7 (2019): e7793.

Ahmad, Shakeel, Muhammad Kamran, Xunbo Zhou, Irshad Ahmad, Xiangping Meng, Tehseen Javed, Anas Iqbal et al. "Melatonin improves the seed filling rate and endogenous hormonal mechanism in grains of summer maize." *Physiologia Plantarum* 172, no. 2 (2021): 1059–1072.

Alharbi, Basmah M., Abeer Hamdy Elhakem, Ghalia S. H. Alnusairi, Mona H. Soliman, Khalid Rehman Hakeem, Md Mahadi Hasan, and Magdi T. Abdelhamid. "Exogenous application of melatonin alleviates salt stress-induced decline in growth and photosynthesis in *Glycine max* (L.) seedlings by improving mineral uptake, antioxidant and glyoxalase system." *Plant, Soil and Environment* 67, no. 4 (2021): 208–220.

Antoniou, Chrystalla, Giannis Chatzimichail, Rafaella Xenofontos, Jan J. Pavlou, Evangelia Panagiotou, Anastasis Christou, and Vasileios Fotopoulos. "Melatonin systemically ameliorates drought stress-induced damage in *Medicago sativa* plants by modulating nitro-oxidative homeostasis and proline metabolism." *Journal of Pineal Research* 62, no. 4 (2017): e12401.

Arnao, Marino B., and Josefa Hernández-Ruiz. "Chemical stress by different agents affects the melatonin content of barley roots." *Journal of Pineal Research* 46, no. 3 (2009): 295–299.

Arnao, Marino B., and Josefa Hernández-Ruiz. "Growth conditions determine different melatonin levels in *Lupinus albus* L." *Journal of Pineal Research* 55, no. 2 (2013): 149–155.

Arnao, Marino B., and Josefa Hernández-Ruiz. "Melatonin as a plant biostimulant in crops and during post-harvest: A new approach is needed." *Journal of the Science of Food and Agriculture* 101, no. 13 (2021): 5297–5304.

Asif, Mahnoor, Arshid Pervez, Usman Irshad, Qaisar Mehmood, and Rafiq Ahmad. "Melatonin and plant growth-promoting rhizobacteria alleviate the cadmium and arsenic stresses and increase the growth of *Spinacia oleracea* L." *Plant, Soil and Environment* 66, no. 5 (2020): 234–241.

Ayyaz, Ahsan, Muhammad Ahsan Farooq, Aneela Kanwal, Muhammad Aslam, Muhammad Iqbal, Azra Manzoor, Ayesha Khalid et al. "Differential responses of exogenous melatonin on growth, photosynthesis and antioxidant defence system in two *Brassica napus* L. cultivars under chromium stress." *International Journal of Environment, Agriculture and Biotechnology* 5, no. 2 (2020).

Back, Kyoungwhan, Dun-Xian Tan, and Russel J. Reiter. "Melatonin biosynthesis in plants: Multiple pathways catalyze tryptophan to melatonin in the cytoplasm or chloroplasts." *Journal of Pineal Research* 61, no. 4 (2016): 426–437.

Bajwa, Vikramjit S., Mukund R. Shukla, Sherif M. Sherif, Susan J. Murch, and Praveen K. Saxena. "Role of melatonin in alleviating cold stress in *Arabidopsis thaliana*." *Journal of Pineal Research* 56, no. 3 (2014): 238–245.

Banerjee Aditya., and Aryadeep. Roychoudhury. "Melatonin as a regulator of abiotic stress tolerance in plants." In Singh V. P., Singh S., Prasad S. M. (Eds.) *Mechanisms behind Phytohormonal Signaling and Crop abiotic Stress Tolerance*, pp. 47–60. Nova Science Publishers, New York, 2017.

Banerjee, Aditya, and Aryadeep Roychoudhury. "Melatonin application reduces fluoride uptake and toxicity in rice seedlings by altering abscisic acid, gibberellin, auxin and antioxidant homeostasis." *Plant Physiology and Biochemistry* 145 (2019): 164–173.

Barman, Dipankar, O. Ghimire, V. Chinnusamy, R. Kumar, and Ajay Arora. "Amelioration of heat stress during reproductive stage in rice by melatonin." *The Indian Journal of Agricultural Sciences* 89 (2019): 1151–1156.

Blanchard, Béatrice, Denis Pompon, and Claire Ducrocq. "Nitrosation of melatonin by nitric oxide and peroxynitrite." *Journal of Pineal Research* 29, no. 3 (2000): 184–192.

Buttar, Zeeshan Ali, Sheng Nan Wu, Marino B. Arnao, Chaojie Wang, Ikram Ullah, and Chengshe Wang. "Melatonin suppressed the heat stress-induced damage in wheat seedlings by modulating the antioxidant machinery." *Plants* 9, no. 7 (2020): 809.

Cao, Yun-Yun, Chuan-Dong Qi, Shuangtao Li, Zhirong Wang, Xiaoyun Wang, Jinfang Wang, Shuxin Ren, Xingsheng Li, Na Zhang, and Yang-Dong Guo. "Melatonin alleviates copper toxicity via improving copper sequestration and ROS scavenging in cucumber." *Plant and Cell Physiology* 60, no. 3 (2019): 562–574.

Çelik, Ömer, and Mustafa Nazıroğlu. "Melatonin modulates apoptosis and TRPM2 channels in transfected cells activated by oxidative stress." *Physiology & Behavior* 107, no. 3 (2012): 458–465.

Chen, Kong, Guo-Jun Li, Ray A. Bressan, Chun-Peng Song, Jian-Kang Zhu, and Yang Zhao. "Abscisic acid dynamics, signaling, and functions in plants." *Journal of Integrative Plant Biology* 62, no. 1 (2020a): 25–54.

Chen, Li, Liantao Liu, Bin Lu, Tongtong Ma, Dan Jiang, Jin Li, Ke Zhang et al. "Exogenous melatonin promotes seed germination and osmotic regulation under salt stress in cotton (*Gossypium hirsutum* L.)." *PLoS One* 15, no. 1 (2020b): e0228241.

Cherono, Sylvia, Charmaine Ntini, MisganawWassie, Mohammad Dulal Mollah, Mohammad A. Belal, Collins Ogutu, and Yuepeng Han. "Exogenous application of melatonin improves drought tolerance in coffee by regulating photosynthetic efficiency and oxidative damage." *Journal of the American Society for Horticultural Science* 146, no. 1 (2021): 24–32.

Choi, Geun-Hee, and Kyoungwhan Back. "Cyclic 3-hydroxymelatonin exhibits diurnal rhythm and cyclic 3-hydroxymelatonin overproduction increases secondary tillers in rice by upregulating MOC1 expression." *Melatonin Research* 2, no. 3 (2019): 120–138.

Cui, Guibin, Xiaoxiao Zhao, Shudong Liu, Fengli Sun, Chao Zhang, and Yajun Xi. "Beneficial effects of melatonin in overcoming drought stress in wheat seedlings." *Plant Physiology and Biochemistry* 118 (2017): 138–149.

Das, K., and A. Roychoudhury. "Reactive oxygen species (ROS) and response of antioxidants as ROS-scavengers during environmental stress in plants." *Frontiers in Environmental Science* 2 (2014): 53

Ding, Fei, Bin Liu, and Shuoxin Zhang. "Exogenous melatonin ameliorates cold-induced damage in tomato plants." *Scientia Horticulturae* 219 (2017a): 264–271.

Ding, Fei, Gang Wang, Meiling Wang, and Shuoxin Zhang. "Exogenous melatonin improves tolerance to water deficit by promoting cuticle formation in tomato plants." *Molecules* 23, no. 7 (2018): 1605.

Ding, Fei, Meiling Wang, Bin Liu, and Shuoxin Zhang. "Exogenous melatonin mitigates photoinhibition by accelerating non-photochemical quenching in tomato seedlings exposed to moderate light during chilling." *Frontiers in Plant Science* 8 (2017b): 244.

Fan, Jibiao, Zhengrong Hu, Yan Xie, Zhulong Chan, Ke Chen, Erick Amombo, Liang Chen, and Jinmin Fu. "Alleviation of cold damage to photosystem II and metabolisms by melatonin in Bermudagrass." *Frontiers in Plant Science* 6 (2015): 925.

Farooq, Muhammad, A. Wahid, N. S. M. A. Kobayashi, D. B. S. M. A. Fujita, and S. M. A. Basra. "Plant drought stress: Effects, mechanisms and management." In *Sustainable Agriculture*, pp. 153–188. Springer, Dordrecht, 2009.

Farouk, Saad, and Salem M. Al-Amri. "Exogenous melatonin-mediated modulation of arsenic tolerance with improved accretion of secondary metabolite production, activating antioxidant capacity and improved chloroplast ultrastructure in rosemary herb." *Ecotoxicology and Environmental Safety* 180 (2019): 333–347.

Feng, Yiqing, Xin Fu, Lujie Han, Chenxiao Xu, Chaoyue Liu, Huangai Bi, and Xizhen Ai. "Nitric oxide functions as a downstream signal for melatonin-induced cold tolerance in cucumber seedlings." *Frontiers in Plant Science* (2021): 1432.

Foyer, Christine H., and Jeremy Harbinson. "Oxygen metabolism and the regulation of photosynthetic electron transport." In *Causes of Photooxidative Stress and Amelioration of Defense Systems in Plants*, pp. 1–42. CRC Press, New York, 2019.

Gao, Wenying, Zheng Feng, Qingqing Bai, Jinjin He, and Yingjuan Wang. "Melatonin-mediated regulation of growth and antioxidant capacity in salt-tolerant naked oat under salt stress." *International Journal of Molecular Sciences* 20, no. 5 (2019): 1176.

Gong, Xiaoqing, Shuting Shi, Fangfang Dou, Yi Song, and Fengwang Ma. "Exogenous melatonin alleviates alkaline stress in *Malus hupehensis* Rehd. by regulating the biosynthesis of polyamines." *Molecules* 22, no. 9 (2017): 1542.

Gu, Quan, Ziping Chen, Xiuli Yu, Weiti Cui, Jincheng Pan, Gan Zhao, Sheng Xu, Ren Wang, and Wenbiao Shen. "Melatonin confers plant tolerance against cadmium stress via the decrease of cadmium accumulation and reestablishment of microRNA-mediated redox homeostasis." *Plant Science* 261 (2017): 28–37.

Gu, Quan, Chuyan Wang, Qingqing Xiao, Ziping Chen, and Yi Han. "Melatonin confers plant cadmium tolerance: An update." *International Journal of Molecular Sciences* 22, no. 21 (2021a): 11704.

Gu, Xianbin, Lian Xue, Linghong Lu, Jinping Xiao, Genhua Song, Ming Xie, and Huiqin Zhang. "Melatonin enhances the waterlogging tolerance of *Prunus persica* by modulating antioxidant metabolism and anaerobic respiration." *Journal of Plant Growth Regulation* 40, no. 5 (2021b): 2178–2190.

Han, Qiao-Hong, Bo Huang, Chun-Bang Ding, Zhong-Wei Zhang, Yang-Er Chen, Chao Hu, Li-Jun Zhou et al. "Effects of melatonin on anti-oxidative systems and photosystem II in cold-stressed rice seedlings." *Frontiers in Plant Science* 8 (2017): 785.

Hardeland, Rüdiger. "Melatonin in plants and other phototrophs: Advances and gaps concerning the diversity of functions." *Journal of Experimental Botany* 66, no. 3 (2015): 627–646.

Hasan, Md Kamrul, Golam Jalal Ahammed, Lingling Yin, Kai Shi, Xiaojian Xia, Yanhong Zhou, Jingquan Yu, and Jie Zhou. "Melatonin mitigates cadmium phytotoxicity through modulation of phytochelatins biosynthesis, vacuolar sequestration, and antioxidant potential in *Solanum lycopersicum* L." *Frontiers in Plant Science* 6 (2015): 601.

Hassan, Muhammad Umair, Athar Mahmood, Masood Iqbal Awan, Rizwan Maqbool, Muhammad Aamer, Haifa A. S. Alhaithloul, Guoqin Huang et al. "Melatonin-induced protection against plant abiotic stress: Mechanisms and prospects." *Frontiers in Plant Science* 13 (2022).

He, Jiali, Xiaolei Zhuang, Jiangtao Zhou, Luyang Sun, Huixue Wan, Huifeng Li, and Deguo Lyu. "Exogenous melatonin alleviates cadmium uptake and toxicity in apple rootstocks." *Tree Physiology* 40, no. 6 (2020): 746–761.

Hernández-Ruiz, J., and M. B. Arnao. "Distribution of melatonin in different zones of lupin and barley plants at different ages in the presence and absence of light." *Journal of Agricultural and Food Chemistry* 56, no. 22 (2008): 10567–10573.

Hoque, Md Najmol, Md Tahjib-Ul-Arif, Afsana Hannan, Naima Sultana, Shirin Akhter, Md Hasanuzzaman, Fahmida Akter et al. "Melatonin modulates plant tolerance to heavy metal stress: Morphological responses to molecular mechanisms." *International Journal of Molecular Sciences* 22, no. 21 (2021): 11445.

Hosseini, Marjan Sadat, Davood Samsampour, Seyed Morteza Zahedi, Kazem Zamanian, Md Mezanur Rahman, Mohammad Golam Mostofa, and Lam-Son Phan Tran. "Melatonin alleviates drought impact on growth and essential oil yield of lemon verbena by enhancing antioxidant responses, mineral balance, and abscisic acid content." *Physiologia Plantarum* 172, no. 2 (2021): 1363–1375.

Hu, Zhengrong, Jibiao Fan, Yan Xie, Erick Amombo, Ao Liu, Margaret Mukami Gitau, A. B. M. Khaldun, Liang Chen, and Jinmin Fu. "Comparative photosynthetic and metabolic analyses reveal mechanism of improved cold stress tolerance in bermudagrass by exogenous melatonin." *Plant Physiology and Biochemistry* 100 (2016): 94–104.

Huang, Bo, Yang-Er Chen, Yu-Qing Zhao, Chun-Bang Ding, Jin-Qiu Liao, Chao Hu, Li-Jun Zhou, Zhong-Wei Zhang, Shu Yuan, and Ming Yuan. "Exogenous melatonin alleviates oxidative damages and protects photosystem II in maize seedlings under drought stress." *Frontiers in Plant Science* 10 (2019): 677.

Huang, Yan-Hua, Si-Jia Liu, Shan Yuan, Cong Guan, Dan-Yang Tian, Xin Cui, Yun-Wei Zhang, and Fu-Yu Yang. "Overexpression of ovine AANAT and HIOMT genes in switchgrass leads to improved growth performance and salt-tolerance." *Scientific Reports* 7, no. 1 (2017): 1–13.

Ibrahim, Mohamed F. M., Ola H. Abd Elbar, Reham Farag, Mohamed Hikal, Amr El-Kelish, Ahmed Abou El-Yazied, Jawaher Alkahtani, and Hany G. Abd El-Gawad. "Melatonin counteracts drought induced oxidative damage and stimulates growth, productivity and fruit quality properties of tomato plants." *Plants* 9, no. 10 (2020): 1276.

Imran, Muhammad, Abdul Latif Khan, Raheem Shahzad, Muhammad Aaqil Khan, Saqib Bilal, Adil Khan, Sang-Mo Kang, and In-Jung Lee. "Exogenous melatonin induces drought stress tolerance by promoting plant growth and antioxidant defence system of soybean plants." *AoB Plants* 13, no. 4 (2021): plab026.

Iqbal, Noushina, Mehar Fatma, Harsha Gautam, Shahid Umar, Adriano Sofo, Ilaria D'ippolito, and Nafees A. Khan. "The crosstalk of melatonin and hydrogen sulfide determines photosynthetic performance by regulation of carbohydrate metabolism in wheat under heat stress." *Plants* 10, no. 9 (2021): 1778.

Jahan, Mohammad Shah, Shirong Guo, Abdul Raziq Baloch, Jin Sun, Sheng Shu, Yu Wang, Golam Jalal Ahammed, Khairul Kabir, and Rana Roy. "Melatonin alleviates nickel phytotoxicity by improving photosynthesis, secondary metabolism and oxidative stress tolerance in tomato seedlings." *Ecotoxicology and Environmental Safety* 197 (2020): 110593.

Jahan, Mohammad Shah, Shirong Guo, Jin Sun, Sheng Shu, Yu Wang, Ahmed Abou El-Yazied, Nadiyah M. Alabdallah et al. "Melatonin-mediated photosynthetic performance of tomato seedlings under high-temperature stress." *Plant Physiology and Biochemistry* 167 (2021a): 309–320.

Jahan, Mohammad Shah, Sheng Shu, Yu Wang, Zheng Chen, Mingming He, Meiqi Tao, Jin Sun, and Shirong Guo. "Melatonin alleviates heat-induced damage of tomato seedlings by balancing redox homeostasis and modulating polyamine and nitric oxide biosynthesis." *BMC Plant Biology* 19, no. 1 (2019): 1–16.

Jahan, Mohammad Shah, Sheng Shu, Yu Wang, Md Mahadi Hasan, Ahmed Abou El-Yazied, Nadiyah M. Alabdallah, Dina Hajjar, Muhammad Ahsan Altaf, Jin Sun, and Shirong Guo. "Melatonin pretreatment confers heat tolerance and repression of heat-induced senescence in tomato through the modulation of ABA-and GA-mediated pathways." *Frontiers in Plant Science* 12 (2021b): 650955.

Jia, Chunhua, Xiaojing Yu, Min Zhang, Zhiguang Liu, Peng Zou, Jun Ma, and Yachun Xu. "Application of melatonin-enhanced tolerance to high-temperature stress in cherry radish (*Raphanus sativus* L. var. radculuspers)." *Journal of Plant Growth Regulation* 39, no. 2 (2020): 631–640.

Jiang, Chaoqiang, Quanren Cui, Kun Feng, Dafeng Xu, Chengfeng Li, and Qingsong Zheng. "Melatonin improves antioxidant capacity and ion homeostasis and enhances salt tolerance in maize seedlings." *Acta Physiologiae Plantarum* 38, no. 4 (2016): 1–9.

Kang, Kiyoon, Kyungjin Lee, Sangkyu Park, Young Soon Kim, and Kyoungwhan Back. "Enhanced production of melatonin by ectopic overexpression of human serotonin N-acetyltransferase plays a role in cold resistance in transgenic rice seedlings." *Journal of Pineal Research* 49, no. 2 (2010): 176–182.

Kaur, Harmeet, and Satish C. Bhatla. "Melatonin and nitric oxide modulate glutathione content and glutathione reductase activity in sunflower seedling cotyledons accompanying salt stress." *Nitric Oxide* 59 (2016): 42–53.

Kaya, Cengiz, Mustafa Okant, Ferhat Ugurlar, Mohammed Nasser Alyemeni, Muhammad Ashraf, and Parvaiz Ahmad. "Melatonin-mediated nitric oxide improves tolerance to cadmium toxicity by reducing oxidative stress in wheat plants." *Chemosphere* 225 (2019): 627–638.

Ke, Qingbo, Jun Ye, Bomei Wang, Jianhong Ren, Lina Yin, Xiping Deng, and Shiwen Wang. "Melatonin mitigates salt stress in wheat seedlings by modulating polyamine metabolism." *Frontiers in Plant Science* 9 (2018): 914.

Khan, Mohammad Nauman, Jing Zhang, Tao Luo, Jiahuan Liu, Muhammad Rizwan, Shah Fahad, Zhenghua Xu, and Liyong Hu. "Seed priming with melatonin coping drought stress in rapeseed by regulating reactive oxygen species detoxification: Antioxidant defense system, osmotic adjustment, stomatal traits and chloroplast ultrastructure perseveration." *Industrial Crops and Products* 140 (2019): 111597.

Kobylińska, A., and Małgorzata M. Posmyk. "Melatonin restricts Pb-induced PCD by enhancing BI-1 expression in tobacco suspension cells." *Biometals* 29, no. 6 (2016): 1059–1074.

Kobylińska, Agnieszka, Russel J. Reiter, and Malgorzata M. Posmyk. "Melatonin protects cultured tobacco cells against lead-induced cell death via inhibition of cytochrome c translocation." *Frontiers in Plant Science* 8 (2017): 1560.

Lee, Hyoung Yool, Kyungjin Lee, and Kyoungwhan Back. "Knockout of Arabidopsis serotonin N-acetyltransferase-2 reduces melatonin levels and delays flowering." *Biomolecules* 9, no. 11 (2019): 712.

Lee, Kyungjin, and Kyoungwhan Back. "Overexpression of rice serotonin N-acetyltransferase 1 in transgenic rice plants confers resistance to cadmium and senescence and increases grain yield." *Journal of Pineal Research* 62, no. 3 (2017): e12392.

Lei, Xiao-Yong, Rui-Yu Zhu, Gui-You Zhang, and Yao-Ren Dai. "Attenuation of cold-induced apoptosis by exogenous melatonin in carrot suspension cells: The possible involvement of polyamines." *Journal of Pineal Research* 36, no. 2 (2004): 126–131.

Li, Chao, Dun-Xian Tan, Dong Liang, Cong Chang, Dongfeng Jia, and Fengwang Ma. "Melatonin mediates the regulation of ABA metabolism, free-radical scavenging, and stomatal behaviour in two Malus species under drought stress." *Journal of Experimental Botany* 66, no. 3 (2015): 669–680.

Li, Hao, Yanliang Guo, Zhixiang Lan, Kai Xu, Jingjing Chang, Golam Jalal Ahammed, Jianxiang Ma, Chunhua Wei, and Xian Zhang. "Methyl jasmonate mediates melatonin-induced cold tolerance of grafted watermelon plants." *Horticulture Research* 8 (2021).

Li, Xiangnan, Dun-Xian Tan, Dong Jiang, and Fulai Liu. "Melatonin enhances cold tolerance in drought-primed wild-type and abscisic acid-deficient mutant barley." *Journal of Pineal Research* 61, no. 3 (2016): 328–339.

Li, Xin, Ji-Peng Wei, Eric R. Scott, Jian-Wei Liu, Shuai Guo, Yang Li, Lan Zhang, and Wen-Yan Han. "Exogenous melatonin alleviates cold stress by promoting antioxidant defense and redox homeostasis in *Camellia sinensis* L." *Molecules* 23, no. 1 (2018): 165.

Liang, Bowen, Changqing Ma, Zhijun Zhang, Zhiwei Wei, Tengteng Gao, Qi Zhao, Fengwang Ma, and Chao Li. "Long-term exogenous application of melatonin improves nutrient uptake fluxes in apple plants under moderate drought stress." *Environmental and Experimental Botany* 155 (2018a): 650–661.

Liang, Dong, Fan Gao, Zhiyou Ni, Lijin Lin, Qunxian Deng, Yi Tang, Xun Wang, Xian Luo, and Hui Xia. "Melatonin improves heat tolerance in kiwifruit seedlings through promoting antioxidant enzymatic activity and glutathione S-transferase transcription." *Molecules* 23, no. 3 (2018b): 584.

Liu, Dan-Dan, Xiao-Shuai Sun, Lin Liu, Hong-Di Shi, Sui-Yun Chen, and Da-Ke Zhao. "Overexpression of the melatonin synthesis-related gene SlCOMT1 improves the resistance of tomato to salt stress." *Molecules* 24, no. 8 (2019): 1514.

Liu, Jianlong, Weixiang Wang, Liying Wang, and Yan Sun. "Exogenous melatonin improves seedling health index and drought tolerance in tomato." *Plant Growth Regulation* 77, no. 3 (2015): 317–326.

Liu, Juan, Sergey Shabala, Jing Zhang, Guohui Ma, Dandan Chen, Lana Shabala, Fanrong Zeng et al. "Melatonin improves rice salinity stress tolerance by NADPH oxidase-dependent control of the plasma membrane K+ transporters and K+ homeostasis." *Plant, Cell & Environment* 43, no. 11 (2020): 2591–2605.

Ma, Xiqing, Jing Zhang, Patrick Burgess, Stephanie Rossi, and Bingru Huang. "Interactive effects of melatonin and cytokinin on alleviating drought-induced leaf senescence in creeping bentgrass (Agrostis stolonifera)." *Environmental and Experimental Botany* 145 (2018): 1–11.

Ma, Yaner, Jian Jiao, Xiucai Fan, Haisheng Sun, Ying Zhang, Jianfu Jiang, and Chonghuai Liu. "Endophytic bacterium *Pseudomonas fluorescens* RG11 may transform tryptophan to melatonin and promote endogenous melatonin levels in the roots of four grape cultivars." *Frontiers in Plant Science* 7 (2017): 2068.

Mano, Jun'ichi, Yoshimitsu Torii, Shun-ichiro Hayashi, Koichi Takimoto, Kenji Matsui, Kaoru Nakamura, Dirk Inzé, Elena Babiychuk, Sergei Kushnir, and Kozi Asada. "The NADPH: quinone oxidoreductase

P1-ζ-crystallin in Arabidopsis catalyzes the α, β-hydrogenation of 2-alkenals: Detoxication of the lipid peroxide-derived reactive aldehydes." *Plant and Cell Physiology* 43, no. 12 (2002): 1445–1455.

Meloni, Diego A., Marco A. Oliva, Carlos A. Martinez, and José Cambraia. "Photosynthesis and activity of superoxide dismutase, peroxidase and glutathione reductase in cotton under salt stress." *Environmental and Experimental Botany* 49, no. 1 (2003): 69–76.

Moustafa-Farag, Mohamed, Ahmed Mahmoud, Marino B. Arnao, Mohamed S. Sheteiwy, Mohamed Dafea, Mahmoud Soltan, Amr Elkelish, Mirza Hasanuzzaman, and Shaoying Ai. "Melatonin-induced water stress tolerance in plants: Recent advances." *Antioxidants* 9, no. 9 (2020): 809.

Murch, S. J., S. KrishnaRaj, and P. K. Saxena. "Tryptophan is a precursor for melatonin and serotonin biosynthesis in in vitro regenerated St. John's wort (*Hypericum perforatum* L. cv. Anthos) plants." *Plant Cell Reports* 19, no. 7 (2000): 698–704.

Namdjoyan, Shahram, Ali Abolhasani Soorki, Nazli Elyasi, Nader Kazemi, and Mehdi Simaei. "Melatonin alleviates lead-induced oxidative damage in safflower (*Carthamus tinctorius* L.) seedlings." *Ecotoxicology* 29, no. 1 (2020): 108–118.

Nazir, Munazza, Muhammad Asad Ullah, Sadia Mumtaz, Aisha Siddiquah, Muzamil Shah, Samantha Drouet, Christophe Hano, and Bilal Haider Abbasi. "Interactive effect of melatonin and UV-C on phenylpropanoid metabolite production and antioxidant potential in callus cultures of purple basil (*Ocimum basilicum* L. var purpurascens)." *Molecules* 25, no. 5 (2020): 1072.

Ni, Jun, Qiaojian Wang, Faheem Afzal Shah, Wenbo Liu, Dongdong Wang, Shengwei Huang, Songling Fu, and Lifang Wu. "Exogenous melatonin confers cadmium tolerance by counterbalancing the hydrogen peroxide homeostasis in wheat seedlings." *Molecules* 23, no. 4 (2018): 799.

Okant, Mustafa, and Cengiz Kaya. "The role of endogenous nitric oxide in melatonin-improved tolerance to lead toxicity in maize plants." *Environmental Science and Pollution Research* 26, no. 12 (2019): 11864–11874.

Okazaki, Masateru, Kenji Higuchi, Asma Aouini, and Hiroshi Ezura. "Lowering intercellular melatonin levels by transgenic analysis of indoleamine 2, 3-dioxygenase from rice in tomato plants." *Journal of Pineal Research* 49, no. 3 (2010): 239–247.

Park, Hee-Soon, Elham Ahmed Kazerooni, Sang-Mo Kang, Abdullah Mohammed Al-Sadi, and In-Jung Lee. "Melatonin enhances the tolerance and recovery mechanisms in *Brassica juncea* (L.) Czern. under saline conditions." *Frontiers in Plant Science* 12 (2021): 593717.

Posmyk, Małgorzata M., M. Bałabusta, M. Wieczorek, E. Sliwinska, and K. M. Janas. "Melatonin applied to cucumber (*Cucumis sativus* L.) seeds improves germination during chilling stress." *Journal of Pineal Research* 46, no. 2 (2009): 214–223.

Posmyk, Małgorzata M., Hanna Kuran, Kazimierz Marciniak, and Krystyna M. Janas. "Presowing seed treatment with melatonin protects red cabbage seedlings against toxic copper ion concentrations." *Journal of Pineal Research* 45, no. 1 (2008): 24–31.

Rezania, Shahabaldin, Mohanadoss Ponraj, AmirrezaTalaiekhozani, Shaza Eva Mohamad, Mohd Fadhil Md Din, Shazwin Mat Taib, Farzaneh Sabbagh, and Fadzlin Md Sairan. "Perspectives of phytoremediation using water hyacinth for removal of heavy metals, organic and inorganic pollutants in wastewater." *Journal of Environmental Management* 163 (2015): 125–133.

Rodriguez, Carmen, Juan C. Mayo, Rosa M. Sainz, Isaac Antolín, Federico Herrera, Vanesa Martín, and Russel J. Reiter. "Regulation of antioxidant enzymes: A significant role for melatonin." *Journal of Pineal Research* 36, no. 1 (2004): 1–9.

Roychoudhury, A., S. Basu, S. N. Sarkar., and D. N. Sengupta. "Comparative physiological and molecular responses of a common aromatic indica rice cultivar to high salinity with non-aromatic indica rice cultivars." *Plant Cell Reports* 27, no. 8 (2008): 1395–1410.

Roychoudhury, A., S. Basu, and D. N. Sengupta. "Antioxidants and stress-related metabolites in the seedlings of two indica rice varieties exposed to cadmium chloride toxicity.' *Acta Physiologiae Plantarum* 34, no. 3 (2012): 835–847.

Roychoudhury, A., C. Roy, and D. N. Sengupta. "Transgenic tobacco plants overexpressing the heterologous *lea* gene *Rab16A* from rice during high salt and water deficit display enhanced tolerance to salinity stress." *Plant Cell Reports* 26. no. 10 (2007): 1839–1859.

Sadak, Mervat Sh, Aboelfetoh M. Abdalla, Ebtihal M. Abd Elhamid, and M. I. Ezzo. "Role of melatonin in improving growth, yield quantity and quality of *Moringa oleifera* L. plant under drought stress." *Bulletin of the National Research Centre* 44, no. 1 (2020): 1–13.

Samanta, Santanu, and Aryadeep Roychoudhury. "Heavy metal toxicity in plants: Physiological and molecular adaptations." In *Heavy Metal Toxicity in Plants*, pp. 1–10. Boca Raton: CRC Press, USA, 2021.

Seleiman, Mahmoud F., Shafaqat Ali, Yahya Refay, Muhammad Rizwan, Bushra Ahmed Alhammad, and Salah E. El-Hendawy. "Chromium resistant microbes and melatonin reduced Cr uptake and toxicity, improved physio-biochemical traits and yield of wheat in contaminated soil." *Chemosphere* 250 (2020): 126239.

Sharma, Anket, and Bingsong Zheng. "Melatonin mediated regulation of drought stress: Physiological and molecular aspects." *Plants* 8, no. 7 (2019): 190.

Shi, Haitao, Chuan Jiang, Tiantian Ye, Dun-Xian Tan, Russel J. Reiter, Heng Zhang, Renyi Liu, and Zhulong Chan. "Comparative physiological, metabolomic, and transcriptomic analyses reveal mechanisms of improved abiotic stress resistance in bermudagrass [*Cynodon dactylon* (L). Pers.] by exogenous melatonin." *Journal of Experimental Botany* 66, no. 3 (2015a): 681–694.

Shi, Haitao, Yongqiang Qian, Dun-Xian Tan, Russel J. Reiter, and Chaozu He. "Melatonin induces the transcripts of CBF/DREB1s and their involvement in both abiotic and biotic stresses in Arabidopsis." *Journal of Pineal Research* 59, no. 3 (2015b): 334–342.

Shi, Haitao, Dun-Xian Tan, Russel J. Reiter, Tiantian Ye, Fan Yang, and Zhulong Chan. "Melatonin induces class A1 heat-shock factors (HSFA 1s) and their possible involvement of thermotolerance in Arabidopsis." *Journal of Pineal Research* 58, no. 3 (2015c): 335–342.

Sun, Chengliang, Lijuan Liu, Luxuan Wang, Baohai Li, ChongweiJin, and Xianyong Lin. "Melatonin: A master regulator of plant development and stress responses." *Journal of Integrative Plant Biology* 63, no. 1 (2021): 126–145.

Sun, Shasha, Dan Wen, Wanying Yang, Qiufeng Meng, Qinghua Shi, and Biao Gong. "Overexpression of caffeic acid O-methyltransferase 1 (COMT1) increases melatonin level and salt stress tolerance in tomato plant." *Journal of Plant Growth Regulation* 39, no. 3 (2020): 1221–1235.

Talaat, Neveen B. "Polyamine and nitrogen metabolism regulation by melatonin and salicylic acid combined treatment as a repressor for salt toxicity in wheat (*Triticum aestivum* L.) plants." *Plant Growth Regulation* 95, no. 3 (2021): 315–329.

Tan, Dun-Xian, Rudiger Hardeland, Lucien C. Manchester, Ahmet Korkmaz, Shuran Ma, Sergio Rosales-Corral, and Russel J. Reiter. "Functional roles of melatonin in plants, and perspectives in nutritional and agricultural science." *Journal of Experimental Botany* 63, no. 2 (2012): 577–597.

Tan, Dun-Xian, Lucien C. Manchester, Pat Helton, and Russel J. Reiter. "Phytoremediative capacity of plants enriched with melatonin." *Plant Signaling & Behavior* 2, no. 6 (2007): 514–516.

Tan, Dun-Xian, and Russel J. Reiter. "An evolutionary view of melatonin synthesis and metabolism related to its biological functions in plants." *Journal of Experimental Botany* 71, no. 16 (2020): 4677–4689.

Tang, Yi, Guochao Sun, Mingjun Miao, Yanhong Li, and Huanxiu Li. "Effects of exogenous melatonin on growth and physiological characteristics of Chinese cabbage seedlings under aluminum stress." In *2017 6th International Conference on Energy and Environmental Protection (ICEEP 2017)*, pp. 1119–1123. Zhuhai: Atlantis Press, China, 2017.

Tiwari, Rahul Kumar, Milan Kumar Lal, Kailash Chandra Naga, Ravinder Kumar, Kumar Nishant Chourasia, S. Subhash, Dharmendra Kumar, and Sanjeev Sharma. "Emerging roles of melatonin in mitigating abiotic and biotic stresses of horticultural crops." *Scientia Horticulturae* 272 (2020): 109592.

Turk, Hulya, Serkan Erdal, MucipGenisel, Okkes Atici, Yavuz Demir, and Derya Yanmis. "The regulatory effect of melatonin on physiological, biochemical and molecular parameters in cold-stressed wheat seedlings." *Plant Growth Regulation* 74, no. 2 (2014): 139–152.

Upadhyaya, Gouranga, Arup Das, Chandradeep Basu, Tanushree Agarwal, Chandra Basak, Chandrima Chakraborty, Tanmoy Halder, Gautam Basu, and Sudipta Ray. "Multiple copies of a novel amphipathic α-helix forming segment in *Physcomitrella patens* dehydrin play a key role in abiotic stress mitigation." *Journal of Biological Chemistry* 296 (2021a).

Upadhyaya, Gouranga, Arup Das, and Sudipta Ray. "A rice R2R3-MYB (OsC1) transcriptional regulator improves oxidative stress tolerance by modulating anthocyanin biosynthesis." *Physiologia Plantarum* 173, no. 4 (2021b): 2334–2349.

Vafadar, Farinaz, Rayhaneh Amooaghaie, Parviz Ehsanzadeh, Faezeh Ghanati, and Reza H. Sajedi. "Crosstalk between melatonin and Ca2+/CaM evokes systemic salt tolerance in *Dracocephalum kotschyi*." *Journal of Plant Physiology* 252 (2020): 153237.

Van Zelm, Eva, Yanxia Zhang, and Christa Testerink. "Salt tolerance mechanisms of plants." *Annual Review of Plant Biology* 71 (2020): 403–433.

Wang, Lin, Yu Zhao, Russel J. Reiter, Changjiu He, Guoshi Liu, Qiong Lei, Bixiao Zuo, Xiao Dong Zheng, Qingtian Li, and Jin Kong. "Changes in melatonin levels in transgenic 'Micro-Tom' tomato overexpressing ovine AANAT and ovine HIOMT genes." *Journal of Pineal Research* 56, no. 2 (2014): 134–142.

Wang, Lin-Feng, Ting-Ting Li, Yu Zhang, Jia-Xing Guo, Kai-Kai Lu, and Wen-Cheng Liu. "CAND2/PMTR1 is required for melatonin-conferred osmotic stress tolerance in Arabidopsis." *International Journal of Molecular Sciences* 22, no. 8 (2021): 4014.

Wang, Meng, Shuhui Duan, Zhicheng Zhou, Shibao Chen, and Duo Wang. "Foliar spraying of melatonin confers cadmium tolerance in *Nicotiana tabacum* L." *Ecotoxicology and Environmental Safety* 170 (2019): 68–76.

Wang, Ping, Xun Sun, Chao Li, Zhiwei Wei, Dong Liang, and Fengwang Ma. "Long-term exogenous application of melatonin delays drought-induced leaf senescence in apple." *Journal of Pineal Research* 54, no. 3 (2013): 292–302.

Wang, Xiaoyun, Haijun Zhang, Qian Xie, Ying Liu, Hongmei Lv, Ruyue Bai, Rui Ma et al. "SlSNAT interacts with HSP40, a molecular chaperone, to regulate melatonin biosynthesis and promote thermotolerance in tomato." *Plant and Cell Physiology* 61, no. 5 (2020): 909–921.

Wei, Zhiwei, Chao Li, Tengteng Gao, Zhijun Zhang, Bowen Liang, Zuosen Lv, Yangjun Zou, and Fengwang Ma. "Melatonin increases the performance of *Malus hupehensis* after UV-B exposure." *Plant Physiology and Biochemistry* 139 (2019): 630–641.

Wu, Yandi, Xiucai Fan, Ying Zhang, Jianfu Jiang, Lei Sun, Faiz Ur Rahman, and Chonghuai Liu. "VvSNAT1 overexpression enhances melatonin production and salt tolerance in transgenic Arabidopsis." *Plant Physiology and Biochemistry* 166 (2021): 485–494.

Xie, Chengcheng, Xi Xiong, Zhuo Huang, Lingxia Sun, Jun Ma, Shizhen Cai, Fei Yu, Woxiu Zhong, Shuyu Chen, and Xi Li. "Exogenous melatonin improves lead tolerance of bermudagrass through modulation of the antioxidant defense system." *International Journal of Phytoremediation* 20, no. 14 (2018): 1408–1417.

Xie, Ziyan, Juan Wang, Wensheng Wang, Yanru Wang, Jianlong Xu, Zhikang Li, Xiuqin Zhao, and Binying Fu. "Integrated analysis of the transcriptome and metabolome revealed the molecular mechanisms underlying the enhanced salt tolerance of rice due to the application of exogenous melatonin." *Frontiers in Plant Science* 11 (2021): 618680.

Xu, Lili, Guangqing Xiang, Qinghua Sun, Yong Ni, ZhongxinJin, Shiwei Gao, and Yuxin Yao. "Melatonin enhances salt tolerance by promoting MYB108A-mediated ethylene biosynthesis in grapevines." *Horticulture Research* 6 (2019).

Xu, Wen, Shu-Yu Cai, Yun Zhang, Yu Wang, Golam Jalal Ahammed, Xiao-Jian Xia, Kai Shi et al. "Melatonin enhances thermotolerance by promoting cellular protein protection in tomato plants." *Journal of Pineal Research* 61, no. 4 (2016): 457–469.

Yan, Feiyu, Haimin Wei, Weiwei Li, Zhenghui Liu, She Tang, Lin Chen, Chengqiang Ding, Yu Jiang, Yanfeng Ding, and Ganghua Li. "Melatonin improves K+ and Na+ homeostasis in rice under salt stress by mediated nitric oxide." *Ecotoxicology and Environmental Safety* 206 (2020): 111358.

Yang, Wen-Jing, Yong-Tao Du, Yong-Bin Zhou, Jun Chen, Zhao-Shi Xu, You-Zhi Ma, Ming Chen, and Dong-Hong Min. "Overexpression of TaCOMT improves melatonin production and enhances drought tolerance in transgenic Arabidopsis." *International Journal of Molecular Sciences* 20, no. 3 (2019): 652.

Yang, X. L., H. Xu, D. Li, X. Gao, T. L. Li, and R. Wang. "Effect of melatonin priming on photosynthetic capacity of tomato leaves under low-temperature stress." *Photosynthetica* 56, no. 3 (2018): 884–892.

Yao, Jing-Wen, Zheng Ma, Yan-Qin Ma, Ying Zhu, Meng-Qi Lei, Cheng-Ying Hao, Li-Yu Chen, Zi-Qin Xu, and Xuan Huang. "Role of melatonin in UV-B signaling pathway and UV-B stress resistance in Arabidopsis thaliana." *Plant, Cell & Environment* 44, no. 1 (2021): 114–129.

Ye, Jun, Shiwen Wang, Xiping Deng, Lina Yin, Binglin Xiong, and Xinyue Wang. "Melatonin increased maize (*Zea mays* L.) seedling drought tolerance by alleviating drought-induced photosynthetic inhibition and oxidative damage." *Acta Physiologiae Plantarum* 38, no. 2 (2016): 1–13.

Yin, Zepeng, Jiazhi Lu, Sida Meng, Yiling Liu, Islam Mostafa, Mingfang Qi, and Tianlai Li. "Exogenous melatonin improves salt tolerance in tomato by regulating photosynthetic electron flux and the ascorbate: Glutathione cycle." *Journal of Plant Interactions* 14, no. 1 (2019): 453–463.

Yu, Yanyan, Yong Ni, Tian Qiao, Xiaomin Ji, Jinghao Xu, Bo Li, and Qinghua Sun. "Overexpression of VvASMT1 from grapevine enhanced salt and osmotic stress tolerance in *Nicotiana benthamiana*." *PloS one* 17, no. 6 (2022): e0269028.

Yu, Yicheng, Aimin Wang, Xiang Li, Meng Kou, Wenjun Wang, Xianyang Chen, Tao Xu et al. "Melatonin-stimulated triacylglycerol breakdown and energy turnover under salinity stress contributes to the maintenance of plasma membrane H+-ATPase activity and K+/Na+ homeostasis in sweet potato." *Frontiers in Plant Science* 9 (2018): 256.

Zahedi, Seyed Morteza, Marjan Sadat Hosseini, Javier Abadía, and Mina Marjani. "Melatonin foliar sprays elicit salinity stress tolerance and enhance fruit yield and quality in strawberry (Fragaria× ananassa Duch.)." *Plant Physiology and Biochemistry* 149 (2020): 313–323.

Zeng, Wen, Salma Mostafa, Zhaogeng Lu, and Biao Jin. "Melatonin-mediated abiotic stress tolerance in plants." *Frontiers in Plant Science* 13 (2022).

Zhang, Jiarong, Bingjie Zeng, Yawen Mao, Xiangying Kong, Xinxun Wang, Ye Yang, Jie Zhang, Jin Xu, Zed Rengel, and Qi Chen. "Melatonin alleviates aluminium toxicity through modulating antioxidative enzymes and enhancing organic acid anion exudation in soybean." *Functional Plant Biology* 44, no. 10 (2017a): 961–968.

Zhang, Jing, Yi Shi, Xunzhong Zhang, Hongmei Du, Bin Xu, and Bingru Huang. "Melatonin suppression of heat-induced leaf senescence involves changes in abscisic acid and cytokinin biosynthesis and signaling pathways in perennial ryegrass (*Lolium perenne* L.)." *Environmental and Experimental Botany* 138 (2017b): 36–45.

Zhang, Ji-Yi, Maria H. Cruz De Carvalho, Ivone Torres-Jerez, Y. U. N. Kang, Stacy N. Allen, David V. Huhman, Yuhong Tang, Jeremy Murray, Lloyd W. Sumner, and Michael K. Udvardi. "Global reprogramming of transcription and metabolism in *Medicago truncatula* during progressive drought and after rewatering." *Plant, Cell & Environment* 37, no. 11 (2014): 2553–2576.

Zhang, Lai Jun, Jing Fen Jia, Yao Xu, Yingli Wang, Jian Guo Hao, and Tian Ke Li. "Production of transgenic Nicotiana sylvestris plants expressing melatonin synthetase genes and their effect on UV-B-induced DNA damage." *In Vitro Cellular & Developmental Biology-Plant* 48, no. 3 (2012): 275–282.

Zhang, Liang, Dongmei Xi, Shanwei Li, Zheng Gao, Shuoli Zhao, Jing Shi, Changai Wu, and Xingqi Guo. "A cotton group C MAP kinase gene, GhMPK2, positively regulates salt and drought tolerance in tobacco." *Plant Molecular Biology* 77, no. 1 (2011): 17–31.

Zhang, Na, Qianqian Sun, Haijun Zhang, Yunyun Cao, Sarah Weeda, Shuxin Ren, and Yang-Dong Guo. "Roles of melatonin in abiotic stress resistance in plants." *Journal of Experimental Botany* 66, no. 3 (2015): 647–656.

Zhang, Na, Bing Zhao, Hai-Jun Zhang, Sarah Weeda, Chen Yang, Zi-Cai Yang, Shuxin Ren, and Yang-Dong Guo. "Melatonin promotes water-stress tolerance, lateral root formation, and seed germination in cucumber (*Cucumis sativus* L.)." *Journal of Pineal Research* 54, no. 1 (2013): 15–23.

Zhang, Pengfei, Lei Liu, Xin Wang, Ziyang Wang, He Zhang, Jingting Chen, Xinyu Liu, Yubo Wang, and Caifeng Li. "Beneficial effects of exogenous melatonin on overcoming salt stress in sugar beets (*Beta vulgaris* L.)." *Plants* 10, no. 5 (2021a): 886.

Zhang, Qiang, Xiaofei Liu, Zhifei Zhang, Ningfang Liu, Danzhu Li, and Longxing Hu. "Melatonin improved waterlogging tolerance in alfalfa (*Medicago sativa*) by reprogramming polyamine and ethylene metabolism." *Frontiers in Plant Science* 10 (2019): 44.

Zhang, Tengguo, Zhongfei Shi, Xiaohua Zhang, Sheng Zheng, Juan Wang, and Jiangnan Mo. "Alleviating effects of exogenous melatonin on salt stress in cucumber." *Scientia Horticulturae* 262 (2020): 109070.

Zhang, Tengguo, Juan Wang, Yuanpei Sun, Lu Zhang, and Sheng Zheng. "Versatile roles of melatonin in growth and stress tolerance in plants." *Journal of Plant Growth Regulation* (2021b): 1–17.

Zhao, Gan, Yingying Zhao, Xiuli Yu, Felix Kiprotich, Han Han, Rongzhan Guan, Ren Wang, and Wenbiao Shen. "Nitric oxide is required for melatonin-enhanced tolerance against salinity stress in rapeseed (*Brassica napus* L.) seedlings." *International Journal of Molecular Sciences* 19, no. 7 (2018): 1912.

Zheng, Xiaodong, Jingzhe Zhou, Dun-Xian Tan, Na Wang, Lin Wang, Dongqian Shan, and Jin Kong. "Melatonin improves waterlogging tolerance of *Malus baccata* (Linn.) Borkh. seedlings by maintaining aerobic respiration, photosynthesis and ROS migration." *Frontiers in Plant Science* 8 (2017): 483.

Zhuang, Weibing, Tianyu Liu, Xiaochun Shu, Hongxue Wang, Zhong Wang, Tao Wang, Fengjiao Zhang, and Shenchun Qu. "Overexpression of MzASMT 1, a gene from malus zumi mats, enhances salt tolerance in transgenic tobacco." *Frontiers in Plant Science* 11 (2020): 561903.

Zuo, Bixiao, Xiaodong Zheng, Pingli He, Lin Wang, Qiong Lei, Chao Feng, Jingzhe Zhou, Qingtian Li, Zhenhai Han, and Jin Kong. "Overexpression of MzASMT improves melatonin production and enhances drought tolerance in transgenic *Arabidopsis thaliana* plants." *Journal of Pineal Research* 57, no. 4 (2014): 408–417.

13 Melatonin-Induced Tolerance against Salt Stress

Shashi Ranjan, Vikas Mangal, Aman Prakash, Muhammad Ahsan Altaf, Soumya Kumar Sahoo, Biswajit Lenka, Biswaranjan Behera, Ravinder Kumar, Awadhesh Kumar, Rahul Kumar Tiwari, and Milan Kumar Lal

13.1 INTRODUCTION

Environmental factors, including drought, salinity, and global warming significantly reduce agricultural crop production in many regions around the world. The world's agricultural land loss is 2 Mha (around 1 %) per year due to salinity, resulting in decreased crop yield (Chourasia u. a. 2022). Salt stress (salinity) is one of the most damaging abiotic stresses to plant processes. Salt stress can inhibit plant growth and development, seed germination, senescence, cellular metabolism, gene expression, crop yield, and so on (Zhu 2001; Chen, Song, und Wang 2010; Song und Wang 2015; Guo u. a. 2018; Kumar u. a. 2022). Essentially, stress in plants causes an adverse condition in which plants struggle to survive, resulting in an unnatural response from plants (Roychoudhury et al. 2008). Climate, human activities, and the composition of the soil content of that particular land are all factors that contribute to salt stress. It is estimated that 900 Mha are affected by salt, which is a major problem for crop productivity (Flowers und Yeo 1995; Munns 2002; Munns und Tester 2008b). The main reason for salinity in soil is the accumulation of NaCl, Na_2CO_3, and trace amounts of $CaCl_2$. Both nature and humans are responsible for the rise in salinity. Natural salinity occurs as a result of capillary water level elevation and continuous evaporation of salty water from the ground. When we talk about man-made salinity, it is due to irrigated land in barren areas, which is highly susceptible to salinity. A large amount of irrigation causes ground water to rise, which increases evaporation (Nawaz u. a. 2010). An increase in salinity or salt level raises the soil pH, degrading the soil structure and affecting the essential air-water balance for root processes (Nawaz u. a. 2010).

Salinity causes an increase in the concentrations of Na^+ and Cl- ions in leaves while decreasing the concentrations of N^+, PO_4^{-3}, K^+, Mg^{+2}, and Ca^{2+} ions. Plant growth is hampered by reduced water uptake, malnutrition, and accumulation of toxic sodium and chloride ions as a result of these factors (Roychoudhury et al. 2007). Plants use chloride as a mineral nutrient to fulfil micronutrient requirements, but if its accumulation increases, it results in low biodiversity, less growth, and production of salt-sensitive plants. Because of the higher concentrations of sodium and chloride ions in the cytosol, enzymes do not function properly. There are two stages of salinity stress-related responses in plants. These are both shoot ion-dependent and shoot ion-independent responses. The former phase is characterized by a shoot ion-dependent response that occurs within a few minutes or over a few days and is caused by sodium ions, which cause stomatal closure as well as leaf expansion inhibition. Salinity has a negative impact on plant water combinations during this phase. In the later stages of salinity stress, toxic ion accumulation occurs, particularly in older leaves, resulting in premature senescence, lower yield, and plant death (Kotagiri und Kolluru 2017). To overcome the situation of tolerance, plants have some mechanisms, such as ion exclusion and shoot and ion-independent tolerance. Plants respond to salt stress through a variety of biochemical, molecular, and cellular changes (Chen u. a. 2018). At the cellular level, plant stress responses

 DOI: 10.1201/9781003343752-13

include changes in cell membrane composition, ion differentiation, and selective ion absorption and exclusion. In plants, the authority of the absorption of ions by the roots, control of transportation from roots to shoots, ion distribution in their different parts, deviation in their photosynthetic pathway, modification of the activity of antioxidant enzymes, changing the levels of phytohormones (salicylic acid, strigolactones, nitric oxide, jasmonic acid, hydrogen sulfide, calcium (Daudi u. a. 2012; Iqbal, Masood, und Khan 2014) contribute to salt stress response (Fahad u. a. 2015; Campos u. a. 2019; Roychoudhury u. a. 2013).

Numerous studies conducted recently have revealed that MT is crucial for plants under salt stress (Wang u. a. 2016; Chen u. a. 2017). *N-acetyl*-5-methoxy-tryptamine, also known as melatonin, was first discovered in plants in 1995, and it was given that name because of its ability to counteract the darkening effects of melanocyte-stimulating hormone (Lerner u. a. 1958). In plants, MT is found in different plant species and in various organs or parts, such as bulb, skin, coleoptiles, roots, stems, seeds, fruits, and leaf (Altaf u. a. 2021a). Previous literature has shown that MT action enhances tolerance to a wide range of abiotic stresses, including salinity, drought, cold, heat, light, and heavy metals (Tiwari u. a. 2020). MT also regulates seed germination, seedling growth, root morphology, leaf senescence, photosynthesis, flowering, and organogenesis (Tiwari u. a. 2020; Lal u. a. 2022; Altaf u. a. 2022). Recently, many studies on MT have focused on the functioning and regulation of MT in plants adapted to salt stress. Altaf et al. (2021b) reported that MT application significantly enhanced the growth status of tomato seedlings under salt stress. In addition, MT treatment enhanced the root architecture system in tomato, improved photosynthetic capacity in cucumber (Wang u. a. 2016), increased osmolyte content in *Citrus* (Kostopoulou u. a. 2015), improved gas exchange elements in watermelon, promoted pigment content in tomato, enhanced antioxidant enzyme system, and reduced ROS production in apple under salt stress (Li u. a. 2019).

13.2 EFFECT OF SALT STRESS ON PLANTS AND THEIR MORPHOLOGY

As a plant suffers from salt stress, it might face difficulty in its growth, metabolism, productivity, reproducibility, and other physiological processes. Due to salinity, a plant loses their water-absorbing capacity through roots, which leads to their diminishing growth. To avoid this, a plant makes some adjustments to avoid stress. Due to lack of water, a plant faces another stress called drought stress. The production of a new leaf depends on the water potential of soil solution. Salt does not directly take part in affecting the growth of tissues; as the cells of the plant grow, the salt that travels through the xylem expands within the vacuoles. That is why the amount of salt uptake by the plant does not directly affect the growth of new plant leaves (Soltabayeva u. a. 2021; Lin u. a. 2018). During plant growth, there comes a stage which is very sensitive, that is, at germination stage and at the seedling stage. During this stage, plants are expected to regulate these steps at a normal rate, slow rate or up regulated condition (Shah 2014).

Water stress causes a slower rate of new leaf or root growth than salt stress (Munns und Tester 2008c). To support this theory, there is evidence that the concentrations of Na^+ and Cl- in growing cells are lower than the actual toxic-causing concentrations (Hu und Schmidhalter 2005; Roychoudhury et al. 2011). The faster-growing cell contributes to an increase in salt concentration. According to experimental results of shoot water relations, hormonal signals are most likely induced by the osmotic effect of salt outside roots, controlling the rate of cell growth (Munns und Gilliham 2015). In comparison to roots, the growth of leaves and stems is greatly influenced by mild osmotic stress (Ho u. a. 2009). The rate of growth is largely determined by the tissue and species' response time, as well as whether the stress is treated slowly or quickly (Ashraf, Ashraf, und Ali 2010).

In plants, toxicity is mainly due to the uptake and utilization of toxic ions from soil after irrigation. The main toxic substance is sodium chloride and sulphate, and these components are capable of reducing productivity. Most woody plants and crops are sensitive to sodium chloride and sulphate

(FAO 1976). In plants, the main reason of death of leaves in salinity conditions is uptake of plant concentrate by old leaves, and, as a result, salt continues to transpire over a long period of time, which eventually results in an increase in the concentration of Na^+ and Cl- (Munns 2002). The injury caused in the leaves is probably due to excess load of salt in cells to compartmentalize in vacuoles (Munns 2002). The amount of availability of salt in soil varies from place to place, affecting the growth of plants or crops according to the concentration of salt availability (Munns und Tester 2008a). In general, sodium chloride is widely used for the study of salt stress (Munns und Tester 2008a). The salinity stress hits the morphology of plant in a different way like fresh or dry biomass of whole plant, height of the plant, etc. In the growing environment, the risk of salt stress also increases with the growth of a plant. However, this type of response will vary from plant to plant. Salinity does not only affect plants but also trees; very few studies are available on salinity's effect on trees. Some markers are used to trace the salinity stress, such as how salinity affects germination. Germination of seeds is the most prominent morphological process used to indicate the availability of stress. Germination takes around two to 14 days to trace any kind of stress, which depends upon plant species. Change in total plant biomass such as weight of the different parts of a plant is used more often to trace salinity stress. The parameters which are set for the growth of a plant in salinity are root architecture, shoot diameter, number of secondary branches, etc. For the examination of salinity effect, young leaf generation rate is also being used. Along with these parameters, the flowering time of a plant and reproductive stage of a plant also help to analyze the salinity effect (Soltabayeva u. a. 2021). Due to salinity stress, a plant loses its growth and producibility capacity because of badly affected physiological processes, especially photosynthesis. In the salt stress condition, plants accumulate Na^+, which changes the ratio of K:Na; the change in the ratio of K:Na is responsible for affecting the process of photosynthesis (Sudhir und Murthy 2004).

Salt stress leads to variation in ROS, causing a disruption or disturbance in electron transport chain during photoinhibition of photosynthesis (Harir und Mittler 2009). If there is a higher level of ROS, it will be toxic for the plant cells, leading to cellular damage or cell death. ROS level is believed to be regulated by enzyme activity of ROS producers and scavengers (Soltabayeva u. a. 2021). Some of the proteomic and transcriptomic studies reveal that stress leads to the change in the proteome and transcript in plants (Soltabayeva u. a. 2021). Some of the genes change their expression in a different salt concentration when a plant is undergoing salt stress, as analysed from transcript analysis. The genes involved in evaluating the impact of salt stress are mostly seen in ROS regulation enzymes, signal transductions (transcription factors and protein kinase), cell protection, and membrane transporters. It is also seen that some of the genes mentioned earlier are associated with senescence such as, NAC, NAP, WRKY53, WRKY25, SAG12, GDH. On the basis of transcriptional changes of genes, changes at the physiological and morphological level can be predicted (Das and Roychoudhury 2014; Awlia u. a. 2021).

13.3 MOLECULAR INTERACTION OF SALINITY STRESS

To overcome salt stress, plant resorts to some mechanism, viz., biochemical and molecular. It was seen that the biochemical pathways lead to some of the compounds which help to improve the ability to tolerate against salt stress (Iyengar und Reddy 1996). The level of complexity of a salt-tolerant plant may be high or low (Parida und Das 2005).

13.3.1 Classification and Control of Ions

While the plants are under a high level of salt stress, the salt concentration increases rapidly up to excessive levels. It can also affect the cytoplasm, enzymes present for carbohydrate metabolism or it can affect chloroplast, leading to toxicity in photosynthetic process (Dajic 2006). A huge amount of salt is hazardous for both halophytes and glycophytes, and to overcome that situation, plant channelizes the excess salt inside the vacuole to enable their metabolic functions (Manchanda and Garg

2008; Iyengar and Reddy 1996; Zhu 2003). Uptake of ions is a very important process not only for normal plant growth but also for the growth under saline conditions (Adams u. a. 1992). The process of vacuolar sequestration of sodium ions takes place with the help of an enzyme known as salt-inducible Na^+/H^+ antiporter (Apse u. a. 1999). By regulating the expression and activity of K^+ and Na^+ transporters and H^+ pumps, plants maintain high concentrations of K^+ and low concentrations of Na^+ in the cytosol under salt stress (Zhu u. a. 1993).

13.3.2 Compatible Solutes

Compounds with lower molecular mass regulate the ionic balance in vacuoles. Such compounds are known as compatible solutes because they do not interrupt in normal biochemical reactions (Ashihara u. a. 1997; Burg, Kwon, und Kültz 1997; Ford 1984; Manchanda und Garg 2008; Hasegawa u. a. 2000; Zhifang und Loescher 2003). In plants, some suitable solutes include proline, glycine betaine, sugars, and polyols (Torabi u. a. 2013; McNeil u. a. 1999). Some of the nitrogen-containing compounds (NCC) accumulate in plants when they are under salt stress, like amino acids, imino acids, amides, proteins, polyamines, and quaternary ammonium compounds (Roychoudhury u. a. 2015). They play a role in detoxification, cellular macromolecule protection, osmotic adjustment, nitrogen storage, maintaining the cellular pH, and free radical scavenging (Mansour 2000).

13.3.3 Alteration in Photosynthetic Pathways

Reduced rate of photosynthesis under salinity stress is an inevitable condition. In this condition, plant may have to shift their C_3 mode of photosynthesis to CAM (Cushman u. a. 1989; Parida und Das 2005).

13.3.4 Initiation of Antioxidative Enzymes

The effect of salt stress and water deficiency are very much similar to osmotic effect (Cheeseman 1988; Greenway und Munns 1980; He u. a. 2007; Imlay 2003). As there is increase in osmotic potentials in cells, it causes increase in ROS such as singlet oxygen, superoxide, hydroxyl radical, and hydrogen peroxide (Imlay 2003). To reduce the level of reactive oxygen species, plants have a very effective system that protects them from hazardous oxidative reactions (Mittova u. a. 2004). Some antioxidative enzymes are involved in this task such as glutathione reductase, glutathione-S-transferase, catalase, and superoxide dismutase (Garratt u. a. 2002; Mittova u. a. 2003). On the other hand, superoxide dismutase metabolizes oxygen radicals to H_2O_2 and protects the plant cells from damage (Luhua et al. 2008). Catalase and ascorbate peroxidase facilitate the breakdown of H_2O_2 to water and oxygen (Garratt u. a. 2002; Gossett u. a. 1994; Mittova u. a. 2002, 2003).

13.3.5 Plant Hormones

Abscisic acid (ABA) plays an important role during salt stress; ABA activates salt stress–induced genes (De Bruxelles u. a. 1996; Gómez-Cadenas u. a. 2002) to ease the inhibitory effect of NaCl in photosynthesis, translocation, and growth (Popova, Stoinova, und Maslenkova 1995), to promote switch from C_3 to CAM, and to promote stomatal closure by rapidly regulating ion flux in guard cells (Thomas, McElwain, und Bohnert 1992).

13.4 ROLE OF MELATONIN IN PLANTS AND ON SALINITY STRESS

Melatonin is often known as *N-acetyl*-5-methoxy-tryptamine, first discovered in an animal in 1958 and later in a plant in 1995 (Hattori u. a. 1995). It was named due to its reversing function of darkening the effect of melanocyte-stimulating hormone (Lerner u. a. 1958). Melatonin was first discovered in α-proteobacteria and photosynthetic cyanobacteria (Vivien-Roels und Pévet 1993). It is an indoleamine compound with low molecular weight that was first identified as a secretary neurohormone in the pineal gland of vertebrates (Tiwari u. a. 2021), and it has since emerged as a useful animal hormone that regulates circadian rhythms, body temperature, appetite, sleep immunological system, seasonal reproduction, and tumorigenesis (Arnao und Hernández-Ruiz 2018). In the early studies, it was found that melatonin works only in animals. After several experiments in animals, it was found that it plays an important role in regulating several activities like antioxidant enzymes (Rodriguez u. a. 2004), circadian rhythms (Brainard u. a. 2001), emotional status, and physical conditions, including some diseases like coronary and Alzheimer's disease (Brugger, Marktl, und Herold 1995; Dollins u. a. 1994; Cardinali u. a. 2002).

As a unique growth regulator, this compound has outstanding environmental protection capabilities. Under salt stress, MT is a widely used powerful antioxidant that helps plants grow and develop (Altaf u. a. 2021a). MT supplementation considerably improved tomato growth traits, mineral nutrient content, and yield traits under salt stress (Wang u. a. 2016). Salt stress reduced leaf gas exchange elements, enhanced ROS production, and caused oxidative damage in the leaves of watermelon plants. Conversely, MT application efficiently improved leaf photosynthesis and maintained redox homeostasis. In addition, MT treatment increased antioxidant enzyme mechanism in watermelon seedlings under salt stress (Li u. a. 2017b). NaCl toxicity may cause changes in root architecture. In a recent study, Altaf et al. (2021b) observed that salt stress dramatically reduced the root architect system of tomato; however, MT treatment effectively increased root morphological traits of tomato seedlings under salt stress. Photosynthesis is one of the most important physiological processes for a plant to survive. Salinity decreased chlorophyll content, reduced leaf photosynthesis, enhanced ROS production, and damaged PSII activity. In contrast, MT application profoundly improved chlorophyll content, reduced overproduction of ROS, and regulated antioxidant enzyme pool in tomato under salt stress (Yin u. a. 2019). Furthermore, MT may protect plants from oxidative damage by directly enhancing the antioxidant enzyme activity, scavenging free radicals, and excessive production of ROS. MT attenuated the effects of salt stress by reducing the levels of ROS and minimizing the oxidative damage in tomato and apple (Zhou u. a. 2016; Banerjee and Roychoudhury 2017). Antioxidant enzymes play a leading role in the plant defence system.

MT is a naturally occurring molecule with broad-spectrum antioxidant properties. MT supplementation efficiently enhanced antioxidant enzymes in cucumber (Zhang u. a. 2013), tomato (Ali u. a. 2021), and pepper (Kaya u. a. 2020) under salt stress.

13.4.1 Melatonin Is a Bio-Stimulatory Molecule

Although plants are sessile, they can change their physiological functions in response to stressful environmental conditions. When a plant is exposed to a harsh environment, its cells undergo rapid and significant changes. As a result of the stressful environment, several bio-stimulators are activated to stimulate the prevailing bioremediation capacities. Melatonin stimulates plant growth, especially in stressful environments (Debnath u. a. 2019). During the last decades, it was discovered that MT plays a role regularly as a free radical scavenger (Samanta et al. 2021). Melatonin research has revealed that the stimulatory effects of melatonin on biomass and antioxidative defence machinery are corroborated by an adaptable primary and secondary metabolism, as well as stress responses by phytohormone (Arnao und Hernández-Ruiz 2019). Melatonin stimulates and/or protects photosynthesis and the stomatal machinery by upregulating many photosystem components, thylakoid electron transporters, and ATPase genes (Arnao and Hernández-Ruiz 2019). Generally, MT produces numerous changes in gene expression in plants, resulting in a bio-stimulatory response (Arnao and Hernández-Ruiz 2019). MT has been considered as a plant master regulator as well as a novel plant hormone due to a wide range of its properties, and its receptor (PMTR1) has been found in *Arabidopsis* (Wei u. a. 2018). MT is the primary regulator of the redox network in plants.

It regulates the ROS/RNS levels and the gene expression of many variables through the nitric oxide signalling cascade and other mechanisms (Tang u. a. 2018). Melatonin protects by scavenging free radicals, ROS and RNS, acting as a signalling molecule in cells, and increasing the activity of many antioxidant enzymes (Shi u. a. 2015). Phytomelatonin may be used as a bio-stimulator in crops, as phytomelatonin-rich extracts from aromatic and/or medicinal plants have been shown to be natural biostimulants against biotic and abiotic stress.

The molecules of plant growth regulators are able to perform multiple tasks such as the growth and development of plants (Asgher u. a. 2015; Iqbal u. a. 2021a, 2021b). In most of the plant families, the formation of melatonin is well established amidst monocot and dicot (Paredes u. a. 2009; Back, Tan, und Reiter 2016). The production of melatonin takes place in mitochondria and chloroplast of the roots and leaves (de la Puerta u. a. 2007; Wang u. a. 2016; Arnao und Hernández-Ruiz 2017)

13.4.2 Synthesis of Melatonin in Plants

Tryptophan is the precursor of melatonin. The whole synthesis procedure includes four steps: The first step includes the conversion of tryptophan to tryptamine by the help of tryptophan decarboxylase (TDC) (Figure 13.1). After that, tryptamine is converted to serotonin with the help of tryptamine 5-hydroxylase (T5H) (Posmyk und Janas 2009). In plants, it is the primary synthetic pathway for serotonin. Following the completion of this pathway, serotonin is converted to N-acetyl-serotonin via the catalytic reaction of N-acetyltransferase (SNAT) or arylalkylamine N-acetyltransferase (AANAT), and then N-acetyl-serotonin is converted to melatonin via the catalytic reaction of N-acetyl-serotonin (HIOMT). Furthermore, SNAT can facilitate the conversion of tryptamine to N-acetyl-tryptamine, but N-acetyl-tryptamine cannot convert T5H into N-acetyl-serotonin (Yang u. a. 2019).

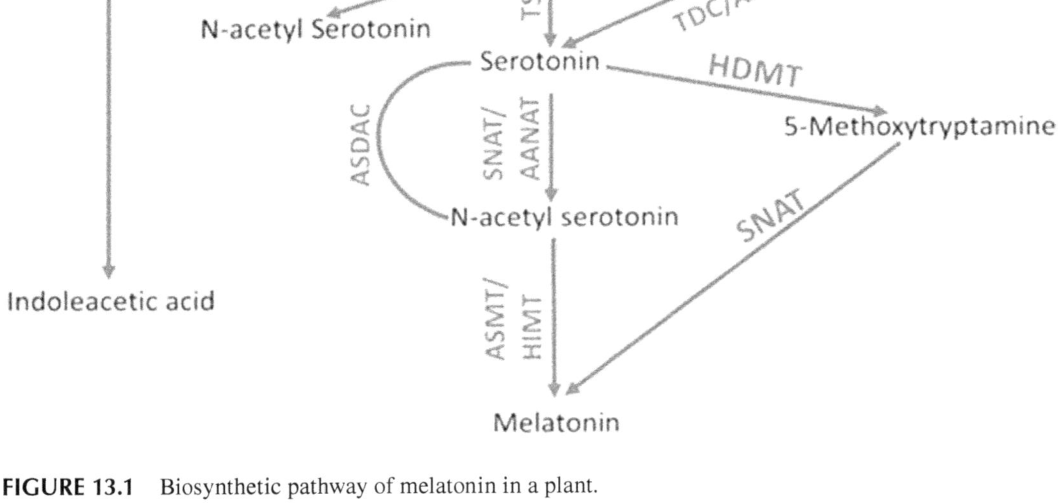

FIGURE 13.1 Biosynthetic pathway of melatonin in a plant.

13.4.3 Melatonin Accumulation Due to Stress

Melatonin is an antioxidant for plants. It interacts with ROS and reduces its level during the stress period (Arnao und Hernández-Ruiz 2013). Thus, we can say that during stress condition, increase in melatonin content is related to increase in the level of ROS (Arnao und Hernández-Ruiz 2013). The gene expression and the enzyme activity are associated with the biosynthesis of melatonin and its catabolic pathways and are closely related to the accumulation of melatonin (Byeon u. a. 2015). Melatonin concentration generally depends on its precursor availability (Rehaman u. a. 2021). Under salt stress, endogenous serotonin and melatonin content increases, indicating their participation in salinity stress. As the analysis goes further, it was found that NaCl stress regulates the activity of N-acetyl-serotonin O-methyltransferase (ASMT), the enzyme which is responsible for the synthesis of melatonin from N-acetyl-serotonin. However, it is unclear how the melatonin pathway is regulated in response to other stress (Zhao u. a. 2022).

Melatonin can regulate plant growth and development, similar to IAA. Kolář, Johnson, und Macháčkov (2003) stated that in the early stages of flower development which is regulated by photoperiod, melatonin plays an important role (Kolář, Johnson, und Macháčková 2003). Due to salt stress, the plant physiological process gets affected, and it reduces large amounts of ROS (Sui u. a. 2010; Sui und Han 2014; Wang u. a. 2015; Lin u. a. 2018). So, to reduce the ROS injury to plants, it is necessary to up regulate melatonin biosynthesis (Shi u. a. 2015). Melatonin has the capability to initiate root development, which makes sure that the plant has a strong root system. The strength of the root is directly related to plant stress resistance (Arnao und Hernández-Ruiz 2007). Some precursors of melatonin, their metabolites, and their intermediates have been identified to engage in plant stress tolerance response (Reddy und Raghavendra 2006). Melatonin plays an important role in photosynthesis and photoprotection (Li u. a. 2012).

TABLE 13.1
Effect of Biochemical Response due to Exogenous Application of Melatonin in Different Crops under Salinity Stress

Crop	Salt stress/melatonin concentration	Biochemical response	Reference
Cucumis Sativus	Seed treatment (1 μM)	Energy production by the activation of different cycles like glycolysis, glyoxylate, citric acid, and cell elongation	(N. Zhang u. a. 2017; S. Zhang u. a. 2017)
	Seed treatment (1 μM)	Synthesis and catabolism of abscisic and gibberellic acid	(Zhang u. a. 2014)
	Seed treatment (50–150 μM)	Activity of antioxidant enzyme enhanced, and chlorophyll content improved	(Wang u. a. 2016)
	Seedling treatment (50–300 μM)	Signalling of mitogen-activated protein kinase (MPAK) enhanced due to increased expression of MAPK3, MAPK4, MAPK6	(Zhang u. a. 2020)
Ipomoea batata	Root dip (1 μM) and leaf spray (100 μM)	Homeostasis of K^+ and Na^+ because there is inhibition of β-oxidation pathway under salt stress	(Zhang u. a. 2013)
Solanum lycopersicum	Seedling treatment at the three-leaf stage (150 μM)	The quantum yield of photosystem 2 and D1 protein improved and is controlled by ROS levels	(Zhou u. a. 2016)

TABLE 13.1 ***(Continued)***
Effect of Biochemical Response due to Exogenous Application of Melatonin in Different Crops under Salinity Stress

Crop	Salt stress/melatonin concentration	Biochemical response	Reference
	Seedling treatment at the three-leaf stage (150 μM)	Photosynthetic electron flux and ascorbate-glutathione cycle balanced	(Yin u. a. 2019)
	100 μM of melatonin (treatment through root irrigation)	Less titer of tobacco mosaic virus (TMV)	(Zhao u. a. 2019)
Malus. hupehensis	Leaf treatment (0.1 μM)	Photosynthetic efficiency improved, and H_2O_2 scavenging is better	(Li u. a. 2012)
Citrus aurantium L.	Seedling treatment (1μM)	Electrolyte leakage reduced, and lipid peroxidation, ion homeostasis, and transcription regulated	(Kostopoulou u. a. 2015)
Citrullus lanatus	Seedlings treated at the three-leaf stage (500 μM)	Improved photosynthesis and ion homeostasis	(Li u. a. 2017a; 2017b)
Gossypium hirsutum L.	Seedlings treated with 25 μM melatonin	Improved efficiency of photosynthesis with ability to scavenge ROS	(Zhang u. a. 2021)
Ulmus americana L.	0.1–0.5 μM of melatonin	Promotes the growth of frozen shoots	(Uchendu u. a. 2013)
Brassica Juncea (L.)	0.1 mM of melatonin	Increases root growth	(Chen u. a. 2009)
Solanum tuberosum	Exogenous treatment of 0, 1, 3, 6, 8, 10 mM of melatonin	Reduces the chances of late blight in potato by reducing the mycelial growth and deformation of cell structure of the pathogen *Phytophthora infestans*	(Zhang u. a. 2017)
Capsicum annum	100 μM of melatonin	Reduces 76% of mycelial growth in phytopathogenic fungi	(Ali u. a. 2021)
Nicotiana glutinosa	100 μM of melatonin (treatment through root irrigation)	Less titer of tobacco mosaic virus (TMV)	(Zhao u. a. 2019)
Zea mays	1 μM melatonin	Elevates growth, enhances photosynthetic rate (P_n) by approximately 19% than normal, increases activities of antioxidant enzymes, decreases electrolyte leakage	(Jiang u. a. 2016)

13.5 CONCLUSION

Altogether it can be concluded that salinity stress negatively impacts plant growth and development. Salinity stress causes an imbalance in ionic strength, higher ROS level, electrolyte leakage, disrupted photosynthesis due to stomatal closure, osmotic imbalance, less productivity. The phytohormone called melatonin, also known as *N-acetyl*-5-methoxy-tryptamine, plays a vital role in overcoming stress. Melatonin is a naturally occurring molecule without any hazardous or toxic effect and hence can be used safely for organic farming. Nowadays, synthetic melatonin is also available at a low cost and hence can be used in agriculture to enhance crop productivity against adverse situations.

REFERENCES

Adams, Patricia, John C. Thomas, Daniel M. Vernon, Hans J. Bohnert, und Richard G. Jensen. 1992. "Distinct Cellular and Organismic Responses to Salt Stress." *Plant and Cell Physiology* 33 (8): 1215–1223. doi:10.1093/oxfordjournals.pcp.a078376.

Ali, Mohamed M., Kaouthar Jeddi, Mohamed S. Attia, Salah M. Elsayed, Mohammad Yusuf, Mahmoud S. Osman, Mona H. Soliman, und Kamel Hessini. 2021. "Wuxal Amino (Bio Stimulant) Improved Growth and Physiological Performance of Tomato Plants under Salinity Stress through Adaptive Mechanisms and Antioxidant Potential." *Saudi Journal of Biological Sciences* 28 (6). Elsevier: 3204–3213. doi:10.1016/j.sjbs.2021.04.040.

Altaf, Muhammad Ahsan, Rabia Shahid, Muhammad Mohsin Altaf, Ravinder Kumar, Safina Naz, Awadhesh Kumar, Pravej Alam, Rahul Kumar Tiwari, Milan Kumar Lal, und Parvaiz Ahmad. 2022. "Melatonin: First-Line Soldier in Tomato under Abiotic Stress Current and Future Perspective." *Plant Physiology and Biochemistry* 185 (August). Elsevier Masson: 188–197. doi:10.1016/j.plaphy.2022.06.004.

Altaf, Muhammad Ahsan, Rabia Shahid, Ming Xun Ren, Muhammad Mohsin Altaf, Mohammad Shah Jahan, und Latif Ullah Khan. 2021a. "Melatonin Mitigates Nickel Toxicity by Improving Nutrient Uptake Fluxes, Root Architecture System, Photosynthesis, and Antioxidant Potential in Tomato Seedling." *Journal of Soil Science and Plant Nutrition* 21 (3). Springer: 1842–1855. doi:10.1007/s42729-021-00484-2.

Altaf, Muhammad Ahsan, Rabia Shahid, Ming Xun Ren, Muhammad Mohsin Altaf, Latif Ullah Khan, Sidra Shahid, und Mohammad Shah Jahan. 2021b. "Melatonin Alleviates Salt Damage in Tomato Seedling: A Root Architecture System, Photosynthetic Capacity, Ion Homeostasis, and Antioxidant Enzymes Analysis." *Scientia Horticulturae* 285 (July). Elsevier B.V.: 110145. doi:10.1016/j.scienta.2021.110145.

Apse, Maris P., Gilad S. Aharon, Wayne A. Snedden, und Eduardo Blumwald. 1999. "Salt Tolerance Conferred by Overexpression of a Vacuolar Na+/H+ Antiport in Arabidopsis." *Science* 285 (5431). American Association for the Advancement of Science: 1256–1258. doi:10.1126/science.285.5431.1256.

Arnao, Marino B., und Josefa Hernández-Ruiz. 2007. "Melatonin Promotes Adventitious- and Lateral Root Regeneration in Etiolated Hypocotyls of Lupinus Albus L." *Journal of Pineal Research* 42 (2). John Wiley & Sons, Ltd: 147–152. doi:10.1111/j.1600-079X.2006.00396.x.

Arnao, Marino B., und Josefa Hernández-Ruiz. 2013. "Growth Conditions Determine Different Melatonin Levels in Lupinus Albus L." *Journal of Pineal Research* 55 (2): 149–155. doi:10.1111/jpi.12055.

Arnao, Marino B., und Josefa Hernández-Ruiz. 2017. "Growth Activity, Rooting Capacity, and Tropism: Three Auxinic Precepts Fulfilled by Melatonin". *Acta Physiologiae Plantarum* 39 (6): 127. doi:10.1007/s11738-017-2428-3.

Arnao, Marino B., und Josefa Hernández-Ruiz. 2018. "Melatonin and Its Relationship to Plant Hormones." *Annals of Botany*. Oxford University Press. doi:10.1093/aob/mcx114.

Arnao, Marino B., und Josefa Hernández-Ruiz. 2019. "Melatonin and Reactive Oxygen and Nitrogen Species: A Model for the Plant Redox Network." *Melatonin Research* 2 (3). ST Bio-life: 152–168. doi:10.32794/11250036.

Asgher, Mohd, M. Iqbal R. Khan, Naser A. Anjum, und Nafees A. Khan. 2015. "Minimising Toxicity of Cadmium in Plants: Role of Plant Growth Regulators." *Protoplasma* 252 (2): 399–413. doi:10.1007/s00709-014-0710-4.

Ashihara, Hiroshi, Kyoko Adachi, Miho Otawa, Eri Yasumoto, Yuko Fukushima, Misako Kato, Hiroshi Sano, Hamako Sasamoto, und Shigeyuki Baba. 1997. "Compatible Solutes and Inorganic Ions in the Mangrove Plant Avicennia Marina and Their Effects on the Activities of Enzymes." *Zeitschrift Fur Naturforschung Section C: Journal of Biosciences* 52 (7–8): 433–440. doi:10.1515/znc-1997-7-804.

Ashraf, Muhammad Arslan, Muhammad Ashraf, und Qasim Ali. 2010. "Response of Two Genetically Diverse Wheat Cultivars to Salt Stress at Different Growth Stages: Leaf Lipid Peroxidation and Phenolic Contents." *Pakistan Journal of Botany* 42 (1): 559–565.

Awlia, Mariam, Nouf Alshareef, Noha Saber, Arthur Korte, Helena Oakey, Klára Panzarová, Martin Trtílek, Sónia Negrão, Mark Tester, und Magdalena M. Julkowska. 2021. "Genetic Mapping of the Early Responses to Salt Stress in Arabidopsis Thaliana." *Plant Journal* 107 (2). John Wiley & Sons, Ltd: 544–563. doi:10.1111/tpj.15310.

Back, Kyoungwhan, Dun Xian Tan, und Russel J. Reiter. 2016. "Melatonin Biosynthesis in Plants: Multiple Pathways Catalyze Tryptophan to Melatonin in the Cytoplasm or Chloroplasts." *Journal of Pineal Research* 61 (4): 426–437. doi:10.1111/jpi.12364.

Banerjee A., und A. Roychoudhury. (2017). "Melatonin as a Regulator of Abiotic Stress Tolerance in Plants." In Singh V. P., Singh S., Prasad S. M. (Eds.) *Mechanisms Behind Phytohormonal Signaling and Crop Abiotic Stress Tolerance*, New York: Nova Science Publishers, pp. 47–60.

Brainard, George C., John R. Hanifin, Jeffrey M. Greeson, Brenda Byrne, Gena Glickman, Edward Gerner, und Mark D. Rollag. 2001. "Action Spectrum for Melatonin Regulation in Humans: Evidence for a Novel Circadian Photoreceptor". *Journal of Neuroscience* 21 (16): 6405–6412. doi:10.1523/jneurosci.21-16-06405.2001.

Brugger, P., W. Marktl, und M. Herold. 1995. "Impaired Nocturnal Secretion of Melatonin in Coronary Heart Disease." *The Lancet* 345 (8962): 1408. doi:10.1016/S0140-6736(95)92600-3.

Burg, Maurice B., Eugene D. Kwon, und Dietmar Kültz. 1997. "Regulation of Gene Expression by Hypertonicity." *Annual Review of Physiology* 59 (1). Annual Reviews: 437–455. doi:10.1146/annurev.physiol.59.1.437.

Byeon, Yeong, Hyoung Yool Lee, Ok Jin Hwang, Hye Jung Lee, Kyungjin Lee, und Kyoungwhan Back. 2015. "Coordinated Regulation of Melatonin Synthesis and Degradation Genes in Rice Leaves in Response to Cadmium Treatment." *Journal of Pineal Research* 58 (4): 470–478. doi:10.1111/jpi.12232.

Campos, Fernanda v., Juraci A. Oliveira, Mayara G. Pereira, und Fernanda S. Farnese. 2019. "Nitric Oxide and Phytohormone Interactions in the Response of Lactuca Sativa to Salinity Stress." *Planta* 250 (5). Springer Verlag: 1475–1489. doi:10.1007/s00425-019-03236-w.

Cardinali, Daniel P., Luis I. Brusco, Cynthia Liberczuk, und Analía M. Furio. 2002. "The Use of Melatonin in Alzheimer's Disease." *Neuroendocrinology Letters* 23 (SUPPL. 1). Departament of Physiology, Faculty of Medicine, University of Buenos Aires, Argentina. cardinal@mail.retina.ar: 20–23.

Cheeseman, John M. 1988. "Mechanisms of Salinity Tolerance in Plants." *Plant Physiology* 87 (3): 547–550. doi:10.1104/pp.87.3.547.

Chen, Min, Jie Song, und Bao Shan Wang. 2010. "NaCl Increases the Activity of the Plasma Membrane H+-ATPase in C3 Halophyte Suaeda Salsa Callus." *Acta Physiologiae Plantarum* 32 (1): 27–36. doi:10.1007/s11738-009-0371-7.

Chen, Min, Zhen Yang, Jing Liu, Tingting Zhu, Xiaocen Wei, Hai Fan, und Baoshan Wang. 2018. "Adaptation Mechanism of Salt Excluders under Saline Conditions and Its Applications." *International Journal of Molecular Sciences*. doi:10.3390/ijms19113668.

Chen, Qian, Wen bo Qi, Russel J. Reiter, Wei Wei, und Bao min Wang. 2009. "Exogenously Applied Melatonin Stimulates Root Growth and Raises Endogenous Indoleacetic Acid in Roots of Etiolated Seedlings of Brassica Juncea." *Journal of Plant Physiology* 166 (3): 324–328. doi:10.1016/j.jplph.2008.06.002.

Chen, Ziping, Yanjie Xie, Quan Gu, Gan Zhao, Yihua Zhang, Weiti Cui, Sheng Xu, Ren Wang, und Wenbiao Shen. 2017. "The AtrbohF-Dependent Regulation of ROS Signaling Is Required for Melatonin-Induced Salinity Tolerance in Arabidopsis." *Free Radical Biology and Medicine* 108 (Juli). Elsevier Inc.: 465–477. doi:10.1016/j.freeradbiomed.2017.04.009.

Chourasia, Kumar Nishant, Sanket Jijabrao More, Ashok Kumar, Dharmendra Kumar, Brajesh Singh, Vinay Bhardwaj, Awadhesh Kumar, u. a. 2022. "Salinity Responses and Tolerance Mechanisms in Underground Vegetable Crops: An Integrative Review." *Planta* 255 (3). Springer: 1–25. doi:10.1007/S00425-022-03845-Y.

Cushman, J. C., G. Meyer, C. B. Michalowski, J. M. Schmitt, und H. J. Bohnert. 1989. "Salt Stress Leads to Differential Expression of Two Isogenes of Phosphoenolpyruvate Carboxylase during Crassulacean Acid Metabolism Induction in the Common Ice Plant." *The Plant Cell* 1 (7): 715–725. doi:10.1105/tpc.1.7.715.

Dajic, Zora. 2006. "Salt Stress." In herausgegeben von K V Madhava Rao, A S Raghavendra, und K Janardhan Reddy (Eds.) *Physiology and Molecular Biology of Stress Tolerance in Plants*, Dordrecht: Springer Netherlands, pp. 41–99. doi:10.1007/1-4020-4225-6_3.

Das, K, und A. Roychoudhury. (2014). "Reactive Oxygen Species (ROS) and Response of Antioxidants as ROS-Scavengers During Environmental Stress in Plants." *Frontiers in Environmental Science* 2: 53.

Daudi, Arsalan, Zhenyu Cheng, Jose A. O'Brien, Nicole Mammarella, Safina Khan, Frederick M. Ausubel, und G. Paul Bolwell. 2012. "The Apoplastic Oxidative Burst Peroxidase in Arabidopsis Is a Major Component of Pattern-Triggered Immunity." *Plant Cell* 24 (1): 275–287. doi:10.1105/tpc.111.093039.

Debnath, Biswojit, Waqar Islam, Min Li, Yueting Sun, Xiaocao Lu, Sangeeta Mitra, Mubasher Hussain, Shuang Liu, und Dongliang Qiu. 2019. "Melatonin Mediates Enhancement of Stress Tolerance in Plants." *International Journal of Molecular Sciences* 20 (5). MDPI AG: 1040. doi:10.3390/ijms20051040.

De Bruxelles, Guy L., W. James Peacock, Elizabeth S. Dennis, und Rudy Dolferus. 1996. "Abscisic Acid Induces the Alcohol Dehydrogenase Gene in Arabidopsis." *Plant Physiology* 111 (2): 381–391. doi:10.1104/pp.111.2.381.

de la Puerta, Cristina, María P. Carrascosa-Salmoral, Pedro P. García-Luna, Patricia J. Lardone, Juan L. Herrera, Rafael Fernández-Montesinos, Juan M. Guerrero, und David Pozo. 2007. "Melatonin Is a Phytochemical in Olive Oil." *Food Chemistry* 104 (2): 609–612. doi:10.1016/j.foodchem.2006.12.010.

Dollins, Andrew B., Irina V. Zhdanova, Richard J. Wurtman, Harry J. Lynch, und Mei H. Deng. 1994. "Effect of Inducing Nocturnal Serum Melatonin Concentrations in Daytime on Sleep, Mood, Body Temperature, and Performance." *Proceedings of the National Academy of Sciences of the United States of America* 91 (5). Proceedings of the National Academy of Sciences: 1824–1828. doi:10.1073/pnas.91.5.1824.

Fahad, Shah, Saddam Hussain, Amar Matloob, Faheem Ahmed Khan, Abdul Khaliq, Shah Saud, Shah Hassan, u. a. 2015. "Phytohormones and Plant Responses to Salinity Stress: A Review". *Plant Growth Regulation* 75 (2). Springer: 391–404. doi:10.1007/s10725-014-0013-y.

FAO. 1976. "3. Saline Soils and Their Management." *Soils Bulletin*, 39.

Flowers, T. J., und A. R. Yeo. 1995. "Breeding for Salinity Resistance in Crop Plants: Where Next?" *Australian Journal of Plant Physiology* 22 (6): 875–884. doi:10.1071/PP9950875.

Ford, Clive W. 1984. "Accumulation of Low Molecular Weight Solutes in Water-Stressed Tropical Legumes." *Phytochemistry* 23 (5): 1007–1015. doi:10.1016/S0031-9422(00)82601-1.

Garratt, Lee C., Basangouda S. Janagoudar, Kenneth C. Lowe, Paul Anthony, J. Brian Power, und Michael R. Davey. 2002. "Salinity Tolerance and Antioxidant Status in Cotton Cultures." *Free Radical Biology and Medicine* 33 (4): 502–511. doi:10.1016/S0891-5849(02)00838-9.

Gómez-Cadenas, Aurelio, Vicent Arbona, Josep Jacas, Eduardo Primo-Millo, und Manuel Talon. 2002. "Abscisic Acid Reduces Leaf Abscission and Increases Salt Tolerance in Citrus Plants." *Journal of Plant Growth Regulation* 21 (3): 234–240. doi:10.1007/s00344-002-0013-4.

Gossett, Dalton R., Eddie P. Millhollon, M. Cran Lucas, Stephen W. Banks, und Marye Michelle Marney. 1994. "The Effects of NaCl on Antioxidant Enzyme Activities in Callus Tissue of Salt-Tolerant and Salt-Sensitive Cotton Cultivars (Gossypium Hirsutum L.)." *Plant Cell Reports* 13 (9): 498–503. doi:10.1007/BF00232944.

Greenway, H, und R Munns. 1980. "Mechanisms of Salt Tolerance in Nonhalophytes." *Annual Review of Plant Physiology* 31 (1): 149–190. doi:10.1146/annurev.pp. 31.060180.001053.

Guo, Jianrong, Yandi Li, Guoliang Han, Jie Song, und Baoshan Wang. 2018. "NaCl Markedly Improved the Reproductive Capacity of the Euhalophyte Suaeda Salsa." *Functional Plant Biology* 45 (3): 350–361. doi:10.1071/FP17181.

Harir, Yael, und Ron Mittler. 2009. "The ROS Signaling Network of Cells." In Luis Alfonso Rio und Alain Puppo (Eds.) *Reactive Oxygen Species in Plant Signaling*, Berlin, Heidelberg: Springer, pp. 165–174. doi:10.1007/978-3-642-00390-5_10.

Hasegawa, Paul M., Ray A. Bressan, Jian Kang Zhu, und Hans J. Bohnert. 2000. "Plant Cellular and Molecular Responses to High Salinity." *Annual Review of Plant Biology* 51: 463–499. doi:10.1146/annurev.arplant.51.1.463.

Hattori, A., H. Migitaka, M. Iigo, M. Itoh, K. Yamamoto, R. Ohtani-Kaneko, M. Hara, T. Suzuki, und R. J. Reiter. 1995. "Identification of Melatonin in Plants and Its Effects on Plasma Melatonin Levels and Binding to Melatonin Receptors in Vertebrates." *Biochemistry and Molecular Biology International* 35 (3): 627–634.

He, Zhong Qun, Chao Xing He, Zhi Bin Zhang, Zhi Rong Zou, und Huai Song Wang. 2007. "Changes of Antioxidative Enzymes and Cell Membrane Osmosis in Tomato Colonized by Arbuscular Mycorrhizae under NaCl Stress." *Colloids and Surfaces B: Biointerfaces* 59 (2): 128–133. doi:10.1016/j.colsurfb.2007.04.023.

Ho, Ying-Ning, Chun-Hao Shih, Shu-Chuan Hsiao, und Chieh-Chen Huang. 2009. "A Novel Endophytic Bacterium, Achromobacter Xylosoxidans, Helps Plants against Pollutant Stress and Improves Phytoremediation." *Journal of Bioscience and Bioengineering* 108 (1): 11.

Hu, Yuncai, und Urs Schmidhalter. 2005. "Drought and Salinity: A Comparison of Their Effects on Mineral Nutrition of Plants." *Journal of Plant Nutrition and Soil Science* 168 (4). John Wiley & Sons, Ltd: 541–549. doi:10.1002/JPLN.200420516.

Imlay, James A. 2003. "Pathways of Oxidative Damage." *Annual Review of Microbiology* 57: 395–418. doi:10.1146/annurev.micro.57.030502.090938.

Iqbal, Noushina, Asim Masood, und Nafees A. Khan. 2014. "Phytohormones in Salinity Tolerance: Ethylene and Gibberellins Cross Talk." In *Phytohormones and Abiotic Stress Tolerance in Plants*, April. Springer, Berlin, Heidelberg, pp. 77–98. doi:10.1007/978-3-642-25829-9_3.

Iqbal, Noushina, Shahid Umar, Nafees A. Khan, und Francisco J. Corpas. 2021a. "Crosstalk between Abscisic Acid and Nitric Oxide under Heat Stress: Exploring New Vantage Points." *Plant Cell Reports* 40 (8): 1429–1450. doi:10.1007/s00299-021-02695-4.

Iqbal, Noushina, Shahid Umar, Nafees A. Khan, und Francisco J. Corpas. 2021b. "Nitric Oxide and Hydrogen Sulfide Coordinately Reduce Glucose Sensitivity and Decrease Oxidative Stress via Ascorbate-Glutathione Cycle in Heat-Stressed Wheat (Triticum Aestivum l.) Plants." *Antioxidants*. doi:10.3390/antiox10010108.

Iyengar, E. R. R. and Reddy, M. P. 1996. "Photosynthesis in Highly Salt Tolerant Plants." *Handbook of photosynthesis*. Marshal Dekar, Baten Rose, USA, 909.

Jiang, Chaoqiang, Quanren Cui, Kun Feng, Dafeng Xu, Chengfeng Li, und Qingsong Zheng. 2016. "Melatonin Improves Antioxidant Capacity and Ion Homeostasis and Enhances Salt Tolerance in Maize Seedlings." *Acta Physiologiae Plantarum* 38 (4): 82. doi:10.1007/s11738-016-2101-2.

Kaya, Cengiz, David Higgs, Muhammad Ashraf, Mohammed N. Alyemeni, und Parvaiz Ahmad. 2020. "Integrative Roles of Nitric Oxide and Hydrogen Sulfide in Melatonin-Induced Tolerance of Pepper (Capsicum Annuum L.) Plants to Iron Deficiency and Salt Stress Alone or in Combination." *Physiologia Plantarum* 168 (2). Blackwell Publishing Ltd: 256–277. doi:10.1111/ppl.12976.

Kolář, Jan, Carl Hirschie Johnson, und Ivana Macháčková. 2003. "Exogenously Applied Melatonin (*N*-Acetyl-5-Methoxytryptamine) Affects Flowering of the Short-Day Plant Chenopodium Rubrum." *Physiologia Plantarum* 118 (4). John Wiley & Sons, Ltd: 605–612. doi:10.1034/j.1399-3054.2003.00114.x.

Kostopoulou, Zacharoula, Ioannis Therios, Efstathios Roumeliotis, Angelos K. Kanellis, und Athanassios Molassiotis. 2015. "Melatonin Combined with Ascorbic Acid Provides Salt Adaptation in Citrus Aurantium L. Seedlings." *Plant Physiology and Biochemistry* 86 (January). Elsevier Masson SAS: 155–165. doi:10.1016/j.plaphy.2014.11.021.

Kotagiri, Divya, und Viswanatha Chaitanya Kolluru. 2017. "Effect of Salinity Stress on the Morphology & Physiology of Five Different Coleus Species." *Biomedical and Pharmacology Journal* 10 (4): 1639–1649. doi:10.13005/bpj/1275.

Kumar, Ravinder, Priyanka Kaundal, Rahul Kumar Tiwari, Sundaresha Siddappa, Hema Kumari, Milan Kumar Lal, Kailash Chandra Naga, Sanjeev Sharma, Vinay Sagar, und Manoj Kumar. 2022. "Establishment of a One-Step Reverse Transcription Recombinase Polymerase Amplification Assay for the Detection of Potato Virus S." *Journal of Virological Methods* 307 (September). Elsevier: 114568. doi:10.1016/J.JVIROMET.2022.114568.

Lal, Milan Kumar, Nitin Sharma, Sandeep B. Adavi, Eshita Sharma, Muhammad Ahsan Altaf, Rahul Kumar Tiwari, Ravinder Kumar, u. a. 2022. "From Source to Sink: Mechanistic Insight of Photoassimilates Synthesis and Partitioning under High Temperature and Elevated [CO2]." *Plant Molecular Biology 2022* 1 (Mai). Springer: 1–20. doi:10.1007/S11103-022-01274-9.

Lerner, Aaron B., James D. Case, Yoshiyata Takahashi, Teh H. Lee, und Wataru Mori. 1958. "Isolation of Melatonin, the Pineal Gland Factor That Lightens Melanocytes." *Journal of the American Chemical Society* 80 (10): 2587. doi:10.1021/ja01543a060.

Li, Chao, Ping Wang, Zhiwei Wei, Dong Liang, Changhai Liu, Lihua Yin, Dongfeng Jia, Mingyang Fu, und Fengwang Ma. 2012. "The Mitigation Effects of Exogenous Melatonin on Salinity-Induced Stress in Malus Hupehensis." *Journal of Pineal Research* 53 (3). John Wiley & Sons, Ltd: 298–306. doi:10.1111/j.1600-079X.2012.00999.x.

Li, Hao, Jingjing Chang, Hejie Chen, Zhongyuan Wang, Xiurong Gu, Chunhua Wei, Yong Zhang, Jianxiang Ma, Jianqiang Yang, und Xian Zhang. 2017a. "Exogenous Melatonin Confers Salt Stress Tolerance to Watermelon by Improving Photosynthesis and Redox Homeostasis." *Frontiers in Plant Science* 8 (März). Frontiers Research Foundation. doi:10.3389/fpls.2017.00295.

Li, Hao, Jingjing Chang, Junxian Zheng, Yuchuan Dong, Qiyan Liu, Xiaozhen Yang, Chunhua Wei, Yong Zhang, Jianxiang Ma, und Xian Zhang. 2017b. "Local Melatonin Application Induces Cold Tolerance in Distant Organs of Citrullus Lanatus L. via Long Distance Transport." *Scientific Reports* 7 (1): 40858. doi:10.1038/srep40858.

Li, Junpeng, Jing Liu, Tingting Zhu, Chen Zhao, Lingyu Li, und Min Chen. 2019. "The Role of Melatonin in Salt Stress Responses." *International Journal of Molecular Sciences* 20 (7). Multidisciplinary Digital Publishing Institute: 1735. doi:10.3390/IJMS20071735.

Lin, J., J. P. Li, F. Yuan, Z. Yang, B. S. Wang, und M. Chen. 2018. "Transcriptome Profiling of Genes Involved in Photosynthesis in Elaeagnus Angustifolia L. under Salt Stress." *Photosynthetica* 56 (4): 998–1009. doi:10.1007/s11099-018-0824-6.

Luhua, Song, Sultan Ciftci-Yilmaz, Jeffery Harper, John Cushman, und Ron Mittler. 2008. "Enhanced Tolerance to Oxidative Stress in Transgenic Arabidopsis Plants Expressing Proteins of Unknown Function." *Plant Physiology* 148 (1): 280–292. doi:10.1104/pp.108.124875.

Manchanda, Geetanjali, und Neera Garg. 2008. "Salinity and Its Effects on the Functional Biology of Legumes." *Acta Physiologiae Plantarum* 30 (5): 595–618. doi:10.1007/s11738-008-0173-3.

Mansour, M. M. F. 2000. "Nitrogen Containing Compounds and Adaptation of Plants to Salinity Stress." *Biologia Plantarum* 43 (4): 491–500. doi:10.1023/A:1002873531707.

McNeil, Scott D., Michael L. Nuccio, und Andrew D. Hanson. 1999. "Betaines and Related Osmoprotectants: Targets for Metabolic Engineering of Stress Resistance." *Plant Physiology* 120 (4): 945–949. doi:10.1104/pp.120.4.945.

Mittova, Valentina, Micha Guy, Moshe Tal, und Micha Volokita. 2004. "Salinity Up-Regulates the Antioxidative System in Root Mitochondria and Peroxisomes of the Wild Salt-Tolerant Tomato Species Lycopersicon Pennellii." *Journal of Experimental Botany* 55 (399): 1105–1113. doi:10.1093/jxb/erh113.

Mittova, Valentina, Moshe Tal, Micha Volokita, und Micha Guy. 2002. "Salt Stress Induces Up-Regulation of an Efficient Chloroplast Antioxidant System in the Salt-Tolerant Wild Tomato Species Lycopersicon Pennellii but Not in the Cultivated Species." *Physiologia Plantarum* 115 (3). John Wiley & Sons, Ltd: 393–400. doi:10.1034/j.1399-3054.2002.1150309.x.

Mittova, Valentina, Moshe Tal, Micha Volokita, und Micha Guy. 2003. "Up-Regulation of the Leaf Mitochondrial and Peroxisomal Antioxidative Systems in Response to Salt-Induced Oxidative Stress in the Wild Salt-Tolerant Tomato Species Lycopersicon Pennellii." *Plant, Cell and Environment* 26 (6). John Wiley & Sons, Ltd: 845–856. doi:10.1046/j.1365-3040.2003.01016.x.

Munns, Rana. 2002. "Comparative Physiology of Salt and Water Stress." *Plant, Cell and Environment* 25 (2). John Wiley & Sons, Ltd: 239–250. doi:10.1046/j.0016-8025.2001.00808.x.

Munns, Rana, und Matthew Gilliham. 2015. "Salinity Tolerance of Crops: What Is the Cost?" *New Phytologist* 208 (3). Blackwell Publishing Ltd: 668–673. doi:10.1111/nph.13519.

Munns, Rana, und Mark Tester. 2008a. "Mechanisms of Salinity Tolerance." *Annual Review of Plant Biology* 59: 651–681. doi:10.1146/annurev.arplant.59.032607.092911.

Munns, Rana, und Mark Tester. 2008b. "Mechanisms of Salinity Tolerance." *Annual Review of Plant Biology* 59: 651–681. doi:10.1146/annurev.arplant.59.032607.092911.

Munns, Rana, und Mark Tester. 2008c. "Mechanisms of Salinity Tolerance." *Annual Review of Plant Biology* 59 (1). Annual Reviews: 651–681. doi:10.1146/annurev.arplant.59.032607.092911.

Nawaz, Khalid, Khalid Hussain, Abdul Majeed, Farah Khan, Shahid Afghan, und Kazim Ali. 2010. "Fatality of Salt Stress to Plants: Morphological, Physiological and Biochemical Aspects." *African Journal of Biotechnology* 9 (34): 5475–5480.

Paredes, Sergio D., Ahmet Korkmaz, Lucien C. Manchester, Dun Xian Tan, und Russel J. Reiter. 2009. "Phytomelatonin: A Review." *Journal of Experimental Botany* 60 (1): 57–69. doi:10.1093/jxb/ern284.

Parida, Asish Kumar, und Anath Bandhu Das. 2005. "Salt Tolerance and Salinity Effects on Plants: A Review." *Ecotoxicology and Environmental Safety* 60 (3): 324–349. doi:10.1016/j.ecoenv.2004.06.010.

Popova, Losanka P., Zhivka G. Stoinova, und Liliana T. Maslenkova. 1995. "Involvement of Abscisic Acid in Photosynthetic Process in Hordeum Vulgare L. during Salinity Stress." *Journal of Plant Growth Regulation* 14 (4): 211–218. doi:10.1007/BF00204914.

Posmyk, Małgorzata M., und Krystyna M. Janas. 2009. "Melatonin in Plants." *Acta Physiologiae Plantarum* 31 (1): 1–11. doi:10.1007/s11738-008-0213-z.

Reddy, Attipalli R., und Agepati S. Raghavendra. 2006. "Photooxidative Stress." In herausgegeben von K. V. Madhava Rao, A. S. Raghavendra, und K. Janardhan Reddy (Eds.) *Physiology and Molecular Biology of Stress Tolerance in Plants.* Dordrecht: Springer Netherlands, pp. 157–186. doi:10.1007/1-4020-4225-6_6.

Rehaman, Abdul, Awdhesh Kumar Mishra, Asma Ferdose, Tasir S. Per, Mohd Hanief, Arif Tasleem Jan, und Mohd Asgher. 2021. "Melatonin in Plant Defense against Abiotic Stress." *Forests* 12 (10): 1–23. doi:10.3390/f12101404.

Rodriguez, Carmen, Juan C. Mayo, Rosa M. Sainz, Isaac Antolín, Federico Herrera, Vanesa Martín, und Russel J. Reiter. 2004. "Regulation of Antioxidant Enzymes: A Significant Role for Melatonin." *Journal of Pineal Research* 36 (1): 1–9. doi:10.1046/j.1600-079X.2003.00092.x.

Roychoudhury, A., A. Banerjee, und V. Lahiri. 2015. "Metabolic and Molecular-Genetic Regulation of Proline Signaling and its Cross-Talk With Major Effectors Mediates Abiotic Stress Tolerance in Plants. *Turkish Journal of Botany* 39 (6): 887–910.

Roychoudhury, A, S. Basu, S. N. Sarkar, und D. N. Sengupta. 2008. "Comparative Physiological and Molecular Responses of a Common Aromatic Indica Rice Cultivar to High Salinity with Non-Aromatic Indica Rice Cultivars." *Plant Cell Reports* 27(8): 1395–1410.

Roychoudhury, A, S. Basu, and D. N. Sengupta. 2011. "Amelioration of Salinity Stress by Exogenously Applied Spermidine or Spermine in Three Varieties of Indica Rice Differing in Their Level of Salt Tolerance." *Journal of Plant Physiology* 168: 317–328.

Roychoudhury, A, S. Paul, und S. Basu. 2013. "Cross-Talk Between Abscisic Acid-Dependent and Abscisic Acid-Independent Pathways During Abiotic Stress." *Plant Cell Reports* 32 (7): 985–1006.

Roychoudhury, A., C. Roy, und D. N. Sengupta. 2007. "Transgenic Tobacco Plants Overexpressing the Heterologous *lea* Gene *Rab16A* From Rice During High Salt and Water Deficit Display Enhanced Tolerance to Salinity Stress." *Plant Cell Reports* 26(10): 1839–1859.

Samanta, S, A. Banerjee, und A. Roychoudhury. 2021. "Melatonin Application Differentially Modulates the Enzymes Associated with Antioxidative Machinery and Ascorbate Glutathione Cycle During Arsenate Exposure in Indica Rice Varieties." *Plant Biology* 23 (S1): 193–201.

Shah, S. 2014. "Morphological, Anatomical and Physiological Responses of Plants to Salinity [Internet]."

Shi, Haitao, Chuan Jiang, Tiantian Ye, Dun Xian Tan, Russel J. Reiter, Heng Zhang, Renyi Liu, und Zhulong Chan. 2015. "Comparative Physiological, Metabolomic, and Transcriptomic Analyses Reveal Mechanisms of Improved Abiotic Stress Resistance in Bermudagrass [Cynodon Dactylon (L). Pers.] by Exogenous Melatonin." *Journal of Experimental Botany* 66 (3): 681–694. doi:10.1093/jxb/eru373.

Shi, Haitao, Dun Xian Tan, Russel J. Reiter, Tiantian Ye, Fan Yang, und Zhulong Chan. 2015. "Melatonin Induces Class A1 Heat-Shock Factors (HSFA1s) and Their Possible Involvement of Thermotolerance in Arabidopsis." *Journal of Pineal Research* 58 (3). Blackwell Publishing Ltd: 335–342. doi:10.1111/jpi.12219.

Soltabayeva, Aigerim, Assel Ongaltay, John Okoth Omondi, und Sudhakar Srivastava. 2021. "Morphological, Physiological and Molecular Markers for Salt-Stressed Plants." *Plants* 10 (2): 1–18. doi:10.3390/plants10020243.

Song, Jie, und Baoshan Wang. 2015. "Using Euhalophytes to Understand Salt Tolerance and to Develop Saline Agriculture: Suaeda Salsa As a Promising Model." *Annals of Botany* 115 (3): 541–553. doi:10.1093/aob/mcu194.

Sudhir, P., und S. D. S. Murthy. 2004. "Effects of Salt Stress on Basic Processes of Photosynthesis." *Photosynthetica* 42 (4): 481–486. doi:10.1007/S11099-005-0001-6.

Sui, N., M. Li, K. Li, J. Song, und B. S. Wang. 2010. "Increase in Unsaturated Fatty Acids in Membrane Lipids of Suaeda Salsa L. Enhances Protection of Photosystem II under High Salinity." *Photosynthetica* 48 (4): 623–629. doi:10.1007/s11099-010-0080-x.

Sui, Na, und Guoliang Han. 2014. "Salt-Induced Photoinhibition of PSII Is Alleviated in Halophyte Thellungiella Halophila by Increases of Unsaturated Fatty Acids in Membrane Lipids." *Acta Physiologiae Plantarum* 36 (4): 983–992. doi:10.1007/s11738-013-1477-5.

Tang, Ruimin, Suyan Niu, Guodong Zhang, Guanshui Chen, Muhammad Haroon, Qing Yang, Om P. Rajora, und Xiu Qing Li. 2018. "Physiological and Growth Responses of Potato Cultivars to Heat Stress." *Botany* 96 (12): 897–912. doi:10.1139/cjb-2018-0125.

Thomas, John C., Elizabeth F. McElwain, und Hans J. Bohnert. 1992. "Convergent Induction of Osmotic Stress-Responses: Abscisic Acid, Cytokinin, and the Effects of NaCl." *Plant Physiology* 100 (1): 416–423. doi:10.1104/pp.100.1.416.

Tiwari, Rahul Kumar, Milan Kumar Lal, Ravinder Kumar, Vikas Mangal, Muhammad Ahsan Altaf, Sanjeev Sharma, Brajesh Singh, und Manoj Kumar. 2021. "Insight into Melatonin-Mediated Response and Signaling in the Regulation of Plant Defense under Biotic Stress." *Plant Molecular Biology* 1 (November). Springer Science and Business Media B.V.: 1–15. doi:10.1007/s11103-021-01202-3.

Tiwari, Rahul Kumar, Milan Kumar Lal, Kailash Chandra Naga, Ravinder Kumar, Kumar Nishant Chourasia, Subhash S., Dharmendra Kumar, und Sanjeev Sharma. 2020. "Emerging Roles of Melatonin in Mitigating Abiotic and Biotic Stresses of Horticultural Crops." *Scientia Horticulturae* 272 (July). Elsevier: 109592. doi:10.1016/j.scienta.2020.109592.

Torabi, Masoud, Ridzwan a Halim, Aliakbar Mokhtarzadeh, und Yasamin Miri. 2013. "Physiological and Biochemical Responses of Plants in Saline Environment." *Crop Biology and Agriculture in Harsh Environments*, Nr. (September): 47–80.

Uchendu, Esther E., Mukund R. Shukla, Barbara M. Reed, und Praveen K. Saxena. 2013. "Melatonin Enhances the Recovery of Cryopreserved Shoot Tips of American Elm (Ulmus Americana L.)." *Journal of Pineal Research* 55 (4). John Wiley & Sons, Ltd: 435–442. doi:10.1111/jpi.12094.

Vivien-Roels, B., und P. Pévet. 1993. "Melatonin: Presence and Formation in Invertebrates." *Experientia* 49. Birkhäuser-Verlag: 642–647. doi:10.1007/BF01923945.

Wang, Fengxia, Yan ge Xu, Shuai Wang, Weiwei Shi, Ranran Liu, Gu Feng, und Jie Song. 2015. "Salinity Affects Production and Salt Tolerance of Dimorphic Seeds of Suaeda Salsa." *Plant Physiology and Biochemistry* 95: 41–48. doi:10.1016/j.plaphy.2015.07.005.

Wang, L. Y., J. L. Liu, W. X. Wang, und Y. Sun. 2016. "Exogenous Melatonin Improves Growth and Photosynthetic Capacity of Cucumber under Salinity-Induced Stress." *Photosynthetica* 54 (1). Kluwer Academic Publishers: 19–27. doi:10.1007/s11099-015-0140-3.

Wei, Jian, Dong Xu Li, Jia Rong Zhang, Chi Shan, Zed Rengel, Zhong Bang Song, und Qi Chen. 2018. "Phytomelatonin Receptor PMTR1-Mediated Signaling Regulates Stomatal Closure in Arabidopsis Thaliana." *Journal of Pineal Research* 65 (2). Blackwell Publishing Ltd: e12500. doi:10.1111/jpi.12500.

Yang, Wen Jing, Yong Tao Du, Yong bin Zhou, Jun Chen, Zhao Shi Xu, You Zhi Ma, Ming Chen, und Dong Hong Min. 2019. "Overexpression of TaCOMT Improves Melatonin Production and Enhances Drought Tolerance in Transgenic Arabidopsis." *International Journal of Molecular Sciences* 20 (3). doi:10.3390/ijms20030652.

Yin, Zepeng, Jiazhi Lu, Sida Meng, Yiling Liu, Islam Mostafa, Mingfang Qi, und Tianlai Li. 2019. "Exogenous Melatonin Improves Salt Tolerance in Tomato by Regulating Photosynthetic Electron Flux and the Ascorbate: Glutathione Cycle." *Journal of Plant Interactions* 14 (1). Taylor and Francis Ltd.: 453–463. doi:10.1080/17429145.2019.1645895.

Zhang, Hai Jun, Na Zhang, Rong Chao Yang, Li Wang, Qian Qian Sun, Dian Bo Li, Yun Yun Cao, u. a. 2014. "Melatonin Promotes Seed Germination under High Salinity by Regulating Antioxidant Systems, ABA and GA4 Interaction in Cucumber (Cucumis Sativus L.)." *Journal of Pineal Research* 57 (3). John Wiley & Sons, Ltd: 269–279. doi:10.1111/jpi.12167.

Zhang, Na, Hai Jun Zhang, Qian Qian Sun, Yun Yun Cao, Xingsheng Li, Bing Zhao, Ping Wu, und Yang Dong Guo. 2017. "Proteomic Analysis Reveals a Role of Melatonin in Promoting Cucumber Seed Germination under High Salinity by Regulating Energy Production." *Scientific Reports* 7 (1): 503. doi:10.1038/s41598-017-00566-1.

Zhang, Na, Bing Zhao, Hai Jun Zhang, Sarah Weeda, Chen Yang, Zi Cai Yang, Shuxin Ren, und Yang Dong Guo. 2013. "Melatonin Promotes Water-Stress Tolerance, Lateral Root Formation, and Seed Germination in Cucumber (Cucumis Sativus L.)." *Journal of Pineal Research* 54 (1). John Wiley & Sons, Ltd: 15–23. doi:10.1111/j.1600-079X.2012.01015.x.

Zhang, Shumin, Xianzhe Zheng, Russel J. Reiter, Shun Feng, Ying Wang, Sen Liu, Liang Jin, Zhengguo Li, Raju Datla, und Maozhi Ren. 2017. "Melatonin Attenuates Potato Late Blight by Disrupting Cell Growth, Stress Tolerance, Fungicide Susceptibility and Homeostasis of Gene Expression in Phytophthora Infestans." *Frontiers in Plant Science* 8 (November): 1–19. doi:10.3389/fpls.2017.01993.

Zhang, Tengguo, Zhongfei Shi, Xiaohua Zhang, Sheng Zheng, Juan Wang, und Jiangnan Mo. 2020. "Alleviating Effects of Exogenous Melatonin on Salt Stress in Cucumber." *Scientia Horticulturae* 262: 109070. doi:10.1016/j.scienta.2019.109070.

Zhang, Yi, Xiujuan Zhou, Yating Dong, Fan Zhang, Qiuling He, Jinhong Chen, Shuijin Zhu, und Tianlun Zhao. 2021. "Seed Priming with Melatonin Improves Salt Tolerance in Cotton through Regulating Photosynthesis, Scavenging Reactive Oxygen Species and Coordinating with Phytohormone Signal Pathways." *Industrial Crops and Products* 169: 113671. doi:10.1016/j.indcrop.2021.113671.

Zhao, Chenxu, Ghazala Nawaz, Qinghe Cao, und Tao Xu. 2022. "Melatonin Is a Potential Target for Improving Horticultural Crop Resistance to Abiotic Stress." *Scientia Horticulturae* 291: 110560. doi:10.1016/j.scienta.2021.110560.

Zhao, L., L. Chen, P. Gu, X. Zhan, Y. Zhang, C. Hou, Z. Wu, Y. F. Wu, und Q. C. Wang. 2019. "Exogenous Application of Melatonin Improves Plant Resistance to Virus Infection." *Plant Pathology* 68 (7). John Wiley & Sons, Ltd: 1287–1295. doi:10.1111/ppa.13057.

Zhifang, G., und Wayne H. Loescher. 2003. "Expression of a Celery Mannose 6-Phosphate Reductase in Arabidopsis Thaliana Enhances Salt Tolerance and Induces Biosynthesis of Both Mannitol and a Glucosyl-Mannitol Dimer." *Plant, Cell and Environment* 26 (2). John Wiley & Sons, Ltd: 275–283. doi:10.1046/j.1365-3040.2003.00958.x.

Zhou, Xiaoting, Hailiang Zhao, Kai Cao, Lipan Hu, Tianhao Du, František Baluška, und Zhirong Zou. 2016. "Beneficial Roles of Melatonin on Redox Regulation of Photosynthetic Electron Transport and Synthesis of D1 Protein in Tomato Seedlings under Salt Stress." *Frontiers in Plant Science* 7 (NOVEMBER 2016). Frontiers Media S.A. doi:10.3389/fpls.2016.01823.

Zhu, Jian Kang. 2001. "Plant Salt Tolerance." *Trends in Plant Science* 6 (2): 66–71. doi:10.1016/S1360-1385(00)01838-0.

Zhu, Jian Kang. 2003. "Regulation of Ion Homeostasis under Salt Stress." *Current Opinion in Plant Biology* 6 (5): 441–445. doi:10.1016/S1369-5266(03)00085-2.

Zhu, Jian Kang, Jun Shi, Utpal Singh, Sarah E. Wyatt, Ray A. Bressan, Paul M. Hasegawa, und Nicholas C. Carpita. 1993. "Enrichment of Vitronectin- and Fibronectin-like Proteins in NaCl-Adapted Plant Cells and Evidence for Their Involvement in Plasma Membrane-Cell Wall Adhesion." *Plant Journal* 3 (5). John Wiley & Sons, Ltd: 637–646. doi:10.1111/j.1365-313X.1993.00637.x.

14 Biochemical and Molecular Mechanism of Melatonin-Induced Tolerance against Salt Stress

Anup Kumar Sarkar and Sanjoy Sadhukhan

14.1 INTRODUCTION

Saline soil renders most of the cultivable areas unproductive or even inefficient (Egamberdieva et al., 2017). Salt stress is pervasive abiotic stress that adversely affects plants by producing initially osmotic stress and, subsequently, ionic toxicity and oxidative stress, leading to growth retardation, yield losses, and plant mortality (Majeed & Muhammad, 2019). In addition to having an immediate effect on growth, osmotic stress has a greater effect on growth rate in the early stages of salt stress than ionic toxicity (Roychoudhury et al., 2007; Munns & Tester, 2008). Salinity can damage plants from the seedling stage to the mature stage. Most glycophytic plants are seriously affected by salt stress, while halophytes and some salt-tolerant glycophytes acclimatize to this stressful circumstance through morphological, biochemical, and physiological changes (Roychoudhury & Chakraborty, 2013; Sarkar & Sadhukhan, 2022a). Salts on the exterior of roots make it more difficult to draw water and have an immediate impact on cell growth and associated metabolic functions; however, toxicity only occurs after a certain amount of salt accumulation inside plants, and, subsequently, it takes time to manifest (Zhang et al., 2014). When plants are subjected to stress, they trigger signalling pathways that allow them to adjust even to minimal environmental changes (Sarkar & Sadhukhan, 2022b). Managing harmful ion intake by roots, controlling the transport of ions from roots to shoots, adjusting photosynthetic pathways, altering the activity of antioxidant enzymes, and regulating hormone levels are some common defence mechanisms in plants against salt stress (Roychoudhury et al., 2008; Liu et al., 2018). Melatonin (*N-acetyl*-5-metoxytriptamine) is appreciated as a biological hormone which is essential for stress responses. Numerous scientific researches are being undertaken at the molecular level to unravel the role of melatonin in influencing the expression patterns of numerous genes across varied plant species. Diverse investigations revealed that melatonin-treated plants grew better and were better competent to maintain reasonably low sodium levels under salt-induced stress conditions than untreated seedlings (Xie et al., 2021). Several researches in this area have focused on the use of melatonin to enhance the relatively low salt tolerance of glycophytes by encouraging development and boosting antioxidant capacity. One of the primary causes of these negative effects of salinity is the excessive production of ROS and methylglyoxal (MG), both of which destroy essential macromolecules such as proteins, lipids, nucleic acids, and so on (Ahanger & Agarwal, 2017; Jahan et al., 2019). Plant tolerance to salinity stress is extremely complex at the whole plant and cellular levels, involving changes in morphology, physiology, and metabolism. It is generally established that plant responses to salt stress vary depending on the severity and duration of the stress, plant variant or species, and stage of growth. Plants have developed a lot of ways to deal with stress, which lets them complete their life cycles even when there is salt stress (Roychoudhury et al., 2011). In 1958, researchers

 DOI: 10.1201/9781003343752-14

found an indoleamine molecule called melatonin in the pineal gland (Lerner et al., 1958). Using mass spectrometry combined with fluorescence detection and certain other approaches, researchers have identified the presence of melatonin in a few plants (Dubbels et al., 1995; Hattori et al., 1995). Various subcellular organelles, such as the cytoplasm, endoplasmic reticulum, and chloroplasts, create numerous melatonin intermediates, which either facilitate or inhibit the succeeding enzymatic processes. Because of the large number of melatonin isomers, at least two analytical techniques, such as high-performance liquid chromatography (HPLC) and liquid chromatography-tandem mass spectrometry (LC-MS/MS), should be employed to quantify melatonin. Melatonin is a primitive chemical that is believed to have arisen concurrently with aerobic respiration, due to its presence in the majority of living species, including archaea, bacteria, algae, protozoa, fungus, plants, invertebrates, and vertebrates (Tan & Reiter, 2019; Tan et al., 2010; Zhao et al., 2019). About 300 plant species, ranging from monocots to dicots, contain melatonin, which has the chemical formula *N-acetyl*-5-methoxytryptamine and is popularly known as phytomelatonin (Arnao, 2014; Tripathi et al., 2021). In most plants, melatonin can move from the mitochondria and chloroplasts to other organs and tissues (Tan et al., 2013). Although melatonin was initially identified as a strong antioxidant, it was later proven to have numerous positive characteristics in numerous plant species against salinity and allied stressors. N-acetyl and 5-methoxy functional groups of melatonin regulate its lipophilicity, hydrophilicity, and receptor selectivity (Arnao & Hernández-Ruiz, 2015; Banerjee & Roychoudhury, 2017).

Several investigations have shown that amino acids, organic acids, and sugars are implicated in melatonin-associated responses to biological as well as abiotic stressors (Hu et al., 2016; Shi et al., 2015a). Under stressful conditions, the activation of synthetic genes rapidly increases endogenous melatonin levels in plants (Chen et al., 2018, 2017; Gu et al., 2017; Sun et al., 2020). Numerous research demonstrates that melatonin improves salinity tolerance in a variety of plants including *Arabidopsis*, *Brassica napus*, *Oryza sativa*, *Triticum aestivum*, *Lycopersicon esculentum*, *Cucumis sativus*, *Malus domestica*, *Limonium bicolor*, *Helianthus annuus*, *Olea* and many more (Chen et al., 2017; Li et al., 2012, 2022; Siddiqui et al., 2019; Su et al., 2021; Tan et al., 2021; Yan et al., 2021; Zahedi et al., 2021; Zhang et al., 2014; Zhao et al., 2018). Various studies have found that melatonin regulates genes encoding transcription factors (TFs) which are intimately associated with salinity resistance, such as those coding drought-responsive element-binding (DREB), WRKY, MYB, bHLH, NAC, and HSF TFs in plants (Li et al., 2016a; Shi et al., 2015a; Weeda et al., 2014).

14.2 MELATONIN IN PLANTS

Melatonin (*N-acetyl*-5-methoxytryptamine) is an indoleamine, a small molecular substance that was initially identified in mammals and later in plants. It has been found and measured in the roots, shoots, leaves, fruits, and seeds of the vast majority of plant species. Dubbels et al. (1995) utilized gas chromatography–mass spectrometry to confirm that melatonin is present in the leaf tissues of tomatoes, bananas, cucumbers, and tobacco. Hattori et al. (1995) employed radioimmunoassay technology to assess the melatonin levels of 24 edible plants from 12 angiospermic families. This indoleamine is mostly found in the roots, leaves, flowers, seeds, and bulbs of plant species from the families Poaceae, Rosaceae, Apiaceae, Vitaceae, and Brassicaceae (Mir et al., 2020). Radioimmunoassay and high-performance liquid chromatography with electrochemical or fluorescence detection are the most used techniques for measuring melatonin. Furthermore, high-performance liquid chromatography coupled with mass spectrometry identification is a strong and indispensable instrument for the accurate measurement of melatonin in plant samples. Melatonin concentrations in plant organs range from picograms to micrograms per gram of plant material. Higher levels of melatonin can be found in plant cells in contrast to animal cells (Tan et al., 2012). Melatonin metabolism in plants generates many essential bioactive compounds that can be categorized as (i) products of oxidative processes, (ii) products of catabolism, and (iii) conjugates and derivatives (Murch & Erland, 2021). Melatonin, like other phytohormones, primarily benefits plants in moderate quantities and, at very high amounts, may diminish plant fitness, while excessively low concentrations may have no impact (Chen et al., 2009).

14.3 FUNCTIONING OF MELATONIN IN PLANTS

Melatonin may exert its effects via receptor-dependent and receptor-independent methods (Tripathi et al., 2021). Since the discovery, researchers have initiated their investigation into specialized binding sites/receptors and associated signalling pathways for the proper understanding of the activities of the molecule (Zhao et al., 2019). Wei et al. (2018) recently discovered a hypothetical phytomelatonin receptor, CAND2/PMTR1, in *Arabidopsis thaliana*. It is located in the plasma membrane with a receptor-like topology and interacts with the G-protein subunit (GPA1), while expression in different tissues is induced by melatonin. The receptor was also found to have a crucial function in the melatonin-mediated osmotic stress response of plants (Bychkov et al., 2021).

With the advancement of metabolome-profiling systems, significant numbers of plant researchers fix their objective to metabolites in response to the treatment of melatonin (Xie et al., 2021). Melatonin is widely distributed, indicating that it is crucial to the existence and activity of organisms. Melatonin functions as a universal regulator in a range of activities, including plant growth and development, root morphology, floral transition, leaf senescence, and fruit setting (Sun et al., 2021). Melatonin is appreciated both as a growth stimulator and as a growth inhibitor, and its stimulating and/or inhibitory effects are proportional to its concentration (Sarropoulou et al., 2012). Melatonin is produced naturally and is necessary for plant growth and development. Melatonin promotes the formation of lateral and adventitious roots, increases seed germination, improves crop production, and delays plant flowering (Arnao & Hernández-Ruiz, 2009; Wang et al., 2012; Zhang et al., 2013). Melatonin was observed at higher concentrations in chloroplast and mitochondria in comparison to other cell components, and because of its antioxidant property, it can shield these two organelles from ROS damage (Shen et al., 2018). In modest quantities, melatonin inhibits photooxidation of the photosynthetic machinery and safeguards chlorophyll (Chl) throughout senescence. Melatonin increases photosynthetic efficiency and boosts PS I and PS II, yielding more nitrogen and chlorophyll content as well as enhancing the level of solubilized proteins and RuBisCo (Arnao & Hernández-Ruiz, 2018; Nawaz et al., 2020). Melatonin is considered to boost the effectiveness of the electron transport chain in mitochondria, and, therefore, it is crucial for plant cell integrity and protection against oxidative damage (Neamah & Jdayea, 2022). Melatonin modifies the cell layer permeability mediated by ion transporters, which controls the opening and closing of stomata in plants (Iqbal et al., 2021). Melatonin did not influence the concentrations of sucrose, fructose, or proline under normal development conditions (Ren et al., 2020).

Some biological functions of melatonin in plants are analogous to those in animals, such as the control of circadian rhythms, the removal of free radicals by sequestrating reactive nitrogen species (RNS) and reactive oxygen species (ROS), and the modulation of innate immunity (Calvo et al., 2013; Qian et al., 2015; Reiter et al., 2015; Shi et al., 2015b, 2015c). The melatonin biosynthesis pathways of animals and plants share some similarities and some differences. The mode of action of melatonin in plants remains unknown, whereas researchers have confirmed that all changes in plants occur as a result of its antioxidant activity, as well as its role in membrane stability and gene expression control (Nawaz et al., 2016). According to Manchester et al. (2015), melatonin can play a protective role against oxidative damage by boosting the activity of antioxidative enzymes.

The beneficial effects of melatonin in plant salt stress resistance have gradually evolved by two methods: exogenous melatonin administration or genetic alteration of the enzymes involved in melatonin production (Kanwar et al., 2018). Furthermore, it has been shown to alter the gene expression of essential proteins like ion channels, antioxidants, and so on (Jahan et al., 2019; Zhao et al., 2017). Melatonin, at both low and high dosages, regulates the expression of a vast number of genes associated with stress response, antioxidant activities, intracellular signalling processes, and hormone regulation (Wan et al., 2018; Weeda et al., 2014). Melatonin also plays a crucial role as a signalling molecule in the mitogen-activated protein kinase (MAPK) pathway that mediates the defence response in plants (Lee et al., 2014, 2015).

14.4 IMPACT OF MELATONIN IN MITIGATION OF SALT STRESS IN PLANTS

Melatonin, as a master regulator, plays a crucial function in plant tolerance to salinity. Extensive research has demonstrated the critical and necessary roles that melatonin plays in enhancing salt tolerance in a wide range of plant species. Melatonin helps plants recover from salt stress by regulating antioxidants, protecting subcellular structures from damage, increasing photosynthesis, modulating hormonal cross-talk, and/or reprogramming secondary metabolism. Basic strategies for mitigating the negative effects of salt stress include metabolic adjustment, which involves the synthesis, accumulation, and temporary increase of compatible solutes, such as proline, betaine, glycine, mannitol, soluble sugar, and soluble protein. These solutes either protect vital enzymes and cellular macromolecular constructions or detoxify free radicals. Melatonin is a pleiotropic molecule that acts as an amphiphilic agent, and thus it can rapidly traverse the cell membrane and enter intracellular compartments (Altaf et al., 2020). Plants can cope with salt-induced osmotic stress by employing a strategy called the osmotic adjustment (Yin et al., 2013). According to a study, melatonin enhanced the proportion of soluble sugar in Arabidopsis under salt-induced osmotic stress (Shi et al., 2015a). In a study, it was discovered that plants treated with melatonin used to have higher relative water content than those not treated (Ren et al., 2020).

14.4.1 Melatonin-Mediated Activation of Antioxidant Systems in Response to Salt Stress

Melatonin helps to improve antioxidant systems and eliminate free radicals when exposed to salt stress (Mohamadi & Karimi, 2020). Arnao & Hernández-Ruiz (2009) reported that NaCl generated a time-dependent rise in the melatonin levels in barley roots. Melatonin can boost chlorophyll levels in addition to enhancing electron transport and stomatal conductance to mitigate the negative impacts of abiotic stressors on photosynthesis (Wang et al., 2018; Samanta et al., 2022). The antioxidant potential of melatonin is postulated as a mechanism for stress responses, as well as the protection of photosynthetic apparatus and reproductive development/germ tissues (Murch & Erland, 2021). The capacity to neutralize ROS also plays an important role in the control of ROS and reactive nitrogen species (RNS) signalling networks within plants, in particular, hydrogen peroxide (H_2O_2) and nitric oxide (NO) signalling networks which allow for rapid signalling in response to the stimulus. Reduced ROS in melatonin-treated plants safeguarded membrane function by inhibiting lipid peroxidation, electrolyte leakage, and lipoxygenase activity (Alharbi et al., 2021). Melatonin can directly scavenge ROS and then generate at least three metabolites viz., 2-hydroxymelatonin, cyclic-3-hydroxymelatonin, and AFMK (Yu et al., 2018). Under salinity stress, the oxidative interaction of free radicals in membrane lipids induces the deposition of more ROS and hydrogen peroxide in plants, resulting in cross-linking polymerization of polypeptides, nucleic acids, and some other biomolecules as well as simultaneously damaging the membrane system, which, in turn, increases malondialdehyde (MDA) content, electrolyte leakage (EL), as well as lipid peroxidation. Due to its amphipathic nature, melatonin may flow across membranes, making it a vital chemical in agriculture, and cross-talk with various plant hormones is a key mechanism for stress tolerance in crops (Mukherjee, 2018). As a result of the rise in melatonin in leaves, the overall amount of melatonin in plants is higher (Tripathi et al., 2021). Transcriptome analysis indicated that exogenous melatonin provided antioxidant protection through influencing genes involved in photosynthesis, ROS metabolism, flavonoids, and melatonin production (Yang et al., 2020). In addition to reacting directly with

ROS, it can also trigger the antioxidant system of plants. Exogenous melatonin raises the activity of SOD, POD, CAT, and APX, hence increasing antioxidant capacity and decreasing the effects of abiotic stresses (Li et al., 2018). Under stress conditions, melatonin increases the concentration of SOD, APX, CAT, and GPX in plants by controlling the expression of genes involved in antioxidant response (Samanta et al., 2021). Melatonin was shown to boost the activity of APX, CAT, and SOD in *Arabidopsis* by up regulating APX1/2, CAT1, and FSD1 transcripts in response to salt stress (Chen et al., 2017). Melatonin and Ca^{+2} administrations in cotton reduce ROS-producing enzymes such as NADPH oxidase in the plasma membrane and GOX in peroxisomes (Siddiqui et al., 2019).

14.4.2 Melatonin-Mediated Promotion of Ion Homeostasis under Salt Stress

The ability of living creatures to maintain constant ion concentrations in a specified space is referred to as ion homeostasis. Many studies have found a favourable relationship between Na^+ and K^+ levels in response to melatonin treatment for salt-induced stress. Exogenous melatonin ($\mu mol \cdot L^{-1}$) maintains a reduced Na^+/K^+ ratio, equivalent to non-stressed plants. Melatonin increases salt tolerance in *Oryza sativa* by lowering K^+ efflux in roots triggered by high NaCl concentrations (Liu et al., 2020). Liu et al. (2020) found that melatonin elicited a ROS burst in rice during salinity-induced stress conditions, which, in turn, boosted the activity of K^+ uptake transporters, allowing K^+ retention. Under salinity, melatonin also lowers osmotic pressure by decreasing salt-induced Na^+ deposition and boosting K^+ and proline levels (Jiang et al., 2020). Na^+ can penetrate plant cells in salt-stress conditions, where it can be damaging to cytosolic enzymes in high quantities (Fukuda et al., 2011). Melatonin increased K^+ and lowered Na^+ levels in maize shoots, resulting in a considerably higher K^+/Na^+ ratio in shoots exposed to melatonin-mediated salinity (Ren et al., 2020). Exogenous melatonin treatments had a significant effect on lipid metabolism and K^+/Na^+ balance in a salinity-stressed potato (Yu et al., 2018). The binding of phytomelatonin receptors induces Gγb and Gα activation of NADPH oxidase-dependent H_2O_2 synthesis (RBOH), increasing Ca^{2+} inflow, promoting K^+ outflow, and, finally, stomatal closure (Arnao & Hernández-Ruiz, 2019). Yang et al. (2020) showed that melatonin can directly or indirectly increase salt stress tolerance in *Hevea brasiliensis* trees by inhibiting H_2O_2 deposition.

14.4.3 Melatonin-Mediated Improvement of Photosynthesis under Salt Stress

Lower photosynthetic levels are frequently associated with a loss in yield in many plant species that are exposed to salt stress. Melatonin significantly regulates the photosynthetic process in response to salt stress. Melatonin can improve chlorophyll concentration, electron transport, and stomatal conductance, which can make up for the negative consequences of salinity on photosynthesis (Arnao & Hernández-Ruiz, 2018). Liang et al. (2015) have reported that melatonin can effectively enhance salt tolerance in rice seedlings by slowing chlorophyll breakdown. Under salt-stress conditions, the addition of melatonin substantially mitigated the reduction in net photosynthetic rate, the highest quantum efficiency of PSII, and total amounts of chlorophyll in *Cucumis sativus* L. (Wang et al., 2018). It has been revealed that exogenous melatonin prevents chlorophyll degradation and delays senescence via modulating a few TFs, notably senescence-related SGR and NAC TFs, and preventing chlorophyll degradation (Liang et al., 2015). According to Li et al. (2017), exogenous melatonin treatment increased the ability of watermelon to enhance photosynthesis during salt stress. Ye et al. (2016) also reported that melatonin administration under stressful conditions enhanced transpiration because it may be mediated by an increase in stomatal conductance and maintained a steady state of photosynthetic rate. Exogenous melatonin treatment increased tolerance to salinity stressors in soybean by alteration of photosynthetic metabolism and starch/sucrose metabolic activity (Wei et al., 2015).

Melatonin improves tolerance by increasing or decreasing downstream regulatory components in the physiological contexts of different plants (Gu et al., 2022). During stress, exogenous melatonin increased the activities of sucrose phosphate synthase (SPS) and acid invertase (AI) enzymes, which are favourable to the enhancement of sucrose content and metabolic capacity, hence preserving the dynamic equilibrium between sucrose production and metabolism (Zhou et al., 2019). Melatonin was discovered to influence SPS as well as sucrose synthase by protein activation or gene transcriptions and to boost invertase activity by decreasing invertase inhibitor gene expression, thus affecting growth (Yang et al., 2019).

14.4.4 Melatonin-Induced Reprogramming of Plant Hormone Metabolism

Research published in the last few years suggests that melatonin plays a role in controlling sugar accumulation and metabolism in some plants, with shifts in sugar status triggering the production of sugar signals that are then incorporated into several hormone signalling pathways (Kobylińska et al., 2018; Ljung et al., 2015; Yang et al., 2019). Melatonin widely participates in the metabolism of most plant hormones, such as indole-3-acetic acid (IAA), abscisic acid (ABA), gibberellic acids (GA), cytokinins (CK), and ethylene (Arnao & Hernández-Ruiz, 2018). Melatonin therapy of tomato seedlings dramatically increased stress tolerance in a dose-dependent manner (Altaf et al., 2020). Under salinity, melatonin enhances the catabolism of ABA as well as biosynthesis of GA and regulates seed germination by enhancing nutrient use and biogenesis in seeds (Li et al., 2019a). In germinating seed, melatonin works primarily by boosting antioxidant enzymes, influencing phytohormone production as well as catabolism, and controlling storage protein breakdown to essential amino acids during seed germination to counteract the inhibitory effects of NaCl stress (Zhang et al., 2014; Zhang et al., 2017). During salt stress, amylase activity in germinating seeds reduced significantly in comparison to the control (Li et al., 2019a). This suggests that melatonin shields enzymes from damage caused by excessive salt. Melatonin increased ethylene synthesis, and ethylene contributed to melatonin-induced salt tolerance. Melatonin greatly stimulated MYB108A expression, and MYB108A could immediately bind to the promoter of ACS1, triggering its expression and boosting ethylene production (Xu et al., 2019). Some researchers opine that role of melatonin in controlling ethylene biosynthesis might be indirect, and its positive or negative impact on ethylene production might depend on other signal molecules (Xu et al., 2019). Melatonin increases seed germination under high-salt conditions via modulating the interaction between ABA and GA4 in cucumber (Zhang et al., 2014).

14.4.5 Melatonin-Induced Alteration of the Genetic Circuit and Signalling Pathway

Under salt stress, the accumulation of phytomelatonin is correlated with gene expression and enzymatic activity. Melatonin therapy protects plant cells against the damaging effects of salt by boosting the expression of the gene that governs the stress-responsive gene. External melatonin could make plants more tolerant by modulation of transcriptional cascade through activating unique TFs, multiple antioxidant pathways, and reprogramming the homeostasis of metabolites (Xie et al., 2021). The most crucial mechanism of the phytomelatonin-mediated stress response is the control of TFs. Zinc finger protein 6 (ZAT6), heat shock factors (HSFA1s), and C-Repeat Binding Factor (CBF)/Drought response element binding 1 factors (DREB1s) were shown to be three melatonin-mediated abiotic stress TFs in plants (Yu et al., 2018). Under abiotic stress like salinity, transcription factors stimulated by melatonin play crucial roles in controlling the transcription of stress-responsive genes. According to Wei et al. (2015), melatonin presumably increased the expression of genes related to sugar metabolism to produce its stimulatory effects in soybean. Melatonin application significantly boosts the levels of proline, polyamines, and sucrose, as well as up regulates the expression of proline biosynthesis genes (*P5CS*) in seedlings of tomatoes under abiotic-stress conditions (Jahan et al., 2019; Zhang et al., 2014).

14.5 CROSS-TALK BETWEEN MELATONIN WITH OTHER SALT STRESS MITIGATING MOLECULES

It is not a fully explained universal mechanism when or how melatonin initiates cross-talk with essential regulators under salinity. There is no doubt that it can interact with several molecules associated with salt tolerance. Different concentrations of melatonin may have differential effects on stress-mitigating molecules as well as on salt signals. In addition to the control of gene expression, interplay with other phytohormones is the most critical aspect in its contribution to an abundance of physiological processes under normal and stressed situations. Under adversity, plants increase the biosynthesis and accumulation of osmotic adjustment molecules such as proline, soluble protein, soluble sugar, betaine, etc. (Khan et al., 2019). Melatonin, as a major regulator of metabolism, collaborates with a variety of signalling molecules and hormones such as auxin, ethylene, salicylic acid, GA, and ABA. Melatonin is related to auxin in many respects, and its structure is almost identical to that of auxin (indole-3-acetic acid). Comparable to auxin, melatonin can affect plant development processes such as rhizogenesis in pericycle from *Lupinus albus*. Despite the molecular similarities between melatonin and IAA, the melatonin molecule lacks the structural criteria necessary for the auxin-like activity (Shibaeva et al., 2018). The primary purpose of melatonin is to stimulate the production of IAA and to collaborate with IAA to enhance the elongation and expansion of cells, which is manifested in the induction of root growth, lateral root formation, adventitious root formation, and fruit enlargement (Yan et al., 2020). Melatonin does not appear to have the same mode of action as auxins, but it can be processed to IAA or alter auxin levels (Park, 2011). Melatonin therapy increases auxin concentration in *Brassica juncea* and *Solanum lycopersicon* by 1.4–2.0 times (Chen et al., 2009; Wen et al., 2016). It should be emphasized that while the auxin-like effect of melatonin is widely recognized, data indicates that this may be due to a reduction of the substrate pool—because tryptophan functions as the precursor for both IAA and melatonin production pathways (Arnao & Hernández-Ruiz, 2015; Pérez-Llorca et al., 2019). While melatonin and IAA are both indoleamine chemicals, they do not share the same signal transduction networks. Instead, melatonin prevents auxin signalling by suppressing the production of the auxin antagonistic transcription factor IAA17 (Pelagio-Flores et al., 2012; Shi et al., 2015a). Several studies have demonstrated the interaction of melatonin with signalling molecules such as nitric oxide (NO), hydrogen sulphide (H_2S), hydrogen gas (H_2), and many more (Mukherjee, 2019; Su et al., 2021; Zhu et al., 2019). These molecules often stimulate the activity of melatonin against salinity stress. Melatonin and NO influence glutathione levels and glutathione reductase activity in the seedlings of *Helianthus annuus* under salt stress (Kaur & Bhatla, 2016). Melatonin accumulation in the plant may be positively modulated by NO. NO might limit ROS generation by modifying the activities of many ROS-scavenging enzymes and act as an antioxidant to improve plant stress tolerance; however, an excess of NO could promote nitrosative stress, leading to toxic plant cell damage (Corpas & Palma, 2018). Melatonin and NO interaction mitigated the negative effects of salt stress by differently altering the expression of Cu/Zn SOD and MnSOD (Arora & Bhatla, 2017). Melatonin's ability to restore redox and ion equilibrium into plant tissues increased NO levels and hence increased salt tolerance in seedling tolerance of rapeseed (Zhao et al., 2018). Exogenous NO raises the level of endogenous melatonin levels, while exogenous melatonin enhances endogenous NO levels (He & He, 2020). Mukherjee & Bhatla (2021) showed that both H_2S and melatonin facilitated the growth of tomatoes under salinity, and also that exogenous melatonin regulates early H_2S signalling. According to them, exogenous melatonin therapy under salt stress modified endogenous H_2S levels and significantly increased L-cysteine desulfhydrase efficiency in tomato seedlings. Overproduction of various kinds of suitable solutes is one of the most important responses that plants have to abiotic stressors like salinity. These include some low-molecular-weight, water-soluble chemicals that are generally safe at high cellular concentrations, which are also called compatible solutes. As a result of their ability to shield cellular components from dehydration damage, some of these solutes have received the name "osmoprotectants". According to Siddiqui et al. (2019), melatonin eased salt stress via osmoregulation by raising the quantity of total soluble sugar and proline as well as up regulating carbohydrate metabolism. It is also reported that melatonin speeds up the conversion of arginine and methionine into polyamines in seedlings of *T. aestivum* as they grow (Ke et al., 2018). Melatonin was also discovered to interact with sugar-based signalling pathways during numerous plant growth processes. In research, it was discovered that low concentrations of melatonin increase maize seedling growth by improving sugar metabolism, photosynthetic activity, and phloem loading of sucrose (Zhao et al., 2015). Through osmotic control, plants primarily withstand salt stress–induced damage, and several studies have confirmed that soluble sugar and soluble protein are crucial osmotic regulatory components for boosting cell fluid concentration as well as enhancing salt tolerance in plants (Hu et al., 2018; Park et al., 2016).

Melatonin is classified as a biostimulator because it is a non-toxic natural substance that can strengthen and stimulate plant health (Janas & Posmyk, 2013). Melatonin has been linked to a variety of physiological responses in diverse plant tissues studied over a wide range of angiosperm groups (Park & Back, 2012; Pelagio-Flores & López-Bucio, 2016; Pelagio-Flores et al., 2012). Melatonin has been reported to act like auxin for the promotion of vegetative growth in several plant species (Paredes et al., 2009).

A large number of salt-responsive genes and other stress-responsive genes are activated and/or modulated through complex signal transduction networks and synthesize many functional proteins to improve the ability of plants to resist salinity and allied stresses. Li et al. (2020) demonstrated that melatonin may up regulate the expression of genes producing ion transporters and vesicle transport proteins to increase salt secretion from the salt glands in some halophytes. Transcriptomic research demonstrated that during salinity stress, melatonin can increase the expression of *MIPS*, *SLAH1*, and *MYB73* genes in *Citrus aurantium*, which are associated with sugar metabolism, ion homeostasis, and transcriptional control (Kostopoulou et al., 2015). In *T. aestivation*, it is important to note that melatonin can be regulated via a positive-feedback loop and an increase in endogenous melatonin levels by the transcription of the *SNAT* transcript, which encodes a crucial enzyme in melatonin production (Ke et al., 2018).

14.6 EXOGENOUS APPLICATION OF MELATONIN FOR PROTECTION OF CROP PLANT

Treatment of exogenous melatonin modifies the hormonal profile, and physiological cross-talk enhances the growth, physiological efficiency, and yield of many crops under abiotic-stress conditions such as drought and salinity (Ahmad et al., 2022). Exogenous melatonin boosts endogenous melatonin synthesis and increases antioxidant capacity (Kaya et al., 2019; Zhao et al., 2017). According to the findings, salt exposure decreased the vegetative development of the plant, whereas exogenous melatonin administration reversed growth suppression under salt stress (Table 14.1). It has been observed that different melatonin application methods such as pretreated coating or soaking, leaf spray, root or rhizospheric application, mixed application with other growth regulators, etc., appear to have multiple regulatory consequences during salinity-induced stress (Kostopoulou et al., 2015). Mohamadi & Karimi (2020) conducted an experiment on *Rosmarinus officinalis* L. and reported that melatonin has an overall protective impact on the plant under salinity-induced stress conditions. Castañares & Bouzo (2019) reported that exogenous melatonin induces biochemical changes in *Cucumis melo* L., increasing salt stress tolerance. Exogenously administered melatonin enhanced endogenous amounts and stimulated the expression of genes involved in melatonin production in the wild type (Bychkov et al., 2021). The roots of plants can readily absorb exogenously treated melatonin. Although concentrations of melatonin up to 1000 μM have been evaluated, the majority of findings indicate that concentrations below 100 μM are more helpful for most plants (Weeda et al., 2014; Yin et al., 2013; Zhang et al., 2013; Zhao et al., 2015). Even though very little is known about the site of melatonin synthesis in plants, there is proof that they can absorb

TABLE 14.1
Application of Exogenous Melatonin and Its Impact on Salt Stress Mitigation in Some Plants

Name of the plant	Amount	Mode	Impact during salt-stress condition	Reference
Abelmoschus esculentus L.	50 μM	Seed treatment	Decreased the levels of lipid peroxidation and H_2O_2 content and scavenged O2•⁻; enhanced the level of defence enzyme	(Zhan et al., 2021)
Actinidia deliciosa	0.1 μM	Application on seedlings	Enhancement of antioxidant capacity under salinity stress	(Xia et al., 2017)
Arabidopsis thaliana	–	Exogenous application	Reestablishment of ion and redox homeostasis through triggering the AtrbohF-dependent ROS signalling	(Chen et al., 2017)
Brassica napus	–	Exogenous application	Recovered the protein damages in the leaves by increasing their soluble sugar contents	(Zeng et al., 2018)
Cucumis sativus	100 μM	Exogenous application	Overall growth	(Wang et al., 2016)
Citrus aurantium	1 mM	Application on the plant in combination with ascorbic acid (0.5 mM)	Protective effect of up to 100 mM NaCl	(Kostopoulou et al., 2015)
Gossypium hirsutum L.	200 μM	Administration in medium	Improves the ROS scavenging capacity by enhancing their antioxidant capacity	(Jiang et al., 2020)
Helianthus annuus	15 μM	Exogenous application	Regulates root growth and hypocotyl elongation under salt stress	(Mukherjee et al., 2014)
Ipomoea batatas	100 μM	Foliar spray	Maintenance of Plasma Membrane H^+-ATPase Activity and K^+/Na^+ Homeostasis	(Yu et al., 2018)
Limonium bicolor	50–300 μM	Application in medium	Promotes ABA catabolism and GA biosynthesis and regulates seed germination by promoting nutrient utilization and protein synthesis in seeds	(Li et al., 2019b)
Malus hupehensis	1 mM	Pretreatment of seed	Alleviated the effects caused by salinity stress such as growth retardation, decreased photosynthesis rates, chlorophyll degradation, and impaired potassium uptake	(Li et al., 2012, 2016b)
Oryza sativa	25–400 μM	Foliar spray	Significantly improved the salt tolerance in seedlings by increasing the activity of antioxidant enzymes, scavenging ROS, and reducing membrane lipid peroxidation and cell damage	(Wei et al., 2022)
	20 μM	Application on seedling	Delayed leaf senescence and cell death in rice	(Liang et al., 2015)

TABLE 14.1 *(Continued)*
Application of Exogenous Melatonin and Its Impact on Salt Stress Mitigation in Some Plants

Name of the plant	Amount	Mode	Impact during salt-stress condition	Reference
Raphanus sativus	100 μM	Application on seedling	Increased chlorophyll a, chlorophyll b and total chlorophyll contents	(Park et al., 2021)
Solanum lycopersicum	100 μM	Application on seedling	Improved photosynthesis and regulation of photosynthetic electron transport	(Yin et al., 2019; Zhou et al., 2016)
	100 μM	Application on seedling	Induced antioxidant enzymes activity and better photosynthetic performance	(Martinez et al., 2018)
Triticum aestivum L.	500 μM	Foliar application	Enhancement of antioxidant enzymes with an alternate decrease in malondialdehyde content	(Zafar et al., 2019)
Vicia faba L.	100–500μM	Seed priming	Increased relative water content; enhanced growth parameters; increased total carbohydrate, photosynthetic pigments, phenolic content, IAA, K, Ca^{+2}; down regulation of Na^+ and Cl- the content of the leaf; decreased the concentrations of compatible solutes	(Dawood & El-Awadi, 2015)
Zea mays	60 μM	Exogenous application	Increased the strength of the antioxidant defence system by lowering ROS generation under salt-stress conditions	(Ren et al., 2020)

it directly from the soil or substrate in which they are cultivated (Paredes et al., 2009). Melatonin pretreatment in tomato seedlings resulted in improved growth and biomass development, increased antioxidant enzyme activity, decreased H_2O_2, O_2^{-}•, electrolyte leakage, and malondialdehyde, promoted root growth, and increased photosynthetic assimilation and photosynthetic pigment content (Altaf et al., 2020). Under abiotic stress, administered melatonin greatly decreased malondialdehyde levels and significantly boosted the expression of antioxidant genes in melon, *Triticum aestivum*, *Carya cathayensis*, and *Camellia sinensis* (Li et al., 2019a; Sun et al., 2018; Wang et al., 2019; Zhang et al., 2017). Soluble protein levels under exogenous melatonin initially increased and then decreased (Jiang et al., 2020). Exogenous melatonin is frequently associated with the accumulation of flavonoids and improved antioxidant activity. Some of the researchers opined that melatonin promotes the biosynthesis of flavonol and flavanol components; whereas some others opined that the application of exogenous melatonin accelerates the metabolism of phenolic substances (Gao et al., 2019; Yang et al., 2020). Globally, researchers are undertaking systematic and deliberate efforts to thoroughly investigate the prospects of melatonin for beneficial synthetic application to combat abiotic stresses, most importantly against salinity.

14.7 LIMITATION AND FUTURE PERSPECTIVE

Melatonin-rich plants that have already been genetically engineered will likely be capable of growing in saline soil (Byeon & Back, 2014; Okazaki & Ezura, 2009; Reiter et al., 2015). Although intriguing results have been gained over the past few years, existing information on melatonin in plants is inadequate, and additional research is required to determine the function of this molecule in plants. Certain studies have documented the regulatory mechanisms of plants connected to melatonin. However, further research is required to determine the regulation and activity of melatonin in salt stress tolerance in plants. Though the importance of melatonin in protecting plants from salinity has been widely described, not all reported data are equally legitimate, and a few of them require advanced methods of verification. For example, the level of melatonin in oat (*Avena sativa* L.) and ginger (*Zingiber officinale*) were reported much higher in the study by Hattori et al. (1995), but later research by Van Tassel et al. (2001) revealed that such plants contain trace amounts of melatonin. Its concentrations in ipomoea and tomato plants as measured by GC-MS were reported lower than those measured by RIA (Van Tassel et al., 2001). The collective knowledge concerning the melatonin production pathways in plants continues to expand. The participation of tryptophan in phytomelatonin biosynthesis is a recognized fact; however, other indole molecules may play a role, given the existence of non-tryptophan routes of IAA biosynthesis under normal and salinity-induced stress conditions.

Present communication reveals that there is still scope for more research into the role of melatonin in physiological processes that deal with the salt stress response.

14.8 CONCLUSION

Melatonin has an important role in the plant kingdom because of its ability to generate at the molecular scale as well as to trigger plant development by improving its hydroxyl antioxidant capability under normal as well as salinity-stress conditions. In most of the region, field-grown plants are exposed to salinity stress. It is envisaged that a good effect of melatonin on salt tolerance will create new potential for its usage in agriculture. Treatment with melatonin ameliorated the deleterious effects of salt stress on the majority of physiological parameters. As melatonin is safe for the environment and may be administered to plants in a variety of methods, including seed coating, foliar spraying, and soil treatment, it may play an important role in future agricultural techniques for crop production as well as crop protection against salinity stress.

Some photosynthetic genes can be down regulated by salinity, and the majority of their expressions were up regulated by melatonin therapy.

REFERENCES

Ahanger, M. A., & Agarwal, R. (2017). Salinity stress induced alterations in antioxidant metabolism and nitrogen assimilation in wheat (*Triticum aestivum* L) as influenced by potassium supplementation. *Plant Physiology and Biochemistry*, *115*, 449–460. https://doi.org/10.1016/j.plaphy.2017.04.017

Ahmad, S., Wang, G. Y., Muhammad, I., Farooq, S., Kamran, M., Ahmad, I., . . . Huang, J. H. (2022). Application of melatonin-mediated modulation of drought tolerance by regulating photosynthetic efficiency, chloroplast ultrastructure, and endogenous hormones in maize. *Chemical and Biological Technologies in Agriculture*, *9*(1), 1–14. https://doi.org/10.1186/s40538-021-00272-1

Alharbi, B. M., Elhakem, A. H., Alnusairi, G. S., Soliman, M. H., Hakeem, K. R., Hasan, M. M., & Abdelhamid, M. T. (2021). Exogenous application of melatonin alleviates salt stress-induced decline in growth and photosynthesis in *Glycine max* (L.) seedlings by improving mineral uptake, antioxidant and glyoxalase system. *Plant, Soil and Environment*, *67*(4), 208–220. https://doi.org/10.17221/659/2020-PSE

Altaf, M., Shahid, R., Ren, M., Naz, S., Altaf, M., Qadir, A., . . . Hayat, F. (2020). Exogenous melatonin enhances salt stress tolerance in tomato seedlings. *Biologia plantarum*, *64*, 604–615. https://doi.org/10.32615/bp.2020.090

Arnao, M., & Hernández-Ruiz, J. (2009). Protective effect of melatonin against chlorophyll degradation during the senescence of barley leaves. *Journal of Pineal Research*, *46*(1), 58–63. https://doi.org/10.1111/j.1600-079X.2008.00625.x

Arnao, M., & Hernández-Ruiz, J. (2015). Melatonin: Synthesis from tryptophan and its role in higher plant. In *Amino Acids in Higher Plants* (pp. 390–435). CAB International Wallingford UK. https://doi.org/10.1079/9781780642635.0390

Arnao, M. B. (2014). Phytomelatonin: Discovery, content, and role in plants. *Advances in Botany, 2014*. https://doi.org/10.1155/2014/815769

Arnao, M. B., & Hernández-Ruiz, J. (2015). Functions of melatonin in plants: A review. *Journal of Pineal Research, 59*(2), 133–150. https://doi.org/10.1111/jpi.12253

Arnao, M. B., & Hernández-Ruiz, J. (2018). Melatonin and its relationship to plant hormones. *Annals of Botany, 121*(2), 195–207. https://doi.org/10.1093/aob/mcx114

Arnao, M. B., & Hernández-Ruiz, J. (2019). Melatonin: A new plant hormone and/or a plant master regulator? *Trends in Plant Science, 24*(1), 38–48. https://doi.org/10.1016/j.tplants.2018.10.010

Arora, D., & Bhatla, S. C. (2017). Melatonin and nitric oxide regulate sunflower seedling growth under salt stress accompanying differential expression of Cu/Zn SOD and Mn SOD. *Free Radical Biology and Medicine, 106*, 315–328. https://doi.org/10.1016/j.freeradbiomed.2017.02.042

Banerjee A, & Roychoudhury A (2017) Melatonin as a regulator of abiotic stress tolerance in plants. In V. P. Singh, S. Singh, & S. M. Prasad (Eds.), *Mechanisms Behind Phytohormonal Signaling and Crop Abiotic Stress Tolerance* (pp. 47–60). Nova Science Publishers.

Bychkov, I., Kudryakova, N., Pojidaeva, E., & Kusnetsov, V. (2021). The melatonin receptor CAND2 is involved in the regulation of photosynthesis and chloroplast gene expression in *Arabidopsis thaliana* under photooxidative stress. *Photosynthetica, 59*(4), 683–692. https://doi.org/10.32615/ps.2021.061

Byeon, Y., & Back, K. (2014). An increase in melatonin in transgenic rice causes pleiotropic phenotypes, including enhanced seedling growth, delayed flowering, and low grain yield. *Journal of Pineal Research, 56*(4), 408–414. https://doi.org/10.1111/jpi.12129

Calvo, J. R., Gonzalez-Yanes, C., & Maldonado, M. (2013). The role of melatonin in the cells of the innate immunity: A review. *Journal of Pineal Research, 55*(2), 103–120. https://doi.org/10.1111/jpi.12075

Castañares, J. L., & Bouzo, C. A. (2019). Effect of exogenous melatonin on seed germination and seedling growth in melon (*Cucumis melo* L.) under salt stress. *Horticultural Plant Journal, 5*(2), 79–87. https://doi.org/10.1016/j.hpj.2019.01.002

Chen, Q., Qi, W. b., Reiter, R. J., Wei, W., & Wang, B. m. (2009). Exogenously applied melatonin stimulates root growth and raises endogenous indoleacetic acid in roots of etiolated seedlings of *Brassica juncea*. *Journal of Plant Physiology, 166*(3), 324–328. https://doi.org/10.1016/j.jplph.2008.06.002

Chen, Z., Gu, Q., Yu, X., Huang, L., Xu, S., Wang, R., . . . Shen, W. (2018). Hydrogen peroxide acts downstream of melatonin to induce lateral root formation. *Annals of Botany, 121*(6), 1127–1136. https://doi.org/10.1093/aob/mcx207

Chen, Z., Xie, Y., Gu, Q., Zhao, G., Zhang, Y., Cui, W., . . . Shen, W. (2017). The *AtrbohF*-dependent regulation of ROS signaling is required for melatonin-induced salinity tolerance in *Arabidopsis*. *Free Radical Biology and Medicine, 108*, 465–477. https://doi.org/10.1016/j.freeradbiomed.2017.04.009

Corpas, F. J., & Palma, J. M. (2018). Assessing nitric oxide (NO) in higher plants: An outline. *Nitrogen, 1*(1), 12–20. https://doi.org/10.3390/nitrogen1010003

Dawood, M. G., & El-Awadi, M. E. (2015). Alleviation of salinity stress on *Vicia faba* L. plants via seed priming with melatonin. *Acta Biológica Colombiana, 20*(2), 223–235. https://doi.org/10.15446/abc.v20n2.43291

Dubbels, R., Reiter, R., Klenke, E., Goebel, A., Schnakenberg, E., Ehlers, C., . . . Schloot, W. (1995). Melatonin in edible plants identified by radioimmunoassay and by high performance liquid chromatography-mass spectrometry. *Journal of Pineal Research, 18*(1), 28–31. https://doi.org/10.1111/j.1600-079x.1995.tb00136.x

Egamberdieva, D., Wirth, S., Jabborova, D., Räsänen, L. A., & Liao, H. (2017). Coordination between *Bradyrhizobium* and *Pseudomonas* alleviates salt stress in soybean through altering root system architecture. *Journal of Plant Interactions, 12*(1), 100–107. https://doi.org/10.1080/17429145.2017.1294212

Fukuda, A., Nakamura, A., Hara, N., Toki, S., & Tanaka, Y. (2011). Molecular and functional analyses of rice NHX-type Na^+/H^+ antiporter genes. *Planta, 233*(1), 175–188. https://doi.org/10.1007/s00425-010-1289-4

Gao, W., Feng, Z., Bai, Q., He, J., & Wang, Y. (2019). Melatonin-mediated regulation of growth and antioxidant capacity in salt-tolerant naked oat under salt stress. *International Journal of Molecular Sciences, 20*(5), 1176. https://doi.org/10.3390/ijms20051176

Gu, Q., Chen, Z., Yu, X., Cui, W., Pan, J., Zhao, G., . . . Shen, W. (2017). Melatonin confers plant tolerance against cadmium stress via the decrease of cadmium accumulation and reestablishment of microRNA-mediated redox homeostasis. *Plant Science, 261*, 28–37. https://doi.org/10.1016/j.plantsci.2017.05.001

Gu, Q., Xiao, Q., Chen, Z., & Han, Y. (2022). Crosstalk between melatonin and reactive oxygen species in plant abiotic stress responses: An update. *International Journal of Molecular Sciences*, *23*(10), 5666. https://doi.org/10.3390/ijms23105666

Hattori, A., Migitaka, H., Iigo, M., Itoh, M., Yamamoto, K., Ohtani-Kaneko, R., . . . Reiter, R. J. (1995). Identification of melatonin in plants and its effects on plasma melatonin levels and binding to melatonin receptors in vertebrates. *Biochemistry and Molecular Biology International*, *35*(3), 627–634.

He, H., & He, L. F. (2020). Crosstalk between melatonin and nitric oxide in plant development and stress responses. *Physiologia Plantarum*, *170*(2), 218–226. https://doi.org/10.1111/ppl.13143

Hu, T., Zhang, G., Zheng, F., & Cao, Y. (2018). Research progress in plant salt stress response. *Molecular Plant Breeding*, *16*(09), 3006–3015.

Hu, Z., Fan, J., Xie, Y., Amombo, E., Liu, A., Gitau, M. M., . . . Fu, J. (2016). Comparative photosynthetic and metabolic analyses reveal mechanism of improved cold stress tolerance in bermudagrass by exogenous melatonin. *Plant Physiology and Biochemistry*, *100*, 94–104. https://doi.org/10.1016/j.plaphy.2016.01.008

Iqbal, N., Fatma, M., Gautam, H., Umar, S., Sofo, A., D'ippolito, I., & Khan, N. A. (2021). The crosstalk of melatonin and hydrogen sulfide determines photosynthetic performance by regulation of carbohydrate metabolism in wheat under heat stress. *Plants*, *10*(9), 1778. https://doi.org/10.3390/plants10091778

Jahan, M. S., Shu, S., Wang, Y., Chen, Z., He, M., Tao, M., . . . Guo, S. (2019). Melatonin alleviates heat-induced damage of tomato seedlings by balancing redox homeostasis and modulating polyamine and nitric oxide biosynthesis. *BMC Plant Biology*, *19*(1), 1–16. https://doi.org/10.1186/s12870-019-1992-7

Janas, K. M., & Posmyk, M. M. (2013). Melatonin, an underestimated natural substance with great potential for agricultural application. *Acta Physiologiae Plantarum*, *35*(12), 3285–3292. https://doi.org/10.1007/s11738-013-1372-0

Jiang, D., Lu, B., Liu, L., Duan, W., Chen, L., Li, J., . . . Dong, H. (2020). Exogenous melatonin improves salt stress adaptation of cotton seedlings by regulating active oxygen metabolism. *PeerJ*, *8*, e10486. https://doi.org/10.7717/peerj.10486

Kanwar, M. K., Yu, J., & Zhou, J. (2018). Phytomelatonin: Recent advances and future prospects. *Journal of Pineal Research*, *65*(4), e12526. https://doi.org/10.1111/jpi.12526

Kaur, H., & Bhatla, S. C. (2016). Melatonin and nitric oxide modulate glutathione content and glutathione reductase activity in sunflower seedling cotyledons accompanying salt stress. *Nitric Oxide*, *59*, 42–53. https://doi.org/10.1016/j.niox.2016.07.001

Kaya, C., Okant, M., Ugurlar, F., Alyemeni, M. N., Ashraf, M., & Ahmad, P. (2019). Melatonin-mediated nitric oxide improves tolerance to cadmium toxicity by reducing oxidative stress in wheat plants. *Chemosphere*, *225*, 627–638. https://doi.org/10.1016/j.chemosphere.2019.03.026

Ke, Q., Ye, J., Wang, B., Ren, J., Yin, L., Deng, X., & Wang, S. (2018). Melatonin mitigates salt stress in wheat seedlings by modulating polyamine metabolism. *Frontiers in Plant Science*, *9*, 914. https://doi.org/10.3389/fpls.2018.00914

Khan, M. N., Zhang, J., Luo, T., Liu, J., Rizwan, M., Fahad, S., . . . Hu, L. (2019). Seed priming with melatonin coping drought stress in rapeseed by regulating reactive oxygen species detoxification: Antioxidant defense system, osmotic adjustment, stomatal traits and chloroplast ultrastructure perseveration. *Industrial Crops and Products*, *140*, 111597. https://doi.org/10.1016/j.indcrop.2019.111597

Kobylińska, A., Borek, S., & Posmyk, M. M. (2018). Melatonin redirects carbohydrates metabolism during sugar starvation in plant cells. *Journal of Pineal Research*, *64*(4), e12466. https://doi.org/10.1111/jpi.12466

Kostopoulou, Z., Therios, I., Roumeliotis, E., Kanellis, A. K., & Molassiotis, A. (2015). Melatonin combined with ascorbic acid provides salt adaptation in *Citrus aurantium* L. seedlings. *Plant Physiology and Biochemistry*, *86*, 155–165. https://doi.org/10.1016/j.plaphy.2014.11.021

Lee, H. Y., Byeon, Y., & Back, K. (2014). Melatonin as a signal molecule triggering defense responses against pathogen attack in *Arabidopsis* and tobacco. *Journal of Pineal Research*, *57*(3), 262–268. https://doi.org/10.1111/jpi.12165

Lee, H. Y., Byeon, Y., Tan, D. X., Reiter, R. J., & Back, K. (2015). *Arabidopsis* serotonin *N*-acetyltransferase knockout mutant plants exhibit decreased melatonin and salicylic acid levels resulting in susceptibility to an avirulent pathogen. *Journal of Pineal Research*, *58*(3), 291–299. https://doi.org/10.1111/jpi.12214

Lerner, A. B., Case, J. D., Takahashi, Y., Lee, T. H., & Mori, W. (1958). Isolation of melatonin, the pineal gland factor that lightens melanocytes. *Journal of the American Chemical Society*, *80*(10), 2587–2587. https://doi.org/10.1021/ja01543a060

Li, C., Liang, B., Chang, C., Wei, Z., Zhou, S., & Ma, F. (2016a). Exogenous melatonin improved potassium content in *Malus* under different stress conditions. *Journal of Pineal Research*, *61*(2), 218–229. https://doi.org/10.1111/jpi.12342

Li, C., Wang, P., Wei, Z., Liang, D., Liu, C., Yin, L., . . . Ma, F. (2012). The mitigation effects of exogenous melatonin on salinity-induced stress in *Malus hupehensis*. *Journal of Pineal Research*, *53*(3), 298–306. https://doi.org/10.1111/j.1600-079X.2012.00999.x

Li, H., Chang, J., Chen, H., Wang, Z., Gu, X., Wei, C., . . . Zhang, X. (2017). Exogenous melatonin confers salt stress tolerance to watermelon by improving photosynthesis and redox homeostasis. *Frontiers in Plant Science*, *8*, 295. https://doi.org/10.3389/fpls.2017.00295

Li, H., Dong, Y., Chang, J., He, J., Chen, H., Liu, Q., . . . Yang, J. (2016b). High-throughput microRNA and mRNA sequencing reveals that microRNAs may be involved in melatonin-mediated cold tolerance in *Citrullus lanatus* L. *Frontiers in Plant Science*, *7*, 1231. https://doi.org/10.3389/fpls.2016.01231

Li, J., Arkorful, E., Cheng, S., Zhou, Q., Li, H., Chen, X., . . . Li, X. (2018). Alleviation of cold damage by exogenous application of melatonin in vegetatively propagated tea plant (*Camellia sinensis* (L.) O. Kuntze). *Scientia Horticulturae*, *238*, 356–362.

Li, J., Liu, Y., Zhang, M., Xu, H., Ning, K., Wang, B., & Chen, M. (2022). Melatonin increases growth and salt tolerance of *Limonium bicolor* by improving photosynthetic and antioxidant capacity. *BMC Plant Biology*, *22*(1), 1–14. https://doi.org/10.1186/s12870-021-03402-x

Li, J., Yang, Y., Sun, K., Chen, Y., Chen, X., & Li, X. (2019a). Exogenous melatonin enhances cold, salt and drought stress tolerance by improving antioxidant defense in tea plant (*Camellia sinensis* (L.) O. Kuntze). *Molecules*, *24*(9), 1826. https://doi.org/10.3390/molecules24091826

Li, J., Yuan, F., Liu, Y., Zhang, M., Liu, Y., Zhao, Y., . . . Chen, M. (2020). Exogenous melatonin enhances salt secretion from salt glands by upregulating the expression of ion transporter and vesicle transport genes in *Limonium bicolor*. *BMC Plant Biology*, *20*(1), 1–11. https://doi.org/10.1186/s12870-020-02703-x

Li, J., Zhao, C., Zhang, M., Yuan, F., & Chen, M. (2019b). Exogenous melatonin improves seed germination in *Limonium bicolor* under salt stress. *Plant Signaling & Behavior*, *14*(11), 1659705. https://doi.org/10.1080/15592324.2019.1659705

Liang, C., Zheng, G., Li, W., Wang, Y., Hu, B., Wang, H., . . . Tan, D. X. (2015). Melatonin delays leaf senescence and enhances salt stress tolerance in rice. *Journal of Pineal Research*, *59*(1), 91–101. https://doi.org/10.1111/jpi.12243

Liu, J., Shabala, S., Zhang, J., Ma, G., Chen, D., Shabala, L., . . . Venkataraman, G. (2020). Melatonin improves rice salinity stress tolerance by NADPH oxidase-dependent control of the plasma membrane K^+ transporters and K^+ homeostasis. *Plant, Cell & Environment*, *43*(11), 2591–2605. https://doi.org/10.1111/pce.13759

Liu, Q., Liu, R., Ma, Y., & Song, J. (2018). Physiological and molecular evidence for Na^+ and Cl^- exclusion in the roots of two *Suaeda salsa* populations. *Aquatic Botany*, *146*, 1–7. https://doi.org/10.1016/j.aquabot.2018.01.001

Ljung, K., Nemhauser, J. L., & Perata, P. (2015). New mechanistic links between sugar and hormone signalling networks. *Current Opinion in Plant Biology*, *25*, 130–137. https://doi.org/10.1016/j.pbi.2015.05.022

Majeed, A., & Muhammad, Z. (2019). Salinity: A major agricultural problem: Causes, impacts on crop productivity and management strategies. In M. Hasanuzzaman, K. Hakeem, K. Nahar, & H. Alharby (Eds.), *Plant Abiotic Stress Tolerance* (pp. 83–99). Springer. https://doi.org/10.1007/978-3-030-06118-0_3

Manchester, L. C., Coto-Montes, A., Boga, J. A., Andersen, L. P. H., Zhou, Z., Galano, A., . . . Reiter, R. J. (2015). Melatonin: An ancient molecule that makes oxygen metabolically tolerable. *Journal of Pineal Research*, *59*(4), 403–419. https://doi.org/10.1111/jpi.12267

Martinez, V., Nieves-Cordones, M., Lopez-Delacalle, M., Rodenas, R., Mestre, T. C., Garcia-Sanchez, F., . . . Rivero, R. M. (2018). Tolerance to stress combination in tomato plants: New insights in the protective role of melatonin. *Molecules*, *23*(3), 535. https://doi.org/10.3390/molecules23030535

Mir, A. R., Faizan, M., Bajguz, A., Sami, F., Siddiqui, H., & Hayat, S. (2020). Occurrence and biosynthesis of melatonin and its exogenous effect on plants. *Acta Societatis Botanicorum Poloniae*, *89*(2). https://doi.org/10.5586/asbp.8922

Mohamadi, M., & Karimi, M. (2020). Effect of exogenous melatonin on growth, electrolyte leakage and antioxidant enzyme activity in rosemary under salinity stress. *Journal of Plant Process and Function*, *9*(37), 60.

Mukherjee, S. (2018). Novel perspectives on the molecular crosstalk mechanisms of serotonin and melatonin in plants. *Plant Physiology and Biochemistry*, *132*, 33–45. https://doi.org/10.1016/j.plaphy.2018.08.031

Mukherjee, S. (2019). Insights into nitric oxide: Melatonin crosstalk and *N*-nitrosomelatonin functioning in plants. *Journal of Experimental Botany*, *70*(21), 6035–6047. https://doi.org/10.1093/jxb/erz375

Mukherjee, S., & Bhatla, S. C. (2021). Exogenous melatonin modulates endogenous H_2S homeostasis and L-cysteine desulfhydrase activity in salt-stressed tomato (*Solanum lycopersicum* L. var. cherry) seedling cotyledons. *Journal of Plant Growth Regulation*, *40*(6), 2502–2514. https://doi.org/10.1007/s00344-020-10261-7

Mukherjee, S., David, A., Yadav, S., Baluška, F., & Bhatla, S. C. (2014). Salt stress-induced seedling growth inhibition coincides with differential distribution of serotonin and melatonin in sunflower seedling roots and cotyledons. *Physiologia Plantarum*, *152*(4), 714–728. https://doi.org/10.1111/ppl.12218

Munns, R., & Tester, M. (2008). Mechanisms of salinity tolerance. *Annual Review of Plant Biology*, *59*, 651–681. https://doi.org/10.1146/annurev.arplant.59.032607.092911

Murch, S. J., & Erland, L. A. (2021). A systematic review of melatonin in plants: An example of evolution of literature. *Frontiers in Plant Science*, *12*, 1060. https://doi.org/10.3389/fpls.2021.683047

Nawaz, K., Chaudhary, R., Sarwar, A., Ahmad, B., Gul, A., Hano, C., . . . Anjum, S. (2020). Melatonin as master regulator in plant growth, development and stress alleviator for sustainable agricultural production: Current status and future perspectives. *Sustainability*, *13*(1), 294. https://doi.org/10.3390/su13010294

Nawaz, M. A., Huang, Y., Bie, Z., Ahmed, W., Reiter, R. J., Niu, M., & Hameed, S. (2016). Melatonin: Current status and future perspectives in plant science. *Frontiers in Plant Science*, *6*, 1230. https://doi.org/10.3389/fpls.2015.01230

Neamah, S. I., & Jdayea, N. A. (2022). Positive response of hyoscyamus pusillus callus cultures to exogenous melatonin on biochemical traits and secondary metabolites under drought conditions. *International Journal of Agronomy*, *2022*. https://doi.org/10.1155/2022/7447024

Okazaki, M., & Ezura, H. (2009). Profiling of melatonin in the model tomato (*Solanum lycopersicum* L.) cultivar Micro-Tom. *Journal of Pineal Research*, *46*(3), 338–343. https://doi.org/10.1111/j.1600-079X.2009.00668.x

Paredes, S. D., Korkmaz, A., Manchester, L. C., Tan, D.-X., & Reiter, R. J. (2009). Phytomelatonin: A review. *Journal of Experimental Botany*, *60*(1), 57–69. https://doi.org/10.1093/jxb/ern284

Park, H. J., Kim, W. Y., & Yun, D. J. (2016). A new insight of salt stress signaling in plant. *Molecules and Cells*, *39*(6), 447–459. https://doi.org/10.14348/molcells.2016.0083

Park, H. S., Kazerooni, E. A., Kang, S. M., Al-Sadi, A. M., & Lee, I. J. (2021). Melatonin enhances the tolerance and recovery mechanisms in *Brassica juncea* (L.) czern. under saline conditions. *Frontiers in Plant Science*, *12*, 593717. https://doi.org/10.3389/fpls.2021.593717

Park, S., & Back, K. (2012). Melatonin promotes seminal root elongation and root growth in transgenic rice after germination. *Journal of Pineal Research*, *53*(4), 385–389. https://doi.org/10.1111/j.1600-079X.2012.01008.x

Park, W. J. (2011). Melatonin as an endogenous plant regulatory signal: Debates and perspectives. *Journal of Plant Biology*, *54*(3), 143–149. https://doi.org/10.1007/s12374-011-9159-6

Pelagio-Flores, R., & López-Bucio, J. (2016). Serotonin and melatonin in plant growth and development. In G. A. Ravishankar, & A. Ramakrishna (Eds.), *Serotonin and Melatonin* (pp. 119–132). CRC Press. https://doi.org/10.1201/9781315369334

Pelagio-Flores, R., Muñoz-Parra, E., Ortiz-Castro, R., & López-Bucio, J. (2012). Melatonin regulates Arabidopsis root system architecture likely acting independently of auxin signaling. *Journal of Pineal Research*, *53*(3), 279–288. https://doi.org/10.1111/j.1600-079X.2012.00996.x

Pérez-Llorca, M., Muñoz, P., Müller, M., & Munné-Bosch, S. (2019). Biosynthesis, metabolism and function of auxin, salicylic acid and melatonin in climacteric and non-climacteric fruits. *Frontiers in Plant Science*, *10*, 136. https://doi.org/10.3389/fpls.2019.00136

Qian, Y., Tan, D. X., Reiter, R. J., & Shi, H. (2015). Comparative metabolomic analysis highlights the involvement of sugars and glycerol in melatonin-mediated innate immunity against bacterial pathogen in *Arabidopsis*. *Scientific Reports*, *5*(1), 1–11. https://doi.org/10.1038/srep15815

Reiter, R. J., Tan, D. X., Zhou, Z., Cruz, M. H. C., Fuentes-Broto, L., & Galano, A. (2015). Phytomelatonin: Assisting plants to survive and thrive. *Molecules*, *20*(4), 7396–7437. https://doi.org/10.3390/molecules20047396

Ren, J., Ye, J., Yin, L., Li, G., Deng, X., & Wang, S. (2020). Exogenous melatonin improves salt tolerance by mitigating osmotic, ion, and oxidative stresses in maize seedlings. *Agronomy*, *10*(5), 663. https://doi.org/10.3390/agronomy10050663

Roychoudhury, A., Basu, S., Sarkar, S. N., & Sengupta, D. N. (2008). Comparative physiological and molecular responses of a common aromatic indica rice cultivar to high salinity with non-aromatic indica rice cultivars. *Plant Cell Reports*, *27*(8), 1395–1410.

Roychoudhury, A., & Chakraborty, M. (2013). Biochemical and molecular basis of varietal difference in plant salt tolerance. *Annual Review & Research in Biology*, *3*(4), 422–454.

Roychoudhury, A., Datta, K., & Datta, S. K. (2011). Abiotic stress in plants: From genomics to metabolomics. In N. Tuteja, S. S. Gill, & R. Tuteja. (Eds.). *Omics and Plant Abiotic Stress Tolerance* (pp. 91–120). Bentham Science Publishers.

Roychoudhury, A., Roy, C., & Sengupta, D. N. (2007). Transgenic tobacco plants overexpressing the heterologous *lea* gene *Rab16A* from rice during high salt and water deficit display enhanced tolerance to salinity stress. *Plant Cell Reports*, *26*(10), 1839–1859.

Samanta, S., Banerjee, A., & Roychoudhury, A. (2021). Melatonin application differentially modulates the enzymes associated with antioxidative machinery and ascorbate glutathione cycle during arsenate exposure in indica rice varieties. *Plant Biology*, *23*(S1), 193–201.

Samanta, S., Banerjee, A., & Roychoudhury, A. (2022). Arsenic toxicity is counteracted by exogenous application of melatonin to different extents in arsenic-susceptible and arsenic-tolerant rice cultivars. *Journal of Plant Growth Regulation*, *41*, 2210–2231.

Sarkar, A. K., & Sadhukhan, S. (2022a). Bioremediation of salt-affected soil through plant-based strategies. In J. A. Malik (Ed.), *Advances in Bioremediation and Phytoremediation for Sustainable Soil Management* (pp. 81–100). Springer. https://doi.org/10.1007/978-3-030-89984-4_5

Sarkar, A. K., & Sadhukhan, S. (2022b). Imperative role of trehalose metabolism and trehalose-6-phosphate signaling on salt stress responses in plants. *Physiologia Plantarum*, *174*(1), e13647. https://doi.org/10.1111/ppl.13647

Sarropoulou, V. N., Therios, I. N., & Dimassi-Theriou, K. N. (2012). Melatonin promotes adventitious root regeneration in in vitro shoot tip explants of the commercial sweet cherry rootstocks CAB-6P (*Prunus cerasus* L.), Gisela 6 (*P. cerasus*× *P. canescens*), and MxM 60 (*P. avium*× *P.mahaleb*). *Journal of Pineal Research*, *52*(1), 38–46. https://doi.org/10.1111/j.1600-079X.2011.00914.x

Shen, Y., Xia, H., Lei, Z., & Liang, D. (2018). The effect of exogenous melatonin on kiwifruit phenolic compounds and antioxidant capacity under low light environment. *2018 7th International Conference on Energy, Environment and Sustainable Development (ICEESD 2018).*

Shi, H., Chen, Y., Tan, D. X., Reiter, R. J., Chan, Z., & He, C. (2015a). Melatonin induces nitric oxide and the potential mechanisms relate to innate immunity against bacterial pathogen infection in *Arabidopsis*. *Journal of Pineal Research*, *59*(1), 102–108. https://doi.org/10.1111/jpi.12244

Shi, H., Jiang, C., Ye, T., Tan, D. X., Reiter, R. J., Zhang, H., . . . Chan, Z. (2015b). Comparative physiological, metabolomic, and transcriptomic analyses reveal mechanisms of improved abiotic stress resistance in bermudagrass [*Cynodon dactylon* (L). Pers.] by exogenous melatonin. *Journal of Experimental Botany*, *66*(3), 681–694. https://doi.org/10.1093/jxb/eru373

Shi, H., Qian, Y., Tan, D. X., Reiter, R. J., & He, C. (2015c). Melatonin induces the transcripts of *CBF/DREB1s* and their involvement in both abiotic and biotic stresses in *Arabidopsis*. *Journal of Pineal Research*, *59*(3), 334–342. https://doi.org/10.1111/jpi.12262

Shibaeva, T., Markovskaya, E., & Mamaev, A. (2018). Phytomelatonin: A review. *Biology Bulletin Reviews*, *8*(5), 375–388. https://doi.org/10.1134/S2079086418050080

Siddiqui, M. H., Alamri, S., Al-Khaishany, M. Y., Khan, M. N., Al-Amri, A., Ali, H. M., . . . Alsahli, A. A. (2019). Exogenous melatonin counteracts NaCl-induced damage by regulating the antioxidant system, proline and carbohydrates metabolism in tomato seedlings. *International Journal of Molecular Sciences*, *20*(2), 353. https://doi.org/10.3390/ijms20020353

Su, J., Yang, X., Shao, Y., Chen, Z., & Shen, W. (2021). Molecular hydrogen-induced salinity tolerance requires melatonin signalling in *Arabidopsis thaliana*. *Plant, Cell & Environment*, *44*(2), 476–490. https://doi.org/10.1111/pce.13926

Sun, C., Liu, L., Wang, L., Li, B., Jin, C., & Lin, X. (2021). Melatonin: A master regulator of plant development and stress responses. *Journal of Integrative Plant Biology*, *63*(1), 126–145. https://doi.org/10.1111/jipb.12993

Sun, L., Li, X., Wang, Z., Sun, Z., Zhu, X., Liu, S., . . . Wang, Y. (2018). Cold priming induced tolerance to subsequent low temperature stress is enhanced by melatonin application during recovery in wheat. *Molecules*, *23*(5), 1091. https://doi.org/10.3390/molecules23051091

Sun, L., Song, F., Guo, J., Zhu, X., Liu, S., Liu, F., & Li, X. (2020). Nano-ZnO-induced drought tolerance is associated with melatonin synthesis and metabolism in maize. *International Journal of Molecular Sciences*, *21*(3), 782. https://doi.org/10.3390/ijms21030782

Tan, D. X., Hardeland, R., Manchester, L. C., Korkmaz, A., Ma, S., Rosales-Corral, S., & Reiter, R. J. (2012). Functional roles of melatonin in plants, and perspectives in nutritional and agricultural science. *Journal of Experimental Botany*, *63*(2), 577–597. https://doi.org/10.1093/jxb/err256

Tan, D. X., Hardeland, R., Manchester, L. C., Paredes, S. D., Korkmaz, A., Sainz, R. M., . . . Reiter, R. J. (2010). The changing biological roles of melatonin during evolution: From an antioxidant to signals of darkness, sexual selection and fitness. *Biological Reviews*, *85*(3), 607–623. https://doi.org/10.1111/j.1469-185X.2009.00118.x

Tan, D. X., Manchester, L. C., Liu, X., Rosales-Corral, S. A., Acuna-Castroviejo, D., & Reiter, R. J. (2013). Mitochondria and chloroplasts as the original sites of melatonin synthesis: A hypothesis related to melatonin's primary function and evolution in eukaryotes. *Journal of Pineal Research*, *54*(2), 127–138. https://doi.org/10.1111/jpi.12026

Tan, D. X., & Reiter, R. J. (2019). Mitochondria: The birth place, battle ground and the site of melatonin metabolism in cells. *Melatonin Research*, *2*(1), 44–66. https://doi.org/10.32794/mr11250011

Tan, K., Zheng, J., Liu, C., Liu, X., Liu, X., Gao, T., . . . Li, C. (2021). Heterologous expression of the melatonin-related gene *HIOMT* improves salt tolerance in *Malus domestica*. *International Journal of Molecular Sciences*, *22*(22), 12425. https://doi.org/10.3390/ijms222212425

Tripathi, G. D., Javed, Z., Mishra, M., Fasake, V., & Dashora, K. (2021). Phytomelatonin in stress management in agriculture. *Heliyon*, *7*(3), e06150. https://doi.org/10.1016/j.heliyon.2021.e06150

Van Tassel, D. L., Roberts, N., Lewy, A., & O'Neill, S. D. (2001). Melatonin in plant organs. *Journal of Pineal Research*, *31*(1), 8–15. https://doi.org/10.1034/j.1600-079X.2001.310102.x

Wan, J., Zhang, P., Wang, R., Sun, L., Ju, Q., & Xu, J. (2018). Comparative physiological responses and transcriptome analysis reveal the roles of melatonin and serotonin in regulating growth and metabolism in *Arabidopsis*. *BMC Plant Biology*, *18*(1), 1–14. https://doi.org/10.1186/s12870-018-1548-2

Wang, J., Chen, J., Sharma, A., Tao, S., Zheng, B., Landi, M., . . . Yan, D. (2019). Melatonin stimulates activities and expression level of antioxidant enzymes and preserves functionality of photosynthetic apparatus in hickory plants (*Carya cathayensis* Sarg.) under PEG-promoted drought. *Agronomy*, *9*(11), 702. https://doi.org/10.3390/agronomy9110702

Wang, L., Liu, J., Wang, W., & Sun, Y. (2016). Exogenous melatonin improves growth and photosynthetic capacity of cucumber under salinity-induced stress. *Photosynthetica*, *54*(1), 19–27. https://doi.org/10.1007/s11099-015-0140-3

Wang, P., Yin, L., Liang, D., Li, C., Ma, F., & Yue, Z. (2012). Delayed senescence of apple leaves by exogenous melatonin treatment: toward regulating the ascorbate: Glutathione cycle. *Journal of Pineal Research*, *53*(1), 11–20. https://doi.org/10.1111/j.1600-079X.2011.00966.x

Wang, Y., Reiter, R. J., & Chan, Z. (2018). Phytomelatonin: A universal abiotic stress regulator. *Journal of Experimental Botany*, *69*(5), 963–974. https://doi.org/10.1093/jxb/erx473

Weeda, S., Zhang, N., Zhao, X., Ndip, G., Guo, Y., Buck, G. A., . . . Ren, S. (2014). Arabidopsis transcriptome analysis reveals key roles of melatonin in plant defense systems. *PloS one*, *9*(3), e93462. https://doi.org/10.1371/journal.pone.0093462

Wei, J., Li, D. X., Zhang, J. R., Shan, C., Rengel, Z., Song, Z. B., & Chen, Q. (2018). Phytomelatonin receptor PMTR1-mediated signaling regulates stomatal closure in *Arabidopsis thaliana*. *Journal of Pineal Research*, *65*(2), e12500. https://doi.org/10.1111/jpi.12500

Wei, L., Zhao, H., Wang, B., Wu, X., Lan, R., Huang, X., . . . Wang, J. (2022). Exogenous melatonin improves the growth of rice seedlings by regulating redox balance and ion homeostasis under salt stress. *Journal of Plant Growth Regulation*, *41*(6), 2108–2121. https://doi.org/10.1007/s00344-021-10417-z

Wei, W., Li, Q.-T., Chu, Y.-N., Reiter, R. J., Yu, X.-M., Zhu, D.-H., . . . Zhang, J.-S. (2015). Melatonin enhances plant growth and abiotic stress tolerance in soybean plants. *Journal of Experimental Botany*, *66*(3), 695–707. https://doi.org/10.1093/jxb/eru392

Wen, D., Gong, B., Sun, S., Liu, S., Wang, X., Wei, M., . . . Shi, Q. (2016). Promoting roles of melatonin in adventitious root development of *Solanum lycopersicum* L. by regulating auxin and nitric oxide signaling. *Frontiers in plant science*, *7*, 718. https://doi.org/10.3389/fpls.2016.00718

Xia, H., Ni, Z., & Pan, D. (2017). Effects of exogenous melatonin on antioxidant capacity in *Actinidia* seedlings under salt stress. *IOP Conference Series: Earth and Environmental Science*.

Xie, Z., Wang, J., Wang, W., Wang, Y., Xu, J., Li, Z., . . . Fu, B. (2021). Integrated analysis of the transcriptome and metabolome revealed the molecular mechanisms underlying the enhanced salt tolerance of rice due to the application of exogenous melatonin. *Frontiers in Plant Science*, *11*, 618680. https://doi.org/10.3389/fpls.2020.618680

Xu, L., Xiang, G., Sun, Q., Ni, Y., Jin, Z., Gao, S., & Yao, Y. (2019). Melatonin enhances salt tolerance by promoting *MYB108A*-mediated ethylene biosynthesis in grapevines. *Horticulture Research*, *6*. https://doi.org/10.1038/s41438-019-0197-4

Yan, F., Wei, H., Ding, Y., Li, W., Liu, Z., Chen, L., . . . Li, G. (2021). Melatonin regulates antioxidant strategy in response to continuous salt stress in rice seedlings. *Plant Physiology and Biochemistry*, *165*, 239–250. https://doi.org/10.1016/j.plaphy.2021.05.003

Yan, Y., Shi, Q., & Gong, B. (2020). Review of melatonin in horticultural crops. In M. Vlachou (Ed.), *Melatonin-the Hormone of Darkness and Its Therapeutic Potential and Perspectives*. IntechOpen. https://doi.org/10.5772/intechopen.90935

Yang, H., Dai, L., Wei, Y., Deng, Z., & Li, D. (2020). Melatonin enhances salt stress tolerance in rubber tree (*Hevea brasiliensis*) seedlings. *Industrial Crops and Products*, *145*, 111990. https://doi.org/10.1016/j.indcrop.2019.111990

Yang, J., Zhang, C., Wang, Z., Sun, S., Zhan, R., Zhao, Y., . . . Li, M. (2019). Melatonin-mediated sugar accumulation and growth inhibition in apple plants involves down-regulation of fructokinase 2 expression and activity. *Frontiers in Plant Science*, *10*, 150. https://doi.org/10.3389/fpls.2019.00150

Ye, J., Wang, S., Deng, X., Yin, L., Xiong, B., & Wang, X. (2016). Melatonin increased maize (*Zea mays* L.) seedling drought tolerance by alleviating drought-induced photosynthetic inhibition and oxidative damage. *Acta Physiologiae Plantarum*, *38*(2), 1–13. https://doi.org/10.1007/s11738-015-2045-y

Yin, L., Wang, S., Li, J., Tanaka, K., & Oka, M. (2013). Application of silicon improves salt tolerance through ameliorating osmotic and ionic stresses in the seedling of *Sorghum bicolor*. *Acta Physiologiae Plantarum*, *35*(11), 3099–3107. https://doi.org/10.1007/s11738-013-1343-5

Yin, Z., Lu, J., Meng, S., Liu, Y., Mostafa, I., Qi, M., & Li, T. (2019). Exogenous melatonin improves salt tolerance in tomato by regulating photosynthetic electron flux and the ascorbate: Gglutathione cycle. *Journal of Plant Interactions*, *14*(1), 453–463. https://doi.org/10.1080/17429145.2019.1645895

Yu, Y., Wang, A., Li, X., Kou, M., Wang, W., Chen, X., . . . Li, Z. (2018). Melatonin-stimulated triacylglycerol breakdown and energy turnover under salinity stress contributes to the maintenance of plasma membrane H^+-ATPase activity and K^+/Na^+ homeostasis in sweet potato. *Frontiers in Plant Science*, *9*, 256. https://doi.org/10.3389/fpls.2018.00256

Zafar, S., Hasnain, Z., Anwar, S., Perveen, S., Iqbal, N., Noman, A., & Ali, M. (2019). Influence of melatonin on antioxidant defense system and yield of wheat (*Triticum aestivum* L.) genotypes under saline condition. *Pakistan Journal of Botany*, *51*(6), 1987–1994. https://doi.org/10.30848/PJB2019-6(5)

Zahedi, S. M., Hosseini, M. S., Fahadi Hoveizeh, N., Gholami, R., Abdelrahman, M., & Tran, L. S. P. (2021). Exogenous melatonin mitigates salinity-induced damage in olive seedlings by modulating ion homeostasis, antioxidant defense, and phytohormone balance. *Physiologia Plantarum*, *173*(4), 1682–1694. https://doi.org/10.1111/ppl.13589

Zeng, L., Cai, J. S., Li, J. J., Lu, G. Y., Li, C.S., Fu, G. P., . . . Cheng, Y. (2018). Exogenous application of a low concentration of melatonin enhances salt tolerance in rapeseed (*Brassica napus* L.) seedlings. *Journal of Integrative Agriculture*, *17*(2), 328–335. https://doi.org/10.1016/S2095-3119(17)61757-X

Zhan, Y., Wu, T., Zhao, X., Wang, Z., & Chen, Y. (2021). Comparative physiological and full-length transcriptome analyses reveal the molecular mechanism of melatonin-mediated salt tolerance in okra (*Abelmoschus esculentus* L.). *BMC Plant Biology*, *21*(1), 1–16. https://doi.org/10.1186/s12870-021-02957-z

Zhang, H. J., Zhang, N., Yang, R. C., Wang, L., Sun, Q. Q., Li, D. B., . . . Ren, S. (2014). Melatonin promotes seed germination under high salinity by regulating antioxidant systems, ABA and GA_4 interaction in cucumber (*Cucumis sativus* L.). *Journal of Pineal Research*, *57*(3), 269–279. https://doi.org/10.1111/jpi.12167

Zhang, N., Zhao, B., Zhang, H. J., Weeda, S., Yang, C., Yang, Z. C., . . . Guo, Y. D. (2013). Melatonin promotes water-stress tolerance, lateral root formation, and seed germination in cucumber (*Cucumis sativus* L.). *Journal of Pineal Research*, *54*(1), 15–23. https://doi.org/10.1111/j.1600-079X.2012.01015.x

Zhang, R., Sun, Y., Liu, Z., Jin, W., & Sun, Y. (2017). Effects of melatonin on seedling growth, mineral nutrition, and nitrogen metabolism in cucumber under nitrate stress. *Journal of Pineal Research*, *62*(4), e12403. https://doi.org/10.1111/jpi.12403

Zhao, D., Wang, R., Meng, J., Li, Z., Wu, Y., & Tao, J. (2017). Ameliorative effects of melatonin on dark-induced leaf senescence in gardenia (*Gardenia jasminoides* Ellis): Leaf morphology, anatomy, physiology and transcriptome. *Scientific Reports*, *7*(1), 1–19. https://doi.org/10.1038/s41598-017-10799-9

Zhao, D., Yu, Y., Shen, Y., Liu, Q., Zhao, Z., Sharma, R., & Reiter, R. J. (2019). Melatonin synthesis and function: Evolutionary history in animals and plants. *Frontiers in Endocrinology*, *10*, 249. https://doi.org/10.3389/fendo.2019.0024

Zhao, G., Zhao, Y., Yu, X., Kiprotich, F., Han, H., Guan, R., . . . Shen, W. (2018). Nitric oxide is required for melatonin-enhanced tolerance against salinity stress in rapeseed (*Brassica napus* L.) seedlings. *International Journal of Molecular Sciences*, *19*(7), 1912. https://doi.org/10.3390/ijms19071912

Zhao, H., Su, T., Huo, L., Wei, H., Jiang, Y., Xu, L., & Ma, F. (2015). Unveiling the mechanism of melatonin impacts on maize seedling growth: Sugar metabolism as a case. *Journal of Pineal Research*, *59*(2), 255–266. https://doi.org/10.1111/jpi.12258

Zhou, X., Zhang, Y., Deng, J., Jiang, Z., He, Z., & Zou, Z. (2019). Effects of melatonin on the sugar metabolism and antioxidative response to excess nitrate stress in lettuce seedlings. *IOP Conference Series: Earth and Environmental Science*.

Zhou, X., Zhao, H., Cao, K., Hu, L., Du, T., Baluška, F., & Zou, Z. (2016). Beneficial roles of melatonin on redox regulation of photosynthetic electron transport and synthesis of D1 protein in tomato seedlings under salt stress. *Frontiers in Plant Science*, *7*, 1823. https://doi.org/10.3389/fpls.2016.01823

Zhu, Y., Gao, H., Lu, M., Hao, C., Pu, Z., Guo, M., . . . Huang, X. (2019). Melatonin-nitric oxide crosstalk and their roles in the redox network in plants. *International Journal of Molecular Sciences*, *20*(24), 6200. https://doi.org/10.3390/ijms20246200

15 Role of Melatonin in Drought Tolerance

Physiological and Molecular Perspectives

Soumya Kumar Sahoo, Biswajit Lenka, Biswaranjan Behera, Muhammad Ahsan Altaf, Ravinder Kumar, Awadhesh Kumar, Rahul Kumar Tiwari, and Milan Kumar Lal

15.1 INTRODUCTION

Melatonin (*N-acetyl*-5-methoxytryptamine) discovered in the bovine group is a multifunctional molecule with several cellular and physiological activities in various kingdoms (Lerner et al. 1958). The pineal gland releases it into the cerebrospinal fluid and the bloodstream, where it functions as a neurohormone in animals involved in the regulation of circadian rhythm, sexual behaviour, appetite, and body temperature (Hardeland et al. 2012; Altaf et al. 2022a; Saqib et al. 2022). In addition, melatonin (MT) was also reported to be engaged in various cellular activities as an antioxidant and free radical scavenger (Reiter et al. 2013). Interestingly, the presence of MT in higher plants was confirmed independently by Dubbels et al. (1995), Hattori et al. in 1995. The successive studies have accumulated details on the occurrence of MT in plants and its potential physiological roles. A number of researchers initially studied phytomelatonin as a fascinating plant-derived antioxidant molecule. Its presence in plant origin foodstuff and its ensuing assimilation into the human bloodstream make it important in human nutrition (Tan et al. 2014). Likewise, as a novel phytoremediation agent, the use of melatonin-rich plants as a recovery bioagent for chemically deteriorated soil has been proposed. Due to the highly diverse functions of this vital molecule, more researchers are interested in working on physiological and molecular functions of phytomelatonin.

One of the primary environmental issues for plants that has a negative impact on their growth and development is the lack of water. Plant efficiency eventually falls due to the physiological imbalances brought about by the decreased water availability to plants (Fahad et al. 2017). The effect of drought stress on the physiological reactions in a plant is reliant upon the duration of dry spell and the development phase of a plant (Okçu, Kaya, and Atak 2005; Basu et al. 2010). Phytotoxicity in plant cells increases as a result of accumulation of reactive oxygen species (ROS) induced by drought stress (Reddy, Chaitanya, and Vivekanandan 2004; Reddy et al. 2004). Enhanced level of ROS in plant cells under drought conditions adversely influences the photosynthetic responses by upsetting the photosynthetic apparatus. Moreover, enhanced ROS accumulation leads to chlorophyll degradation and ultimately impairs performance of photosynthesis in drought-stressed plants (Aryadeep Roychoudhury and Kaushik Das 2014; Shao et al. 2016). Melatonin is one of the plant growth regulators which plays a vital role in imparting resistance to plants through the regulation of the photosynthetic apparatus and the antioxidant defence system when there is a water shortage (Liang et al. 2019). However, the effects of drought stress have received less attention, and further research is needed to understand the precise processes that

DOI: 10.1201/9781003343752-15

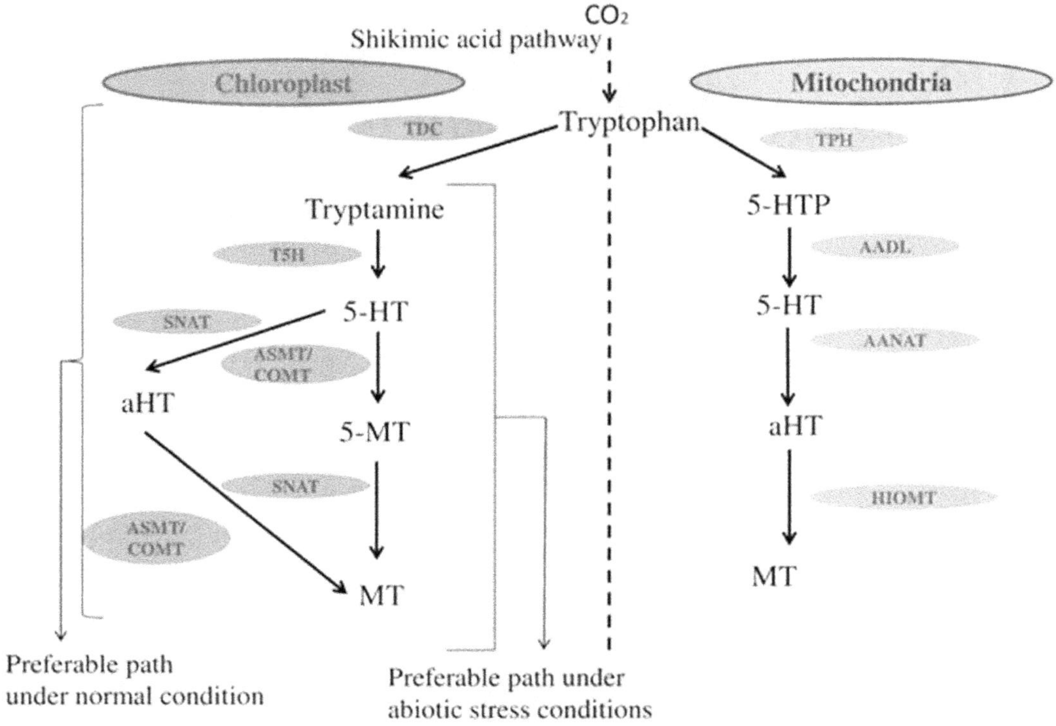

FIGURE 15.1 A schematic representation of steps involved in melatonin biosynthesis in plants.

control melatonin-mediated drought tolerance. This chapter is summarized with recent research efforts that have focused mainly on mechanisms of MT actions at physiological and molecular levels to alleviate drought stress in plants.

15.2 MELATONIN-MEDIATED DROUGHT STRESS REGULATION IN PLANTS

Melatonin is regarded as a growth regulator due to its important role in controlling plant growth and development. It has been thought to have comparable effects to auxins and controls plant developmental processes starting with seed germination (Hernández-Ruiz, Cano, and Arnao 2005). It is an extraordinarily strong, naturally occurring antioxidant that effectively mitigates the negative effects of water scarcity on plants (Altaf et al. 2021, 2022b; Tiwari et al. 2021b, 2022). Melatonin is therefore strongly recommended to reduce the effects of drought stress in a variety of plant species, including model plants (Lee and Back 2019), field crops (Cao, Jin, and Zhang 2019), fruit crops (Xia et al. 2020), and vegetable crops (Kaya and Doganlar 2019) as well as ornamental and medicinal plants (Li et al. 2019). The scientific community has been interested in examining the direct evidence of the link of melatonin with drought resilience because it is commonly used to treat drought stress in plants (Park et al. 2009). Through the melatonin biosynthesis genes N-acetylserotonin methyltransferase (ASMT), tryptophan decarboxylase (TDC), and serotonin N-acetyltransferase, drought stress is reduced (SNAT) (Di Fiore et al. 2002). Melatonin levels change as a result of changes in the environmental factors that affect plant development (Banerjee and Roychoudhury 2017).

15.3 EXOGENOUS MELATONIN APPLICATIONS ENHANCE DROUGHT TOLERANCE IN CROP PLANTS

Along with the endogenous production of melatonin, applications of exogenous melatonin through different pre-treatment techniques have been potentially used in several researches to in drought stress mitigation of plants. In Arabidopsis, melatonin treatment was done along with nutrient solution containing 50 μM melatonin during watering to the 13-day-old soil-grown plants, and sampling was done 0, 1, 3, 6, 12, and 24 h after treatment (Shi et al. 2015). In an experiment on maize, soil amendments/biochar and P were properly mixed with soil, and melatonin (100 μM) solution was sprayed as foliar application; this proved to be effective in drought stress tolerance (Alharby and Fahad 2020; Alharby et al. 2020). Seven-day-old rice seedlings were carefully taken from the media and placed into separate 50 mL Falcon tubes filled with distilled water for a combined cold and drought study. After stabilizing for 24 h, seedlings were treated for an additional day with either MT or 2-hydroxymelatonin solutions (each at a concentration of 100 M) and stored in distilled water. The seedlings were placed in a refrigerator set at 4°C with a 12-h photoperiod (0.1 mol photons $m^{-2}s^{-1}$ during the light period) in order to apply a combination of stressors to rice. After two days of incubation, water was drained from the plant-containing vessels in order to increase drought stress while lowering the temperature (cold/drought combination). The plants were then incubated further for five days. Under the combined stress of cold and drought, this form of melatonin delivery was demonstrated to be beneficial (Lee and Back 2016). In a study on the ability of maize to withstand drought, 1 mm of melatonin was added to irrigation water, or Hoagland nutrient solution, but not to the other (control plants). Samples were collected from plants treated with melatonin and plants that were not treated (control) after eight days of water-restriction-caused stress. The findings of this study demonstrated that melatonin in irrigation water increased PS-II efficiency even in the presence of stress, which, in turn, caused changes in the concentration of phytohormones, antioxidants, and pigments that are associated to stress (Altaf et al. 2022b). In addition, it was found that maize plants exposed to high melatonin concentrations in irrigation water displayed extremely high endogenous melatonin content in leaves 24 h after the application, but endogenous contents rapidly decreased, indicating a very quick consumption of this compound during recovery and the ensuing drought stress (Fleta-Soriano et al. 2017). An experiment with six different treatments was carried out to ascertain the impact of melatonin on germination and antioxidant metabolism in wheat under drought stress. (1) Distilled water (control), (2) PEG-6000 at 20%, and (3) melatonin at 1 M plus 20% PEG-6000. Seed germination was monitored daily for up to seven days using the following concentrations: (4) 10 M melatonin + 20% PEG; (5) 100 M melatonin + 20% PEG; and (6) 300 M melatonin + 20% PEG. In the growth chamber, a different experiment was carried out using liquid MS media, three treatments—control, 20% PEG, and 20% PEG combined with a 300 M melatonin solution—were applied at the three-leaf stage. By increasing the radicle length, radicle number, and plumule length of the germinated seeds, the results showed that 300 M melatonin had a favourable influence on drought stress tolerance (Li et al. 2020a). From the several reports, it can be summarized that the exogenous application of melatonin in plants under abiotic stress condition is mainly done through foliar application, along with irrigation water, mixing with nutrient solution or hydroponics medium; and in some cases, seed priming is effective for better growth and development of plants, but it depends on the crop and growth stage of the crop.

15.4 ROLE OF MELATONIN IN AMELIORATING DROUGHT STRESS

Melatonin is considered as a multifunctional phyto-protectant molecule as it ameliorates effects of several abiotic and biotic stresses. This vital molecule also has potential to improve drought stress tolerance (Liu et al. 2020). The major protective mechanism associated with melatonin for drought stress is due to its highly antioxidative properties. In some systems, melatonin has a greater antioxidant capacity than vitamins E, C, and K (Tan et al. 2007). Melatonin controls several processes in plants under stress to lessen the effects of drought stress, including rhizogenesis, stomatal conductance, carbon assimilation, RuBisCO accumulation, photochemical efficiency of the photosystems, catabolism of chlorophyll molecules, and the ascorbate-glutathione cycle (Liang et al. 2019; Moustafa-Farag et al. 2020). Moreover, as a growth regulator, melatonin performs auxin-like activity that regulates growth of both root and shoot but acting independently of auxin signalling (Pelagio-Flores et al. 2012). It has encouraging influence on fruit development and delaying senescence (Gao et al. 2016). In plants, the amount of endogenous melatonin production is very low, viz. nmol/g DW range in leaves of maize, as compared to other stress-related phytohormones (Ye et al. 2016). Details of the methods of drought treatments and methods of melatonin applications and its drought-ameliorating effects are described in Table 15.1.

TABLE 15.1
Details on Crops, Methods of Drought Imposed, and Melatonin Application and Its Drought Ameliorating Effects

Group	Crop	Drought-imposed methods	Melatonin application methods	Concentration	Effects	References
Cereal crops	Rice	Water drainage from vessels	Pre-treatment with distilled H_2O	100 µM	Plant growth (↑), proline (↑), genes related to stress tolerance (↑), stability in ATP-producing machinery (↑), free radicals (↓), electrolyte leakage (↓)	(Lee and Back 2016)
	Rice	25% and 35% of PEG 6000 treatments	Melatonin seed priming	0–500 µM	100 µM favours seed emergence (↑), seedling development (↑), agronomic traits (↑), antioxidant enzyme activity (↑), soluble protein (↑) and malondialdehyde (↓)	(Li et al. 2022)
	Rice	Prevented irrigation for seven days or leaf roll	foliar spray	50–300 µM	100 µM melatonin was the most effective in drought tolerance in the following parameters: growth of plants (↑), RWC (↑), and chlorophyll pigments (↑), antioxidant enzyme activities such as SOD, CAT, APX, and GPX (↑), proline (↑), and total soluble sugar accumulation (↑), electrolyte leakage (↓), MDA (↓)	(Silalert and Pattanagul 2021)

TABLE 15.1 ***(Continued)***
Details on Crops, Methods of Drought Imposed, and Melatonin Application and Its Drought Ameliorating Effects

Group	Crop	Drought-imposed methods	Melatonin application methods	Concentration	Effects	References
	Rice	Plants grown at 100% field capacity (FC; as control), 80, 60, 40, and 25% FC (drought stress) for seven days	Seedlings of more than 25% FC were sprayed in each two-day interval	0, 50, 100, 200, and 1000 μmol/L	Survival percentage (↑), antioxidant activities (↑), and osmotic balance (↑)	(Li et al. 2022)
	Wheat	20% poly ethylene glycol 6000 (one week)	Seed treatment	10 and 300 M	Germination-related parameters such as its rate, index and potential (↑), radicle and plumule length (↑), germination-related amino acid (i.e. lysine) (↑)	(Li et al. 2020b)
	Tartary Buckwheat	20% field capacity (15 days)	Foliar application	100 M	Water balance and osmoprotection (↑), secondary metabolites (↑), antioxidative protection (↑), photosynthetic rate (↑), ROS (↓)	(Hossain et al. 2020)
	Maize	Water stress applied for eight days,	Supplemented with irrigation before water withholding	1 mM	Photoprotection (PSII effi ciency) (↑)	(Fleta-Soriano et al. 2017)
	Maize	30–60% SWC (eight days)	Foliar application	100 M	Recovering after rehydration (↑), photosynthesis (↑), conductance of stomata (↑), transpiration rates (↑), cell turgidity (↑), enzymatic and non-enzymatic antioxidants (↑), osmotic potential (↓), reactive oxygen species (↓)	(Ye et al. 2016)
	Maize	20% PEG6000 (three days)	Pre-treatment as foliar spray	10–100 M	Photosynthetic rate (↑), antioxidative enzyme content (↑), fixing of carbons (↑), amino acidS and biosynthesis of secondary metabolites (↑), reactive oxygen species (↓)	(Su et al. 2019)

(Continued)

TABLE 15.1 *(Continued)*
Details on Crops, Methods of Drought Imposed, and Melatonin Application and Its Drought Ameliorating Effects

Group	Crop	Drought-imposed methods	Melatonin application methods	Concentration	Effects	References
	Maize	Water withholding (seven days)	Both in foliar application and root irrigation	100 M	Photosynthetic rate (↑), reactive oxygen species (↓)	(Huang et al. 2019)
	Maize	40–45% field capacity (50 days)	Foliar application or soil treatment	Foliar spray (50 M) and soil drenching (100 M)	Photosynthesis (↑), antioxidant enzymes (↑), ROS (↓)	(Ahmad et al. 2019)
Pulses	Soybean	20% field capacity (ten days)	Seed coating	50 M	Seedling growth (↑), biomass (↑), electrolyte leakage (↓)	(Wei et al. 2015)
	Soybean	15% PEG 6000 (seven days)	Supplemented with nutrient solution	100 M	Seedling growth and photosynthetic rate (↑)	(Zou et al. 2019)
	Soy-bean	45% RSWC (15 days)	Foliar application	100 M	Antioxidant enzymes (↑), osmolytes (↑), malondialdehyde (↓)	(Cao, Jin, and Zhang 2019)
	Soy-bean	15% PEG6000 (three days)	Both spray in leaves and root incorporation	100 M	Plant growth and flower formation (↑), yield potential (↑), gas exchange potential (↑), PSII e ffciency (↑), antioxidant enzyme activities (↑), malondialdehyde (↓)	(Zhang et al. 2019)
Oilseeds	Safflower	Stress applied at flowering stage	Melatonin seed priming for 6 h	0.1 and 0.5 mM	Scavenging ROS (↑), chl/carotenoid ratio (↑), antioxidant enzymes activity (↑), degradation of proteins (↓), proline content (↑)	(Heshmati et al. 2021)
	Rape seed	PEG-6000	Melatonin-priming	500 μM	Stomatal number (↑), stomatal length and width (↑), chloroplast structure preserved, cell expansion maintained and strengthened the cell wall, antioxidant enzymes activities (↑), non-enzymatic antioxidants (↑), and osmoprotectant amounts (↑)	(Khan et al. 2019)

TABLE 15.1 *(Continued)*
Details on Crops, Methods of Drought Imposed, and Melatonin Application and Its Drought Ameliorating Effects

Group	Crop	Drought-imposed methods	Melatonin application methods	Concentration	Effects	References
	Rape seed	After melatonin pre-treatment, the seedlings were exposed to drought	Seedlings were irrigated with melatonin solution	100 μM	Root catalase (CAT) activity (↑), taproot and lateral root growth (↑), stomatal opening (↑) resulting in photosynthesis (↑), no effects on the root POD and SOD activities	(Dai et al. 2020)
	Rape seed	4% PEG applied on Hoagland's culture medium	Melatonin added to culture medium	0–100 μmol/L	Leaf area (↑), fresh weight of shoot (↑), fresh weight of root (↑), shoot dry weight (↑), and root dry weight (↑) at 50 μmol/L, peroxidase (↑), catalase (↑), proline (↑)	(Li et al. 2018)

15.5 PHYSIOLOGICAL MECHANISMS OF MELATONIN ASSISTING DROUGHT TOLERANCE

Under the deficiency of soil moisture, plants experience drought stress. Drought stress causes hyperaccumulation of oxidant products, that is, reactive oxygen species (ROS), hydrogen peroxide (H_2O_2), malondialdehyde (MDA), and superoxide ions (O_2^-), which causes oxidative damages of cell contents (Kapoor et al. 2020) and ultimately restricts the overall growth and productivity of plants by alternating their physio-biochemical and molecular machinery (Cherono et al. 2021). As sessile organisms, evolutionary plants have several defence mechanisms to manage the unfavourable environmental conditions like drought stress, to ensure their survival and reproduction successfully. Melatonin is a multifunctional biomolecule and considered as a highly efficient abiotic stress regulator in plants (Wang, Reiter, and Chan 2018). Over the recent years, the job of melatonin as a multifunctional regulator of plant stress under drought conditions, including physical and physiological perspectives, has been continuously studied. Typically, drought stress encourages melatonin biosynthesis (Shi et al. 2015). Further, it has been documented that endogenous melatonin increases during drought stress as melatonin biosynthesis genes (e.g., *TDC*, *ASMT*, *COMT*, and *SNAT*) were up regulated. Drought-induced melatonin biosynthesis has been reported in Arabidopsis, rice, barley, apple, and grapevine (Moustafa-Farag et al. 2020). Application of exogenous melatonin causes an increase in endogenous melatonin level that favours to improve turgidity, stability in cell, chlorophyll recovery, and stability of the photosynthetic system under drought stress (Ding et al. 2018). In an investigation, it was discovered that exogenous application of melatonin can lessen the damage under drought stress, with the normal function of leaf cell membrane, improvement in net photosynthetic rate and water use efficiency, by promoting the formation of osmotic adjustment substances and adaptability of *Dendranthema morifolium* seedlings under dry spell (Yan et al. 2016). In soybean, melatonin-mediated drought tolerance assisted with glucose metabolism. Melatonin treatment also encourages seed germination, hypocotyl elongation, lateral root formation, vegetative growth, and improvement in the quality of fruit under drought (Sun et al. 2021). Physiological parameters, viz., increase in membrane stability, grana compactness in chloroplast, increased photosynthesis rate, higher efficiency of PS-II, cell turgidity, and water-holding capacity, were observed in melatonin-treated wheat seedlings, favouring drought tolerance (Cui et al. 2017). Pre-treatment of melatonin reduced drought stress impacts by increasing photosynthetic efficiency, altering internal solute concentrations (e.g. proline and sugars) and promoting vegetative growth in tomato (Ibrahim et al. 2020). Melatonin seed priming improved stomatal conductance and osmotic balance, promoting drought tolerance in rapeseed (Khan et al. 2019).

The oxidative damage due to drought stress can be mitigated through melatonin treatment, as seen in many crops (Antoniou et al. 2017; Cui et al. 2017). Melatonin potentially activates many antioxidative enzymes such as SOD, CAT, peroxidase, DHAR, and GSH reductase, and these enzymes efficiently suppress ROS accumulation under drought conditions (Alharby and Fahad 2020; Samanta et al. 2021). Many drought stress–tolerant genes (*APX*, *MDHAR*, and *DHAR*) were up regulated, activating the integral antioxidative defence system of the AsA-GSH cycle after melatonin treatment, ultimately enhancing tolerance under drought stress (Cui et al. 2017; Tiwari et al. 2021b). Melatonin treatment also improved drought tolerance by MAPKs and transcription factors (Tiwari et al. 2021a) and up regulated expression of MAPKs (e.g., Asmapl and Aspk11) and transcription factor genes (e.g., WRKY1, DREB2, and MYB) (Gao et al. 2018).

Melatonin is structurally much similar to indole-3-acetic acid (IAA)—a natural auxin. Both have similar planar aromatic ring, a carboxylic acid-binding site, and a hydrophobic transition region. Evidently, some of the physiological functions related to growth and development and antioxidative properties are also similar among melatonin and IAA (Arnao and Hernández-Ruiz 2006). Moreover, melatonin triggers lateral and adventitious root formation, seed germination, hypocotyl elongation, vegetative growth, and fruit quality improvement during drought (Hosseini et al. 2021; Sun et al. 2021). It encourage photosynthesis, CO_2 uptake, and biomass accumulation (Park et al. 2012; Reiter et al. 2015), rhizogenesis and caulogenesis in explant cultures (Murch and Saxena 2002; Jones et al. 2007), and highly associated with flower and fruit development and seed formation (Okazaki and Ezura 2009; Murch et al. 2010). Under drought stress, crosstalk between melatonin and other phytohormones also plays a regulatory mechanism for drought stress tolerance in plants. Exogenous application of melatonin improves flowering events during drought by inducing GA biosynthesis in plants (Tiwari et al. 2020). The interaction of melatonin and ABA also manages stomatal closure (Chen et al. 2020; Qi et al. 2021). Melatonin negatively regulates ABA biosynthesis genes and positively regulates ABA catabolism genes (MdCYP707A1 and MdCYP707A2), which regulate stomatal function during drought stress (Li et al. 2015).

15.6 MOLECULAR MECHANISMS OF MELATONIN ASSISTING DROUGHT TOLERANCE

The defence mechanisms of melatonin have been investigated, and its ability to protect plant cells from redox homeostasis disruption under drought stress has been addressed. Reactive oxygen species (ROS) levels, antioxidant-related genes including SOD, POD, CAT, DHAR, and glutathione reductase (GR) activities are all regulated by melatonin in plant cells (Wei et al. 2015). Pathway and gene ontology (GO) term enrichment analyses in Bermuda grass revealed that genes involved in nitrogen metabolism, major carbohydrate metabolism, tricarboxylic acid (TCA)/org transformation, transport, hormone metabolism, metal handling, redox, and secondary metabolism were overexpressed after melatonin pre-treatment (Shi et al. 2015; Sharma et al. 2020), along with energy production (Cui et al. 2018), carotenoid metabolism and photosynthesis (Sharma et al. 2020), and cuticle wax biosynthesis (Ding et al. 2018). Melatonin is likewise a vital regulator of nitrogen metabolism, as demonstrated by the higher level of expression of N-uptake genes. The characteristics of plant growth, chlorophyll pigments, fluorescence, and gas exchange parameters were all significantly enhanced by the administration of melatonin along with nitrogen (Kobylińska et al. 2018). Additionally, under drought-stress circumstances, melatonin and nitrogen treatment drastically decreased the buildup of reactive oxygen species (ROS) via raising antioxidant and nitrogen metabolism enzyme activities (Zhao et al. 2021).

Plants are protected from oxidative stress by ROS and reactive nitrogen species (RNS), which increases their resistance to a number of environmental abiotic stresses. Melatonin synthesis depends on a specific enzyme called N-acetylserotonin-O-methyltransferase (ASMT) and designated as MzASMT1. The MzASMT1, an ASMT gene cloned from apple rootstock, expression was induced by drought stress in apple leaves. The upregulation of MzASMT1 in the apple leaf positively relates to melatonin production over a 24-h dark/light cycle (Zuo et al. 2014). Caffeic acid 3-O-methyltransferase (COMT) functions at a vital step in the biosynthesis of the phytohormone melatonin. Overexpression of wheat *TaCOMT* (a gene similar to the *COMT* gene) was found to enhance drought tolerance in common wheat (*Triticum aestivum*) and also increased melatonin content in transgenic *Arabidopsis thaliana* (Yang et al. 2019). Overexpression of TaCOMT led to the upregulation of various drought responsive genes, such as RAB18, RD29A, KIN1, and DREB2A by melatonin (Yang et al. 2019). The *oAANAT* transgenic plants were reported with a higher level of melatonin and lower indoleacetic acid (IAA) contents because of the competition for tryptophan, which is a common precursor for both melatonin and IAA. As a result, the oAANAT lines lacked apical dominance, proving that melatonin is likely devoid of auxin action. The critical involvement of ASMT in plant melatonin synthesis is further demonstrated by the considerably higher melatonin content in oHIOMT lines than oAANAT lines (Wang et al. 2014). Serotonin N-acetyltransferase (SNAT) and N-acetylserotonin methyltransferase (ASMT) act as two key enzymes for melatonin biosynthesis. SNAT and ASMT are practically involved in melatonin production, as evident by the reduced melatonin synthesis in transgenic rice plants in which expression of either endogenous SNAT or ASMT was repressed (Byeon and Back 2016). The mechanistic study indicated that the ectopically expressing MzSNAT5 allowed plant mitochondria to increase melatonin synthesis (Wang et al. 2017).

Reduced expression of several stress-related genes (CDPK1, MAPK1, TSPMS, ERF4, HSP80, and ERD15), as well as decreased activity of antioxidative enzymes (SOD, CAT, and APX), which were in charge of the stress tolerance, were the results of RBOH inhibition, which significantly hindered the melatonin-induced defence response (Gong et al. 2017). A study revealed that chloroplastic serotonin N-acetyltransferase (SNAT) expression prevented enhanced production of melatonin following stress, ultimately leading to a steady-state melatonin level inside plant cells (Lee, Choi, and Back 2017). Autophagy plays a role in reducing drought stress, as evident by the greater expression of the autophagosome marker (lipidated [ATG8-PE]) in melatonin-primed drought-stressed cotton plants (Supriya et al. 2022). The overexpression of the caffeic acid O-methyltransferase gene (CrCOMT) from *Carex rigescens*, which triggers the transcription of the drought stress response, confers melatonin-mediated drought resistance in transgenic tobacco (Li et al. 2022).

Increased transcript levels of CER1 (very-long-chain aldehyde decarbonylase 1) were seen together with increased cuticular waxes on leaves, indicating the presence of additional intracuticular waxes that serve as fillers in the cutin matrix (Li et al. 2020a). Cuticle waxes are essential substances for plants when they are under water stress, because they help to control water loss via the leaf surface (Ding et al. 2018). Melatonin stimulates the biosynthesis of cuticular waxes, which results in their deposition on the leaf surface, minimising water loss. Increased transcript levels of genes including KCS1 (ketoacyl-CoA synthase 1), CER3 (ECERIFERUM3), TTS1 (triterpenoid synthase 1), and LTP1 (lipid transfer protein) are responsible for the improved production of cuticular waxes (Ding et al. 2018).

Melatonin promotes the critical physiological functions, including photosynthesis, transpiration, and stomatal conductance, by preventing the breakdown of chlorophyll during dry spells (Karaca and Cekic 2019). Chlorophyll-degrading peroxidase (Chl-PRX), pheophytinase, and other genes are downregulated as a result of exogenous melatonin administration, which reduces the degradation of chlorophyll molecules (Ma et al. 2018). Melatonin has also been linked to reports of accessory photosynthetic pigment regeneration during drought stress, including carotenoids (Liang et al. 2019). The metabolism of chlorophyll also involves the enzyme pheophorbide-a-oxygenase (PAO). Melatonin inhibits the transcription of PAO, which reduces the breakdown of chlorophyll during drought stress (Wang et al. 2013).

Mitogen-activated protein kinase (MAPK) cascade pathways also play a vital role in the regulation of plant biology under drought stress (Zhang et al. 2011). The NAC, WRKY, MYB, and DREB are a few transcription factors that are important parts of the MAPK cascade pathway in stressed plants (Wang et al. 2018). For abiotic stress tolerance, these variables are involved in the regulation of many stress-sensitive genes. Melatonin, a phytohormone, controls the MAPK pathway in response to drought stress by increasing the expression of MAPKs such as Asmapl and Aspk11, which is accompanied by an increase in transcription factors including WRKY1, DREB2, and MYB (Gao et al. 2018). H_2O_2 signalling is thought to control this melatonin-mediated regulation of the MAPK cascade, leading to the enhanced plant resistance against drought stress (Gao et al. 2018).

Nitrogen metabolism in plant cells is negatively impacted by abiotic stress, particularly drought stress. However, melatonin modifies physiological and molecular components of plant biology to control nitrogen metabolism under drought stress (Liang et al. 2018). Under conditions of water scarcity, melatonin also increases the pattern of gene expression in plants for the genes NR, NiR, GS, and GOGAT (Liang et al. 2018). Melatonin also increases the transcript levels of genes involved in nitrogen uptake, including AMT (ammonium transporter) and NRT (nitrate transporter), in plants under drought stress (Liang et al. 2018). Melatonin promotes cytokinin (CK) production by increasing the transcript levels of important CK-signalling genes, including Type-A RRs, Type-B RRs (response regulators), HKs (histidine kinases), and HPs (histidine phosphatases) (Ma et al. 2018).

15.7 CONCLUSION

By promoting ROS scavenging, melatonin imparts resistance to plants under drought stress. Melatonin shields plant cells from oxidative damage and speeds up chloroplast structural recovery, improving the effectiveness of photosynthetic machinery. Melatonin-mediated drought alleviation is accomplished by enhanced cell signalling, which in turn controls a number of physiological processes at the molecular level. Melatonin levels in the field can be useful from an agronomic perspective because drought stress affects crop yield and quality. A few recent research have supported the development of genetically engineered plants with improved melatonin production and increased drought tolerance due to genetic engineering. Additionally, the identification of key genes involved in the control of melatonin-mediated drought alleviation and the investigation of their expression in response to exogenous melatonin administration will open up new possibilities for the development of drought-tolerant crops.

15.8 FUTURE PERSPECTIVE

Future scientific research on the phytohormone melatonin holds promise because it may be accounted for as a significant component of the regulatory network governing the mechanisms behind drought tolerance in plants. Melatonin research has accelerated recently, but more research is needed to understand how it affects the ability to withstand drought stress. Understanding how melatonin affects nutrient uptake will give us the chance to leverage the benefits of this molecule for boosting plant tolerance to water stress. As melatonin contributes to the formation of secondary metabolites such as phenylpropanoids, flavonoids, and steroids, it is important to establish a link between these substances and the propensity of plants to withstand drought using cutting-edge research techniques. Understanding the factors influencing how plants respond to exogenous melatonin applications from a physical, physiological, and biochemical standpoint will be made easier with the aid of microarray studies, transcriptomic, and proteomic analyses. It would also be a commendable strategy to investigate new receptor-mediated melatonin signalling in order to comprehend the physiological basis of melatonin-mediated drought resistance. The understanding of the mechanisms underlying NO and H_2S signalling and how they interact with melatonin has made significant strides in recent decades. Due to the complexity of plants and various environmental conditions, future study should take into account the link between melatonin and various stressors.

REFERENCES

Ahmad, Shakeel, Muhammad Kamran, Ruixia Ding, Xiangping Meng, Haiqi Wang, Irshad Ahmad, Shah Fahad, and Qingfang Han. 2019. "Exogenous Melatonin Confers Drought Stress by Promoting Plant Growth, Photosynthetic Capacity and Antioxidant Defense System of Maize Seedlings." *PeerJ* 2019 (10). doi:10.7717/peerj.7793.

Alharby, Hesham F., and Shah Fahad. 2020. "Melatonin Application Enhances Biochar Efficiency for Drought Tolerance in Maize Varieties: Modifications in Physio-Biochemical Machinery." *Agronomy Journal* 112 (4). doi:10.1002/agj2.20263.

Altaf, Muhammad Ahsan, Rabia Shahid, Muhammad Mohsin Altaf, Ravinder Kumar, Safina Naz, Awadhesh Kumar, Pravej Alam, Rahul Kumar Tiwari, Milan Kumar Lal, and Parvaiz Ahmad. 2022a. "Melatonin: First-Line Soldier in Tomato under Abiotic Stress Current and Future Perspective." *Plant Physiology and Biochemistry* 185 (August). Elsevier Masson: 188–197. doi:10.1016/J.PLAPHY.2022.06.004.

Altaf, Muhammad Ahsan, Rabia Shahid, Ming Xun Ren, Latif Ullah Khan, Muhammad Mohsin Altaf, Mohammad Shah Jahan, Muhammad Azher Nawaz, et al. 2021. "Protective Mechanisms of Melatonin Against Vanadium Phytotoxicity in Tomato Seedlings: Insights into Nutritional Status, Photosynthesis, Root Architecture System, and Antioxidant Machinery." *Journal of Plant Growth Regulation*, (October). Springer, 1–17. doi:10.1007/s00344-021-10513-0.

Altaf, Muhammad Ahsan, Rabia Shahid, Ming Xun Ren, Safina Naz, Muhammad Mohsin Altaf, Latif Ullah Khan, Milan Kumar Lal, Rahul Kumar Tiwari, and Awais Shakoor. 2022b. "Melatonin Mitigates Cadmium Toxicity by Promoting Root Architecture and Mineral Homeostasis of Tomato Genotypes." *Journal of Soil Science and Plant Nutrition* 22 (1). Springer: 1112–1128. doi:10.1007/s42729-021-00720-9.

Antoniou, Chrystalla, Giannis Chatzimichail, Rafaella Xenofontos, Jan J. Pavlou, Evangelia Panagiotou, Anastasis Christou, and Vasileios Fotopoulos. 2017. "Melatonin Systemically Ameliorates Drought Stress-Induced Damage in Medicago Sativa Plants by Modulating Nitro-Oxidative Homeostasis and Proline Metabolism." *Journal of Pineal Research* 62 (4). doi:10.1111/jpi.12401.

Arnao, Marino B., and Josefa Hernández-Ruiz. 2006. "The Physiological Function of Melatonin in Plants." *Plant Signaling and Behavior.* doi:10.4161/psb.1.3.2640.

Banerjee, A., and A. Roychoudhury. 2017. "Melatonin as a Regulator of Abiotic Stress Tolerance in Plants." In: Singh VP, Singh S, Prasad SM (Eds.) *Mechanisms behind Phytohormonal Signaling and Crop abiotic Stress Tolerance*, Nova Science Publishers, New York, pp. 47–60.

Basu S., A. Roychoudhury, P. P. Saha, and D. N. Sengupta. 2010. "Differential Antioxidative Responses of Indica Rice Cultivars to Drought Stress." *Plant Growth Regulation* 60: 51–59.

Byeon, Yeong, and Kyoungwhan Back. 2016. "Low Melatonin Production by Suppression of Either Serotonin N-Acetyltransferase or N-Acetylserotonin Methyltransferase in Rice Causes Seedling Growth Retardation with Yield Penalty, Abiotic Stress Susceptibility, and Enhanced Coleoptile Growth under Anoxic Conditions." *Journal of Pineal Research* 60 (3). doi:10.1111/jpi.12317.

Cao, L., X. J. Jin, and Y. X. Zhang. 2019. "Melatonin Confers Drought Stress Tolerance in Soybean (Glycine Max l.) by Modulating Photosynthesis, Osmolytes, and Reactive Oxygen Metabolism." *Photosynthetica* 57 (3). doi:10.32615/ps.2019.100.

Chen, Kong, Guo Jun Li, Ray A. Bressan, Chun Peng Song, Jian Kang Zhu, and Yang Zhao. 2020. "Abscisic Acid Dynamics, Signaling, and Functions in Plants." *Journal of Integrative Plant Biology.* doi:10.1111/jipb.12899.

Cherono, Sylvia, Charmaine Ntini, Misganaw Wassie, Mohammad Dulal Mollah, Mohammad A. Belal, Collins Ogutu, and Yuepeng Han. 2021. "Exogenous Application of Melatonin Improves Drought Tolerance in Coffee by Regulating Photosynthetic Efficiency and Oxidative Damage." *Journal of the American Society for Horticultural Science* 146 (1). doi:10.21273/JASHS04964-20.

Cui, Guibin, Fengli Sun, Xinmei Gao, Kunliang Xie, Chao Zhang, Shudong Liu, and Yajun Xi. 2018. "Proteomic Analysis of Melatonin-Mediated Osmotic Tolerance by Improving Energy Metabolism and Autophagy in Wheat (Triticum Aestivum L.)." *Planta* 248 (1). doi:10.1007/s00425-018-2881-2.

Cui, Guibin, Xiaoxiao Zhao, Shudong Liu, Fengli Sun, Chao Zhang, and Yajun Xi. 2017. "Beneficial Effects of Melatonin in Overcoming Drought Stress in Wheat Seedlings." *Plant Physiology and Biochemistry* 118. doi:10.1016/j.plaphy.2017.06.014.

Dai, Lulu, Jun Li, Harry Harmens, Xiaodong Zheng, and Chunlei Zhang. 2020. "Melatonin Enhances Drought Resistance by Regulating Leaf Stomatal Behaviour, Root Growth and Catalase Activity in Two Contrasting Rapeseed (Brassica Napus L.) Genotypes." *Plant Physiology and Biochemistry* 149. doi:10.1016/j.plaphy.2020.01.039.

Di Fiore, Stefano, Qiurong Li, Mark James Leech, Flora Schuster, Neil Emans, Rainer Fischer, and Stefan Schillberg. 2002. "Targeting Tryptophan Decarboxylase to Selected Subcellular Compartments of Tobacco Plants Affects Enzyme Stability and in Vivo Function and Leads to a Lesion-Mimic Phenotype." *Plant Physiology* 129 (3). doi:10.1104/pp.010889.

Ding, Fei, Gang Wang, Meiling Wang, and Shuoxin Zhang. 2018. "Exogenous Melatonin Improves Tolerance to Water Deficit by Promoting Cuticle Formation in Tomato Plants." *Molecules* 23 (7). doi:10.3390/molecules23071605.

Dubbels, R., R. J. Reiter, E. Klenke, A. Goebel, E. Schnakenberg, C. Ehlers, H. W. Schiwara, and W. Schloot. 1995. "Melatonin in Edible Plants Identified by Radioimmunoassay and by High Performance Liquid Chromatography-Mass Spectrometry." *Journal of Pineal Research* 18 (1). doi:10.1111/j.1600-079X.1995.tb00136.x.

Fahad, Shah, Ali A. Bajwa, Usman Nazir, Shakeel A. Anjum, Ayesha Farooq, Ali Zohaib, Sehrish Sadia, et al. 2017. "Crop Production Under Drought and Heat Stress: Plant Responses and Management Options." *Frontiers in Plant Science* 8 (June): 1–16. doi:10.3389/fpls.2017.01147.

Fleta-Soriano, E., L. Díaz, E. Bonet, and S. Munné-Bosch. 2017. "Melatonin May Exert a Protective Role against Drought Stress in Maize." *Journal of Agronomy and Crop Science* 203 (4). doi:10.1111/jac.12201.

Gao, Hui, Zheng Ke Zhang, Hong Kang Chai, Ni Cheng, Yue Yang, Dan Na Wang, Ting Yang, and Wei Cao. 2016. "Melatonin Treatment Delays Postharvest Senescence and Regulates Reactive Oxygen Species Metabolism in Peach Fruit." *Postharvest Biology and Technology* 118. Elsevier B.V.: 103–110. doi:10.1016/j.postharvbio.2016.03.006.

Gao, Wenying, Yujing Zhang, Zheng Feng, Qingqing Bai, Jinjin He, and Yingjuan Wang. 2018. "Effects of Melatonin on Antioxidant Capacity in Naked Oat Seedlings under Drought Stress." *Molecules* 23 (7). doi:10.3390/molecules23071580.

Gong, Biao, Yanyan Yan, Dan Wen, and Qinghua Shi. 2017. "Hydrogen Peroxide Produced by NADPH Oxidase: A Novel Downstream Signaling Pathway in Melatonin-Induced Stress Tolerance in Solanum Lycopersicum." *Physiologia Plantarum* 160 (4). doi:10.1111/ppl.12581.

Hardeland, Rüdiger, Juan Antonio Madrid, Dun Xian Tan, and Russel J. Reiter. 2012. "Melatonin, the Circadian Multioscillator System and Health: The Need for Detailed Analyses of Peripheral Melatonin Signaling." *Journal of Pineal Research*. doi:10.1111/j.1600-079X.2011.00934.x.

Hernández-Ruiz, Josefa, Antonio Cano, and Marino Bañón Arnao. 2005. "Melatonin Acts as a Growth-Stimulating Compound in Some Monocot Species." *Journal of Pineal Research* 39 (2). doi:10.1111/j.1600-079X.2005.00226.x.

Heshmati, Siavash, Majid Amini Dehaghi, Muhammad Farooq, Łukasz Wojtyla, Keyvan Maleki, and Sima Heshmati. 2021. "Role of Melatonin Seed Priming on Antioxidant Enzymes and Biochemical Responses of Carthamus Tinctorius L. under Drought Stress Conditions." *Plant Stress* 2. doi:10.1016/j.stress.2021.100023.

Hossain, Md Shakhawat, Jing Li, Ashim Sikdar, Mirza Hasanuzzaman, Ferdinand Uzizerimana, Izhar Muhammad, Yuhao Yuan, Chengjin Zhang, Chenyang Wang, and Baili Feng. 2020. "Exogenous Melatonin Modulates the Physiological and Biochemical Mechanisms of Drought Tolerance in Tartary Buckwheat (Fagopyrum Tataricum (L.) Gaertn)." *Molecules* 25 (12). doi:10.3390/molecules25122828.

Hosseini, Marjan Sadat, Davood Samsampour, Seyed Morteza Zahedi, Kazem Zamanian, Md Mezanur Rahman, Mohammad Golam Mostofa, and Lam Son Phan Tran. 2021. "Melatonin Alleviates Drought

Impact on Growth and Essential Oil Yield of Lemon Verbena by Enhancing Antioxidant Responses, Mineral Balance, and Abscisic Acid Content." *Physiologia Plantarum* 172 (2). doi:10.1111/ppl.13335.

Huang, Bo, Yang Er Chen, Yu Qing Zhao, Chun Bang Ding, Jin Qiu Liao, Chao Hu, Li Jun Zhou, Zhong Wei Zhang, Shu Yuan, and Ming Yuan. 2019. "Exogenous Melatonin Alleviates Oxidative Damages and Protects Photosystem Ii in Maize Seedlings under Drought Stress." *Frontiers in Plant Science* 10. doi:10.3389/fpls.2019.00677.

Ibrahim, Mohamed F. M., Ola H. Abd Elbar, Reham Farag, Mohamed Hikal, Amr El-Kelish, Ahmed Abou El-Yazied, Jawaher Alkahtani, and Hany G. Abd El-Gawad. 2020. "Melatonin Counteracts Drought Induced Oxidative Damage and Stimulates Growth, Productivity and Fruit Quality Properties of Tomato Plants." *Plants* 9 (10). doi:10.3390/plants9101276.

Jones, Maxwell P. A., Jin Cao, Rob O'Brien, Susan J. Murch, and Praveen K. Saxena. 2007. "The Mode of Action of Thidiazuron: Auxins, Indoleamines, and Ion Channels in the Regeneration of Echinacea Purpurea L." *Plant Cell Reports* 26 (9). doi:10.1007/s00299-007-0357-0.

Kapoor, Dhriti, Savita Bhardwaj, Marco Landi, Arti Sharma, Muthusamy Ramakrishnan, and Anket Sharma. 2020. "The Impact of Drought in Plant Metabolism: How to Exploit Tolerance Mechanisms to Increase Crop Production." *Applied Sciences (Switzerland)* 10 (16). doi:10.3390/app10165692.

Karaca, Perihan, and Fazilet Özlem Cekic. 2019. "Exogenous Melatonin-Stimulated Defense Responses in Tomato Plants Treated with Polyethylene Glycol." *International Journal of Vegetable Science* 25 (6). doi:10.1080/19315260.2019.1575317.

Kaya, Armagan, and Zeynep Banu Doganlar. 2019. "Melatonin Improves the Multiple Stress Tolerance in Pepper (Capsicum Annuum)." *Scientia Horticulturae* 256. doi:10.1016/j.scienta.2019.05.036.

Khan, Mohammad Nauman, Jing Zhang, Tao Luo, Jiahuan Liu, Muhammad Rizwan, Shah Fahad, Zhenghua Xu, and Liyong Hu. 2019. "Seed Priming with Melatonin Coping Drought Stress in Rapeseed by Regulating Reactive Oxygen Species Detoxification: Antioxidant Defense System, Osmotic Adjustment, Stomatal Traits and Chloroplast Ultrastructure Perseveration." *Industrial Crops and Products* 140. doi:10.1016/j.indcrop.2019.111597.

Kobylińska, Agnieszka, Sławomir Borek, and Małgorzata M. Posmyk. 2018. "Melatonin Redirects Carbohydrates Metabolism during Sugar Starvation in Plant Cells." *Journal of Pineal Research* 64 (4). doi:10.1111/jpi.12466.

Lee, Hye Jung, and Kyoungwhan Back. 2016. "2-Hydroxymelatonin Promotes the Resistance of Rice Plant to Multiple Simultaneous Abiotic Stresses (Combined Cold and Drought)." *Journal of Pineal Research*. doi:10.1111/jpi.12347.

Lee, Hye-Jung, and Kyoungwhan Back. 2019. "2-Hydroxymelatonin Confers Tolerance against Combined Cold and Drought Stress in Tobacco, Tomato, and Cucumber as a Potent Anti-Stress Compound in the Evolution of Land Plants." *Melatonin Research* 2 (2). doi:10.32794/mr11250020.

Lee, Kyungjin, Geun Hee Choi, and Kyoungwhan Back. 2017. "Cadmium-Induced Melatonin Synthesis in Rice Requires Light, Hydrogen Peroxide, and Nitric Oxide: Key Regulatory Roles for Tryptophan Decarboxylase and Caffeic Acid O-Methyltransferase." *Journal of Pineal Research* 63 (4). doi:10.1111/jpi.12441.

Lerner, Aaron B., James D. Case, Yoshiyata Takahashi, Teh H. Lee, and Wataru Mori. 1958. "Isolation of Melatonin, the Pineal Gland Factor That Lightens Melanocytes." *Journal of the American Chemical Society*. doi:10.1021/ja01543a060.

Li, Chao, Dun Xian Tan, Dong Liang, Cong Chang, Dongfeng Jia, and Fengwang Ma. 2015. "Melatonin Mediates the Regulation of ABA Metabolism, Free-Radical Scavenging, and Stomatal Behaviour in Two Malus Species under Drought Stress." *Journal of Experimental Botany* 66 (3). doi:10.1093/jxb/eru476.

Li, Dongxiao, William D. Batchelor, Di Zhang, Hanxiao Miao, Hongye Li, Shijia Song, and Ruiqi Li. 2020a. "Analysis of Melatonin Regulation of Germination and Antioxidant Metabolism in Different Wheat Cultivars under Polyethylene Glycol Stress." *PLoS ONE* 15 (8 August). doi:10.1371/journal.pone.0237536.

Li, Hao, Yanliang Guo, Qi Cui, Zixing Zhang, Xing Yan, Golam Jalal Ahammed, Xiaozhen Yang, Jianqiang Yang, Chunhua Wei, and Xian Zhang. 2020b. "Alkanes (C29 and C31)-Mediated Intracuticular Wax Accumulation Contributes to Melatonin-and ABA-Induced Drought Tolerance in Watermelon." *Journal of Plant Growth Regulation* 39 (4). doi:10.1007/s00344-020-10099-z.

Li, Jiahao, Yiqing Yang, Kang Sun, Yi Chen, Xuan Chen, and Xinghui Li. 2019. "Exogenous Melatonin Enhances Cold, Salt and Drought Stress Tolerance by Improving Antioxidant Defense in Tea Plant (Camellia Sinensis (L.) O. Kuntze)." *Molecules* 24 (9). doi:10.3390/molecules24091826.

Li, Jingjing, Liu Zeng, Yong Cheng, Guangyuan Lu, Guiping Fu, Haiqing Ma, Qingyun Liu, Xuekun Zhang, Xiling Zou, and Chunsheng Li. 2018. "Exogenous Melatonin Alleviates Damage from Drought Stress in Brassica Napus L. (Rapeseed) Seedlings." *Acta Physiologiae Plantarum* 40 (3). doi:10.1007/s11738-017-2601-8.

Li, Yan, Yan Sun, Huiting Cui, Mingna Li, Guofeng Yang, Zengyu Wang, and Kun Zhang. 2022. "Carex Rigescens Caffeic Acid O-Methyltransferase Gene CrCOMT Confer Melatonin-Mediated Drought Tolerance in Transgenic Tobacco." *Frontiers in Plant Science* 0 (August). Frontiers: 2872. doi:10.3389/FPLS.2022.971431.

Liang, Bowen, Changqing Ma, Zhijun Zhang, Zhiwei Wei, Tengteng Gao, Qi Zhao, Fengwang Ma, and Chao Li. 2018. "Long-Term Exogenous Application of Melatonin Improves Nutrient Uptake Fluxes in Apple Plants under Moderate Drought Stress." *Environmental and Experimental Botany* 155. doi:10.1016/j.envexpbot.2018.08.016.

Liang, Dong, Zhiyou Ni, Hui Xia, Yue Xie, Xiulan Lv, Jin Wang, Lijin Lin, Qunxian Deng, and Xian Luo. 2019. "Exogenous Melatonin Promotes Biomass Accumulation and Photosynthesis of Kiwifruit Seedlings under Drought Stress." *Scientia Horticulturae* 246. doi:10.1016/j.scienta.2018.10.058.

Liu, Juan, Sergey Shabala, Jing Zhang, Guohui Ma, Dandan Chen, Lana Shabala, Fanrong Zeng, et al. 2020. "Melatonin Improves Rice Salinity Stress Tolerance by NADPH Oxidase-Dependent Control of the Plasma Membrane K+ Transporters and K+ Homeostasis." *Plant Cell and Environment* 43 (11). doi:10.1111/pce.13759.

Ma, Xiqing, Jing Zhang, Patrick Burgess, Stephanie Rossi, and Bingru Huang. 2018. "Interactive Effects of Melatonin and Cytokinin on Alleviating Drought-Induced Leaf Senescence in Creeping Bentgrass (Agrostis Stolonifera)." *Environmental and Experimental Botany* 145. doi:10.1016/j.envexpbot.2017.10.010.

Moustafa-Farag, Mohamed, Abdulwareth Almoneafy, Ahmed Mahmoud, Amr Elkelish, Marino B. Arnao, Linfeng Li, and Shaoying Ai. 2020. "Melatonin and Its Protective Role against Biotic Stress Impacts on Plants." *Biomolecules*. doi:10.3390/biom10010054.

Murch, Susan J., Barbara A. Hall, Cuong H. Le, and Praveen K. Saxena. 2010. "Changes in the Levels of Indoleamine Phytochemicals during Véraison and Ripening of Wine Grapes." *Journal of Pineal Research* 49 (1). doi:10.1111/j.1600-079X.2010.00774.x.

Murch, Susan J., and Praveen K. Saxena. 2002. "Melatonin: A Potential Regulator of Plant Growth and Development?" *In Vitro Cellular & Developmental Biology: Plant* 38 (6). Springer: 531–536. doi:10.1079/IVP2002333.

Okazaki, Masateru, and Hiroshi Ezura. 2009. "Profiling of Melatonin in the Model Tomato (Solanum Lycopersicum L.) Cultivar Micro-Tom." *Journal of Pineal Research* 46 (3). doi:10.1111/j.1600-079X.2009.00668.x.

Okçu, Gamze, Mehmet Demir Kaya, and Mehmet Atak. 2005. "Effects of Salt and Drought Stresses on Germination and Seedling Growth of Pea (Pisum Sativum L.)." *Turkish Journal of Agriculture and Forestry* 29 (4). doi:10.3906/tar-0407-2.

Park, Sangkyu, Kiyoon Kang, Kyungjin Lee, Doil Choi, Young Soon Kim, and Kyoungwhan Back. 2009. "Induction of Serotonin Biosynthesis Is Uncoupled from the Coordinated Induction of Tryptophan Biosynthesis in Pepper Fruits (Capsicum Annuum) upon Pathogen Infection." *Planta* 230 (6). doi:10.1007/s00425-009-1015-2.

Park, Sangkyu, Kyungjin Lee, Young Soon Kim, and Kyoungwhan Back. 2012. "Tryptamine 5-Hydroxylase-Deficient Sekiguchi Rice Induces Synthesis of 5-Hydroxytryptophan and N-Acetyltryptamine but Decreases Melatonin Biosynthesis during Senescence Process of Detached Leaves." *Journal of Pineal Research* 52 (2). doi:10.1111/j.1600-079X.2011.00930.x.

Pelagio-Flores, Ramón, Edith Muñoz-Parra, Randy Ortiz-Castro, and José López-Bucio. 2012. "Melatonin Regulates Arabidopsis Root System Architecture Likely Acting Independently of Auxin Signaling." *Journal of Pineal Research* 53 (3). doi:10.1111/j.1600-079X.2012.00996.x.

Qi, Hua, Fan Nv Xia, and Shi Xiao. 2021. "Autophagy in Plants: Physiological Roles and Post-Translational Regulation." *Journal of Integrative Plant Biology*. doi:10.1111/jipb.12941.

Reddy, Attipalli Ramachandra, Kolluru Viswanatha Chaitanya, and Munusamy Vivekanandan. 2004. "Drought-Induced Responses of Photosynthesis and Antioxidant Metabolism in Higher Plants." *Journal of Plant Physiology*. doi:10.1016/j.jplph.2004.01.013.

Reiter, Russel J., Dun-Xian Tan, Sergio Rosales-Corral, and Lucien C. Manchester. 2013. "The Universal Nature, Unequal Distribution and Antioxidant Functions of Melatonin and Its Derivatives." *Mini-Reviews in Medicinal Chemistry* 13 (3). doi:10.2174/1389557511313030006.

Reiter, Russel J., Dun Xian Tan, Zhou Zhou, Maria Helena Coelho Cruz, Lorena Fuentes-Broto, and Annia Galano. 2015. "Phytomelatonin: Assisting Plants to Survive and Thrive." *Molecules*. doi:10.3390/molecules20047396.

Roychoudhury, A., and Das, K. 2014. "Functional Role of Polyamines and Polyamine-Metabolizing Enzymes during Salinity, Drought and Cold Stresses." In: *Plant Adaptation to Environmental Change: Significance of Amino Acids and Their Derivatives*, CABI, Wallingford, UK, pp. 141–156.

Samanta, S., A. Banerjee, and A. Roychoudhury. 2021. "Melatonin Application Differentially Modulates the Enzymes Associated with Antioxidative Machinery and Ascorbate Glutathione Cycle during Arsenate Exposure In Indica Rice Varieties. *Plant Biology* 23 (S1): 193–201.

Saqib, Muhammad, Umbreen Shahzad, Sana Naz, Azhar Abbas, Shah Jahan Bahauddin Zakariya University Sub-Campus Layyah Rabia Shahid, Muhammad Mohsin Altaf, Zakia Abid, Rahul Kumar Tiwari, Milan Kumar Lal, and Muhammad Ahsan Altaf. 2022. "Melatonin Alleviates Cadmium Phytotoxicity through Regulation of Growth, Photosynthesis and Antioxidant Potential in Two Pepper Genotypes." doi:10.21203/rs.3.rs-1651489/v1.

Shao, R. X., L. F. Xin, H. F. Zheng, L. L. Li, W. L. Ran, J. Mao, and Q. H. Yang. 2016. "Changes in Chloroplast Ultrastructure in Leaves of Drought-Stressed Maize Inbred Lines." *Photosynthetica* 54 (1). doi:10.1007/s11099-015-0158-6.

Sharma, Anket, Junfeng Wang, Dongbin Xu, Shenchen Tao, Sunli Chong, Daoliang Yan, Zhen Li, Huwei Yuan, and Bingsong Zheng. 2020. "Melatonin Regulates the Functional Components of Photosynthesis, Antioxidant System, Gene Expression, and Metabolic Pathways to Induce Drought Resistance in Grafted Carya Cathayensis Plants." *Science of the Total Environment* 713. doi:10.1016/j.scitotenv.2020.136675.

Shi, Haitao, Chuan Jiang, Tiantian Ye, Dun Xian Tan, Russel J. Reiter, Heng Zhang, Renyi Liu, and Zhulong Chan. 2015. "Comparative Physiological, Metabolomic, and Transcriptomic Analyses Reveal Mechanisms of Improved Abiotic Stress Resistance in Bermudagrass [Cynodon Dactylon (L). Pers.] by Exogenous Melatonin." *Journal of Experimental Botany* 66 (3). doi:10.1093/jxb/eru373.

Silalert, Parichart, and Wattana Pattanagul. 2021. "Foliar Application of Melatonin Alleviates the Effects of Drought Stress in Rice (Oryza Sativa L.) Seedlings." *Notulae Botanicae Horti Agrobotanici Cluj-Napoca* 49 (3). doi:10.15835/nbha49312417.

Su, Xiaoyu, Xiaocong Fan, Ruixin Shao, Jiameng Guo, Yongchao Wang, Jianping Yang, Qinghua Yang, and Lin Guo. 2019. "Physiological and ITRAQ-Based Proteomic Analyses Reveal That Melatonin Alleviates Oxidative Damage in Maize Leaves Exposed to Drought Stress." *Plant Physiology and Biochemistry* 142. doi:10.1016/j.plaphy.2019.07.012.

Sun, Chengliang, Lijuan Liu, Luxuan Wang, Baohai Li, Chongwei Jin, and Xianyong Lin. 2021. "Melatonin: A Master Regulator of Plant Development and Stress Responses." *Journal of Integrative Plant Biology*. doi:10.1111/jipb.12993.

Supriya, Laha, Pullaiahgari Durgeshwar, Mehanathan Muthamilarasan, and Gudipalli Padmaja. 2022. "Melatonin Mediated Differential Regulation of Drought Tolerance in Sensitive and Tolerant Varieties of Upland Cotton (*Gossypium Hirsutum L.)*." *Frontiers in Plant Science* 13 (April). Frontiers Media S.A.: 821353. doi:10.3389/FPLS.2022.821353.

Tan, Dun Xian, Lucien C. Manchester, Maria P. Terron, Luis J. Flores, and Russel J. Reiter. 2007. "One Molecule, Many Derivatives: A Never-Ending Interaction of Melatonin with Reactive Oxygen and Nitrogen Species?" *Journal of Pineal Research*. doi:10.1111/j.1600-079X.2006.00407.x.

Tan, Dun Xian, Brian M. Zanghi, Lucien C. Manchester, and Russel J. Reiter. 2014. "Melatonin Identified in Meats and Other Food Stuffs: Potentially Nutritional Impact." *Journal of Pineal Research* 57 (2). doi:10.1111/jpi.12152.

Tiwari, Rahul Kumar, Milan Kumar Lal, Ravinder Kumar, Kumar Nishant Chourasia, Kailash Chandra Naga, Dharmendra Kumar, Sourav Kumar Das, and Gaurav Zinta. 2021a. "Mechanistic Insights on Melatonin-Mediated Drought Stress Mitigation in Plants." *Physiologia Plantarum* 172 (2). doi:10.1111/ppl.13307.

Tiwari, Rahul Kumar, Milan Kumar Lal, Ravinder Kumar, Kumar Nishant Chourasia, Kailash Chandra Naga, Dharmendra Kumar, Sourav Kumar Das, and Gaurav Zinta. 2021b. "Mechanistic Insights on Melatonin-Mediated Drought Stress Mitigation in Plants." *Physiologia Plantarum* 172 (2). John Wiley & Sons, Ltd: 1212–1226. doi:10.1111/ppl.13307.

Tiwari, Rahul Kumar, Milan Kumar Lal, Ravinder Kumar, Vikas Mangal, Muhammad Ahsan Altaf, Sanjeev Sharma, Brajesh Singh, and Manoj Kumar. 2022. "Insight into Melatonin-Mediated Response and Signaling in the Regulation of Plant Defense under Biotic Stress." *Plant Molecular Biology* 109 (4–5). Springer Science and Business Media B.V.: 385–399. doi:10.1007/s11103-021-01202-3.

Tiwari, Rahul Kumar, Milan Kumar Lal, Kailash Chandra Naga, Ravinder Kumar, Kumar Nishant Chourasia, Subhash S, Dharmendra Kumar, and Sanjeev Sharma. 2020. "Emerging Roles of Melatonin in Mitigating Abiotic and Biotic Stresses of Horticultural Crops." *Scientia Horticulturae*. doi:10.1016/j.scienta.2020.109592.

Wang, Jinfang, Lei Zhang, Yunyun Cao, Chuandong Qi, Shuangtao Li, Lun Liu, Gongle Wang, Aijun Mao, Shuxin Ren, and Yang Dong Guo. 2018. "CsATAF1 Positively Regulates Drought Stress Tolerance by an ABA-Dependent Pathway and by Promoting ROS Scavenging in Cucumber." *Plant and Cell Physiology* 59 (5). doi:10.1093/pcp/pcy030.

Wang, Lin, Chao Feng, Xiaodong Zheng, Yan Guo, Fangfang Zhou, Dongqian Shan, Xuan Liu, and Jin Kong. 2017. "Plant Mitochondria Synthesize Melatonin and Enhance the Tolerance of Plants to Drought Stress." *Journal of Pineal Research* 63 (3). doi:10.1111/jpi.12429.

Wang, Lin, Yu Zhao, Russel J. Reiter, Changjiu He, Guoshi Liu, Qiong Lei, Bixiao Zuo, Xiao Dong Zheng, Qingtian Li, and Jin Kong. 2014. "Changes in Melatonin Levels in Transgenic 'Micro-Tom' Tomato Overexpressing Ovine AANAT and Ovine HIOMT Genes." *Journal of Pineal Research* 56 (2). doi:10.1111/jpi.12105.

Wang, Ping, Xun Sun, Chao Li, Zhiwei Wei, Dong Liang, and Fengwang Ma. 2013. "Long-Term Exogenous Application of Melatonin Delays Drought-Induced Leaf Senescence in Apple." *Journal of Pineal Research* 54 (3). doi:10.1111/jpi.12017.

Wang, Yanping, Russel J. Reiter, and Zhulong Chan. 2018. "Phytomelatonin: A Universal Abiotic Stress Regulator." *Journal of Experimental Botany*. doi:10.1093/jxb/erx473.

Wei, Wei, Qing Tian Li, Ya Nan Chu, Russel J. Reiter, Xiao Min Yu, Dan Hua Zhu, Wan Ke Zhang, et al. 2015. "Melatonin Enhances Plant Growth and Abiotic Stress Tolerance in Soybean Plants." *Journal of Experimental Botany* 66 (3). doi:10.1093/jxb/eru392.

Xia, Hui, Zhiyou Ni, Rongping Hu, Lijin Lin, Honghong Deng, Jin Wang, Yi Tang, et al. 2020. "Melatonin Alleviates Drought Stress by a Non- Enzymatic and Enzymatic Antioxidative System in Kiwifruit Seedlings." *International Journal of Molecular Sciences* 21 (3). doi:10.3390/ijms21030852.

Yan, Wu, Lian Hong Yan, Mu Xue Jiao, Wang Xue Juan, and Zhang Yuan Bing. 2016. "Effects of Foliar Spraying Exogenous Melatonin on Physiological and Biochemical Characteristics of Dendranthema Morifolium 'Chuju' Seedlings under Drought Stress." *Acta Botanica Boreali-Occidentalia Sinica* 36 (11). Science Press: 2241–2246.

Yang, Hong, Longjun Dai, Yongxuan Wei, Zhi Deng, and Dejun Li. 2020. "Melatonin Enhances Salt Stress Tolerance in Rubber Tree (Hevea Brasiliensis) Seedlings." *Industrial Crops and Products* 145. doi:10.1016/j.indcrop.2019.111990.

Yang, Wen Jing, Yong Tao Du, Yong Bin Zhou, Jun Chen, Zhao Shi Xu, You Zhi Ma, Ming Chen, and Dong Hong Min. 2019. "Overexpression of TaCOMT Improves Melatonin Production and Enhances Drought Tolerance in Transgenic Arabidopsis." *International Journal of Molecular Sciences* 20 (3). doi:10.3390/ijms20030652.

Ye, Jun, Shiwen Wang, Xiping Deng, Lina Yin, Binglin Xiong, and Xinyue Wang. 2016. "Melatonin Increased Maize (Zea Mays L.) Seedling Drought Tolerance by Alleviating Drought-Induced Photosynthetic Inhibition and Oxidative Damage." *Acta Physiologiae Plantarum* 38 (2). doi:10.1007/s11738-015-2045-y.

Zhang, Liang, Dongmei Xi, Shanwei Li, Zheng Gao, Shuoli Zhao, Jing Shi, Changai Wu, and Xingqi Guo. 2011. "A Cotton Group C MAP Kinase Gene, GhMPK2, Positively Regulates Salt and Drought Tolerance in Tobacco." *Plant Molecular Biology* 77 (1). doi:10.1007/s11103-011-9788-7.

Zhang, Mingcong, Songyu He, Yingce Zhan, Bin Qin, Xijun Jin, Mengxue Wang, Yuxian Zhang, Guohua Hu, Zhanlin Teng, and Yaokun Wu. 2019. "Exogenous Melatonin Reduces the Inhibitory Effect of Osmotic Stress on Photosynthesis in Soybean." *PLoS ONE* 14 (12). doi:10.1371/journal.pone.0226542.

Zhao, Chengfeng, Mei Yang, Xi Wu, Yifan Wang, and Renhe Zhang. 2021. "Physiological and Transcriptomic Analyses of the Effects of Exogenous Melatonin on Drought Tolerance in Maize (Zea Mays L.)." *Plant Physiology and Biochemistry* 168. doi:10.1016/j.plaphy.2021.09.044.

Zou, J. N., X. J. Jin, Y. X. Zhang, C. Y. Ren, M. C. Zhang, and M. X. Wang. 2019. "Effects of Melatonin on Photosynthesis and Soybean Seed Growth during Grain Filling under Drought Stress." *Photosynthetica* 57 (2). doi:10.32615/ps.2019.066.

Zuo, Bixiao, Xiaodong Zheng, Pingli He, Lin Wang, Qiong Lei, Chao Feng, Jingzhe Zhou, Qingtian Li, Zhenhai Han, and Jin Kong. 2014. "Overexpression of MzASMT Improves Melatonin Production and Enhances Drought Tolerance in Transgenic Arabidopsis Thaliana Plants." *Journal of Pineal Research* 57 (4). doi:10.1111/jpi.12180.

16 Melatonin-Induced Plant Tolerance against Temperature Stress

Triptesh Mondal, Chandini, Upasana Sahoo, Masina Sairam, Tanmoy Shankar, Sagar Lalichetti, Dinkar J. Gaikwad, Akbar Hossain, Aryadeep Roychoudhury, Debojyoti Moulick, and Sagar Maitra

16.1 INTRODUCTION

The accelerating increase of greenhouse gas emissions is raising the earth's temperature. The effect of high temperature includes the melting of glaciers, higher precipitation, extreme weather calamities, and the delaying of the seasons. The increasing pace of climate change, along with the alarming rate of population rise, threatens food security throughout the world. The global average temperature has increased by 1°C, whereas the temperature in India has increased around 0.7°C during 1901–2018. Also, the temperature is projected to be increased by 4.4°C by the end of the twenty-first century. By the end of the twenty-first century, the frequency of summer heat waves is also anticipated to rise by 3–4% above the prior base data (Krishnan et al., 2020). Plant growth is constrained by such unfavourable conditions. Stress is defined as those situations that prohibit plants from growing and developing. Stress can be categorized broadly into two types, biotic stress and abiotic stress. Among them, abiotic stress alone causes around 71% of the losses, and the remaining 29% is due to other stresses (Boyer, 1982). According to reports, abiotic stresses are the primary global constraints on crop production, which reduce yield by more than 50% (Mahajan and Tuteja, 2005; Roychoudhury et al., 2011a).

Abiotic stress constitutes the adverse effects of the nonliving components of the environment on plants (Manepalli et al., 2022). As the plants are sessile in nature, they ubiquitously face abiotic stresses. Abiotic stress constitutes about 70% of the total crop losses in agriculture (Canter, 2018; Martinez et al., 2018; Zörb et al., 2018). Temperature extremes, water shortage, waterlogging, and salinity are the major abiotic stresses (Waqas et al., 2017; Vaughan et al., 2017; Zafar et al., 2018). The abiotic stresses have affected almost 90% of the land. Only 10% of the land is free from any type of stress (Dos Reis et al., 2012). Biotic stress is defined as the damage caused by fungi, bacteria, viruses, nematodes, insects, and weeds. They normally cause diseases and damage to various plant parts. The nematodes normally affect the roots and eventually affect the overall growth.

Global climate change has led to various changes, but the most prominent among them is temperature change. Temperature affects plant growth and its functions starting from seed germination to harvest. The optimum temperature range for different crops is the range in which the crop performs its best. Both extremes in temperature limit water and nutrient uptake capacity, which, in turn, affects flowering, fruiting, and seed formation (Pierce, 1987). The global temperature has increased over a period of time, leading to heat stress. Due to the enormous amounts of reactive oxygen species that are produced and accumulated during heat stress, it can cause

DOI: 10.1201/9781003343752-16

cellular homeostasis to become disrupted. High temperature also affects protein production and protein structure (Das and Roychoudhury, 2014). Due to the lower water content, extremely high temperature generally impedes cell division and growth (Lazar et al., 2013; Salvucci and Crafts-Brandner, 2004). Superoxide radicals, singlet oxygen, hydroxyl radical (OH•), and hydrogen peroxide (H_2O_2) are formed in excess, resulting in heat shock in plants (Nadeem et al., 2018). These molecules produce excessive amounts of malondialdehyde (MDA) and cause severe oxidative damage. MDA reduces the permeability, stability, and mobility of the membranes (Nadeem et al., 2019; Bhat et al., 2020).

Low-temperature abnormality is a vital abiotic burden for crop yield (Esim and Atici, 2014). Two types of injury are observed in plants due to low temperature, namely chilling injury and freezing injury. Temperature less than the minimum temperature is often harmful to a plant and negatively affects the growth as well as the development of the plant. It affects the physiological, metabolic, biochemical, and different molecular processes (Repkina et al., 2021). It slows down seed germination and emergence. Cold stress is most prevalent in temperate areas where extremely low temperature limits the productivity of crops (Aazami et al., 2021; Feng et al., 2011). Plant type belonging to hot climates have greater sensitivity to lower temperatures as they are already adapted to the higher temperature (Decoteau, 2000).

Metabolism slows down at a lower temperature and often leads to water stress (Kratsch and Wise, 2000). It is because of the generation of massive amounts of reactive oxygen species (ROS). Heat stress can cause cellular homeostasis to become disrupted. The effects of high temperature also affect protein synthesis and structure. Cold stress damages cell membranes because of variations in the generation and metabolism of phospholipids, phenolics, proteins, sugars, and ATP (Lyons, 1973). Cold conditions also encourage the enhanced generation of ROS inclusive of H_2O_2, superoxide radical, and hydroxyl radical in the cells, causing lipid peroxidation and oxidative alterations of proteins (Fan et al., 2014). Crop loss can be overcome by using several stimulators which help the plant resist or fight against the adverse abiotic conditions. One of them is melatonin (MT), which is ubiquitous in all living organisms.

Both plants and animals produce MT, an amphiphilic indole amine that is generated from tryptophan (Shida et al., 1994) and produced in the chloroplast and mitochondria of plants. By directly preventing the buildup of ROS and reactive nitrogen species, MT controls plant stress responses. The amphiphilic character of MT helps the easy movement of the MT and thus allows better response to the exogenous application. It alleviates stress by virtue of its excellent antioxidant property (Kołodziejczyk et al., 2015, 2016, 2021). Besides, MT takes part in other basic physiological procedures of the plant (Zhang et al., 2015). MT helps plants to tolerate heat stress by removing ROS from the environment (Zhang et al., 2021). MT also impacts the expression of genes that control antioxidative actions (Li et al., 2018a; Samanta et al., 2022). By inhibiting photosynthesis, degrading starch, and enhancing electron transport, MT increases cold tolerance (Li et al., 2018b). In this chapter, we have systematically discussed the characteristics of MT, its metabolic pathway, and its role in thermo-stress tolerance along with its future implications.

16.2 FUNCTIONS OF MELATONIN

MT is responsible for inducing plant defense mechanisms by virtue of its ability to produce secondary metabolites, helping the plant or animal to sustain against adverse environmental conditions such as abiotic and biotic stresses (Figure 16.1). MT performs a significant role in growth (Cleland, 2004; Hernandez-Ruiz et al., 2004) and reproductive stages (Wolf et al., 2001) and cell protection from free radicals (Lei et al., 2004). Various functions of MT such as auxin level regulation, cytokinin-like actions, abscisic acid regulation, ethylene-like activity, enhanced photosynthesis, and secondary metabolite production are observed (Table 16.1). Moreover, MT plays a major role in relieving abiotic and biotic stresses in plants and in the process of combating stresses (Arnao and Hernández-Ruiz, 2006; Banerjee and Roychoudhury, 2017).

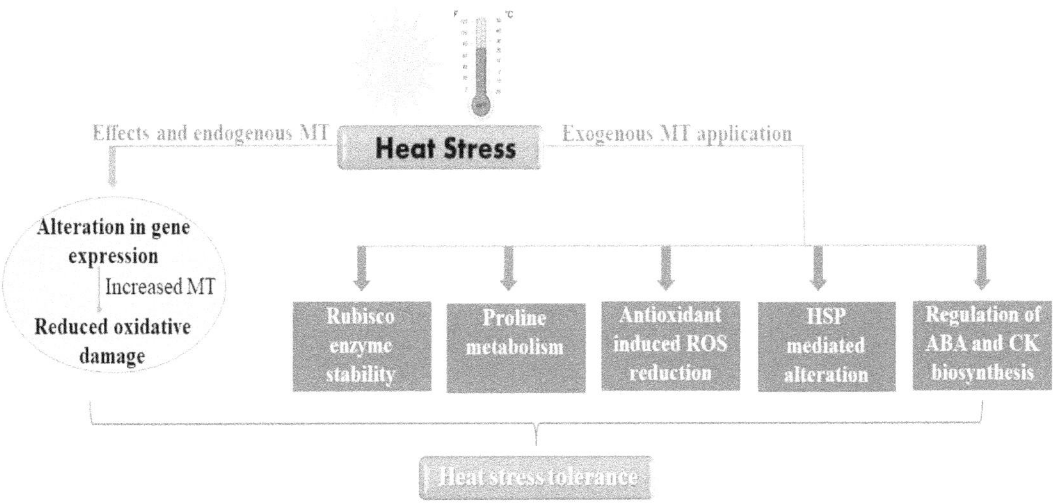

FIGURE 16.1 Impacts of exogenous MT application during heat stress.

[MT = Melatonin; ROS = Reactive oxygen species; HSP = Heat shock protein; ABA = Absicic acid; CK = Cytokinin]

TABLE 16.1
Different Functions of MT in Plants

	Functions	Types of activity	Mechanism involved	References
1.	Hormone-like functions	Auxin agonists	• Approximately three- to four-fold enhanced growth of the aerial and rhizosphere plant parts at threshold MT concentrations of > 10 µm while 1.4- to 2.0-fold increase in auxin-like compound production during stress environment • Suppressed apical dominance and an increased lateral branching and enhanced production of lateral and adventitious roots	Chen et al. (2009); Pelagio-Flores et al. (2012); Zhang et al. (2013); Mukherjee et al. (2014); Zuo et al. (2014)
		Cytokinin-like actions	• Reduction in stressor-induced senescence • Reduced chlorophyll loss • Induced nutrient mobilization with increased MT accumulation • Down regulation of chlorophyll-degrading enzymes (pheophorbide oxygenase and chlorophyllase) • Enhanced CO_2 assimilation, preserving chloroplasts, maintaining osmolyte balance, increased stomatal conductance, inducing proline biosynthesis, and maintenance of cell turgidity • Enhanced expression of carbohydrate and fatty acid metabolism due to corresponding gene up regulation	Wang et al. (2013); Zhang et al. (2013); Arnao and Hernández-Ruiz (2014); Weeda et al. (2014); Gao et al. (2016)

(Continued)

TABLE 16.1 ***(Continued)***
Different Functions of MT in Plants

	Functions	Types of activity	Mechanism involved	References
		Abscisic acid (ABA) metabolism	• High MT concentration results in enhanced expression of CYP707 monooxygenases (ABA catabolic genes) and suppressed expression of 9-cis-epoxycarotenoid dioxygenase (ABA synthesis genes) • MT is also responsible for the suppression of the heat-induced enhanced expression of LpABI5 and LpABI3 (ABA transcription factor)	Zhang et al. (2014b); Li et al. (2016); Zhang et al. (2017a)
		Ethylene like activity	• Enhanced expression of ACC (1-aminocyclopropane-1-carboxylic acid) synthase, resulting in high ethylene production • Blocks the expression of ACC oxidase • Up regulates post-harvest fruit ripening effect • Up regulation of expression of genes responsible for protein accumulation related to oxidative phosphorylation, fatty acid metabolism, cell wall metabolism, carbohydrate, and flavonoid biosynthesis	Kim et al. (2001); Sun et al. (2015); Zhang et al. (2016)
2.	Growth and development	Increases in biomass	• Improved growth efficiency of plants • Developed defense responses against stressors • Improve the establishment and germination of seedlings and seeds, respectively	Tan et al. (2012); Posmyk et al. (2008)
		Photosynthetic activity	• Enhanced chlorophyll preservation • Scavenges ROS produced as a result of temperature stress that may damage the D1 protein binding reaction center	Sharif et al. (2018); Xu (2010); Yamamoto et al. (2008)
		Crop improvement	• MT, as a bio-stimulator, results in enhanced crop yield, increased shelf life, and reduced peroxidation	Hardeland (2009); Asif et al. (2019); Tiwari et al. (2020)
		Secondary metabolites	• Enhanced production of secondary metabolites such as salicylic acid and jasmonic acid responsible for antioxidant, anti-inflammatory, neuroprotective, and immunomodulatory effects	Carrillo-Vico et al. (2013); Bonilla et al. (2004)
3.	Resistance to abiotic stress	Drought or excess moisture tolerance	• Down regulation of *MdNCED3* gene (ABA synthesis gene) and up regulation of *MdCYP707A2* and *MdCYP707A1* (ABA catabolic gene) • Enhanced H_2O_2 scavenging activity and increased accumulation of antioxidants	Li et al. (2015); Meng et al. (2014)
		Cold or heat tolerance	• Up regulation of genes for increased polyamine content • Enhanced production of putrescine and spermidine levels • Acts as a modulator of redox balance	Lei et al. (2004); Hu et al. (2016); Parida and Das (2005)

TABLE 16.1 *(Continued)*
Different Functions of MT in Plants

	Functions	Types of activity	Mechanism involved	References
		Salt tolerance	• Modulating gene expression responsible for catabolism and biosynthesis of ABA and GA • Compartmentalization of ions, altering photosynthetic pathways, selective ion exclusion, alteration in membrane structure, antioxidant enzyme induction, compatible solute production to cope with the effects of salt stressors	Arnao and Hernández-Ruiz (2009a); Wang et al. (2013); Zhang et al. (2014a); Li et al. (2015)
		Heavy metal tolerance	• Inhibition of the transportation and translocation of toxic metals via xylem and phloem • Reduction in accumulation of heavy metals. Acts as a free-radical scavenger against hazardous contaminants. • Induced mobilization of toxic metal chelation	Okazaki and Ezura (2009); Wang et al. (2013)
4.	Resistance to biotic stress	Disease resistance	• Enhanced photosystem efficiency, increased production of plant defense-related enzymes, promotes antioxidant action post MT treatment • Induced production of the defense-signaling molecule	Yin et al. (2013); Lee et al. (2014)

16.3 GENETIC MODIFICATION MODULATES STRESS TOLERANCE IN PLANTS

Generally, crops never face one single stress at a time, and it is difficult for plants to overcome multiple stresses together. Genome editing and cis- or trans-genesis to develop stress-tolerant crops are popular techniques to enhance tolerance. Gene pyramiding or stacking of multiple genes appears to be a prominent method to withstand different environmental stresses simultaneously. MT is one of the pleiotropic signaling molecules which participates in the tolerance against stresses (Jahan et al., 2019). Polyamines (PAs), viz., putrescine (Put), spermidine (Spd), and spermine (Spm), are low-molecular-weight nitrogen compounds exhibiting many functions such as morphogenesis of plants, reproductive stimulation, and reducing the rate of leaf senescence and involvement for adaptation against different abiotic stress conditions like extreme temperature (high and low), salinity, dry spell, heavy metal ion concentration, changes in osmotic pressure, ultraviolet light stress, and waterlogging stress (Roychoudhury et al., 2011b; Ke et al., 2018). PAs have a positive charge that acts in maintaining the integrity of cell membranes, easing the function of enzymes, and safeguarding the structure of nucleic acids and various proteins.

The *P5CS* gene expression is responsible for the biological synthesis of another osmolyte, proline (Roychoudhury et al., 2015). MT is responsible for enhancing *P5CS* expression in plants affected by high temperature stress (Jahan et al., 2019). The levels of expression of some genes that encode antioxidative enzymes showed that superoxide dismutase (SOD), catalase (CAT), ascorbate peroxidase (APX), and glutathione S-transferase (GST) expressions were raised by 1.6, 1.2, 3.2 and 1.3, respectively, than their control sets. MT can up regulate the expression of all these genes during heat stress (Jahan et al., 2019).

The glutathione pool (glutathione [GSH] + glutathione disulfide [GSSG]) was up regulated by MT in tea plants that face chilling stress. Exposure to low temperature resulted in a significant reduction in Fv/Fm by more than 50%. However, MT brought about significant improvement in Fv/Fm during low-temperature conditions (Li et al., 2018a, 2021a). In case of drought priming-induced cold tolerance (DPICT) process, the early drought priming induced an increase in internal MT level, which showed no connection with abscisic acid (ABA). An increment in MT level induced by drought priming ultimately increased the ability of the antioxidants in energy storage and preserving the photosynthetic apparatus. Transportation of electrons in chloroplasts of the barley plants under stressful conditions due to low temperature was thus sustained (Li et al., 2016). MT is also able to reinstate the circadian rhythmic oscillation of circadian clock genes such as the morning gene (*HvCCA1*) and evening loop gene (*HvTOC1*), whose circadian rhythmic phenotypes were not affected due to cold stress (Chang et al., 2021). The involvement of MT was found in grafting-induced low temperature tolerance by encouraging the accumulation of methyl jasmonate (MeJA) and H_2O_2. MeJA subsequently enhanced the accumulation of MT, developing a self-augmenting feedback loop that ultimately led to low temperature tolerance in watermelon (Li et al., 2021b). MT regulates a set of miRNAs under stressful conditions due to low temperature. MT-induced lesser expression of a few miRNAs like miR159–5p, miR858, miR8029–3p, and novel-m0048–3p correlated with the overexpression of cold responsive genes [calcium-dependent protein kinase (CDPK), basic helix-loop-helix (bHLH), dehydration-responsive element binding gene (DREB, *etc.*), and LEA] under freezing temperature (Li et al., 2016).

MT decreased oxidative stress as it diminished thiobarbituric acid reactive substances (TBARS) and H_2O_2 content, increased the activity of antioxidative enzymes, as well as elevated the rate of photosynthesis in wheat plants *vis-à-vis* metabolism of carbohydrate that was required for energy supply and providing carbon skeleton to the wheat seedlings affected by stresses. Hydrogen sulfide (H_2S)-mediated MT enhancement adjusted many physiological responses under heat stress, directing seed germination and development of plants and reducing leaf senescence in wheat plants (Li et al., 2021b).

Antioxidant enzyme activity, hydrogen peroxide (H_2O_2), electrolyte leakage, deterioration of chlorophyll, soluble sugar, proline, and γ-aminobutyric acid (GABA) content increase in Pistachio (*Pistacia vera* L.) seedlings under low temperature stress. Phenolic content, phenylalanine ammonia lyase (PAL) activity, and the expression of fatty acid desaturase (*FAD2*) gene is up regulated, but the concentration of ascorbate and glutathione is distinctly lowered as compared to control. MT remarkably decreased the antioxidant enzyme activity, malondialdehyde (MDA), H_2O_2, degeneration of chlorophyll, γ-aminobutyric acid (GABA) concentration, soluble sugar, and proline content under stress conditions. Along with these, electrolyte leakage, PAL activity, and *FAD2* gene expression were decreased in presence of MT, reducing the oxidative injuries and preserving the structure and mobility of cell membrane in pistachio seedlings under different stresses (Barandet al., 2020). MT enhanced the accumulation of C-repeat binding factors (CBFs)/drought response element binding factors (DREBs). A low-temperature responsive gene, *COR15a*, codes for an important cryoprotective protein that increases the freezing tolerance of chloroplast and protoplast. Calmodulin-binding transcription activators (CAMTAs) are other important transcription factors implicated in freezing tolerance in *Arabidopsis thaliana* which play a positive role as regulators of CBF2 (Bajwa et al., 2014). MT boosts up phenol accumulation which is related to 2, 2-diphenyl-1 picryl hydrazyl (DPPH) scavenging capacity and decrement of oxidative damages following harvest of fruits and vegetables (Madebo et al., 2021).

CRISPR/Cas9-based genome editing plays a major role in the creation of novel varieties with essential traits like yield and quality improvement as well as plants resistant to stressful situations. Presently, the CRISPR/Cas9 system of MT biosynthetic gene-dependent genome editing is largely utilized to develop thermo-stress-tolerant plant species (Zeng et al., 2020; Raza et al., 2022).

16.4 MELATONIN LEVELS IN STRESSED PLANTS

In response to external variables like rapid alteration in light, temperature, or any other climatic fluctuations, the concentration of MT within a plant significantly increases. The amount of MT in plants varies depending on the stress level in various environment. For instance, when barley seedlings were exposed to chemical agents like NaCl, $ZnSO_4$, and H_2O_2, the amount of MT remarkably increased. Lupin (*Lupinus albus*) also showed similar outcomes (Arnao and Hernández-Ruiz, 2013).

The functions of MT biosynthesis enzymes were found to be strictly regulated by stress stimulation. For instance, elevated ASMT and SNAT activity improved MT production in rice seedlings under heat stress and abnormally dark environment (Byeon and Back, 2015). In sunflower (*Helianthus annuus*) plants undergoing salt stress, HIOMT activity was elevated in the cotyledons, resulting in increased levels of MT concentration in the seedlings (Mukherjee et al., 2014). According to studies, these rate-limiting enzymes or isoenzymes were regulated by upstream transcription factors, which might explain how MT is produced and elevated in response to stress. MT levels rose as a result of the overexpression of enzymes like HIOMT or AANAT, which improved tolerance in tomato against soil moisture stress (Wang et al., 2014). In an experiment, Arnao and Hernández-Ruiz (2009b) exposed a test plant to unfavourable environmental circumstances that were either created intentionally or naturally. The MT content in barley roots was measured in the presence of chemical stress agents such as chlorides of sodium, sulphates of zinc, or peroxides of hydrogen (Arnao and Hernández-Ruiz, 2009b), and in another experiment, lupin was exposed to several stresses like drought, flood, abnormal pH, or other stresses like heat and cold stress (Arnao and Hernández-Ruiz, 2013). Levels of MT were somewhat elevated when the root oxygen supply was reduced. High acidity caused a slight rise in MT concentration, while an abnormal pH and alkali stress in the medium diminished it in the roots. Moreover, the responses in barley and lupin were generally dose as well as time dependent. Almost all stresses could initiate an elevation in MT levels in the examined plants. When compared to levels grown under artificial light, the roots and leaves of crops grown in natural sunshine produced 3.0 and 2.5 times more MT content, respectively (Tan et al., 2007).

The MT content in shaded pepper fruits was found to be lower than that in the non-shaded fruit, indicating the direct relation of solar radiation in enhancing MT production (Riga et al., 2014). The MT content in *Glycyrrhyiza auralensis* was also assessed with various spectral quality lights, and it was discovered that plants could synthesize more MT under exposure to UV-B radiation (Afreen et al., 2006). MT could reduce heat stress damage to plants, as observed from the increase in MT level in plants subjected to heat stress (Byeon and Back, 2015; Tal et al., 2011). Accumulation of heavy metals such as Cd, Pb, and Zn in algae resulted in enhanced MT production; likewise, exogenous MT application enhanced Cd stress tolerance in algae, *Solanum lycopersicum*, and in many more crops by improving the processes like photosynthesis and synthesis of antioxidant enzymes (Hasan et al., 2015).

MT plays a key role in the processes of photosynthesis or photo-protection. The amount of MT produced by plants in the Mediterranean and Alpine regions is significantly higher than that in regions with less frequent exposure to intense UV radiation (Simopoulos et al., 2005). The comparatively higher level of MT synthesis in ozone-resistant plant species compared to ozone-susceptible species suggests yet another potential function for this hormone (Dubbels et al., 1995). The pollutant-tolerant plant, such as the water hyacinth, produces MT and some of its significant metabolites, such as N1-acetyl-N2-formyl-5-methoxy kynuramine (AFMK). Plants like water hyacinth can withstand waste water contamination caused by a variety of sources, including agrochemical and industrial wastes (Munavalli and Saler, 2009; Trivedy and Pattanshetty, 2002), nitrogen and phosphate (Jayaweera and Kasturiarachchi, 2004), residue of plant protection chemicals (Xia and Ma, 2006), and heavy metal wastes like lead (Misbahuddin and Fariduddin, 2002). Artificial supplementation with MT ensures a higher phytoremediation in addition to the naturally occurring MT, because of its ability to produce antioxidants for elevating tolerance to pollution stress and removing other harmful contaminants (Tan et al., 2007). High-temperature stress in conjunction with the heavy metal loads caused a rise in MT levels in algal species like *Ulva* sp., supporting its role in promoting heat stress tolerance (Tal et al., 2011).

In contrast to lead and zinc exposure, which only slightly increased MT levels in algae, exposure to Cd caused a considerable enhancement in MT concentration in plants. Rice seedlings in the nursery, when exposed to abnormally high-temperature stress or low-light stress, revealed an increased MT synthesis due to enhanced *N*-acetylserotonin methyltransferase as well as serotonin *N*-acetyltransferase activities (Byeon and Back, 2015). All the evidence showed that MT takes part in adaptive roles against environmental abnormality and/or any chemical stress. High MT levels also have a role in protecting the seeds during sprouting and germination, as an increased MT concentration was recorded in sprouting seeds of sunflowers (Cho et al., 2008). The MT content varied among plant organs and was reported to be the highest in reproductive portions, particularly seeds (Arnao and Hernández-Ruiz, 2009c).

16.5 MECHANISMS OF MELATONIN-MEDIATED STRESS TOLERANCE AND DETOXIFICATION

MT is a molecule that contains both polar and non-polar portions and can very smoothly diffuse through plasmalemma into cell cytoplasm and reach the sub-cellular compartments. MT acts as a plant growth regulator for the management of different environmental stresses (Marta et al., 2016; Zhang et al., 2017a, 2017b, 2017c, 2017d). The main mechanism of action against abiotic stresses in crop plants is the enhancement of photosynthetic capacity by decreasing chlorophyll deterioration. ROS production was decreased by MT through balance in the distribution of photosynthetic electron flux as well as stabilizing of photosystem II in abiotic stress conditions in tomato (Yin et al., 2019). Earlier application of MT in heat-stressed seedlings produced 212% higher proline content than control (Jahan et al., 2019). Exogenous application of MT improved the water retention and presented 7% more relative water content than the untreated high-temperature stressed seedlings (Jahan et al., 2019). MT increased chlorophyll content as well as the rate of photosynthesis under low-temperature stress situations in watermelon (Zhang et al. 2017a, 2017b, 2017c, 2017d) and high-temperature stress situations in perennial ryegrass (Zhang et al., 2017a, 2017b, 2017c, 2017d).

16.6 EXOGENOUS APPLICATION OF MELATONIN AGAINST THERMO-STRESS

Among the thermo-stresses, heat stress is more prominent in the era of climate change where global warming is a major concern. MT has enough potential against high-temperature stress. In heat-stressed conditions, some alterations are noted in physiological activities as well as in gene expression; however, increased endogenous MT level under heat-stressed conditions (Ahammed et al., 2019) enhances improved stress tolerance by reducing oxidative damages (Figure 16.2). Exogenous application of MT is known to provide thermotolerance (Jahan et al., 2019; Zeng et al., 2022). The defense mechanisms include RuBisCo enzyme stability, proline metabolism, antioxidant-induced ROS reduction, heat shock protein (HSP) mediated alteration, and regulation of abscisic acid and cytokinin (Xu et al., 2016; Zhang et al., 2017e; Imran et al., 2021). Research evidence revealed the importance of the exogenous application of MT in plants in combating heat stress, as represented in Table 16.2.

Similarly, low-temperature conditions are harmful to plants, and under low-temperature conditions, MT plays a pivotal role in stress alleviation (Fig. 3). The plants suffering from low temperature show redox imbalance, excess ROS accumulation, and enhancement in ABA synthesis that activate C-repeat binding factor pathway (Zeng et al., 2022). Low temperature induces endogenous MT accumulation in plants for protecting them against severe stresses (Kang et al., 2010). MT can provide long-distance signals, and the signals are transported from roots to the aboveground parts (Li et al., 2021a). In general, under low-temperature stress conditions, exogenous application of MT is beneficial as it modulates phytohormone biosynthesis, reduces lignin content, triggers the accumulation of phenols, produces beneficial secondary metabolites, and alleviates ROS accumulation and lipid peroxidation (Lee and Back, 2019; Zeng et al., 2022). Based on the earlier studies, impacts of exogenous application of MT have been tabulated in Table 16.3.

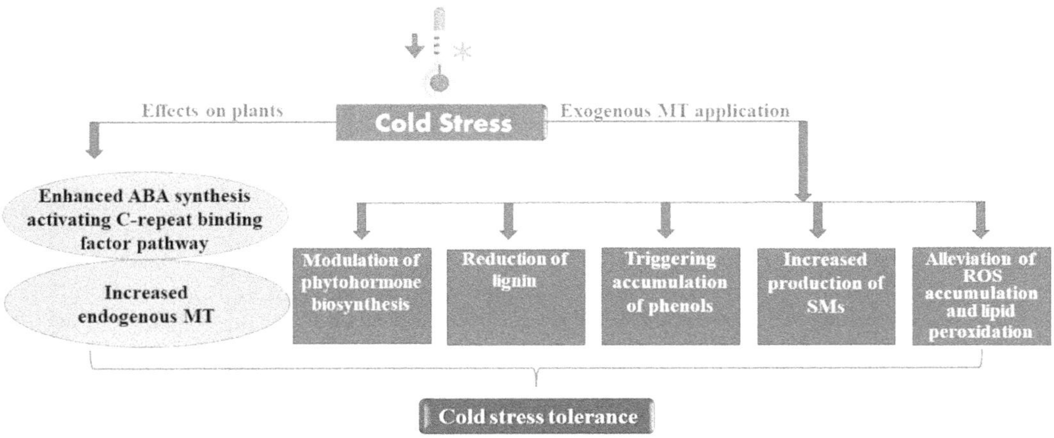

FIGURE 16.2 Impact of exogenous MT during cold stress.

[SMs = Secondary metabolites]

TABLE 16.2
Effect of Exogenous Melatonin Application during High Temperature

Sl. No	Crop	Effect	Dose	References
1.	Tomato	• Reduced generation of ROS (H_2O_2, MDA), increased antioxidant enzyme actions; altered AsA-GSH cycle, and elevated antioxidant-mediated gene expression	100 µM	Jahan et al. (2019)
2.	Wheat	• Increasing the plant dry matter and leaf area by 43.4% and 21.7%, respectively, increase in stomatal conductance, photosynthetic and transpiration rates, and chlorophyll contents	100 µM	
3.	Creeping bent grass	• Significant effect on the antioxidant activity of SOD and POD	200 µM	Merewitz and Liu (2019)
4.	Cherry Radish	• MT foliar spraying significantly enhanced biomass, chlorophyll content, RuBisCo activity	67.0 mg L^{-1} 7 days after imposition of stress.	Jia et al. (2019)
5.	Rye Grass	• Increase in chlorophyll content, plant height, dry weight, photosynthetic rate, membrane stability, endogenous cytokinin, and reduction in ABA accumulation	20 µM	Zhang et al. (2017f)
6.	Soybean	• Reduction of H_2O_2 production, lipid peroxidation, MDA accumulation, and electrolyte leakage	100 µM	Imran et al. (2021)
7.	Crysanthemum	• Reduction in MDA and H_2O_2 production and rate of superoxide anion production	200 µM	Xing et al. (2021)
8.	Strawberry	• Improvement in proline content and higher activity of APX, CAT, GPX, and GSH and expression of heat shock proteins	100 µM	Manafi et al. (2022)

(Continued)

TABLE 16.2 *(Continued)*
Effect of Exogenous Melatonin Application during High Temperature

Sl. No	Crop	Effect	Dose	References
9.	Kiwifruit	• MT increases carotenoid biosynthesis and regulates the expression of heat shock proteins	200 μM	Xia et al. (2021)
10.	Peppermint	• Increase in the activity of CAT, SOD, GST, and POX	30 M	Haydari et al. (2019)
11.	Arabidopsis	• Up regulation of heat-shock factors	20 μM	Shi et al. (2015)
12.	Maize	• Increase in photosynthesis and reduction in oxidative damage	100 μM	Ye et al. (2016)

TABLE 16.3
Effect of Exogenous Melatonin Application during Low Temperature

Sl. No	Crop	Effect	Dose	References
1.	Barley	• Increase in germination and the growth of seedlings; up regulation of genes like *HvCCA1* and HvTOC1	1 μM	Chang et al. (2021)
2.	Watermelon	• Increase in the concentration of MT in the plant and accumulation of MeJA, chlorophyll fluorescence, up regulation of genes like *ClCBF1* and *ClCBF2*	150 μM	Li et al. (2021a)
3.	Tea	• The net photosynthetic rate increased by 99.3%. SOD showed an increase in enzyme activity by 105.6%, followed by CAT (81.6%) and GR (80.2%), decrease in MDA by 78.1%	100 μM	Li et al. (2018a)
4.	Wheat	• Increased nitrate reductase and glutamine synthetase activity, dry matter accumulation, root/shoot ratio, and nutrient uptake. Additionally, it increased stomatal conductance, photosynthetic efficiency, and Cu/Zn SOD expression, all of which aid in improving cold tolerance.	1 μM (20°C) 1mM (10/4°C DNT)	Qiao et al. (2019) Sun et al. (2018)
5.	Maize	• MT application boosts overall growth, chlorophyll content, relative water content, and higher concentrations of magnesium, selenium, potassium, boron, iron, and zinc	1 mM	Turk and Erdal (2015)
6.	Sapota	• MT application reduced electrolyte leakage, MDA contents, and production of free radicals	90 μM	Mirshekari et al. (2020)
7.	Tomato	• Its supplementation reduced chilling injury, ion leakage, MDA accumulation, and H_2O_2 production	200 μM (5°C)	Azadshahraki et al. (2018)
8.	Tomato	• Exogenous MT reduced the electrolyte leakage, MDA, and H_2O_2 accumulation under cold stress	100 μM (4°C)	Ding et al. (2017)
9.	Pepper	• It improved membrane stability and reduced the MDA and H_2O_2 accumulation	25 μM (15°C)	Korkmaz et al. (2017)

TABLE 16.3 ***(Continued)***
Effect of Exogenous Melatonin Application in Low Temperature

Sl. No	Crop	Effect	Dose	References
10.	Melon	• Its application reduced the MDA contents and ROS production	400 μM (12/6°C DNT)	Zhang et al. (2017f)
11.	Cucumber	• Increased activity of the enzymes glutathione reductase, ascorbate peroxidase, and mono dehydro ascorbate reductase (MDHAR)	200 μM (15/8°C DNT)	Zhao et al. (2016)
12.	Cucumber	• Enhances catalase, peroxidase, glutathione peroxidase, and superoxide dismutase activities	500 μM (10°C)	Marta et al. (2016)

16.7 CONCLUSION

MT biosynthesis under thermo-stress is a natural process of crop plants. Exogenous application of MT shows a prominent impact in thermo-stress alleviation, mainly through alteration in gene expression and reduction in ROS accumulation. Change in gene expression also results in the production of polyamines, proline, and other protectants. Overall, melatonin plays a major role in maintaining the chlorophyll in leaves and photosynthetic rate of the crops as well as energy storage in mitochondria. This molecule definitely has a lot of potential to improve crop protection during temperature stress for enhanced yield and productivity.

REFERENCES

Aazami, M. A., Asghari-Aruq, M., Hassanpouraghdam, M. B., Ercisli, S., Baron, M. and Sochor, J. 2021. Low-temperature stress mediates the antioxidants pool and chlorophyll fluorescence in *Vitis vinifera* L. cultivars. *Plants.*, *10*(9), 1877.

Afreen, F., Zobayed, S. and Kozai, T. 2006. Melatonin in *Glycyrrhiza uralensis*: Response of plant roots to spectral quality of light and UV-B radiation. *J. Pineal Res.*, *41*, 108–115.

Ahammed, G. J., Xu, W., Liu, A. and Chen, S. 2019. Endogenous melatonin deficiency aggravates high temperature-induced oxidative stress in *Solanum lycopersicum* L. *Environ. Exp. Bot.*, *161*, 303–311. doi: 10.1016/j. envexpbot.2018.06.006

Arnao, M. B. and Hernández-Ruiz, J. 2006. The physiological function of melatonin in plants. *Plant Signaling Behav.*, *1*(3), 89–95.

Arnao, M. B. and Hernández-Ruiz, J. 2009a. Protective effect of melatonin against chlorophyll degradation during the senescence of barley leaves. *J. Pineal Res.*, *46*(1), 58–63.

Arnao, M. B. and Hernández-Ruiz, J. 2009b. Assessment of different sample processing procedures applied to the determination of melatonin in plants. *Phytochem. Analysis.*, *20*(1), 14–18.

Arnao, M. B. and Hernández-Ruiz, J. 2009c. Chemical stress by different agents affects the melatonin content of barley roots. *J. Pineal Res.*, *46*, 295–299.

Arnao, M. B. and Hernández-Ruiz, J. 2013. Growth conditions determine different melatonin levels in *Lupinus albus* L. *J. Pineal Res.*, *55*, 149–155.

Arnao, M. B. and Hernández-Ruiz, J. 2014. Melatonin: Plant growth regulator and or biostimulator during stress. *Trends Plant Sci.*, *19*(12), 789–797.

Asif, M., Pervez, A. and Ahmad, R. 2019. Role of melatonin and plant-growth-promoting rhizobacteria in the growth and development of plants. *Clean: Soil, Air, Water*, *47*(6), 1800459.

Azadshahraki, F., Jamshidi, B. and Mohebbi, S. 2018. Postharvest melatonin treatment reduces chilling injury and enhances antioxidant capacity of tomato fruit during cold storage. *Adv. Hort. Sci.*, *32*, 299–309.

Bajwa, V. S., Shukla, M. R., Sherif, S. M., Murch, S. J. and Saxena, P. K. 2014. Role of melatonin in alleviating cold stress in *Arabidopsis thaliana*. *J. Pineal Res.*, *56*(3), 238–245.

Banerjee, A. and Roychoudhury, A. 2017. Melatonin as a regulator of abiotic stress tolerance in plants. In: Singh VP, Singh S, Prasad SM (Eds.) *Mechanisms Behind Phytohormonal Signaling and Crop Abiotic Stress Tolerance*. New York: Nova Science Publishers, pp. 47–60.

Barand, A., Nasibi, F., Kalantari, K. M. and Moradi, M. 2020. The effects of foliar application of melatonin on some physiological and biochemical characteristics and expression of fatty acid desaturase gene in pistachio seedlings (*Pistacia vera* L.) under freezing stress *J. Plant Interactions*., *15*, 257–265, doi:10.1080/17429145.2020.1781271

Bhat, M. A., Kumar, V., Bhat, M. A., Wani, I. A. Dar, F. L., Farooq, I., Bhatti, F., Koser, R., Rahman, S., and Jan, A. T. 2020. Mechanistic insights of the interaction of plant growth-promoting rhizobacteria (PGPR) with plant roots toward enhancing plant productivity by alleviating salinity stress. *Front. Microbiol*., *11*, 1952.

Bonilla, E., Valero, N., Chacín-Bonilla, L. and Medina-Leendertz, S. 2004. Melatonin and viral infections. *J. Pineal Res*., *36*(2), 73–79.

Boyer, J. S. 1982. Plant productivity and environment. *Science*, *218*(4571), 443–448.

Byeon, Y. and Back, K. W. 2015. Molecular cloning of melatonin 2-hydroxylase responsible for 2 hydroxymelatonin production in rice (*Oryza sativa*). *J. Pineal Res*., *58*, 343–351.

Canter, L. W. 2018. *Environmental Impact of Agricultural Production Activities*. Boca Raton: CRC Press, p. 400.

Carrillo-Vico, A., Lardone, P. J., Álvarez-Sánchez, N., Rodríguez-Rodríguez A. and Guerrero, J. M. 2013. Melatonin: Buffering the immune system. *Int. J. Mol. Sci*., *14*(4), 8638–8683.

Chang, T., Zhao, Y., He, H., Xi, Q., Fu, J. and Zhao, Y. 2021. Exogenous melatonin improves growth in hulless barley seedlings under cold stress by influencing the expression rhythms of circadian clock genes. *Peer J*, *9*, e10740.

Chen, Q., Qi, W. B., Reiter, R. J., Wei, W. and Wang, B. M. 2009. Exogenously applied melatonin stimulates root growth and raises endogenous indoleacetic acid in roots of etiolated seedlings of *Brassica juncea*. *J. Plant Physiol*., *166*(3), 324–328.

Cho, M. H., No, H. K. and Prinyawiwatkul, W. 2008. Chitosan treatments affect growth and selected quality of sunflower sprouts. *J. Food Sci*., *73*, 70–77.

Cleland, R. E. 2004. Auxin and cell elongation. In: Davies PJ (Ed.). *Plant Hormones: Physiology, Biochemistry and Molecular Biology*. 3rd ed. Dordrecht: Kluwer Academic Publishers, pp. 204–220.

Das, K. and Roychoudhury, A. 2014. Reactive oxygen species (ROS) and response of antioxidants as ROS-scavengers during environmental stress in plants. *Frontiers in Environmental Science*, *2*, 53.

Decoteau, D. R. 2000. *Vegetable Crops* (No. 635 D3589v Ej. 1 025327). Madison: Prentice Hall.

Ding, F., Liu, B. and Zhang, S. 2017. Exogenous melatonin ameliorates cold induced damage in tomato plants. *Sci. Horticul*., *219*, 264–271.

Dos Reis, S. P.; Lima, A. M. and De Souza, C. R. B. 2012. Recent molecular advances on downstream plant responses to abiotic stress. *Int. J. Mol. Sci*., *13*, 8628–8647.

Dubbels, R., Reiter, R., Klenke, E., Goebel, A., Schnakenberg, E., Ehlers, C., Schiwara, H. and Schloot, W. 1995. Melatonin in edible plants identified by radioimmunoassay and by high performance liquid chromatography-mass spectrometry. *J. Pineal Res*., *18*, 28–31.

Esim, N. and Atici, O. 2014. Nitric oxide improves chilling tolerance of maize by affecting apoplastic antioxidative enzymes in leaves. *Plant Growth Reg*., *72*(1), 29–38.

Fan, J., Ren, J., Zhu, W., Amombo, E., Fu, J. and Chen, L. 2014. Antioxidant responses and gene expression in bermudagrass under cold stress. *J. Am. Soc. Horticul. Sci*., *139*(6), 699–705.

Feng, H., Yan, M., Fan, X., Li, B., Shen, Q., Miller, A. J. and Xu, G. 2011. Spatial expression and regulation of rice high-affinity nitrate transporters by nitrogen and carbon status. *J. Exp. Bot*., *62*(7), 2319–2332.

Gao, H., Zhang, Z. K., Chai, H. K., Cheng, N., Yang, Y., Wang, D. N and Cao, W. 2016. Melatonin treatment delays postharvest senescence and regulates reactive oxygen species metabolism in peach fruit. *Postharvest Biol. Technol*., *118*, 103–110.

Hardeland, R. 2009. Melatonin: Signaling mechanisms of a pleiotropic agent. *Biofactors*., *35*(2), 183–192.

Hasan, M., Ahammed, G. J., Yin, L., Shi, K., Xia, X., Zhou, Y., Yu, J. and Zhou, J. 2015. Melatonin mitigates cadmium phytotoxicity through modulation of phytochelatins biosynthesis, vacuolar sequestration, and antioxidant potential in *Solanum lycopersicum* L. *Front. Plant Sci*., *6*, 60.

Haydari, M., Maresca, V., Rigano, D., Taleei, A., Shahnejat-Bushehri, A. A., Hadian, J., Sorbo, S., Guida, M., Manna, C., Piscopo, M. and Notariale, R. 2019. Salicylic acid and melatonin alleviate the effects of heat stress on essential oil composition and antioxidant enzyme activity in *Mentha × piperita* and *Mentha arvensis* L. *Antioxidants*, *8*(11), 547.

Hernandez-Ruiz, J., Cano, A. and Arnao, M. B. 2004. Melatonin: A growth-stimulating compound present in lupin tissues. *Plants.*, *220*, 140–144.

Hu, Z., Fan, J., Xie, Y., Amombo, E., Liu, A., Gitau, M. M., Khaldun, A. B. M., Chen, L. and Fu, J. 2016. Comparative photosynthetic and metabolic analyses reveal mechanism of improved cold stress tolerance in bermudagrass by exogenous melatonin. *Plant Physiol. Biochem.*, *100*, 94–104.

Imran, M., Aaqil Khan, M., Shahzad, R., Bilal, S., Khan, M., Yun, B. W., Khan, A. L. and Lee, I. J. 2021. Melatonin ameliorates thermotolerance in soybean seedling through balancing redox homeostasis and modulating antioxidant defense, phytohormones and polyamines biosynthesis. *Molecules*, *26*, 5116. doi: 10.3390/molecules26175116

Jahan, M. S., Shu, S., Wang, Y., Chen, Z., He, M., Tao, M. and Guo, S. 2019. Melatonin alleviates heat-induced damage of tomato seedlings by balancing redox homeostasis and modulating polyamine and nitric oxide biosynthesis. *BMC Plant Biol.*, *19*(1), 1–16.

Jayaweera, M. and Kasturiarachchi, J. 2004. Removal of nitrogen and phosphorus from industrial wastewaters by phytoremediation using water hyacinth (*Eichhornia crassipes* (Mart.) Solms). *Water Sci. Technol.*, *50*, 217–225.

Jia, C., Yu, X., Zhang, M., Liu, Z., Zou, P., Ma, J. and Xu, Y. 2019 Application of melatonin-enhanced tolerance to high-temperature stress in cherry radish (*Raphanus sativus* L. var. radculuspers). *J. Plant Growth Reg.*, *39*, 631–640.

Kang, K., Lee, K., Park, S., Kim, Y. S. and Back, K. 2010. Enhanced production of melatonin by ectopic overexpression of human serotonin N-acetyltransferase plays a role in cold resistance in transgenic rice seedlings. *J. Pineal Res.*, *49*, 176–182. doi: 10.1111/j.1600-079X.2010.00783.x

Ke, Q., Ye, J., Wang, B., Ren, J., Yin, L., Deng, X. and Wang, S. 2018. Melatonin mitigates salt stress in wheat seedlings by modulating polyamine metabolism. *Front. Plant Sci.*, *9*, 914.

Kim, J. H., Kim, W. T. and Kang, B. G. 2001. IAA and N 6-benzyladenine inhibit ethylene-regulated expression of *ACC oxidase* and *ACC synthase* genes in mungbean hypocotyls. *Plant Cell Physiol.*, *42*(10), 1056–1061.

Kołodziejczyk, I., Bałabusta, M., Szewczyk, R. and Posmyk, M. M. 2015. The levels of melatonin and its metabolites in conditioned corn (*Zea mays* L.) and cucumber (*Cucumis sativus* L.) seeds during storage. *Acta Physiologiae Plantarum.*, *37*(6), 1–11.

Kołodziejczyk, I., Dzitko, K., Szewczyk, R. and Posmyk, M. M. 2016. Exogenous melatonin improves corn (*Zea mays* L.) embryo proteome in seeds subjected to chilling stress. *J. Plant Physiol.*, *193*, 47–56.

Kołodziejczyk, I., Kazmierczak, A. and Posmyk, M. M. 2021. Melatonin application modifies antioxidant defense and induces endoreplication in maize seeds exposed to chilling stress. *Int. J. Mol. Sci.*, *22*, 8628. doi: 10.3390/ijms22168628

Korkmaz, A., Karaca, A., Kocacinar, F. and Cuci, Y. 2017. The effects of seed treatment with melatonin on germination and emergence performance of pepper seeds under chilling stress. *J. Agric. Sci.*, *23*, 167–176.

Kratsch, H. A. and Wise, R. R. 2000. The ultrastructure of chilling stress. *Plant, Cell & Environment*, *23*(4), 337–350.

Krishnan, R., Sanjay, J., Gnanaseelan, C., Mujumdar, M., Kulkarni, A. and Chakraborty, S. 2020. *Assessment of Climate Change Over the Indian Region: A Report of the Ministry of Earth Sciences (MOES), Government of India*. Singapore: Springer, p. 226.

Lazar, D., Murch, S. J., Beilby, M. J. and Al Khazaaly, S. 2013. Exogenous melatonin affects photosynthesis in characeae *Chara australis*. *Plant Signaling Behav.*, *8*(3), e23279.

Lee, H. J. and Back, K. 2019. 2-Hydroxymelatonin confers tolerance against combined cold and drought stress in tobacco, tomato, and cucumber as a potent anti-stress compound in the evolution of land plants. *Melatonin Res.*, *2*, 35–46. doi: 10.32794/mr11250020

Lee, H. Y., Byeon, Y. and Back, K. 2014. Melatonin as a signal molecule triggering defence responses against pathogen attack in Arabidopsis and tobacco. *J. Pineal Res.*, *57*(3), 262–268.

Lei, X. Y., Zhu, R. Y., Zhang, G. Y. and Dai, Y. R. 2004. Attenuation of cold-induced apoptosis by exogenous melatonin in carrot suspension cells: The possible involvement of polyamines. *J. Pineal Res.*, *36*, 126–131.

Li, C., Tan, D. X., Liang, D., Chang, C., Jia, D. and Ma, F. 2015. Melatonin mediates the regulation of ABA metabolism, free-radical scavenging, and stomatal behaviour in two *Malus* species under drought stress. *J. Exp. Bot.*, *66*(3), 669–680.

Li, H., Guo, Y., Lan, Z., Xu, K., Chang, J. and Ahammed, G. J. 2021a. Methyl jasmonate mediates melatonin-induced cold tolerance of grafted watermelon plants. *Horticul. Res.*, *8*, 57. doi: 10.1038/s41438-021-00496-0

Li, X., Brestic, M., Tan, D. X., Zivcak, M., Zhu, X., Liu, S., Song, F., Reiter, R. J. and Liu, F., 2018b. Melatonin alleviates low PSI-limited carbon assimilation under elevated CO_2 and enhances the cold tolerance of offspring in chlorophyll b-deficient mutant wheat. *J. Pineal Res.*, *64*(1), e12453.

Li, X., Tan, D. X., Jiang, D. and Liu, F. 2016. Melatonin enhances cold tolerance in drought-primed wild-type and abscisic acid-deficient mutant barley. *J. Pineal Res.*, *61*(3), 328–339.

Li, X., Wei, J. P., Scott, E. R., Liu, J. W., Guo, S., Li, Y., Zhang, L. and Han, W. Y. 2018a. Exogenous melatonin alleviates cold stress by promoting antioxidant defense and redox homeostasis in *Camellia sinensis* L. *Molecules*, *23*(1), 165.

Li, Z. G., Xiang, R. H. and Wang, J. Q. 2021b. Hydrogen sulfide: Phytohormone interaction in plants under physiological and stress conditions. *J. Plant Growth Reg*, 1–9.

Lyons, J. M. 1973. Chilling injury in plants. *Ann. Review Plant Physiol.*, *24*(1), 445–466.

Madebo, M. P., Hu, S., Zheng, Y. and Jin, P. 2021. Mechanism of chilling tolerance in melatonin treated postharvest fruits and vegetables: A review. *J. Future Foods.*, *1–2*, 56–167.

Mahajan, S. and Tuteja, N. 2005. Cold, salinity and drought stresses: An overview. *Arch. Biochem. Biophy.*, *444*(2), 139–158.

Manafi, H., Baninasab, B., Gholami, M., Talebi, M. and Khanizadeh, S. 2022. Exogenous melatonin alleviates heat-induced oxidative damage in strawberry (*Fragaria* × *ananassa* Duch. cv. Ventana) Plant. *J. Plant Growth Reg.*, *41*(1), 52–64.

Manepalli, S. B., Tomar, S., Gaikwad, D. J. and Maitra, S. 2022. Abiotic stress signaling in plants and transgenic technology as a triumph: A review. *J. Appl. Biol. Biotechnol.*, *10*(5), 5–13.

Marta, B., Szafranska, K. and Posmyk, M. M. 2016. Exogenous melatonin improves antioxidant defense in cucumber seeds (*Cucumis sativus* L.) germinated under chilling stress. *Front. Plant Sci.*, *7*, 575. doi: 10.3389/fpls.2016.00575.

Martinez, V., Nieves-Cordones, M., Lopez-Delacalle, M., Rodenas, R., Mestre, T. C., Garcia-Sanchez, F. and Rivero, R. M. 2018. Tolerance to stress combination in tomato plants: New insights in the protective role of melatonin. *Molecules*, *23*(3), 535.

Meng, J. F., Xu, T. F., Wang, Z. Z., Fang, Y. L., Xi, Z. M. and Zhang, Z. W. 2014. The ameliorative effects of exogenous melatonin on grape cuttings under water-deficient stress: antioxidant metabolites, leaf anatomy, and chloroplast morphology. *J. Pineal Res.*, *57*(2), 200–212.

Merewitz, E. B. and Liu, S. 2019. Improvement in heat tolerance of creeping Bentgrass with melatonin, rutin, and silicon. *J. Am. Soc. Horticul. Sci.*, *144*(2), 141–148.

Mirshekari, A., Madani, B., Yahia, E. M., Golding, J. B. and Vand, S. H. 2020. Postharvest melatonin treatment reduces chilling injury in sapota fruit. *J. Sci. Food Agricul.*, *100*, 1897–1903. doi: 10.1002/jsfa.10198.

Misbahuddin, M. and Fariduddin, A. 2002. Water hyacinth removes arsenic from arsenic-contaminated drinking water. *Arch. Environ. Health.*, *57*, 516–518.

Mukherjee, S., David, A., Yadav, S., Baluška, F. and Bhatla, S. C. 2014. Salt stress-induced seedling growth inhibition coincides with differential distribution of serotonin and melatonin in sunflower seedling roots and cotyledons. *Physiologia Plantarum.*, *152*(4), 714–728.

Munavalli, G. R. and Saler, P. S. 2009. Treatment of dairy wastewater by water hyacinth. *Water Sci. Technol.*, *59*, 713–722.

Nadeem, M., Li, J., Wang, M., Shah, L., Lu, S., Wang, X. and Ma, C. 2018. Unraveling field crops sensitivity to heat stress: Mechanisms, approaches, and future prospects. *Agron.*, *8*(7), 128.

Nadeem, M., Li, J., Yahya, M., Wang, M., Ali, A., Cheng, A. and Ma, C. 2019. Grain legumes and fear of salt stress: Focus on mechanisms and management strategies. *Int. J. Mol. Sci.*, *20*(4), 799.

Okazaki, M. and Ezura, H. 2009. Profiling of melatonin in the model tomato (*Solanum lycopersicum* L.) cultivar Micro-Tom. *J. Pineal Res.*, *46*(3), 338–343.

Parida, A. K. and Das, A. B. 2005. Salt tolerance and salinity effects on plants: A review. *Ecotoxicol. Environm. Saf.*, *60*(3), 324–349.

Pelagio-Flores, R., Muñoz-Parra, E., Ortiz-Castro, R. and López-Bucio, J. 2012. Melatonin regulates Arabidopsis root system architecture likely acting independently of auxin signaling. *J. Pineal Res.*, *53*(3), 279–288.

Pierce, L. C. 1987. *Vegetables: Characteristics, Production and Marketing.* Hoboken: John Willey and Sons Inc, p. 433.

Posmyk, M. M., Kuran, H., Marciniak, K. and Janas, K. M. 2008. Presowing seed treatment with melatonin protects red cabbage seedlings against toxic copper ion concentrations. *J. Pineal Res.*, *45*(1), 24–31.

Qiao, Y., Yin, L., Wang, B., Ke, Q., Deng, X. and Wang, S. 2019. Melatonin promotes plant growth by increasing nitrogen uptake and assimilation under nitrogen deficient condition in winter wheat. *Plant Physiol. Biochem.*, *139*, 342–349. doi: 10.1016/j.plaphy.2019.03.037

Raza, A., Charagh, S., García-Caparrós, P., Rahman, M. A., Vincent, H. O., Faisal, S. and Wanmei, J. 2022. Melatonin-mediated temperature stress tolerance in plants, *GM Crops Food.*, *13*(1), 196–217. doi:10.1080/21645698.2022.2106111

Repkina, N., Ignatenko, A., Holoptseva, E., MiszalskI, Z., Kaszycki, P. and Talanova, V. 2021. Exogenous methyl jasmonate improves cold tolerance with parallel induction of two cold-regulated (COR) genes expression in *Triticum aestivum* L. *Plants.*, *10*, 1421. doi: 10.3390/plants10071421

Riga, P., Medina, S., Garcia-Flores, L. A. and Gil-Izquierdo, A. 2014. Melatonin content of pepper and tomato fruits: Effects of cultivar and solar radiation. *Food Chem.*, *156*, 347–352.

Roychoudhury, A., Banerjee, A. and Lahiri, V. 2015. Metabolic and molecular-genetic regulation of proline signaling and its cross-talk with major effectors mediates abiotic stress tolerance in plants. *Turkish Journal of Botany*, *39*(6), 887–910.

Roychoudhury, A., Basu, S. and Sengupta, D. N. 2011b. Amelioration of salinity stress by exogenously applied spermidine or spermine in three varieties of indica rice differing in their level of salt tolerance. *Journal of Plant Physiology*, *168*, 317–328.

Roychoudhury, A., Datta, K., and Datta S. K. 2011a. Abiotic stress in plants: From genomics to metabolomics. In: Tuteja N, Gill SS, Tuteja R (Eds.). *Omics and Plant Abiotic Stress Tolerance*. Sharjah, UAE: Bentham Science Publishers, pp. 91–120.

Salvucci, M. E. and Crafts-Brandner, S. J. 2004. Relationship between the heat tolerance of photosynthesis and the thermal stability of RuBisCo activase in plants from contrasting thermal environments. *Plant Physiol.*, *134*(4), 1460–1470.

Samanta, S., Banerjee, A. and Roychoudhury, A. 2022. Arsenic toxicity is counteracted by exogenous application of melatonin to different extents in arsenic-susceptible and arsenic-tolerant rice cultivars. *Journal of Plant Growth Regulation*, *41*, 2210–2231.

Sharif, R., Xie, C., Zhang, H., Arnao, M. B., Ali, M., Ali, Q., Muhammad, I., Shalmani, A., Nawaz, M. A., Chen, P. and Li, Y. 2018. Melatonin and its effects on plant systems. *Molecules*, *23*(9), 2352.

Shi, H., Reiter, R. J., Tan, D. X. and Chan, Z. 2015. Indole-3-acetic acid inducible 17 positively modulates natural leaf senescence through melatonin-mediated pathway in Arabidopsis. *J. Pineal Res.*, *58*(1), 26–33.

Shida, C., Castrucci, A. and Lamyfreund, M. 1994. High melatonin solubility in aqueous-medium. *J. Pineal Res.*, *16*, 198–201.

Simopoulos, A., Tan, D. X., Manchester, L. C. and Reiter, R. J. 2005. Purslane: A plant source of omega-3 fatty acids and melatonin. *J. Pineal Res.*, *39*, 331–332.

Sun, L., Li, X., Wang, Z., Sun, Z., Zhu, X. and Liu, S. 2018. Cold priming induced tolerance to subsequent low temperature stress is enhanced by melatonin application during recovery in wheat. *Molecules*, *23*, 1091. doi: 10.3390/molecules23051091.

Sun, Q., Zhang, N., Wang, J., Zhang, H., Li, D., Shi, J., Li, R., Weeda, S., Zhao, B., Ren, S. and Guo, Y.D. 2015. Melatonin promotes ripening and improves quality of tomato fruit during postharvest life. *J. Exp. Bot.*, *66*(3), 657–668.

Tal, O., Haim, A., Harel, O. and Gerchman, Y. 2011. Melatonin as an antioxidant and its semi-lunar rhythm in green macroalga *Ulva* sp. *J. Exp. Bot.*, *62*, 1903–1910.

Tan, D. X., Hardeland, R., Manchester, L. C., Korkmaz, A., Ma, S., Rosales-Corral, S. and Reiter, R. J. 2012. Functional roles of melatonin in plants, and perspectives in nutritional and agricultural science. *J. Exp. Bot.*, *63*(2), 577–597.

Tan, D. X., Manchester, L. C., Di, Mascio. P, Martinez G. R., Prado, F. M. and Reiter, R. J. 2007. Novel rhythms of N-1-acetyl-N-2-formyl-5- methoxy kynuramine and its precursor melatonin in water hyacinth: Importance for phytoremediation. *FASEB J.*, *21*, 1724–1729.

Tiwari, R. K., Lal, M. K., Naga, K. C., Kumar, R., Chourasia, K. N., Subhash, S., Kumar, D. and Sharma, S. 2020. Emerging roles of melatonin in mitigating abiotic and biotic stresses of horticultural crops. *Sci. Horticul.*, *272*, 109592.

Trivedy, R. and Pattanshetty, S. 2002. Treatment of dairy waste by using water hyacinth. *Water Sci. Technol.*, *45*, 329–334.

Turk, H. and Erdal, S. 2015. Melatonin alleviates cold-induced oxidative damage in maize seedlings by up-regulating mineral elements and enhancing antioxidant activity. *J. Plant Nut. Soil Sci., 178*, 433–439.

Vaughan, M. M., Block, A., Christensen, S. A., Allen, L. H. and Schmelz, E. A. 2017. The effects of climate change associated abiotic stresses on maize phytochemical defences. *Phytochem. Rev., 17*, 37–49.

Wang, L., Zhao, Y., Reiter, R. J., He, C., Liu, G., Lei, Q., Zuo, B., Zheng, X., Li, Q. and Kong, J. 2014. Changes in melatonin levels in transgenic 'Micro-Tom'tomato overexpressing ovine *AANAT* and ovine *HIOMT* genes. *J. Pineal Res., 56*(2), 134–142.

Wang, P., Sun, X., Li, C., Wei, Z., Liang, D. and Ma, F. 2013. Long-term exogenous application of melatonin delays drought-induced leaf senescence in apple. *J. Pineal Res., 54*(3), 292–302.

Waqas, M. A., Khan, I., Akhter, M. J., Noor, M. A. and Ashraf, U. 2017. Exogenous application of plant growth regulators (PGRs) induces chilling tolerance in short-duration hybrid maize. *Environ. Sci. Poll. Res., 24*, 11459–11471.

Weeda, S., Zhang, N., Zhao, X., Ndip, G., Guo, Y., Buck, G. A., Fu, C. and Ren, S. 2014. Arabidopsis transcriptome analysis reveals key roles of melatonin in plant defense systems. *PloS one, 9*(3), e93462.

Wolf, K., Kolar, J., Witters, van Dongen, W., van Onckelen, H. and Machackova, I. 2001. Daily profile of melatonin levels in *Chenopodium rubrum* depends on photoperiod. *J. Plant Physiol., 158*, 1491–1493.

Xia., H. and Ma, X. 2006. Phytoremediation of ethion by water hyacinth (*Eichhornia crassipes*) from water. *Biores. Technol., 97*, 1050–1054.

Xia, H., Zhou, Y., Deng, H., Lin, L., Deng, Q., Wang, J., Lv, X., Zhang, X. A. and Liang, D. 2021. Melatonin improves heat tolerance in *Actinidia deliciosa* via carotenoid biosynthesis and heat shock proteins expression. *Physiologia Plantarum., 172*(3), 1582–1593.

Xing, X., Ding, Y., Jin, J., Song, A., Chen, S., Chen, F., Fang, W. and Jiang, J. 2021. Physiological and transcripts analyses reveal the mechanism by which melatonin alleviates heat stress in chrysanthemum seedlings. *Front. Plant Sci.*, 2012.

Xu, W., Cai, S. Y., Zhang, Y., Wang, Y., Ahammed, G. J., Xia, X. J., et al. 2016. Melatonin enhances thermotolerance by promoting cellular protein protection in tomato plants. *J. Pineal Res., 61*, 457–469. doi: 10.1111/jpi.12359

Xu, X. D. 2010. Effects of exogenous melatonin on physiological response of cucumber seedlings under high temperature stress. *Master's Degree Thesis, Northwest A and F University.*

Yamamoto, Y., Aminaka, R., Yoshioka, M., Khatoon, M., Komayama, K., Takenaka, D., Yamashita, A., Nijo, N., Inagawa, K., Morita, N. and Sasaki, T. 2008. Quality control of photosystem II: Impact of light and heat stresses. *Photosynt. Res., 98*(1), 589–608.

Ye, J., Wang, S., Deng, X., Yin, L., Xiong, B. and Wang, X. 2016. Melatonin increased maize (*Zea mays* L.) seedling drought tolerance by alleviating drought-induced photosynthetic inhibition and oxidative damage. *Acta Physiol. Plant., 38*(2), 1–13.

Yin, L., Wang, P., Li, M., Ke, X., Li, C., Liang, D., Wu, S., Ma, X., Li, C., Zou, Y. and Ma, F., 2013. Exogenous melatonin improves *Malus* resistance to *M arssonina* apple blotch. *J. Pineal Res., 54*(4), 426–434.

Yin, Z., Lu, J., Meng, S., Liu, Y., Mostafa, I., Qi, M. and Li, T. 2019. Exogenous melatonin improves salt tolerance in tomato by regulating photosynthetic electron flux and the ascorbate: Glutathione cycle. *J. Plant Interaction., 14*(1), 453–463.

Zafar, S. A., Noor, M. A., Waqas, M. A., Wang, X., Shaheen, T., Raza, M. and Rahman, M. U. 2018. Temperature extremes in cotton production and mitigation strategies. *Past, Present and Future Trends in Cotton Breeding*, 65–91.

Zeng, W., Mostafa, S., Lu, Z., Jin, B. 2022. Melatonin-mediated abiotic stress tolerance in plants. *Front Plant Sci.*, May 9; *13*, 847175. doi: 10.3389/fpls.2022.847175.

Zeng, Y., Wen, J., Zhao, W., Wang, Q. and Huang, W. 2020. Rational improvement of rice yield and cold tolerance by editing the three genes *OsPIN5b, GS3*, and *OsMYB30* with the CRISPR-Cas9 system. *Front Plant Sci., 10*, 1663. doi:10.3389/fpls.2019.01663

Zhang, H., Liu, L., Wang, Z., Feng, G., Gao, Q. and Li, X. 2021. Induction of low temperature tolerance in wheat by pre-soaking and parental treatment with melatonin. *Molecules, 26*, 1192. doi: 10.3390/molecules26041192

Zhang, H. J., Zhang, N. A., Yang, R. C., Wang, L., Sun, Q. Q., Li, D. B., Cao, Y. Y., Weeda, S., Zhao, B., Ren, S. and Guo, Y. D. 2014a. Melatonin promotes seed germination under high salinity by regulating antioxidant systems, ABA and GA 4 interaction in cucumber (*Cucumis sativus* L.). *J. Pineal Res., 57*(3), 269–279.

Zhang, J., Shi, Y., Zhang, X., Du, H., Xu, B. and Huang, B. 2017a. Melatonin suppression of heat-induced leaf senescence involves changes in abscisic acid and cytokinin biosynthesis and signaling pathways in perennial ryegrass (*Lolium perenne* L.). *Environ. Exp. Bot.*, *138*, 36–45. doi: 10.1016/j.envexpbot.2017.02.012

Zhang, N., Sun, Q., Li, H., Li, X., Cao, Y., Zhang, H., Li, S., Zhang, L., Qi, Y., Ren, S. and Zhao, B. 2016. Melatonin improved anthocyanin accumulation by regulating gene expressions and resulted in high reactive oxygen species scavenging capacity in cabbage. *Front. Plant Sci.*, *7*, 197.

Zhang, N., Sun, Q., Zhang, H., Cao, Y., Weeda, S., Ren, S. and Guo, Y. D. 2015. Roles of melatonin in abiotic stress resistance in plants. *J. Exp. Bot.*, *66*, 647–656.

Zhang, N., Zhang, H. J., Sun, Q. Q., Cao, Y. Y., Li, X., Zhao, B., Wu, P. and Guo, Y. D. 2017b. Proteomic analysis reveals a role of melatonin in promoting cucumber seed germination under high salinity by regulating energy production. *Sci. Rep.*, *7*(1), 503.

Zhang, N., Zhang, H. J., Zhao, B., Sun, Q. Q., Cao, Y. Y., Li, R., Wu, X. X., Weeda, S., Li, L., Ren, S. and Reiter, R. J. 2014b. The RNA-seq approach to discriminate gene expression profiles in response to melatonin on cucumber lateral root formation. *J. Pineal Res.*, *56*(1), 39–50.

Zhang, N., Zhao, B., Zhang, H. J., Weeda, S., Yang, C., Yang, Z. C., Ren, S. and Guo, Y. D. 2013. Melatonin promotes water-stress tolerance, lateral root formation, and seed germination in cucumber (*Cucumis sativus* L.). *J. Pineal Res.*, *54*(1), 15–23.

Zhang, R., Sun, Y., Liu, Z., Jin, W., and Sun, Y. 2017e. Effects of melatonin on seedling growth, mineral nutrition, and nitrogen metabolism in cucumber under nitrate stress. *J. Pineal Res.*, *62*(4), e12403.

Zhang, Y., Xu, S., Yang, S. and Chen, Y. 2017c. Melatonin alleviates cold-induced oxidative damage by regulation of ascorbate: Glutathione and proline metabolism in melon seedlings (*Cucumis melo* L.). *J. Horticul. Sci. Biotechnol.*, *92*(3), 313–324.

Zhang, Y., Yang, S. and Chen, Y. 2017d. Effects of melatonin on photosynthetic performance and antioxidants in melon during cold and recovery. *Biol. Plantarum.*, *61*(3), 571–578. doi: 10.1007/s10535-017-0717-8

Zhao, H., Ye, L., Wang, Y., Zhou, X., Yang, J., Wang, J., Cao, K. and Zou, Z. 2016. Melatonin increases the chilling tolerance of chloroplast in cucumber seedlings by regulating photosynthetic electron flux and the ascorbate-glutathione cycle. *Front. Plant Sci.*, *7*, 1814.

Zörb, C., Geilfus, C. and Dietz, K. 2018. Salinity and crop yield. *Plant Biol.*, *21*, 31–38.

Zuo, B., Zheng, X., He, P., Wang, L., Lei, Q., Feng, C., Zhou, J., Li, Q., Han, Z. and Kong, J. 2014. Overexpression of MzASMT improves melatonin production and enhances drought tolerance in transgenic Arabidopsis thaliana plants. *J. Pineal Res.*, *57*(4), 408–417.

17 Melatonin-Induced Plant Tolerance to Heavy Metal Stress

A. Sher, A. Sattar, S. Ul-Allah, M. Ijaz, B. Sarwar, R. Calone and L. Barbanti

LIST OF ABBREVIATIONS

Abbreviation	Full name
APX	Ascorbate peroxidase
AsA	Ascorbic acid
ASMT	Acetylserotonin methyltransferase
CA	Carbonic anhydrase
CAT	Catalase
Chl	Chlorophyll
CHS	Chalcone synthase
COMT	Caffeoyl-O-methyltransferase
CWI	Cell wall invertase
GOx	Glycolate oxidase
GPX	Glutathione peroxidase
GR	Glutathione reductase
GS	Glutamine synthetase
GSH	Reduced glutathione
H^+-ATPase	Proton pump ATPase
MDA	Malondialdehyde
P5CS	1-pyrroline-5-carboxylate synthetase
PAL	Phenylalanine ammonia-lyase
PDH	Proline dehydrogenase
POD	Peroxidase
PRX	Peroxiredoxin
RNS	Reactive nitrogen species
ROS	Reactive oxygen species
RWC	Relative water content
SNAT	Serotonin N-acetyltransferase
SOD	Superoxide dismutase
T5H	Tryptamine 5-hydroxylase
TDC	Tryptophan decarboxylase
TSC	Total soluble carbohydrate
WSS	Water-soluble sugar

DOI: 10.1201/9781003343752-17

δ-ALA	δ-aminolevulinic acid
δ-ALAD	δ-aminolevulinic acid dehydratase

17.1 INTRODUCTION

Plants are exposed to abiotic and biotic stresses, and both have an adverse effect on plant ability to grow, develop and produce. Living organisms such as viruses, weeds, bacteria, insects, fungi and nematodes are the reasons for occurrence of biotic stresses (Mertens et al., 2021). Abiotic stresses are exclusively determined by inanimate or environmental elements such as salt, ultraviolet light, heavy metals (HMs), temperature and water availability (Godoy et al., 2021). Abiotic stressors, which affect plant growth and productivity, include disturbed homeostasis in cell metabolism, oxidative stress, ionic imbalance and osmotic stress (Roychoudhury et al., 2008). Some HMs, such as nickel (Ni), molybdenum (Mo), iron (Fe), copper (Cu), cobalt (Co), manganese (Mn) and zinc (Zn) are necessary for plants at specific quantities but become hazardous at higher levels (Hossain et al., 2012). On the contrary, plants do not need arsenic (As), mercury (Hg), cadmium (Cd) or lead (Pb), all of which are extremely detrimental to plants.

When plants are going through higher HM levels, their primary feedback mechanism is the generation of ROS (reactive oxygen species). ROS in the mitochondria, peroxisomes and chloroplasts are continuously produced by the majority of heavy metals that can lead to unanticipated side effects of heavy metal toxicity and oxidative stress (Chen and Su, 2018). HM stress frequently promotes the photorespiratory pathway activities, increases the closure of stomata, disrupts the antioxidant system, causes the generation of ROS and slows down plant metabolism and electron transport chain (Edelstein and Ben-Hur, 2018) as well. Lipid peroxidation, a harmful condition brought on by HM-induced ROS, impairs the integrity and functionality of the cell membrane (Roychoudhury et al., 2012).

N-acetyl-5-methoxytrytamine, often known as melatonin (MT), is a well-known pleiotropic signaling molecule and works as an antioxidant that promotes plant tolerance to a variety of abiotic and biotic stimuli (Siddiqui et al., 2019). From seed germination to biological yield, MT stimulates a variety of morphological, physiological and biochemical features through the up regulation of stress-related genes (Li et al., 2021; Arnao and Hernández-Ruiz, 2019) that scavenge ROS and enhance plant antioxidant defenses against abiotic stress. MT reduces the harmful impacts of HMs on plants, perhaps by promoting the production of tiny molecules that regulate the ion balance and redox state to prevent the relocation of ions of toxic elements by acting as chelating agents in crops. The preventive effect of MT against heavy metal stress has not been systematically demonstrated despite multiple research trials (Samanta et al., 2022). This chapter will therefore focus upon the exogenous MT application for reducing HM stress in plants and examine the detailed processes underlying MT-facilitated toxic metal stress tolerance in plants.

17.2 HEAVY METAL STRESS TOLERANCE BY MELATONIN

By reducing the negative impact of HM stress, exogenously applied MT improves plant attributes including biomass of root and shoot. Exogenous MT treatments have been shown to reduce the harmful effects of many HMs, including zinc (Zn), nickel (Ni), chromium (Cr), iron (Fe), cadmium (Cd), aluminium (Al) and copper (Cu) (Table 17.1).

Literature survey revealed that exogenous MT increases the stress tolerance of various plant species to HMs such Pb, Cu, Ni, Al, As, Fe, Cd and Cr by a direct channel, directly scavenging ROS and indirectly scavenging ROS by enhancing photosynthetic efficiency, metabolite content and antioxidant activities of enzymes (Table 17.1). Optimal MT concentrations also reduce the harmful effects of Al on soybean plants. As a result of enhanced antioxidant activity of CAT, POD and

TABLE 17.1
Effects of Melatonin in Alleviation of Heavy Metal Toxicity in Crop Plants

Plant species	Heavy metal concentration	Melatonin treatment	Effect of melatonin in alleviation of toxicity		Reference
			Increase	Decrease	
Oryza sativa	Cd (10 and 50 µM)	10 and 50 µM	POD and SOD activity, plant biomass in underground and aboveground areas	Transcription of Cd uptake and transport-related genes, accumulation of Cd	(Lv et al., 2019)
	150 µm arsenate	20 µM	Root and shoot length, fresh and dry weight, chlorophyll A and B content and overall antioxidant capacity	Ascorbate peroxidase and H_2O_2 content	(Samanta et al., 2021)
Medicago sativa	800 mg kg^{-1} Pb	10 µM	Growth and increased antioxidant response	Pb uptake	(Zhang et al., 2020)
Nicotiana tabacum	200 µM Cd	250 µM	Antioxidant capacity and cell wall elongation	Expression of genes related to Cd uptake	(Wang et al., 2019)
Solanum lycopersicum	100 µM Cd	100 µM	Antioxidant activities and biosynthesis of secondary metabolites	Oxidative stress	(Li et al., 2016)
Triticum aestivum	50 µM Cd	50 µM	ASA-GSH metabolism	Cd-induced metal transporter genes TaNramp1, TaNramp5, TaHMA2, TaHMA3 and TaLCT1 in roots	(Li et al., 2022)
Zea mays	100 µM $PbCl_2$	50 and 100 µM	Plant biomass, PSII maximum efficiency, chlorophyll content, potassium, calcium contents of leaf and leaf water potential	Growth and development	(Okant and Kaya, 2019)

SOD, as well as increased citrate and malate exudation, MT treatment in soybean roots decreased Al-induced H_2O_2 content (Zhang et al., 2017). Exogenous foliar MT boosted phytochelatins (PCs) and non-protein thiols in rosemary plants under As stress and promoted the osmoregulatory accrual of chemicals, which upgraded water status and increased plant development and tolerance to As stress (Farouk and Al-Amri, 2019). MT reduced the accumulation of Cd and improved ROS homeostasis, resulting in reduced inhibition of alfalfa (*Medicago sativa*) seedling growth (Gu et al., 2017). Cd tolerance in tobacco plants was increased by foliar spraying of MT by reducing growth inhibition and Cd accumulation (Wang et al., 2019). Additionally, MT decreased the oxidative damage caused by Cd by directly scavenging ROS using antioxidative enzymes POD, CAT and APX in tobacco plants (Wang et al., 2019). By strengthening the antioxidant system, sequestering Cd and raising H^+-ATPase activity, MT increased tomato tolerance against Cd stress (Li et al., 2016).

By enhancing K homeostasis, boosting active Fe content, intracellular H_2S content, intracellular NO content, lowering H_2O_2 and MDA content and enhancing antioxidative enzyme (SOD, CAT and POD) activities, MT improved resistance of *Capsicum annum* (pepper) plants to Fe deficiency (Kaya et al., 2020). When MT was applied to rice seedlings, both the subterranean and surface biomass of the plants increased. Foliar MT treatment boosted SOD and POD while decreasing MDA in rice plants under Cd stress (Lv et al., 2019). In safflower and maize plants, MT enhanced chlorophyll synthesis and the composition of plant nutrients, as well as strengthened antioxidant defense, overpowering MDA and ROS formation and reducing leakage of electrolytes under Pb stress (Namdjoyan et al., 2020; Okant and Kaya, 2019). *Brassica napus* (canola) subjected to external MT treatment showed improved growth features by increasing plastoquinone (PQ) redox rate and efficiency of photosystem II under Cr stress.

Exogenous MT exerts, therefore, antagonist effects for heavy metal stress, with significant advantages for justifiable production of crops. Exogenously administered MT helps to increase tolerance against HM stress, but more studies are needed in order to fully comprehend and utilize the functions of this molecule (Samanta et al., 2021).

17.3 MECHANISMS OF HEAVY METAL STRESS TOLERANCE MEDIATED BY MELATONIN

By regulating the antioxidant system, preserving nutritional and metabolic equilibrium, enhancing secondary metabolism, adjusting osmotic balance, boosting photosynthetic capacity and regulating stress-resistant gene expression, MT strengthens plant defense mechanisms (Al-Huqail et al., 2020). In this section, we discuss the heavy metal stress tolerance through MT-mediated mechanism.

17.3.1 Antioxidant Up Regulation and Reactive Species Detoxification Modulated by Melatonin

At lower quantities, reactive species like RNS, RCS and ROS serve as signaling molecules. When plants are exposed to HM stress, free radicals in large amounts, including NO, O_{2-} and H_2O_2, are formed (Savvides et al., 2016). These HM-produced free radicals regulate lipid peroxidation, damage cellular membranes and boost reactive carbon species like MDA (Kaciene et al., 2017). However, the use of MT, which has received a lot of attention in recent years, can reduce oxidative stress. By controlling the activity of antioxidants including POX, GPX and Cu/Zn-SOD, as well as by scavenging too many reactive species, MT can reduce the toxicity of HM in plants (Nawaz et al., 2018; Nabaei and Amooaghaie, 2019).

It was found (Jahan et al., 2020) that tomato seedlings treated with MT and exposed to Ni stress exhibited up regulation in genes like chlorophyll a oxygenase (CAO) and cytochrome P450 oxidoreductase (POR) as well as elevation in genes associated with MT biosynthesis including T5H, ASMT, SNAT and TDC. Higher amounts of endogenous MT were produced as a result of positive regulation of the genes involved in biosynthesis of chlorophyll and MT, which is principally connected to scavenging excessive ROS by increasing antioxidative enzyme activities (Nawaz et al., 2018). Under Zn, Pb, Fe, Cd, Al, Cu and V stress, ROS scavenging and antioxidative activity were triggered by MT, and up regulation was also noted in mustard, watermelon, wheat, cucumber and tomato plants (Sami et al., 2020; Xalxo and Keshavkant, 2019; Ahammed et al., 2020). Redox balance, which is affected by stress, is the equilibrium between the production and removal of reactive oxygen species in plant cellular components. In plants under stress, antioxidant mechanisms are essential for lowering ROS buildup and overproduction, as well as the severity of oxidative damage (Li et al., 2018). By either directly or indirectly detoxifying the excessive ROS, MT increases the antioxidant activity in plants, improving their ability to withstand stress (Reiter et al., 2016). For example, application of MT can increase the antioxidant enzyme (POD, SOD and CAT) activities under deficiency of iron in severe circumstances. Additionally, increased transcription levels of POD, CAT and Fe-SOD was observed. Oxidative stress in cucumber was mitigated by MT that functions in ROS scavenging and enhances the antioxidant system in plants (Ahammed et al., 2020). Under heavy-metal-stressed environments, such as Cd stress in strawberry (Wu et al., 2021) and wheat and Cu stress in cucumber, a similar increase in scavenging of ROS as a result of MT application was noted.

Additionally, plants treated with exogenous MT had higher levels of non-enzymatic antioxidants as phenolics, proline, anthocyanins, AsA, total thiols, tocopherol and GSH (Al-Huqail et al., 2020; He et al., 2020). To increase plant tolerance against heavy metal toxicity, exogenous application of MT that also works as vigorous ROS scavenger can be used to provide the plants with an antioxidant defense mechanism system.

17.3.2 Different Metabolic Imbalances Are Protected by Melatonin

When HMs enter the plant, the synthesis of several plant metabolites, particularly proteins, may be hindered (Wu et al., 2021). Therefore, the amount of soluble protein in plants may be a good indicator of the physiological state of the plants. The soluble protein level of strawberries under Cd stress was higher in MT-treated plants than in MT-untreated plants. Studies on wheat and tomatoes exposed to Cd stress helped to corroborate this result (Cai et al., 2017). Additionally, HMs have been connected to N metabolism. For instance, nitrate uptake and transport were impeded due to changes in GS and NR activity under Cd stress, eventually impacting the processes of primary N assimilation (Liang et al., 2018; Qiao et al., 2019).

17.3.3 Osmotic Imbalance in Plants Is Adjusted by Melatonin

Heavy metals in excess can alter root morphology and structure, decrease transpiration rates, block aquaporins or disrupt intercellular contacts (plasmodesmata), all leading to a decline in plant water contents. Pb stress, for instance, interfered with the relationship between plants and water in *Sesbania grandiflora* (Malar et al., 2014). Because of water retention and nutrient uptake during Ni stress, tomato root length was shown to be greatly reduced, although pre-treatment with MT preserved adequate root functioning through repair of damages (Jahan et al., 2020). Osmolytes are compatible solutes that improve cellular water retention without impairing normal metabolism. By maintaining turgor pressure, water content and a high water potential, major osmolytes such as sucrose, trehalose, proline, glycine betaine, polyamines and polyols help to protect plants from HM stress. Plant roots typically acquire significantly more metals than aboveground plant portions because they are the first site of interaction with HM ions. Because proline may directly chelate heavy metals, metal harm to plants is diminished. By decreasing PDH activity and increasing P5CS activity, MT enhances proline accumulation (Siddiqui et al., 2019).

Proline accumulation and glucose metabolism are interrelated. Proline accumulation, according to one theory, needs carbohydrates, and MT up regulates glucose metabolism to promote proline synthesis (Al-Huqail et al., 2020). In *Arabidopsis*, production of polyamine is stimulated by exogenous MT under situations of either excessive or deficient iron. Through protein deformation, where the sulfur in S-containing proteins (Met and Cys) is replaced by Se, exogenous MT promotes sugar buildup (Ulhassan et al., 2019).

17.3.4 Homeostasis of Essential Nutrients Maintained by Melatonin

Mineral elements are very important because they maintain both plant development and water relations and are crucial in a range of metabolic processes (Liang et al., 2018). Mineral absorption is crucial for plant survival during specific physiological processes, and any modification to mineral absorption would adversely affect the metabolic functions. Root architectural distortion brought about by excessive HM buildup in roots can interfere with nutrition uptake. For instance, macronutrient (Ca, Mg, S, P, N, K) concentrations and micronutrient (Zn, Mn, Fe) concentrations are noticeably reduced under Ni-stressed circumstances, along with the decline in root activity (Jahan et al., 2020). Due to the fact that Cd and mineral nutrients utilize the same transport pathways, their effects on the equilibrium and stability of the plasma membrane are comparable at the ionic level. By preserving the balance of nutrients and reducing HM stress, MT greatly influences the composition of plant nutrients and benefits plant root architecture. In tomato, MT may be essential for maintaining nutritional homeostasis (Jahan et al., 2020). Deleterious effects of Ni were reversed by MT treatment, which also markedly boosted the concentrations of Mn, P, Mg and N in tomato roots and leaves (Jahan et al., 2020). By promoting lateral root formation and the exudation of organic acid anions, activating auxin-related gene synthesis and suppressing embryonic root growth, MT treatment enhances the architecture of roots, which may reduce the accumulation and translocation of HMs (Nawaz et al., 2018).

MT can be transformed into 5-methoxytryptamine, which improves H^+-ATPase activity in plants. By boosting gene expression, MT can also increase nutrient uptake. In tomato plants, for example, MT increased the expression of the caffeoyl-O-methyltransferase (COMT) gene, which boosts the SUT1 and SUT2 sulfate transporter genes and enhances S-uptake and assimilation. Similar to this, MT boosted the uptake of Fe in *Arabidopsis* by increasing the gene expression involved in Fe acquisition.

17.3.5 Melatonin Regulates Secondary Metabolites

Many compounds are produced by secondary metabolism in plants and serve primarily to increase plant tolerance to various stresses (Li et al., 2017). Flavonoids, phenols and anthocyanins are examples of secondary metabolites that specifically help to counter HM stress by chelating metals (Cd and Fe^{2+}), preventing the production of free radicals and lowering the level of ROS. Anthocyanins, for instance, prevent the production of •OH by chelating Fe^{2+}. The generation of secondary metabolism by plants is thought to be significantly influenced by the concentration of HMs. Higher quantities of HMs hinder the synthesis of secondary metabolites, while lower levels of HMs increase the secondary metabolite production (Ibrahim et al., 2017).

In rice roots, Fe toxicity drastically reduced the expression of genes involved in the shaping of flavonoids and phenolics as chalcone isomerase (Wu et al., 2017). By increasing the activity of the enzymes DFR (dihydroflavonol reductase), PAL (phenylalanine ammonia lyase) and CHS (chalcone synthase), MT controls the formation of secondary metabolites. The phenylpropanoid pathway, a crucial metabolic pathway for the synthesis of secondary metabolites, generates a large number of secondary metabolites, with PAL serving as the rate-limiting enzyme. According to Ahammed et al. (2020), treatment with MT significantly boosted the activity of PAL, while also raising the amounts of phenols and flavonoids in cucumber plants subjected to Fe stress. It was also discovered by Jahan et al. (2020) that MT treatment increased the levels of phenols and flavonoids in Ni-stressed tomato seedlings. Similar findings were noted in rosemary herb under As toxicity (Farouk and Al-Amri, 2019) and pepper plants under B toxicity, which was corroborated by the characteristics of MT. Low amounts of anthocyanins were accumulated by the MT-deficient *Arabidopsis* mutant. The anthocyanin content of rosemary under Cr stress and tomato plants under Ni stress, however, were both benefited by the addition of MT (Jahan et al., 2020).

17.3.6 Melatonin Protects Photosynthetic Attributes

By reducing cell division and chlorophyll biosynthesis, and by weakening cell wall expansion, accumulation of lignin and suberin in plants and carbonic anhydrase (CA) activity, HM toxicity disrupts the plant photosynthetic system. Additionally, too much HM causes oxidative damage, which can change in the structure of thylakoids, cause soft parenchymatic tissue to grow irregularly and harm reaction centers. Exogenous MT accelerates photosynthesis under HM toxicity by activating the enzymes involved in pigment production and the photosynthetic pathway. For instance, foliar application of MT increased the production of photosynthetic pigments by boosting the activity of RuBisCo and the content of δ-ALA, CA and δ-ALAD while decreasing the activity of chlorophyll-ase (Al-Huqail et al., 2020).

In Cd stress, MT can decrease the rate of chloride degradation and increase chloride content to advance plant photosynthesis. Additionally, MT raises ferredoxins, which stop chlorophyll from degrading. Ferredoxins increase the amount of AsA and decrease ROS levels, which ultimately prevents chlorophyll from degrading because of altered photosynthetic electron transport chain by reducing the overproduction of high-energy electrons.

Increased carbon fixation may have been made possible by maintaining homeostasis and CO_2 supply in MT-treated tomato seedlings despite B toxicity (Siddiqui et al., 2019). According to Nawaz et al. (2018), MT treatment significantly boosts photosynthesis and antioxidant enzyme activities in watermelon and tomato under V stress. Additionally, by applying MT under various stress conditions, smooth functioning in tomato plants has been demonstrated by the PS II (Fv/Fm).

17.4 CONCLUSION AND FUTURE PERSPECTIVE

The prime goal of this chapter is to provide the latest information regarding the role of melatonin in alleviation of heavy metal in crop plants and to encourage botanists to further evaluate the tolerance mechanisms of melatonin in plants. Consequently, plant growth and production are enhanced by exogenous melatonin application under heavy metal stress, acting as a strong plant growth regulator and antioxidant. Exogenous melatonin increases photosynthetic pigment level by up regulating chlorophyll and enzyme synthesis while down regulating degrading genes of chlorophyll. Exogenous melatonin reduces the negative impacts of heavy metal toxicity on plant growth, development and photosynthesis. Exogenous melatonin reduces the toxic effects of heavy metals by up regulating a variety of protective genes that are in charge of increased antioxidant activity and metal chelating abilities. By accumulating osmolytes, increasing antioxidant enzyme activity, adjusting osmotic pressure, maintaining integrity of membranes, reducing ROS production and peroxidation of lipids and boosting the activity of both non-enzyme and enzyme-based antioxidants, the application of melatonin improves plant tolerance to heavy metals. Different nutrient homeostasis is also improved by melatonin. Plant secondary metabolites are produced when melatonin is applied exogenously. Despite the abovementioned progress, we still do not fully understand the mechanisms underlying melatonin-regulated heavy metal absorption, sequestration and transport. For sustainable utilization of those pathways and a better understanding of the actions of this molecule to ensure higher crop yield in heavy metal stress situations, additional studies are required.

REFERENCES

Ahammed, G.J.; Wu, M.; Wang, Y.; Yan, Y.; Mao, Q.; Ren, J.; Ma, R.; Liu, A.; Chen, S. 2020. Melatonin Alleviates Iron Stress by Improving Iron Homeostasis, Antioxidant Defense and Secondary Metabolism in Cucumber. Sci. Hortic. 265, 109205.

Al-Huqail, A.A.; Khan, M.N.; Ali, H.M.; Siddiqui, M.H.; Al-Huqail, A.A.; AlZuaibr, F.M.; Al-Muwayhi, M.A.; Marraiki, N.; Al-Humaid, L. 2020. Exogenous Melatonin Mitigates Boron Toxicity in Wheat. Ecotoxicol. Environ. Saf. 201, 110822.

Arnao, M.B.; Hernández-Ruiz, J. 2019. Melatonin: A New Plant Hormone and/or a Plant Master Regulator? Trends Plant Sci. 24, 38–48.

Cai, S.; Zhang, Y.; Xu, Y.; Qi, Z.; Li, M.; Ahammed, G.J.; Xia, X.; Shi, K.; Zhou, Y.; Reiter, R.J. 2017. HsfA1a Upregulates Melatonin Biosynthesis to Confer Cadmium Tolerance in Tomato Plants. J. Pineal Res. 62, e12387.

Chen, T.; Su, Y. 2018. Uptake by Rice Seedlings and In-Plant Degradation of Atrazine as Influenced by the Oxidative Stress Induced by Added Arsenic or Phosphate Deficiency. Hum. Ecol. Risk Assess. Int. J. 24, 1550–1564.

Edelstein, M.; Ben-Hur, M. 2018. Heavy Metals and Metalloids: Sources, Risks and Strategies to Reduce Their Accumulation in Horticultural Crops. Sci. Hortic. 234, 431–444.

Farouk, S.; Al-Amri, S.M. 2019. Exogenous Melatonin-Mediated Modulation of Arsenic Tolerance with Improved Accretion of Secondary Metabolite Production, Activating Antioxidant Capacity and Improved Chloroplast Ultrastructure in Rosemary Herb. Ecotoxicol. Environ. Saf. 180, 333–347.

Godoy, F.; Olivos-Hernández, K.; Stange, C.; Handford, M. 2021. Abiotic Stress in Crop Species: Improving Tolerance by Applying Plant Metabolites. Plants. 10, 186.

Gu, Q.; Chen, Z.; Yu, X.; Cui, W.; Pan, J.; Zhao, G.; Xu, S.; Wang, R.; Shen, W. 2017. Melatonin Confers Plant Tolerance against Cadmium Stress via the Decrease of Cadmium Accumulation and Reestablishment of MicroRNA-Mediated Redox Homeostasis. Plant Sci. 261, 28–37.

He, J.; Zhuang, X.; Zhou, J.; Sun, L.; Wan, H.; Li, H.; Lyu, D. 2020. Exogenous Melatonin Alleviates Cadmium Uptake and Toxicity in Apple Rootstocks. Tree Physiol. 40, 746–761.

Hossain, M.A.; Piyatida, P.; da Silva, J.A.T.; Fujita, M. 2012. Molecular Mechanism of Heavy Metal Toxicity and Tolerance in Plants: Central Role of Glutathione in Detoxification of Reactive Oxygen Species and Methylglyoxal and in Heavy Metal Chelation. J. Bot. 2012, 37. DOI: 10.1155/2012/872875

Ibrahim, M.H.; Chee Kong, Y.; Mohd Zain, N.A. 2017. Effect of Cadmium and Copper Exposure on Growth, Secondary Metabolites and Antioxidant Activity in the Medicinal Plant Sambung Nyawa (*Gynura Procumbens* (Lour.) Merr). Molecules. 22, 1623.

Jahan, M.S.; Guo, S.; Baloch, A.R.; Sun, J.; Shu, S.; Wang, Y.; Ahammed, G.J.; Kabir, K.; Roy, R. 2020. Melatonin Alleviates Nickel Phytotoxicity by Improving Photosynthesis, Secondary Metabolism and Oxidative Stress Tolerance in Tomato Seedlings. Ecotoxicol. Environ. Saf. 197, 110593.

Kaciene, G.; Juknys, R.; Januškaitienė, I. 2017. The Role of Oxidative Stress in Spring Barley Cross-Adaptation to Different Heavy Metals. Arch. Agron. Soil Sci. 63, 1037–1048.

Kaya, C.; Higgs, D.; Ashraf, M.; Alyemeni, M.N.; Ahmad, P. 2020. Integrative Roles of Nitric Oxide and Hydrogen Sulfide in Melatonin Induced Tolerance of Pepper (*Capsicum Annuum* L.) Plants to Iron Deficiency and Salt Stress Alone or in Combination. Physiol. Plant. 168, 256–277.

Li, G.Z.; Wang, Y.Y.; Liu, J.; Liu, H.T.; Liu, H.P.; Kang, G.Z. 2022. Exogenous Melatonin Mitigates Cadmium Toxicity through Ascorbic Acid and Glutathione Pathway in Wheat. Ecotoxicol. Environ. Saf. 237, 113533.

Li, M.; Hasan, M.K.; Li, C.; Ahammed, G.J.; Xia, X.; Shi, K.; Zhou, Y.; Reiter, R.J.; Yu, J.; Xu, M. 2016. Melatonin Mediates Selenium Induced Tolerance to Cadmium Stress in Tomato Plants. J. Pineal Res. 61, 291–302.

Li, X.; Ahammed, G.J.; Zhang, X.-N.; Zhang, L.; Yan, P.; Zhang, L.-P.; Fu, J.-Y.; Han, W.-Y. 2021. Melatonin-Mediated Regulation of Anthocyanin Biosynthesis and Antioxidant Defense Confer Tolerance to Arsenic Stress in *Camellia Sinensis* L. J. Hazard. Mater. 403, 123922.

Li, X.; Ke, M.; Zhang, M.; Peijnenburg, W.; Fan, X.; Xu, J.; Zhang, Z.; Lu, T.; Fu, Z.; Qian, H. 2018. The Interactive Effects of DiclofopMethyl and Silver Nanoparticles on *Arabidopsis Thaliana*: Growth, Photosynthesis and Antioxidant System. Environ. Pollut. 232, 212–219.

Li, X.; Zhang, L.; Ahammed, G.J.; Li, Z.-X.; Wei, J.-P.; Shen, C.; Yan, P.; Zhang, L.-P.; Han, W.-Y. 2017. Stimulation in Primary and Secondary Metabolism by Elevated Carbon Dioxide Alters Green Tea Quality in Camellia Sinensis L. Sci. Rep. 7, 1–12.

Liang, W.; Ma, X.; Wan, P.; Liu, L. 2018. Plant Salt-Tolerance Mechanism: A Review. Biochem. Biophys. Res. Commun. 495, 286–291.

Lv, X.; Fang, Y.; Zhang, L.; Zhang, W.; Xu, L.; Han, J.; Jin, B.; Zhang, X.; Zhang, X.; Xue, D. 2019. Effects of Melatonin on Growth, Physiology and Gene Expression in Rice Seedlings under Cadmium Stress. Phyton. 88, 91.

Malar, S.; Manikandan, R.; Favas, P.J.; Sahi, S.V.; Venkatachalam, P. 2014. Effect of Lead on Phytotoxicity, Growth, Biochemical Alterations and Its Role on Genomic Template Stability in *Sesbania Grandiflora*: A Potential Plant for Phytoremediation. Ecotoxicol. Environ. Saf. 108, 249–257.

Mertens, D.; Boege, K.; Kessler, A.; Koricheva, J.; Thaler, J.S.; Whiteman, N.K.; Poelman, E.H. 2021. Predictability of Biotic Stress Structures Plant Defence Evolution. Trends Ecol. Evol. 36, 444–456.

Nabaei, M.; Amooaghaie, R. 2019. Nitric Oxide Is Involved in the Regulation of Melatonin-Induced Antioxidant Responses in Catharanthus Roseus Roots under Cadmium Stress. Botany. 97, 681–690.

Namdjoyan, S.; Soorki, A.A.; Elyasi, N.; Kazemi, N.; Simaei, M. 2020. Melatonin Alleviates Lead-Induced Oxidative Damage in Safflower (*Carthamus Tinctorius* L.) Seedlings. Ecotoxicology. 29, 108–118.

Nawaz, M.A.; Jiao, Y.; Chen, C.; Shireen, F.; Zheng, Z.; Imtiaz, M.; Bie, Z.; Huang, Y. 2018. Melatonin Pretreatment Improves Vanadium Stress Tolerance of Watermelon Seedlings by Reducing Vanadium Concentration in the Leaves and Regulating Melatonin Biosynthesis and Antioxidant-Related Gene Expression. J. Plant Physiol. 220, 115–127.

Okant, M.; Kaya, C. 2019. The Role of Endogenous Nitric Oxide in Melatonin-Improved Tolerance to Lead Toxicity in Maize Plants. Environ. Sci. Pollut. Res. 26(12), 11864–11874.

Qiao, Y.; Yin, L.; Wang, B.; Ke, Q.; Deng, X.; Wang, S. 2019. Melatonin Promotes Plant Growth by Increasing Nitrogen Uptake and Assimilation under Nitrogen Deficient Condition in Winter Wheat. Plant Physiol. Biochem. 139, 342–349.

Reiter, R.J.; Mayo, J.C.; Tan, D.; Sainz, R.M.; Alatorre-Jimenez, M.; Qin, L. 2016. Melatonin as an Antioxidant: Under Promises but over Delivers. J. Pineal Res. 61, 253–278.

Roychoudhury, A.; Basu, S.; Sarkar, S.N.; Sengupta, D.N. 2008. Comparative Physiological and Molecular Responses of a Common Aromatic Indica Rice Cultivar to High Salinity with Non-Aromatic Indica Rice Cultivars. Plant Cell Rep. 27(8), 1395–1410.

Roychoudhury, A.; Basu, S.; Sengupta, D.N. 2012. Antioxidants and Stress-Related Metabolites in the Seedlings of Two Indica Rice Varieties Exposed to Cadmium Chloride Toxicity. Acta Physiol. Plant. 34(3), 835–847.

Samanta, S.; Banerjee, A.; Roychoudhury, A. 2021. Melatonin Application Differentially Modulates the Enzymes Associated with Antioxidative Machinery and Ascorbate-Glutathione Cycle during Arsenate Exposure in Indica Rice Varieties. Plant Biol. 23, 193–201.

Samanta, S.; Banerjee, A.; Roychoudhury, A. 2022. Arsenic Toxicity Is Counteracted by Exogenous Application of Melatonin to Different Extents in Arsenic-Susceptible and Arsenic-Tolerant Rice Cultivars. J. Plant Growth Regul. 41, 2210–2231.

Sami, A.; Shah, F.A.; Abdullah, M.; Zhou, X.; Yan, Y.; Zhu, Z.; Zhou, K. 2020. Melatonin Mitigates Cadmium and Aluminium Toxicity through Modulation of Antioxidant Potential in *Brassica Napus* L. Plant Biol. 22, 679–690.

Savvides, A.; Ali, S.; Tester, M.; Fotopoulos, V. 2016. Chemical Priming of Plants against Multiple Abiotic Stresses: Mission Possible? Trends Plant Sci. 21, 329–340.

Siddiqui, M.H.; Alamri, S.; Alsubaie, Q.D.; Ali, H.M.; Ibrahim, A.A.; Alsadon, A. 2019. Potential Roles of Melatonin and Sulfur in Alleviation of *Lanthanum* Toxicity in Tomato Seedlings. Ecotoxicol. Environ. Saf. 180, 656–667.

Ulhassan, Z.; Huang, Q.; Gill, R.A.; Ali, S.; Mwamba, T.M.; Ali, B.; Hina, F.; Zhou, W. 2019. Protective Mechanisms of Melatonin against Selenium Toxicity in *Brassica Napus*: Insights into Physiological Traits, Thiol Biosynthesis and Antioxidant Machinery. BMC Plant Biol. 19, 1–16.

Wang, M.; Duan, S.; Zhou, Z.; Chen, S.; Wang, D. 2019. Foliar Spraying of Melatonin Confers Cadmium Tolerance in *Nicotiana Tabacum* L. Ecotoxicol. Environ. Saf. 170, 68–76.

Wu, L.; Ueda, Y.; Lai, S.; Frei, M. 2017. Shoot Tolerance Mechanisms to Iron Toxicity in Rice (*Oryza sativa* L.). Plant Cell Environ. 40, 570–584.

Wu, S.; Wang, Y.; Zhang, J.; Gong, X.; Zhang, Z.; Sun, J.; Chen, X.; Wang, Y. 2021. Exogenous Melatonin Improves Physiological Characteristics and Promotes Growth of Strawberry Seedlings under Cadmium Stress. Hortic. Plant J. 7, 13–22.

Xalxo, R.; Keshavkant, S. 2019. Melatonin, Glutathione and Thiourea Attenuates Lead and Acid Rain-Induced Deleterious Responses by Regulating Gene Expression of Antioxidants in *Trigonella Foenum Graecum* L. Chemosphere. 221, 1–10.

Zhang, H.; Xu, Z.; Guo, K.; Huo, Y.; He, G.; Sun, H.; Guan, Y.; Xu, N.; Yang, W.; Sun, G. 2020. Toxic Effects of Heavy Metal Cd and Zn on Chlorophyll, Carotenoid Metabolism and Photosynthetic Function in Tobacco Leaves Revealed by Physiological and Proteomics Analysis. Ecotoxicol. Environ. Saf. 202, 110856.

Zhang, J.; Zeng, B.; Mao, Y.; Kong, X.; Wang, X.; Yang, Y.; Zhang, J.; Xu, J.; Rengel, Z.; Chen, Q. 2017. Melatonin Alleviates Aluminium Toxicity through Modulating Antioxidative Enzymes and Enhancing Organic Acid Anion Exudation in Soybean. Funct. Plant Biol. 44, 961–968.

18 Ameliorative Response against Herbicide, Pesticide and Stress by Establishing Resistance in Plants via Melatonin

Anamika Paul, Partha Halder, Sayanika Roy, Krishnendu Acharya and Nilanjan Chakraborty

18.1 INTRODUCTION

The damage of agricultural crops on a global scale, caused by plant pathogens (50,000 species), pests (8000 species) and insects and mites (9000 species) is a common matter (Zhang, 2018; Zikankuba et al., 2019). Pesticides are toxic chemical substances that induce a defense response in plants with some side effects on plants and fields (Kaur et al., 2019), whereas loss in horticultural production is a common matter due to the infectious organisms. So the application of pesticides is common to save the plants from external undesired organisms. However, the requirement of pesticide-free foods, amelioration of the side effects caused by the pesticides including herbicide in plants, increasing resistance to pesticides and herbicide is a matter of concern to society (Moustafa-Farag et al., 2019). Rosinger (2014) first introduced the term "safeners" for protection from herbicide. Safeners are herbicide antidotes which help the crop plants in the reduction of herbicidal injurious effects. These chemical substances are applied along with herbicide. However, the farmers have not given so much attention to the use of safeners, because somehow they thought that their main concern is to increase yield; so they have not focused on the effects of herbicides in crops and in production. In 2011, the market value of some safeners was estimated at around €1.7 billion (Rosinger, 2014; Giraldo Acosta et al., 2022). Moreover, melatonin (*N-acetyl*-5-methoxytryptamine), a biomolecule synthesized in nature, is already used as a safener for protection of plants from pesticides (Giraldo Acosta et al., 2022). It plays an important role in many of the stages of plant development (seed germination, crop growth, induction of flowering, ripening of fruits, postharvest preservation, etc.) (Gao et al., 2022). Melatonin is naturally synthesized from tryptophan (Giraldo Acosta et al., 2022). It was first discovered in the pineal gland of cows (Lerner et al., 1958), then in humans (Lerner et al., 1959) and, so many years later, in plants (Dubbels et al., 1995). It has a powerful antioxidant capacity, increases signaling pathways in plants and helps in their growth and defense response (Arnao and Hernández-Ruiz, 2014; Hardeland, 2016; Lee and Back, 2016). Due to having such properties, it protects the plants from several abiotic (Arnao and Hernández-Ruiz, 2014; Hernández-Ruiz and Arnao, 2018) and biotic stresses (Moustafa-Farag et al., 2019) and control transcription factors (Arnao and Hernández-Ruiz, 2021). In this chapter, our main focus is on melatonin and its production and positive impact on plants in offering resistance against pesticide- and herbicide-mediated negative effects on crop plants.

DOI: 10.1201/9781003343752-18

18.2 MELATONIN AND ITS BIOSYNTHESIS

Melatonin (*N-acetyl*-5-methoxytryptamine) is a naturally synthesized biomolecule found in almost all existing organisms like plants and animals. It is derived from tryptophan. It is known as a hormone in an animal and acts as a potent signaling molecule in plants (Back et al., 2016). Melatonin was first identified in the pineal gland of cows in the year 1958 by Lerner et al. (Lerner et al., 1958). Then it was identified in humans, with its chemical structure (Lerner et al., 1959). Lastly in the year of 1995, it was first identified in plants (Dubbels et al., 1995). Almost 10 years after its identification in plants, its functions started to be revealed (Arnao and Hernández-Ruiz, 2015). It acts as a plant growth regulator and biostimulator during stress (Arnao and Hernández-Ruiz, 2014), plays a role as morphogenetic factor, as signaling molecule for ethylene, salicylic acid, jasmonic acid, abscisic acid pathways, as a defensive compound during a pathogen attack by accelerating mitogen-activated potein kinase (MAPK) pathway (Hardeland, 2016; Lee and Back, 2016), and causing delayed senescence (Hardeland, 2016). It also helps the plants to fight against different abiotic stress like salinity, pH, heavy metal, etc. (Moustafa-Farag et al., 2020) and biotic stresses (like fungi, viruses and bacteria) (Moustafa-Farag et al., 2019). Melatonin, applied together with different pesticides (herbicides, fungicides, etc.), induces tolerance against those pesticides in plants (GiraldoAcosta et al., 2022). It increases several biological activities at the cellular level, as it acts as a biostimulator.

As melatonin's importance is already discussed, the biosynthetic pathway related to melatonin synthesis is another matter of concern. Back et al. (2016) proposed that there are multiple (at least four different) routes available for the synthesis of melatonin in plants. These biosynthesis pathways are controlled by chloroplast, cytoplasm and endoplasmic reticulum (ER) in different steps. A total number of six enzymes are found to be involved in these biosynthetic pathways, viz., tryptophan decarboxylase (TDC), tryptophan hydroxylase (TPH), tryptamine 5-hydroxylase (T5H), serotonin *N*-acetyltransferase (SNAT) (also known as arylalkylamine *N*-acetyltransferase or AANAT), *N*-acetylserotonin methyltransferase (ASMT) and caffeic acid *O*-methyltransferase (COMT) (Back et al., 2016; Tan et al., 2016). Four enzymatic steps are involved in each of these four pathways. These enzymes are encoded by six genes, namely *TDC, TPH, T5H, SNAT (or AANAT), ASMT* and *COMT*, respectively. The up regulation and down regulation of these six genes can control the melatonin biosynthesis (Back et al., 2016). Figure 18.1 represents the multiple pathways of melatonin biosynthesis in plants along with the subcellular compartments in which these routes are dependent due to the formation of their intermediates.

Tryptophan is found in the chloroplast of the higher plants. The higher level of tryptophan is found in the leaves (Cho et al., 2000). The first step of the first two pathways (separated by using red and blue arrows for route 1 and route 2, respectively, in Figure 18.1 (A) are catalyzed by TDC that resulted in the formation of tryptamine in cytoplasm. Then tryptamine is converted to serotonin and move into endoplasmic reticulum. This reaction is catalyzed by T5H. In the third step, first route produces *N*-acetylserotonin in the presence of SNAT enzyme in the chloroplast, then it is converted to melatonin that is catalyzed by ASMT or COMT in cytoplasm. In the case of the second route, ASMT or COMT catalyzes the conversion of serotonin to 5-methoxytryptamine in the cytoplasm, and the last step of this route is performed in the chloroplast where melatonin is produced in the presence of SNAT. The next two routes convert tryptophan to 5-hydroxytryptophan, firstly by the catalysis of TPH (route 3 and 4 are marked with black and green arrows in Figure 18.1A) in cytoplasm. In the next step, it converted to serotonin, catalyzed by TDC. These two intermediates are found in the cytoplasm. In the third step, the third route produces *N*-acetylserotonin in the chloroplast, followed by the conversion of melatonin from *N*-acetylserotonin in the cytoplasm. These two steps are catalyzed by SNAT and ASMT or COMT, respectively. The fourth route produces 5-methoxytryptamine from serotonin, catalyzed by ASMT or COMT in the cytoplasm. Lastly, 5-methoxytryptamine is catalyzed by SNAT to melatonin in the chloroplast (Back et al., 2016).

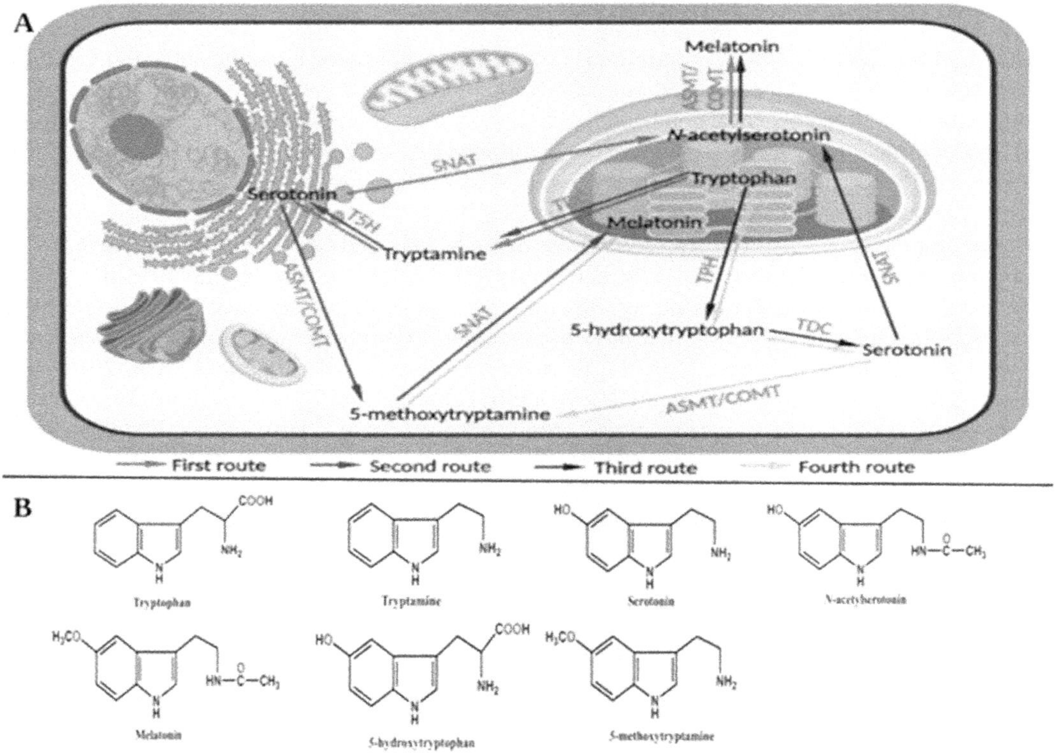

FIGURE 18.1 Multiple pathways of melatonin biosynthesis in plant cell. A. Multiple pathways of melatonin biosynthesis in plant cell with subcellular compartments. B. Chemical structures of the precursor, intermediates and end product.

Source: TDC, Tryptophan decarboxylase; TPH, tryptophan hydroxylase; T5H, tryptamine 5-hydroxylase; SNAT, serotonin *N*-acetyltransferase; ASMT, *N*-acetylserotonin methyltransferase; COMT, caffeic acid *O*-methyltransferase

18.3 HERBICIDE AND PESTICIDE: AT A GLANCE

Loss of billions of dollars occurs due to the damage of plants by the weeds. So it is necessary to control the weeds by killing them in the field, which helps the plants grow with increasing yield. Pesticides are toxic chemical substances or a mixture of substances that are used in the environment to cause harm, destroy, control or kill the insects, rodents, weeds, fungi, pests, etc. (Kaur et al., 2019). These are either biological or synthetic by origin. Pesticides are mainly categorized in three categories, viz. insecticides, herbicides and fungicides (Kumar et al., 2021). On the basis of the target organism, it can be classified as bactericide (target organism is bacteria), algicide (target organism is algae), fungicide (target organism is fungi), insecticide (target organisms are insects), termiticide (target organisms are termites and ants), nematicide (target organisms are nematodes), rodenticide (target organisms are rodents), herbicide (target organisms are weeds), avicide (target organisms are birds), fumigant (target organisms are pests of buildings or soil), slimicide (target organism is slime), molluscicide (target organisms are gastropods like snails, slugs), acaricide (target organisms are acarids), etc. (Britt, 2015; Ali Beg, 2017). Herbicides are defined in a single term: "weed killer". It belongs to a class of pesticide. It helps to manage undesirable plants in gardening, landscaping, agriculture, etc. (Thrall et al., 2011). Herbicides are classified in different groups on the basis of their chemical families, selectivity, site and mode of action, application time, translocation route, etc. Primarily, herbicides are classified on pre-emergent and post-emergent basis. Pre-emergently, they are chosen depending on the morphology of weeds, like either they are grassy or broad-leaved, whereas in the case of post-emergent basis, they are selected based on their selective and non-selective nature. The selectivity of herbicides is the capacity of herbicides to kill the undesirable weeds without harming the desired ones. When selective herbicides are applied to the field, these are directly absorbed by the weeds and directly translocated through xylem and phloem and disrupt the metabolic or biosynthetic pathways. It either kills or injures the weeds (Fernández et al., 2013). On the other hand, non-selective herbicides are mainly used in industries where clearance of the land is mostly required (Green and Owen, 2011). Modes of actions of herbicides are variable due to variability of their chemical compositions. General actions of herbicides on weeds were observed by performing five stages: (1) Contact: herbicides get in touch with the weeds, (2) Absorption: absorbed either by the surface or by the cells of weed, (3) Movement: herbicide translocates through xylem and phloem to the action site of weed, (4) Toxicity: it possesses toxic substances that kills the weeds, and (5) Death: it causes injury that leads to eventual death (DeBoer et al., 2011). Furthermore, herbicides are grouped on the basis of their mode of action in undesired plants. They play regulatory functions like inhibition, disruption and interruption in the enzymatic system, which are related with normal plant growth and development. The number of groups and name depending on their mode of action is given in Figure 18.2. The groups are named aromatic amino acid inhibitors; amino acid synthesis inhibitors (acetolactate synthesis inhibitor or ALS inhibitor); root growth inhibitors (seedling root growth inhibitors); plant growth regulators; shoot growth inhibitors; photosynthesis inhibitors, that is, photosystem I inhibitors and photosystem II inhibitors;

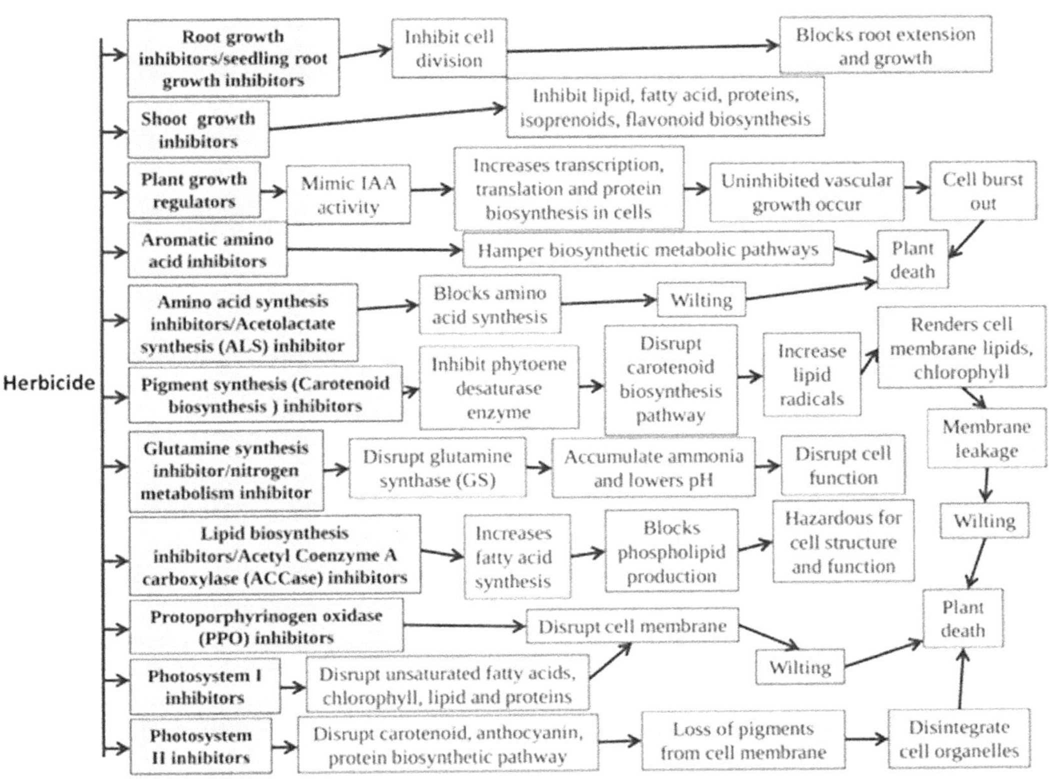

FIGURE 18.2 Groups of herbicides depending on their mode of action.

protoporphyrinogen oxidase (PPO) inhibitors; pigment synthesis inhibitors (carotenoid biosynthesis inhibitors); glutamine synthesis inhibitors (nitrogen metabolism inhibitor); and lipid biosynthesis inhibitors (acetyl coenzyme A carboxylase inhibitors or ACCase inhibitors).

Herbicides named as Lazurit, Zenkor, Dicamba, Tribenuron act on weeds surrounding potato plants to reduce the weed number (Wall, 1994; Galeev et al., 2021). *Phoenix dactylifera* var. Zaghloul (date palm) is a fungal disease-prone plant. It is infected by the attack of *Fusarium oxysporum*, *F. monliforme*, *F. solani*, *Thielaviopsis paradoxa*, *Botryodiplodia theobromae* and *Rhizoctonia solani*. Topsin M70 is a fungicide which helps to reduce the disease severity caused by these fungi in date palm (Baraka et al., 2011). Imidacloprid and acetamiprid are the two potent insecticides used to control the number of canola aphid (*Lipaphis erysimi*) in canola crop (Ahmad et al., 2017). Grasses and broad-leaved weeds of soybeans, sugarcane and numerous other crops are controlled by the application of metribuzin (Fairchild and Sappington, 2002). Moreover, to control the broad-leaved weeds of peanut plants, a number of herbicides are used, like Bentazon (basagran), 2,4-dichlorophenoxybutanoic acid, fluazifop-P-butyl (fusilade), clethodim, acifluorfen, etc. (Wilcut et al., 1994). A list of pesticides including herbicides, insecticides and fungicides are shown in Table 18.1.

18.4 MELATONIN: MECHANISM IN STRESS

Melatonin is well known for its stress-response properties like activator of stressors, plant growth regulator, biostimulation, antioxidant activity and reducing the effects caused by the pesticides in plants (Giraldo Acosta et al., 2022). Melatonin has a capacity to act as a hormonal molecule in a plant for biostimutaion of productivity enhancers, photosynthetic processes in crops, inducing the activity of stressors connected with salinity, pH, heavy metals, drought, cold, heat, pollutants and harmful ultraviolet radiation (Arnao and Hernández-Ruiz, 2014; Hernández-Ruiz and Arnao, 2018; Moustafa-Farag et al., 2020; Altaf et al., 2021; Arnao et al., 2022). It regulates some actions played by auxin, like promoting the growth of roots, shoots, adventitious roots, lateral roots, etc. It prevents the plants from senescence (Altaf et al., 2021). Besides abiotic stress response (Banerjee and Roychoudhury, 2017), it also helps the plants to fight against biotic stresses (fungi, bacteria and viruses) (Moustafa-Farag et al., 2019). Melatonin shows a relation with plant redox network by reestablishing cellular homeostasis, free radical scavenging activity like reactive nitrogen species (RNS), and reactive oxygen species (ROS) (Arnao and Hernández-Ruiz, 2019; Samanta et al., 2021; Arnao et al., 2022). It up-regulates the transcription of many enzymes related to primary metabolism, regulates several secondary metabolite biosynthetic pathways to control the production of polyphenols, terpenoid, glucosinolate and alkaloids (Arnao et al., 2022). It increases the tolerance response of plants by controlling transcription factors, such as CBFs (C-repeat binding factors), DREs (dehydration-responsive elements), WRKY transcription factors (required to response in both abiotic and biotic stress), ERFs (ethylene-responsive elements), NACs, ZATs (a zinc finger element related to the response of ROS), LTI (low-temperature induced factors), RD (factors to dehydration), COR and KIN (response as anti-freezing protein), CIPKs (CBL-interacting protein kinases) and CBLs (calcineurin B-like proteins) regulating salt stress, HSFs (heat shock factors) and heat-sock protein (HSP) response during heat stress (Arnao and Hernández-Ruiz, 2021). An exogenous single application of melatonin acts as a protecting agent that increases expression of genes, enzyme activities associated with pathogen and enhances photosynthetic efficiency. Not only that, it also induces yield of fruit and resistance of cucumber against downy mildew disease (Sun et al., 2019). Another experiment was done on *Fusarium oxysporum* causing *Fusarium* wilt in cucumber, where exogenous application of melatonin increases arbuscular mycorrhizal fungi (AMF) colonization in cucumber roots and reduces disease intensity (Ahammed et al., 2020). Ahammed et al. (2020) also stated that collaboration between AMF and melatonin increases resistance to *Fusarium* wilt in cucumber plant. Melatonin alters the *Fusarium*-induced oxidative stress and attenuates the electrolytic leakage and levels of hydrogen peroxide (H_2O_2) and malondialdehyde in inoculated plant. As

TABLE 18.1
List of Pesticides with Their Impact on Their Target Organism

Sl no.	Pesticide	Name	Plant	Target organism	Impact on target organism	Reference
1.	Herbicide	Zenkor	Potato (*Solanum tuberosum* L.)	Weeds	High rate of weed reduction	(Galeev et al., 2021)
2.	Herbicide	Lazurit	Potato (*Solanum tuberosum* L.)	Weeds	High rate of weed reduction	(Galeev et al., 2021)
3.	Herbicide	Metribuzin	Soybeans (*Glycine max* (L.) Merr.), sugarcane (*Saccharum officinarum* L.) and numerous other crops	Grasses and broad-leaved weeds	Controls weeds	(Fairchild and Sappington, 2002)
4.	Herbicide	Imazethapyr	Chickpea (*Cicer arietinum* L.)	Weeds	Low effect	(Gaur et al., 2013)
5.	Herbicide	Lontrel Grand	Winter wheat (*Triticum aestivum* L.)	Weeds	Reduction of weeds	(Nazarenko et al., 2019)
6.	Herbicide	Lontrel Monitor	Winter wheat (*Triticum aestivum* L.)	Weeds	Reduction of weeds	(Nazarenko et al., 2019)
7.	Herbicide	Picloram	Sunflower (*Helianthus annuus* L), rapeseed (*Brassica napus* L.)	Weeds	Controls weeds	(Hall and Born, 1988)
8.	Herbicide	Dicamba	Potato (*Solanum tuberosum* L.)	Weeds	Reduction in both weeds and in tuber production of potato	(Wall, 1994)
9.	Herbicide	Tribenuron	Potato (*Solanum tuberosum* L.)	Weeds	Reduction in both weeds and in tuber production of potato	(Wall, 1994)
10.	Herbicide	Clopyralid	Sunflower (*Helianthus annuus* L.), rapeseed (*Brassica napus* L.)	Weeds	Controls weeds	(Hall and Born, 1988)
11.	Herbicide	Clopyralid	Strawberry (*Fragaria ananassa* Duchesne)	Broad-leaved weeds	Weed control improved in early season	(Sharpe et al., 2018)
12.	Herbicide	Clopyralid	Plants belong to Asteraceae, Apiaceae	Yellow star-thistle (*Centaurea solstitialis*), native and exotic members	Decline its cover	(Morghan et al., 2003)

(Continued)

TABLE 18.1 *(Continued)*
List of Pesticides with Their Impact on Their Target Organism

Sl no.	Pesticide	Name	Plant	Target organism	Impact on target organism	Reference
13.	Herbicide	Clopyralid	Plants belong to Fabaceae	Exotic members	Decline its cover	(Morghan et al., 2003)
14.	Herbicide	Kusagard	Members of any family	*Avena fatua* and *Agropyron repens*	Show harmful effect	(Pessala, 1981)
15.	Herbicide	DPX-PE350	Cotton (*Gossypium hirsutum* L.)	Pitted and entire leaf morning glory, smooth pigweed and prickly sida	Control of weeds	(Jordan et al., 1993)
16.	Herbicide	Sethoxydim	Cotton (*Gossypium hirsutum* L.)	Weeds	Controls the growth	(Jordan et al., 1993)
17.	Herbicide	Cyclohexene oxime	Broad-leaved crops	Graminaceous weeds	Controls them	(Akyıl et al., 2015)
18.	Herbicide	Paraquat	Peanuts (*Arachis hypogaea* L.)	Broad-leaved weeds	Controls their number	(Wilcut et al., 1994)
19.	Herbicide	2,4-dichlorophenoxybutanoic acid	Peanuts (*Arachis hypogaea* L.)	Broad-leaved weeds	Controls their number	(Wilcut et al., 1994)
20.	Herbicide	Bentazon (basagran)	Peanuts (*Arachis hypogaea* L.)	Broad-leaved weeds	Controls their number	(Wilcut et al., 1994)
21.	Herbicide	Acifluorfen	Peanuts (*Arachis hypogaea* L.)	Broad-leaved weeds	Controls their number	(Wilcut et al., 1994)
22.	Herbicide	Clethodim	Peanuts (*Arachis hypogaea* L.)	Broad-leaved weeds	Controls their number	(Wilcut et al., 1994)
23.	Herbicide	Fluazifop-P-butyl (fusilade)	Peanuts (*Arachishypogaea* L.)	Broad-leaved weeds	Controls their number	(Wilcut et al., 1994)
24.	Herbicide	Sethoxydim	Peanuts (*Arachis hypogaea* L.)	Broad-leaved weeds	Controls their number	(Wilcut et al., 1994)
25.	Fungicide	Carbendazium	Banana (*Musa* sp.)	*Fusarium oxysporum* f. sp. *psidii*	Inhibited radial growth of fungi	(Srivastava et al. 2011)
26.	Fungicide	Topsin M70	Date palm (*Phoenix dactylifera* var. Zaghloul)	*F. oxysporum*, *F. monliforme*, *F. solani*, *Thielaviopsis paradoxa*, *Botryodiplodiatheobromae* and *Rhizoctonia solani*	Reduced disease severity caused by fungi	(Baraka et al., 2011)
27.	Insecticide	Imidacloprid	Canola crop (*Brassica napus* L.)	Canola aphid (*Lipaphis erysimi*)	Reduced the number of aphids	(Ahmad et al., 2017)
28.	Insecticide	Acetamiprid	Canola crop (*Brassica napus* L.)	Canola aphid (*Lipaphis erysimi*)	Reduced the number of aphids	(Ahmad et al., 2017)
29.	Insecticide	Carbosulfan	Canola crop (*Brassica napus* L.)	Canola aphid (*Lipaphis erysimi*)	Reduced the number of aphids	(Ahmad et al., 2017)
30.	Insecticide	Thiamethoxam	Canola crop (*Brassica napus* L.)	Canola aphid (*Lipaphis erysimi*)	Reduced the number of aphids	(Ahmad et al., 2017)
31.	Herbicide	Bentazon	Sweet potato (*Ipomoea batatas* (L.) Lam.)	Yellow nutsedge (*Cyperus esculentus* L.)	Reduced weed number	(Caputo et al., 2020)

green bell peppers (*Capsicum annuum*) are a subtropical fruit, they are vulnerable to cold stress. Treatment of bell pepper fruit with 100 μM melatonin showed a level of reduction of cold stress injury (Kong et al., 2020). Melatonin induces an antioxidative response, blocks the accumulation of ROS, which helps to preserve normal cell structure and function (Liu et al., 2020). Exogenous melatonin treatment decreases the H_2O_2 content and the enzymes associated with H_2O_2 production to increase the resistance capacity in citrus fruit to postharvest green mold (*Penicillium digitatum*) (Lin et al., 2019). In *Arabidopsis thaliana*, an increased glycerol and sugar accumulation was noted after exogenous treatment with melatonin. This helps to increase endogenous nitric oxide (NO) levels by inducing innate immunity against *Pseudomonas syringae* pv. *tomato* DC3000 (Pst DC3000) via salicylic acid and NO-dependent pathway (Zhao et al., 2015). Exogenous application of melatonin reduces the disease intensity of late blight of potato, caused by *Phytophthora infestans*, by inhibiting the mycelia growth, changes in cellular ultrastructure and enhancing the innate immunity of the plant. It also shows an increase in the effectiveness of fungicide against *P. infestans* (Zhang et al., 2017). Exogenous application of 200 mg l^{-1} (~861 μM) melatonin inhibits the growth of *Xanthomonas oryzae* pv. *oryzicola* and reduces bacterial population about 45% during bacterial leaf streak disease of rice (Chen et al., 2018). Exogenous application twice of 100 μM melatonin showed a decrease in the relative RNA level of tobacco mosaic virus (TMV) in tomato and simultaneous increase in *PR1* and *PR5* gene expression (Zhao et al., 2019). Abscisic acid induces melatonin endogenously in melon seeds, which increases gibberellin (GA_3) content and partially reduces abscisic acid (ABA) content that promotes its seed germination (Li et al., 2021). Exogenous application of 20 μM melatonin induces seed germination by increasing the GA_3 accumulation under salt stress in *Gossypium hirsutum* L. (cotton) (Chen et al., 2021). Table 18.2 shows the role of melatonin in response to stresses.

TABLE 18.2
Role of Melatonin to Alleviate Stress Responses

Slno.	Plant	Type of stress	Pathogen/stress	Mode of melatonin application	Effects	Reference
1.	Cucumber (*Cucumis sativus* L.)	Biotic (Fungi)	*Fusarium oxysporum*	Exogenous	Attenuates electrolytic leakage, hydrogen peroxide (H2O2), malondialdehyde level	(Ahammed et al., 2020)
2.	Citrus fruit (*Citrus limon* L.)	Biotic (Fungi)	*Penicillium digitatum*	Exogenous	Decreases H2O2 content	(Lin et al., 2019)
3.	*Arabidopsis thaliana* L.	Biotic (Bacteria)	*Pseudomonas syringae*	Exogenous	Increases glycerol and sugar accumulation, innate immunity increase by acceleration in NO level via NO- and salicylic acid–dependent pathway	(Zhao et al., 2015)
4.	Bell pepper (*Capsicum annuum* L.)	Abiotic	Cold stress	Exogenous	Activates antioxidant defense system and reduces cold stress–induced membrane lipid peroxidation	(Kong et al., 2020)
5.	Potato (*Solanum tuberosum* L.)	Biotic (Fungi)	*Phytophthora infestans*	Exogenous	Inhibits mycelia growth, changes in cellular ultrastructure, enhances innate immunity	(Zhang et al., 2017)

(*Continued*)

TABLE 18.2 *(Continued)*
Role of Melatonin to Alleviate Stress Responses

Slno.	Plant	Type of stress	Pathogen/stress	Mode of melatonin application	Effects	Reference
6.	Cucumber (*Curcumissativus* L.)	Biotic (Fungi)	Downy mildew disease	Exogenous	Increases the expression of genes, enzyme activities associated with pathogen, enhances photosynthetic efficiency, increases the capacity of nitrogen metabolism and antioxidant ability	(Pelagio-Flores et al., 2012; Sun et al., 2019)
7.	Rice (*Oryza sativa* L.)	Biotic (Bacteria)	*Xanthomonas oryzae* pv. *oryzicola*	Exogenous	Inhibits pathogen growth and reduces bacterial population	(Chen et al., 2018)
8.	Tomato (*Solanum lycopersicum* L.)	Biotic (Virus)	Tobacco mosaic virus (TMV)	Exogenous	Decreases relative RNA level of TMV and increases *PR1* and *PR5* gene expression	(Zhao et al., 2019)
9.	Melon (*Cucumismelo* L.)	Abiotic	Abscisic acid stress	Endogenous	Increases GA3 content and partially reduces ABA content that promotes its seed germination	(Li et al., 2021)
10.	Cotton (*Gossypium hirsutum* L.)	Abiotic	Salt stress	Exogenous	Induces seed germination by increasing the GA3 accumulation	(Chen et al., 2021)

18.5 MELATONIN: AS A SAFENER

Pesticides such as herbicides, fungicides, etc. are used to make the plants safe from the foreign organisms. They either cause some cellular or subcellular damages, or help to induce plant growth. Some of the desired but sensitive plants get affected when they come in touch with the herbicides or with other pesticides. New seedling plants are damaged in the vegetative stage, or they show negative effects during seed production. Sprayed herbicides cause a direct effect during seed development (Boutin et al., 2014). Safeners are chemical compounds by nature, helping the plants to reduce the negative, harmful or toxic effects caused by the herbicides. These are either applied singly or together with herbicide. Safeners are also called herbicide antidotes (Rosinger, 2014). The safener term first emerged in 1971, though Giraldo Acosta et al. (2022) made an extension of this term with pesticide like fungicide, insecticide and others. So it was used as an agent for increasing resistance of plants against herbicide and pesticide. Several experiments by different authors had already shown their effects; when paraquat (a herbicide) was applied together with melatonin, they reduced two products of lipid peroxidation (malondialdehyde and 4-hydroxyalkenals) (Melchiorri et al., 1995). In bees, when melatonin level was induced due to the over-expressive activity of T5H (tryptamine 5-hydroxylase, which is an important enzyme of melatonin biosynthesis pathway), it showed oxidative stress tolerance against ultraviolet radiation, low temperature and chemicals like bifenthrin, paraquat, abamectin, cyhalothrin. So survival rates of bees increased (Fan et al., 2020). Table 18.3 shows the effect of co-applications of pesticides and melatonin to increase the defense responses in plants to combat stresses. Fungicide like carbendazim, when used to treat the tomato plant, caused phytotoxicity in the host plant, whereas exogenous application of melatonin was able to alleviate phytotoxicity caused by carbendazim. Yan et al. (2019) showed that the overexpression

TABLE 18.3
Effects of Co-Applications of Pesticides and Melatonin in Plants

Sl no.	Pesticide	Name of pesticide	Mode of application	Plant name	Effect of melatonin on plant	Reference
1.	Fungicide	Carbendazim	Exogenous	Tomato (*Solanum lycopersicum* L.)	Overexpression of *COMT1* (*caffeic acid O-methyltransferase*) gene resulted in ROS scavenging, melatonin increases as well as increases in antioxidant enzymatic activity	(Yan et al., 2019)
2.	Herbicide	Butafenacil	Genetic engineering process	Melatonin-rich transgenic (MRT) rice	Increases chlorophyll levels and decreases hydrogen peroxide and malondialdehyde content, high superoxide dismutase (SOD) and catalase activity	(Park et al., 2013)
3.	Herbicide	Bentazon	Exogenous	Sweet potato (*Ipomoea batatas* (L.) Lam).	Detoxifies herbicide, blocks the interaction of herbicide with target site and alleviates the damage caused due to ROS by oxidative stress	(Caputo et al., 2020)
4.		Paraquat	Exogenous	Pea (*Pisum sativum* L.)	Increases photosynthesis	(Szafranska et al., 2016)
5.	Herbicide	Paraquat	Exogenous	Poplar leaves (*Populus alba* L., *Populus glandulosa* (Uyeki) Uyeki)	Reduces membrane damage, enhances antioxidant enzyme (superoxide dismutase, catalase, ascorbate peroxidase) activity, antioxidant metabolites like glutathione (GSH), proline and ascorbate	(Ding et al., 2018)
6.	Biofungicide	Ethylicin	Exogenous	Tobacco (*Nicotiana tabacum* L.)	Reduced fungal hyphal growth, cell viability and virulence capacity, altered the virulence-related genes, controlled the overexpression of apoptosis-inducing factor, interfered in amino acid metabolism	(Zhang et al., 2018)
7.	Fungicide	Fluopicolide and propamocarb	Exogenous	Potato (*Solanum tuberosum* L.)	Controlled the pathogen growth during late blight disease of potato	(Zhang et al., 2017)

of *COMT1* (*caffeic acid O-methyltransferase*) gene resulted in ROS scavenging, increased melatonin content and an increase in antioxidant enzymatic activity that reduced the phytotoxicity caused by carbendazim. *Pseudoperonospora cubensis* is a mould that causes downy mildew of cucumber (*Cucumis sativus*). ASC-GSH cycle was modulated with increasing antioxidant metabolite (ascorbate, glutathione) content and antioxidative enzyme levels due to the melatonin application in this case (Sun et al., 2019).

Yellow nutsedge (*Cyperus esculentus* L.) is a weed that causes problems to the production of sweet potato (*Ipomoea batatas* (L.) Lam). Bentazon is a herbicide used to control yellow nutsedge. However, it showed a negative alteration in plant development and foliar and root growth (Caputo et al., 2020). Other effects were also observed by many workers, like how it competes for the binding site present on D1 protein with plastoquinone, and furthermore, it blocks electron transport from photosystem II. This causes an inhibition in photosynthesis, and cell damages occur due to oxidative stress ((Dat et al., 1998; Han and Wang, 2002; Caputo et al., 2020). As we know, melatonin acts as a growth regulator to increase photosynthetic ability in different plants (Arnao and Hernández-Ruiz, 2014; Sun et al., 2019). Caputo et al. (2020) also concluded that melatonin is a safener, because even after an exogenous application of melatonin, it increases the detoxification capacity of plants against herbicides. It possibly blocks the interaction of herbicide with the target site and alleviates the damage caused due to ROS by oxidative stress. Paraquat is another broad-spectrum herbicide; it generally interacts with photosystem I and blocks the photosynthesis process and increases excessive amount of ROS generation, which leads to cell damage (Moustaka and Moustakas, 2014). When the pea plants were primarily hydro-primed with melatonin and then treated by paraquat, it showed increase in photosynthesis due to the increase in the functions of photosynthetic pigments. It also induced oxidative stress tolerance as compared to the melatonin-untreated plant (Szafranska et al., 2016). In another case, poplar leaves treated primarily with melatonin and then paraquat showed reduced membrane damage, enhanced activities of some antioxidant enzymes (superoxide dismutase, catalase, ascorbate peroxidase), antioxidant metabolites like glutathione (GSH), proline and ascorbate in the leaves (Ding et al., 2018). Ethylicin is also a broad-spectrum biofungicide. In tobacco black shank disease, cause by *Phytophthora nicotianae*, ethylicin co-treated with melatonin caused great reduction in hyphal growth, cell viability and virulence capacity. Subcellularly, this co-treatment dysregulated the virulence-related genes, interfered in amino acid metabolisms and controlled the overexpression of the apoptosis-inducing factor. This combination treatment led to an eco-friendly method of controlling oomycetic diseases (Zhang et al., 2018). As reported by Zhang et al. (2017), melatonin is able to control pathogen growth in late blight disease of potato. Later it was also proved that similar control strategies were observed when diseased plants were treated with Infinito® fungicide (a mixture of fluopicolide and propamocarb) with melatonin, so that synergistic antifungal effects were noted in combination treatment. Melatonin reduced the dose of this fungicide (Zhang et al., 2018). It served as an eco-friendly formulation to save the plants both from the harmful organism and pesticides. Furthermore, normal rice plants showed many adverse effects due to the application of the herbicide butafenacil. Park et al. (2013) produced melatonin-rich transgenic (MRT) rice plants through genetic engineering. In the experimental setup, they observed that MRT rice plant produces resistant phenotype, and after 0.1 μM butafenacil treatment, chlorophyll level increased, and hydrogen peroxide and malondialdehyde content decreased as compared to the water-treated transgenic plant. This experiment concluded that melatonin scavenged ROS, leading to resistance against oxidative stress under *in vivo* condition.

18.6 CONCLUSION

Melatonin is a naturally synthesized molecule from tryptophan. It shows hormonal response and functions as plant growth regulator and antimicrobial agent either by killing pathogen or by decreasing their number, preservative, safener (by giving resistance to pesticides) and induces stressors molecule. Melatonin is applied either singly or in combination with other pesticides. It has the capacity to induce innate immunity in plants in both types of application. Sometimes it is induced endogenously due to the action of other substances. On the other hand, pesticides including insecticide, herbicide, fungicide, bacteriocide, etc. are used to control the organism upon which they show their action. On one hand, they protect the plant against foreign organisms, while, on the other hand, some of the host plants appear sensitive to these pesticides with harmful effects on plants. In all cases, melatonin showed its role efficiently and effectively by moderating the gene expression and enzyme activities and accelerating antioxidative properties, etc. Hence, melatonin can ameliorate the damage caused by pesticides.

REFERENCES

Ahammed, G.J., Mao, Q., Yan, Y., Wu, M., Wang, Y., Ren, J., Guo, P., Liu, A. and Chen, S., 2020. Role of melatonin in arbuscular mycorrhizal fungi-induced resistance to Fusarium wilt in cucumber. *Phytopathology*, *110*(5), 999–1009.

Ahmad, S., Aftab, A., Sabri, M.A., Ghazal, A., Ullah, Z., Ahmad, M.I., Bilal, H., Mohsin, M. and Naz, S.I., 2017. Efficacy of four different insecticides with different mode of action against canola aphid (Lipaphiserysimi) under field condition. *Journal of Entomology and Zoology Studies*, *5*(3), 626–629.

Akyıl, D., Özkara, A., Erdoğmuş, S.F., Eren, Y., Konuk, M. and Sağlam, E., 2015. Micronucleus assay in human lymphocytes after exposure to alloxydim sodium herbicide in vitro. *Cytotechnology*, *67*(6), 1059–1066.

Ali Beg, M.A., 2017. Classification of pesticides. In: *Pesticides toxicity specificity & politics*, pp. 22–45. DOI: 10.13140/RG.2.2.22592.33280.

Altaf, M.A., Shahid, R., Ren, M.X., Mora-Poblete, F., Arnao, M.B., Naz, S., Anwar, M., Altaf, M.M., Shahid, S., Shakoor, A., Sohail, H., Ahmar, S., Kamran, M. and Chen, J.-T., 2021. Phytomelatonin: An overview of the importance and mediating functions of melatonin against environmental stresses. *Physiology Plant*, *172*, 820–846.

Arnao, M.B., Cano, A. and Hernández-Ruiz, J., 2022. Phytomelatonin: An unexpected molecule with amazing performances in plants. *Journal of Experimental Botany*, erac009.

Arnao, M.B. and Hernández-Ruiz, J., 2014. Melatonin: Plant growth regulator and/or biostimulator during stress? *Trends in Plant Science*, *19*(12), 789–797.

Arnao, M.B. and Hernández-Ruiz, J., 2015. Functions of melatonin in plants: A review. *Journal of Pineal Research*, *59*(2), 133–150.

Arnao, M.B. and Hernández-Ruiz, J., 2019. Melatonin and reactive oxygen and nitrogen species: A model for the plant redox network. *Melatonin Research*, *2*(3), 152–168.

Arnao, M.B. and Hernández-Ruiz, J., 2021. Melatonin against environmental plant stressors: A review. *Current Protein and Peptide Science*, *22*(5), 413–429.

Back, K., Tan, D.X. and Reiter, R.J., 2016. Melatonin biosynthesis in plants: Multiple pathways catalyze tryptophan to melatonin in the cytoplasm or chloroplasts. *Journal of Pineal Research*, *61*(4), 426–437.

Banerjee, A. and Roychoudhury, A., 2017. Melatonin as a regulator of abiotic stress tolerance in plants. In: Singh, V.P., Singh, S., Prasad, S.M. (Eds.) *Mechanisms behind phytohormonal signaling and crop abiotic stress tolerance*, Nova Science Publishers, New York, pp. 47–60.

Baraka, M.A., Radwan, F.M., Shaban, W.I. and Arafat, K.H., 2011. Efficiency of some plant extracts, natural oils, biofungicides and fungicides against root rot disease of date palm. *Journal of Biological Chemistry and Environmental Sciences*, *6*(2), 405–429.

Boutin, C., Strandberg, B., Carpenter, D., Mathiassen, S.K. and Thomas, P.J., 2014. Herbicide impact on non-target plant reproduction: What are the toxicological and ecological implications? *Environmental Pollution*, *185*, 295–306.

Britt, J., 2015. Properties and effects of pesticides. In: Roberts, S.M., James, R.C., Williams, P.L. (Eds.) *Principles of toxicology: Environmental and industrial applications*, Hoboken, NJ: Wiley Publishers, pp. 309–324.

Caputo, G.A., Wadl, P.A., McCarty, L., Adelberg, J., Jennings, K.M. and Cutulle, M., 2020. In vitro safening of bentazon by melatonin in sweetpotato (*Ipomoea batatas*). *Hort Science*, *55*(9), 1406–1410.

Chen, L., Lu, B., Liu, L., Duan, W., Jiang, D., Li, J., Zhang, K., Sun, H., Zhang, Y., Li, C. and Bai, Z., 2021. Melatonin promotes seed germination under salt stress by regulating ABA and GA3 in cotton (*Gossypium hirsutum* L.). *Plant Physiology and Biochemistry*, *162*, 506–516.

Chen, X., Sun, C., Laborda, P., Zhao, Y., Palmer, I., Qing Fu, Z., Qiu, J. and Liu, F., 2018. Melatonin treatment inhibits the growth of *Xanthomonas oryzae* pv. *oryzae*. *Frontiers in Microbiology*, *9*, 2280.

Cho, H.J., Brotherton, J.E., Song, H.S. and Widholm, J.M., 2000. Increasing tryptophan synthesis in a forage legume *Astragalus sinicus* by expressing the tobacco feedback-insensitive anthranilate synthase (ASA2) gene. *Plant Physiology*, *123*(3), 1069–1076.

Dat, J.F., Lopez-Delgado, H., Foyer, G.H. and Scott, I.M., 1998. Parallel changes in H_2O_2 and catalase during thermos tolerance induced by salicylic acid or heat acclimation in mustard seedlings. *Plant Physiology*, 116, 1351–1357.

DeBoer, G.J., Thornburgh, S., Gilbert, J. and Gast, R.E., 2011. The impact of uptake, translocation and metabolism on the differential selectivity between blackgrass and wheat for the herbicide pyroxsulam. *Pest Management Science*, *67*(3), 279–286.

Ding, F., Wang, G. and Zhang, S., 2018. Exogenous melatonin mitigates methyl viologen-triggered oxidative stress in poplar leaf. *Molecules*, *23*, 2852.

Dubbels, R., Reiter, R.J., Klenke, E., Goebel, A., Schnakenberg, E., Ehlers, C., Schiwara, H.W. and Schloot, W., 1995. Melatonin in edible plants identified by radioimmunoassay and by high performance liquid chromatography-mass spectrometry. *Journal of Pineal Research*, *18*(1), 28–31.

Fairchild, J.F. and Sappington, L.C., 2002. Fate and effects of the triazinone herbicide metribuzin in experimental pond mesocosms. *Archives of Environmental Contamination and Toxicology*, *43*(2), 198–202.

Fan, W., Li, G., Zhang, X., Wang, Y., Wang, C., Xu, B., Guo, X. and Li, H., 2020. The role of melatonin and tryptophan-5-hydroxylase-1 in different abiotic stressors in apisceranacerana. *Journal of Insect Physiology*, *128*, 104180.

Fernández, V., Sotiropoulos, T. and Brown, P.H., 2013. *Foliar fertilization: Scientific principles and field pratices*. International Fertilizer Industry Association.

Galeev, R.R., Shulga, M.S. and Kovalev, E.A., 2021. Effectiveness of herbicide application on potatoes in the northern forest-steppe of the novosibirskob region. Вестникнгау, *3*(60), 8.

Gao, T., Liu, X., Tan, K., Zhang, D., Zhu, B., Ma, F. and Li, C., 2022. Introducing melatonin to the horticultural industry: Physiological roles, potential applications, and challenges. *Horticulture Research*, *9*, uhac094.

Gaur, P.M., Jukanti, A.K., Samineni, S., Chaturvedi, S.K., Singh, S., Tripathi, S., Singh, I., Singh, G., Das, T.K., Aski, M. and Mishra, N., 2013. Large genetic variability in chickpea for tolerance to herbicides imazethapyr and metribuzin. *Agronomy*, *3*(3), 524–536.

Giraldo Acosta, M., Cano, A., Hernández-Ruiz, J. and Arnao, M.B., 2022. Melatonin as a possible natural safener in crops. *Plants*, *11*(7), 890.

Green, J.M. and Owen, M.D., 2011. Herbicide-resistant crops: Utilities and limitations for herbicide-resistant weed management. *Journal of Agricultural and Food Chemistry*, *59*(11), 5819–5829.

Hall. J.C. and Born, W.H.V., 1988. The absence of a role of absorption, translocation, or metabolism in the selectivity of picloram and clopyralid in two plant species. *Weed Science*, *36*(1), 9–14.

Han, Y.C. and Wang, C.Y., 2002. Physiological basis of bentazon tolerance in rice (*Oryza sativa* L.) lines. *Weed Biology and Management*, *4*, 186–193.

Hardeland, R., 2016. Melatonin in plant-diversity of levels and multiplicity of functions. *Frontiers in Plant Science*, *7*, 198.

Hernández-Ruiz, J. and Arnao, M.B., 2018. Relationship of melatonin and salicylic acid in biotic/abiotic plant stress responses. *Agronomy*, *8*(4), 33.

Jordan, D.L., Frans, R.E. and McClelland, M.R., 1993. Total post emergence herbicide programs in cotton (*Gossypium hirsutum*) with sethoxydim and DPX-PE350. *Weed Technology*, *7*(1), 196–201.

Kaur, R., Mavi, G.K., Raghav, S. and Khan, I., 2019. Pesticides classification and its impact on environment. *International Journal of Current Microbiology and Applied Sciences*, *8*(3), 1889–1897.

Kong, X.M., Ge, W.Y., Wei, B.D., Zhou, Q., Zhou, X., Zhao, Y.B. and Ji, S.J., 2020. Melatonin ameliorates chilling injury in green bell peppers during storage by regulating membrane lipid metabolism and antioxidant capacity. *Postharvest Biology and Technology*, *170*, 111315.

Kumar, R., Sankhla, M.S., Kumar, R. and Sonone, S.S., 2021. Impact of pesticide toxicity in aquatic environment. *Biointerface Research in Applied Chemistry*, *11*(3), 10131–10140.

Lee, H.Y. and Back, K., 2016. Mitogen-activated protein kinase pathways are required for melatonin-mediated defense responses in plants. *Journal of Pineal Research*, *60*(3), 327–335.

Lerner, A.B., Case, J.D. and Heinzelmann, R.V., 1959. Structure of Melatonin. *Journal of the American Chemical Society*, *81*, 6084–6085.

Lerner, A.B., Case, J.D., Takahashi, Y., Lee, T.H. and Mori, W., 1958. Isolation of melatonin, the pineal gland factor that lightens melanocyte S1. *Journal of the American Chemical Society*, *80*(10), 2587–2587.

Li, H., Guo, Y., Lan, Z., Zhang, Z., Ahammed, G.J., Chang, J., Zhang, Y., Wei, C. and Zhang, X., 2021. Melatonin antagonizes ABA action to promote seed germination by regulating Ca2+ efflux and H_2O_2 accumulation. *Plant Science*, *303*, 110761.

Lin, Y., Fan, L., Xia, X., Wang, Z., Yin, Y., Cheng, Y. and Li, Z., 2019. Melatonin decreases resistance to postharvest green mold on citrus fruit by scavenging defense-related reactive oxygen species. *Postharvest Biology and Technology*, *153*, 21–30.

Liu, G.S., Zhang, Y. and Yun, Z., 2020. Melatonin enhances cold tolerance by regulating energy and proline metabolism in litchi fruit. *Foods*, *9*, 454.

Melchiorri, D., Reiter, R.J., Attia, A.M., Hara, M., Burgos, A. and Nistico, G., 1995. Potent protective effect of melatonin on in vivo paraquat-induced oxidative damage in rats. *Life Science*, *56*, 83–89.

Morghan, K.J.R., Leger, E.A. and Rice, K.J., 2003. Clopyralid effects on yellow starthistle (*Centaurea solstitialis*) and nontarget species. *Weed Science*, *51*(4), 596–600.

Moustafa-Farag, M., Almoneafy, A., Mahmoud, A., Elkelish, A., Arnao, M.B., Li, L. and Ai, S., 2019. Melatonin and its protective role against biotic stress impacts on plants. *Biomolecules*, *10*(1), 54.

Moustafa-Farag, M., Elkelish, A., Dafea, M., Khan, M., Arnao, M.B., Abdelhamid, M.T., El-Ezz, A.A., Almoneafy, A., Mahmoud, A., Awad, M. and Li, L., 2020. Role of melatonin in plant tolerance to soil stressors: Salinity, pH and heavy metals. *Molecules*, *25*(22), 5359.

Moustaka, J. and Moustakas, M., 2014. Photoprotective mechanism of the non-target organism *Arabidopsis thaliana* to paraquat exposure. *Pesticide Biochemistry and Physiology*, 111, 1–6.

Nazarenko, M., Khromykh, N., Matyukha, V., Lykholat, Y., Bezus, R., Alexeeva, A., Lykholat, T. and Shupranova, L., 2019. Chemical plant protection agents change the yield structure and the grain quality of winter wheat ("*Triticum Aestivum* L."). *Bulletin of the Transilvania University of Brasov. Series II: Forestry• Wood Industry• Agricultural Food Engineering*, 97–106.

Park, S., Lee, D.E., Jang, H., Byeon, Y., Kim, Y.S. and Back, K., 2013. Melatonin-rich transgenic rice plants exhibit resistance to herbicide-induced oxidative stress. *Journal of Pineal Research*, *54*(3), 258–263.

Pelagio-Flores, R., Muñoz-Parra, E., Ortiz-Castro, R. and López-Bucio, J., 2012. Melatonin regulates Arabidopsis root system architecture likely acting independently of auxinsignaling. *Journal of Pineal Research*, *53*(3), 279–288.

Pessala, B., 1981. *Alloxydim-Na (Kusagard) against avena fatua and agropyron repens.*

Rosinger, C., 2014. Herbicide safeners: An overview. *Julius-Kühn-Archiv*, (443), 516–525.

Samanta, S., Banerjee, A. and Roychoudhury, A., 2021. Melatonin application differentially modulates the enzymes associated with antioxidative machinery and ascorbate glutathione cycle during arsenate exposure in indica rice varieties. *Plant Biology*, *23*(S1), 193–201.

Sharpe, S.M., Boyd, N.S., Dittmar, P.J., MacDonald, G.E. and Darnell, R.L., 2018. Effect of temperature on clopyralid safety in strawberry. *Weed Technology*, *32*(3), 347–351.

Srivastava, S., Singh, V.P., Kumar, R., Srivastava, M., Sinha, A. and Simon, S., 2011. In vitro evaluation of carbendazim 50% WP, antagonists and botanicals against *Fusarium oxysporum* f. sp. *psidii* associated with rhizosphere soil of guava. *Asian Journal of Plant Pathology*, *5*(1), 46–53.

Sun, Y., Liu, Z., Lan, G., Jiao, C. and Sun, Y., 2019. Effect of exogenous melatonin on resistance of cucumber to downy mildew. *Scientia Horticulturae*, *255*, 231–241.

Szafranska, K., Reiter, R.J. and Posmyk, M.M., 2016. Melatonin application to *Pisum sativum* L. seeds positively influences the function of the photosynthetic apparatus in growing seedlings during paraquat-induced oxidative stress. *Frontiers in Plant Science*, *7*, 1663.

Tan, D.X., Hardeland, R., Back, K., Manchester, L.C., Alatorre-Jimenez, M.A. and Reiter, R.J., 2016. On the significance of an alternate pathway of melatonin synthesis via 5-methoxytryptamine: Comparisons across species. *Journal of Pineal Research*, *61*(1), 27–40.

Thrall, P.H., Oakeshott, J.G., Fitt, G., Southerton, S., Burdon, J.J., Sheppard, A., Russell, R.J., Zalucki, M., Heino, M. and Ford Denison, R., 2011. Evolution in agriculture: The application of evolutionary approaches to the management of biotic interactions in agro-ecosystems. *Evolutionary Applications*, *4*(2), 200–215.

Wall, D.A., 1994. Potato (*Solanum tuberosum*) response to simulated drift of dicamba, clopyralid, and tribenuron. *Weed Science*, *42*(1), 110–114.

Wilcut, J.W., York, A.C. and Wehtje, G.R., 1994. The control and interaction of weeds in peanut. *Reviews of Weed Science*, *6*, 177–205.

Yan, Y., Sun, S., Zhao, N., Yang, W., Shi, Q. and Gong, B., 2019. COMT1 overexpression resulting in increased melatonin biosynthesis contributes to the alleviation of carbendazim phytotoxicity and residues in tomato plants. *Environmental Pollution*, *252*, 51–61.

Zhang, S., Liu, S., Zhang, J., Reiter, R.J., Wang, Y., Qiu, D., Luo, X., Khalid, A.R., Wang, H., Feng, L., Lin, Z. and Ren, M., 2018. Synergistic anti-oomycete effect of melatonin with a biofungicide against oomycetic black shank disease. *Journal of Pineal Research*, *65*, e12492.

Zhang, S., Zheng, X., Reiter, R.J., Feng, S., Wang, Y., Liu, S., Jin, L., Li, Z., Datla, R. and Ren, M., 2017. Melatonin attenuates potato late blight by disrupting cell growth, stress tolerance, fungicide susceptibility and homeostasis of gene expression in *Phytophthora infestans*. *Frontiers in Plant Science*, *8*, 1993.

Zhang, W., 2018. Global pesticide use: Profile, trend, cost/benefit and more. *Proceedings of the International Academy of Ecology and Environmental Sciences*, *8*(1).

Zhao, H., Xu, L., Su, T., Jiang, Y., Hu, L. and Ma, F., 2015. Melatonin regulates carbohydrate metabolism and defenses against *Pseudomonas syringae* pv. tomato DC 3000 infection in Arabidopsis thaliana. *Journal of Pineal Research*, *59*(1), 109–119.

Zhao, L., Chen, L., Gu, P., Zhan, X., Zhang, Y., Hou, C., Wu, Z., Wu, Y.F. and Wang, Q.C., 2019. Exogenous application of melatonin improves plant resistance to virus infection. *Plant Pathology*, *68*(7), 1287–1295.

Zikankuba, V.L., Mwanyika, G., Ntwenya, J.E. and James, A., 2019. Pesticide regulations and their malpractice implications on food and environment safety. *Cogent Food & Agriculture*, *5*(1), 1601544.

19 Prospective Role of Melatonin in Pathogen Suppression and Host Resistance

Retwika Ganguly, Anik Sarkar, Krishnendu Acharya, and Nilanjan Chakraborty

19.1 INTRODUCTION

Melatonin is a small naturally occurring low-molecular-weight molecule that exists in animals as well as in plants (Zhu et al. 2021). The synthesis pathway of melatonin in plants has been discussed by many authors (Back et al. 2016; Nawaz et al. 2020). Reports have confirmed that serotonin N-acetyltransferase (Kang et al. 2013) and N-acetylserotonin methyltransferase (Park et al. 2013) are the two main enzymes that are involved in the synthesis pathways of melatonin. Various functions of this indole-based metabolite (Mandal et al. 2018) have been described in many reports (Arnao and Hernández-Ruiz 2015; Fan et al. 2018; Mathur and Pramanik 2020; Nawaz et al. 2020).

In plants, melatonin was first discovered in the year 1993 in the plant *Pharbitis nil*, which is commonly known as Japanese morning glory. From then on, the presence of melatonin has been studied widely in every part of plants, such as roots, leaves, stems, and fruits (Debnath et al. 2019; Zhao et al. 2019b). Extensive studies with melatonin over the years revealed that its concentration is different in different plant parts, and it also depends on different species and cultivars. High-quality studies have been reported by many authors about the contribution of melatonin in growth, development, signaling, and biochemistry of plants (Erland and Saxena 2018; Sharif et al. 2018; Tan et al. 2012; Zhao et al. 2019b). Some of the metabolites of melatonin, such as 3-orthohydoxymelatonin, 2-hydoxymelatonin, AFMK, are considered as potential antioxidants (Debnath et al. 2019). They can show the effectiveness to decrease the oxidative effects of reactive oxygen species (ROS) and reactive nitrogen species (RNS) (Arnao and Hernández-Ruiz 2019b).

The presence of melatonin has been noticed in a huge number and high value of nutritional crops and medicinal crops (Salehi et al. 2019). According to many studies, it interacts with various plant hormones, such as auxins, cytokinins, gibberellins, abscisic acid, ethylene, etc. It not only interacts with these phytohormones but also provokes many signaling-related physiological and biochemical responses in plants (Arnao and Hernández-Ruiz 2018; Kanwar et al. 2018; Samanta et al. 2021a). It responds to various stress factors, such as biotic (Mandal et al. 2018; Zhao et al. 2021) as well as abiotic stress factors (Chen et al. 2017; Yao et al. 2021; Samanta et al. 2022). Participation of some signaling molecules and transmission of their intracellular and intercellular signals is required for the immune system of melatonin in plants. Such important signaling molecules are reactive oxygen species or ROS (Pardo-Hernández et al. 2020) and nitric oxide or NO (Zhu et al. 2019). Melatonin has shown powerful resistance against numerous fungi (Moustafa-Farag et al. 2019), such as *Phytophthora infestans*, *Botrytis sp.*, *Penicillium sp.*, *Fusarium sp.*, *Penicillium nicotianae*, *and Alternaria* sp., etc.

When the level of melatonin is increased in plants, it enhances the defensive action against *Sphaerotheca fuliginea*, as well as promotes sensitivity to the pathogen *Phytophthora infestans* (Zhang et al. 2017; Mandal et al. 2018). By triggering the chitinase gene expression and

DOI: 10.1201/9781003343752-19

controlling hydrogen peroxide and pathogenesis-related proteins (PRPs), the pre-treatment of melatonin strengthens the resistance against Marssonina apple blotch symptom caused by *Diplocarpon mali* in apple (Yin et al. 2013).

Apart from inhibiting biotic stresses, melatonin is also involved in modulating several physiological functions. For instance, it promotes plant growth and development in numerous dicots (Hernàndez-Ruiz et al. 2004) as well as monocots (Hernàndez-Ruiz et al. 2005). It induces the reformation and elongation of both the lateral and adventitious roots (Arnao and Hernández-Ruiz 2007; Sarropoulou et al. 2012; Zhang et al. 2013). In addition, it adds an important contribution to growth enhancement of the seedlings (Byeon et al. 2014), delay in the process of flowering (Byeon et al. 2014; Kolář et al. 2003), and maturation and quality improvement of fruit (Sun et al. 2015).

19.2 MELATONIN: ITS VARIABILITY IN THE MODE OF FUNCTION IN PLANTS

In plants, melatonin is of great importance. It acts as a versatile factor, starting with its in vitro effects such as exogenous treatment-induced plant growth to in vivo growth and development of plants in response to endogenously produced melatonin (Nawaz et al. 2020). Studies have confirmed about the natural endogenous production of melatonin in several medicinal herbaceous plants, crops, and fruits; however, the amount of melatonin found in various plants is significantly different from each other (Fan et al. 2018). Melatonin is known to regulate the gene expression of phytohormones such as auxin carrier proteins and plays significant roles in the metabolism of IAA (indole-3-acetic acid), ABA (abscisic acid), cytokinin, GAs (gibberellins), and ethylene. Just like auxin, melatonin enhances the growth of lateral as well as adventitious roots (Arnao and Hernández-Ruiz 2018).

Melatonin is unique for its antioxidant nature, by which it can exhibit a defensive response against biotic and abiotic stress, and protection against bacterial pathogens. In this way, it has become a very useful bio-stimulant in the field of agriculture (Huang et al. 2019a). As a result of such defensive actions by melatonin, a satisfied amount of improvement has been found in plant quality, plant characters and traits, and in plant biomass. Melatonin is immensely helpful in promoting root and seed germination, growth of flower, development of fruit, and enhancing the rate of survival under certain adverse conditions. Presence of melatonin in plants in higher concentration checks chlorophyll degradation and improves the photosynthetic activity, which eventually results in the delay in senescence. It is involved in the resistance mechanism of plants against several biotic factors, such as bacteria, virus and fungi, which will be discussed later in detail (Nawaz et al. 2020).

Apart from these, melatonin has some other variabilities in function. For instance, melatonin produces light and dark signals in plants, which in turn signals flowering response during the daytime light. Reiter et al. (1980) revealed that *Chenopodium renbur* L. shows diurnal cadence taking place at night. Transition in flowering is modulated by exogenous melatonin in many plants, such as *Chenopodium rubrum* L., *Pterygpra californica* Ruprecht, and *Arabidopsis thaliana* (Wolf et al. 2001). It helps in maintaining the reproductive physiology and floral development of *Hypericum perforatum* L. plant. This is because the concentration of melatonin remains high during the stage of uninucleate sporogenesis, which results into an enhanced regenerative power of isolated anthers (Kolář et al. 2003). Presence of increased melatonin content in juicy fruits is the indication of the ripening of fruits. It controls the growing stage in plants, because of which the state of dormancy is well conserved (Murch et al. 2001). Melatonin can either be extracted from food or produced endogenously. It maintains its antioxidant properties both in foods and many endogenous compounds. So melatonin can directly be obtained from food in the correct amount and also from various living organisms. However, excessive melatonin production can be dangerous for both animals and plants as it promotes aging (Paredes et al. 2009). On the other hand, regular intake of melatonin-rich food through rice, tea, coffee, corn, etc., improves health condition (Tan et al. 2012). Several versatile roles and activities of melatonin are mentioned in Figure 19.1.

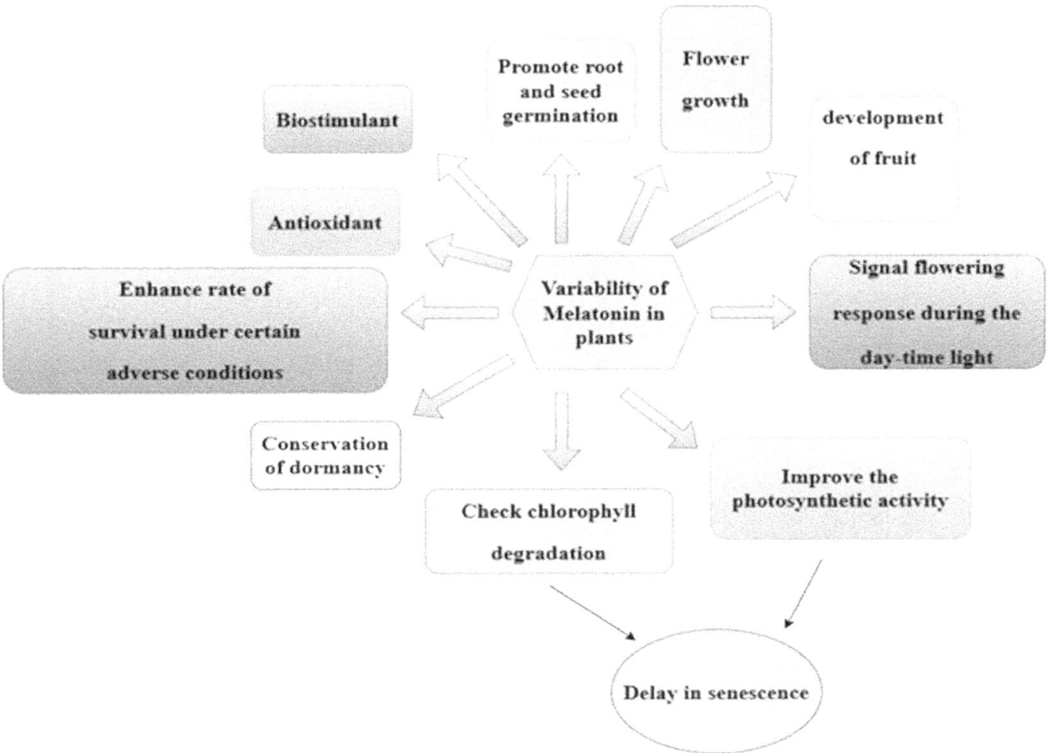

FIGURE 19.1 The flowchart shows the variability of melatonin in plants. It plays numerous roles in plants, such as seed germination, growth of flowers and fruits, stopping degradation of chlorophyll, improving photosynthesis rate, and so on.

19.3 MELATONIN: A PERFECT CHOICE FOR PLANTS IN PLANT STRESS RESPONSE

N-acetyl-5-methoxytryptamine or melatonin is a multifunctional molecule that performs various functions in plants (Figure 19.3). These functions include growth and development of different plant parts, germination of seed, fruit ripening, aging, productivity of crops, and stress response (Hardeland 2016; Hernández et al. 2015; Sun et al. 2015). Melatonin is highly active against numerous abiotic factors, such as cold, heat, drought, salt, heavy metal, etc., and biotic stresses, such as virus, bacteria, and fungi. Production of reactive oxygen species or ROS is very essential for regular cellular metabolism in plants. However, ROS generation is triggered under any abiotic or biotic stresses in plants. At this point, melatonin plays a significant role. During the stressed conditions, it suppresses and terminates the ROS generation, which leads to activation of stress resistance in plants (Foyer et al. 1997; Hardeland 2016; Manchester et al. 2015; Prasad et al. 1994; Reiter et al. 2018).

Moreover, the stimulation of plant antioxidant system, reduction of photosynthesis inhibition, alteration of the transcription factors involved in stress resistance, and triggering heavy metal transport, or renewal of stress-linked hormones, such as salicylic acid, jasmonic acid, and ethylene boost the stress tolerance power in plants in the presence of melatonin (Hernández-Ruiz and Arnao 2018; Limson et al. 1998; Tan et al. 2015). It is well known that tryptophan is the precursor to the biosynthesis of melatonin. It requires the generation and participation of several intermediates, such as tryptamine, 5-hydroxytryptophan, serotonin, N-acetylserotonin, and 5-methoxytryptamine (Kang

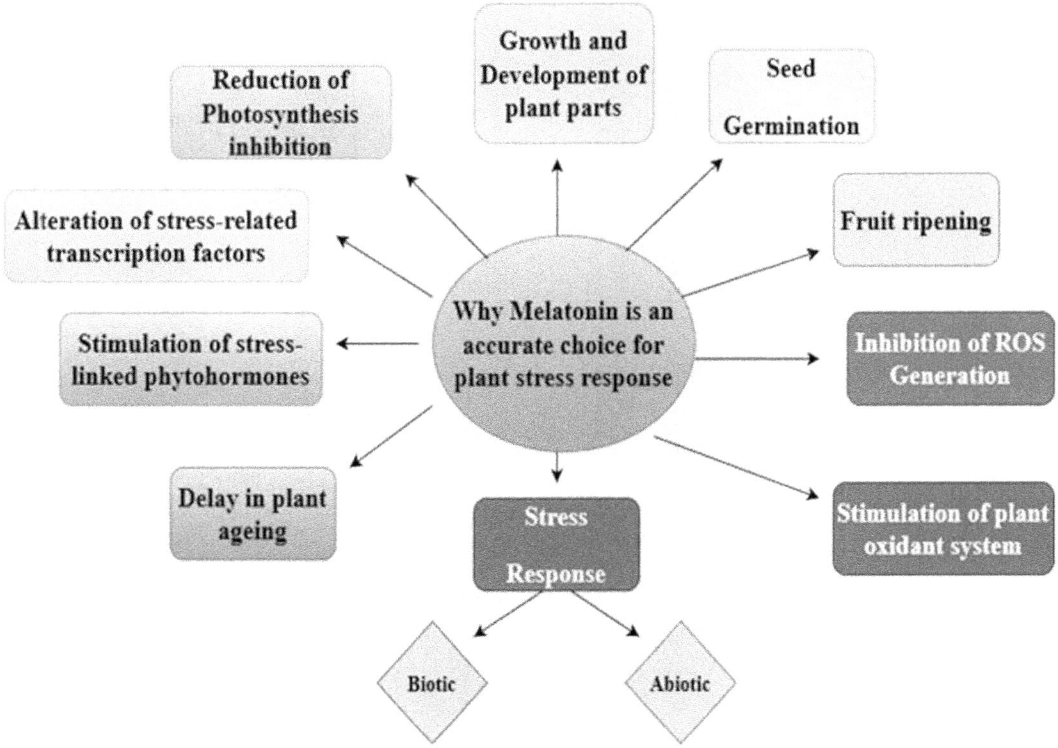

FIGURE 19.2 The flowchart shows some specific functions of melatonin that makes it a perfect choice against plant stress response. It causes delay in senescence, activates the oxidative system of plants, restricts the unwanted excessive production of ROS (reactive oxygen species), induces growth, development of different plant parts, seed germination, and fruit ripening, and so on.

et al. 2011). Subsequently, melatonin is transformed into some other metabolites, which include 2-hydroxymelatonin, cyclic-3-hydroxymelatonin, and N1-acetyl-N2-formyl-5-methoxyknuramine or AFMK (Hardeland 2017; Kołodziejczyk et al. 2015). On top of all the functionality of melatonin, its precursors and metabolites are also responsible for modulating stress resistance in plants (Yu et al. 2018). In this way, melatonin could be regarded as a perfect choice for plants in exhibiting stress response.

19.4 RELATIONSHIP BETWEEN MELATONIN AND REACTIVE OXYGEN SPECIES (ROS) IN REGULATING REGULAR FUNCTIONING IN PLANTS AND ENHANCING STRESS TOLERANCE

Plants confront several biotic stresses, such as bacteria, virus, and fungi during growth and development that can be harmful for the yield of the crops. Apart from the alterations in physiological, biochemical, and molecular features, one very significant change in plants noted as a result of pathogen encounter is the production of ROS. Although the primary sites of ROS are chloroplast, mitochondria, and peroxisome, during the period of stress in plants, ROS are generally found to be generated at the secondary sites, such as cell wall, cell membrane, endoplasmic reticulum, and apoplast. ROS play a pivotal role in maintaining normal plant growth and also improve stress tolerance (Huang et al. 2019b). Low level of ROS, when produced naturally, are required for the plants to maintain basic biological processes (Tsukagoshi et al. 2010; Zafra et al. 2010; Das and Roychoudhury 2014), whereas when ROS are generated at higher levels in response to pathogen attack in plants, it may subsequently result in chlorophyll damage, disparity in redox count, lipid peroxidation, degradation of protein and nucleic acid, and DNA damage (Xie et al. 2014).

ROS are generated as by-products of aerobic respiration in the primary sites (chloroplast, mitochondria, and peroxisome) during a favorable condition, but under a biotic-stressed condition, the secondary sites (cell wall, cell membrane, endoplasmic reticulum, and apoplast) become the platform of ROS generation (Das and Roychoudhury 2014). In many plants, it is noticed that the involvement of ROS acts in different modes in different cases and conditions. For instance, sometimes plants are found to use hypersensitive reaction (HR), or sometimes systemic acquired resistance (SAR) becomes the ideal choice of defense for the plants; moreover, sometimes they apply phytohormone-mediated defense as well (Apel and Hirt 2004; Tripathy and Oelmüller 2012). During biotic stress or pathogen encounter, ROS production is mainly regulated by NADPH oxidases (Bailey-Serres and Mittler 2006; Fridovich 1995; Li et al. 2016; Torres and Dangl 2005), cell wall peroxidases (Bolwell et al. 1998; Grant et al. 2000), and copper-containing amine oxidase (Jasso-Robles et al. 2020). The generation of ROS in the apoplast results in oxidative burst.

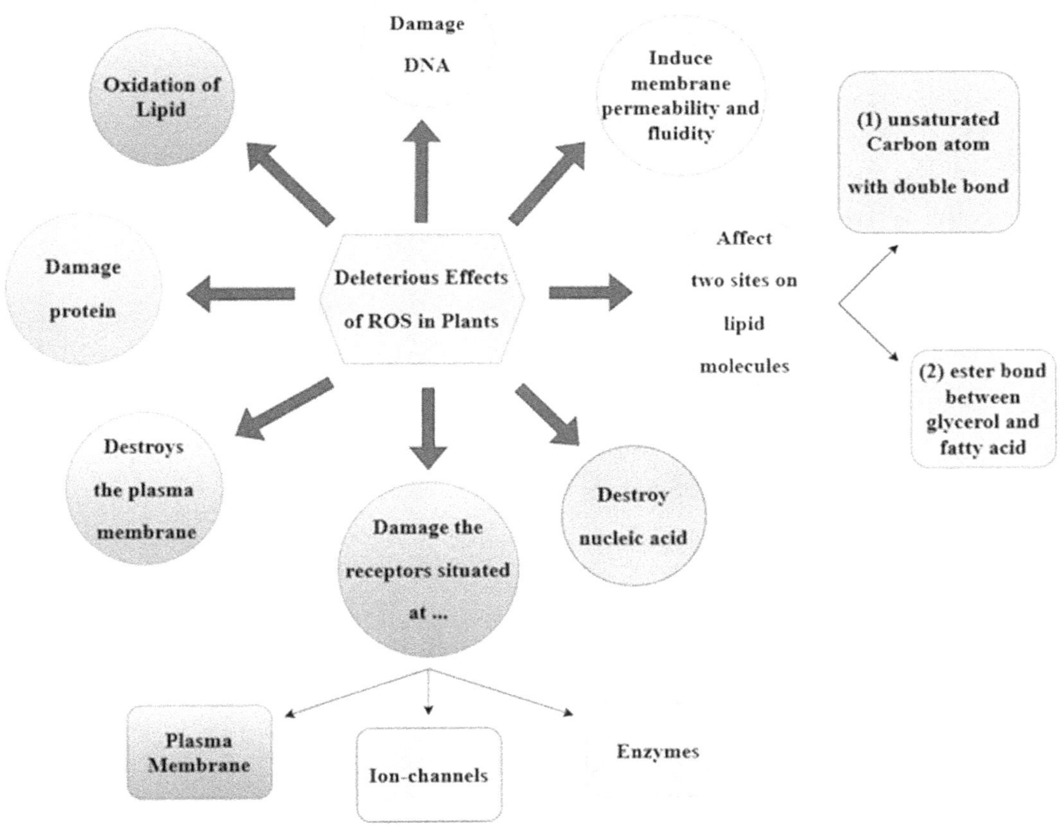

FIGURE 19.3 The flowchart represents the destructive effects of ROS (reactive oxygen species) in plants at the time of biotic stress. During the pathogen attack, ROS is produced in higher amounts, and it causes severe damages to the structure of cell membrane or plasma membrane, protein, lipid, and nucleic acid. Besides, it promotes membrane fluidity and permeability, by which the basic nature of the plasma membrane gets disrupted.

It is known that the cell membrane or plasma membrane mainly consists of lipids. However, the structure and nature of lipid gets altered under biotic stress in plants, as higher concentration of ROS causes oxidation of lipid. Lipid radicals formed after lipid peroxidation damage protein and nucleic acid later via chain reactions. Lipid peroxidation results in destruction of the plasma membrane, followed by damage in the receptors situated at the membranes, ion channels, and enzymes. On top of that, it also induces membrane permeability and fluidity, which, overall, destroys the nature and characteristics of the membrane. Thereafter, ROS target two sites on lipid molecules: (1) unsaturated carbon atom with double bond and (2) ester bond between glycerol and fatty acid. Several destructive effects of ROS in plants during biotic-stressed condition are listed in the following flowchart (Figure 19.4). At such point of time, it becomes necessary for the plants to resist the negative outcomes caused by excessive ROS generation under biotic stress. So melatonin, a ubiquitous, multifunctional signaling molecule, comes in action (Zhu et al. 2021).

The antioxidant property of melatonin helps it to regulate ROS levels by controlling the redox enzymes and metabolites (Arnao and Hernández-Ruiz 2019b; Chen et al. 2017; Gu et al. 2017, 2021; Lee et al. 2016; Tousi et al. 2020). Furthermore, any deflection from ROS balance may be considered as a response to ROS signaling (Choudhury et al. 2017; Mittler 2017). The interaction

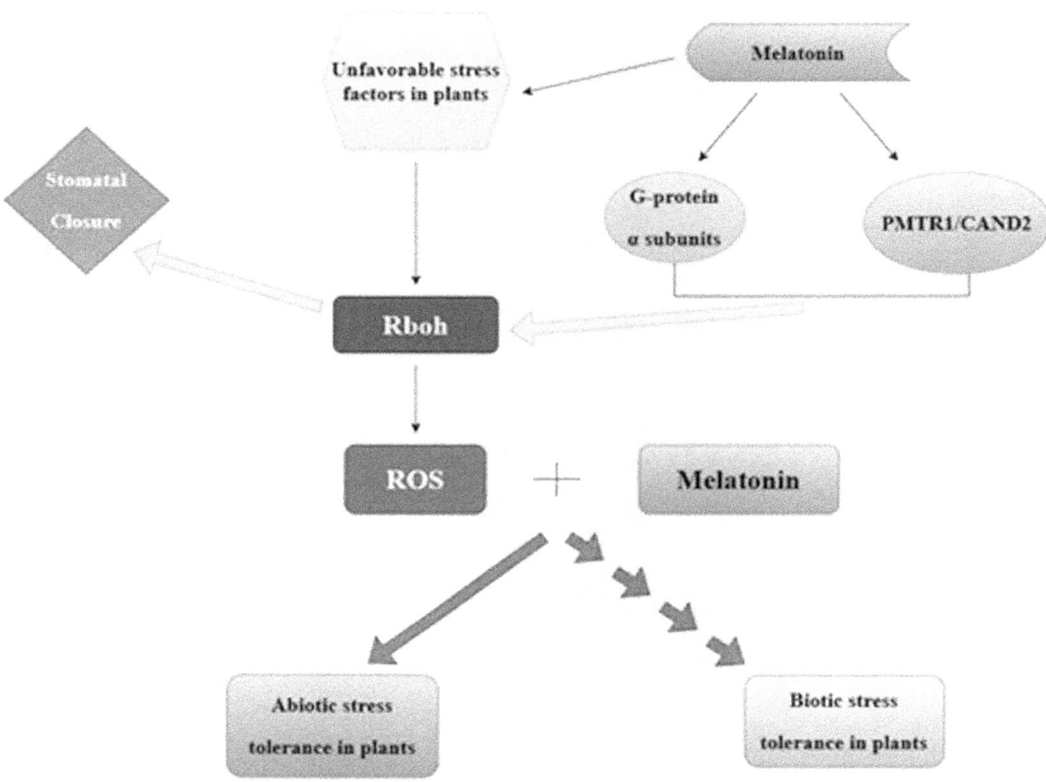

FIGURE 19.4 A possible model of interaction between melatonin and ROS, how they work together to promote stress resistance in plants. PMTR1/CAND2 is the transmembrane receptor of melatonin. These transmembrane receptors work alongside the α subunits of G-protein and promotes Rboh (respiratory burst oxidase homolog). As a result, it shows stomatal closure in *Arabidopsis*. ROS and melatonin together results in enhancing abiotic stress tolerance in plants (shown as continuous dark-green arrow), while their role in defending biotic stress in plants are still unknown (shown as discontinuous red arrow), and so further studies are required.

between melatonin and ROS against stress tolerance in plants is well represented in Figure 19.4. The functioning of melatonin with other signal molecules (such as ROS) with the aim of controlling several stress factors has been depicted well in many studies (Samanta et al. 2021b; Gu et al. 2022). As reported by many studies, the transmembrane receptors of melatonin, such as PMTR1/CAND2, have been found in *Arabidopsis*, alfalfa, maize, and tobacco (Kong et al. 2021; Wang et al. 2021a, 2021b; Wei et al. 2018a). These transmembrane receptors are sometimes also found in the plasma membrane and works alongside α subunits of G-protein, which, in turn, induces Rbohs (respiratory burst oxidase homolog) genes to trigger stomatal closure in *Arabidopsis* (Wang et al. 2021b; Wei et al. 2018a). It decreases the S-nitrosylation of Rboh in order to stimulate ROS signaling (Gong et al. 2017). Although reports have shown that transmembrane receptors of melatonin PMTR1/CAND2 and α subunits of G-protein interact with each other, it still needs to be decoded how this interaction manages Rbohs to stand against unfavorable stresses in plants (Gu et al. 2022).

19.5 HOW MELATONIN STANDS AGAINST PATHOGEN STRESS IN PLANTS

Melatonin has been reported to have a defensive mechanism against several bacteria, fungi, and viruses. In addition to its excellent antioxidant nature against both reactive oxygen species (ROS) and reactive nitrogen species (RNS), the protective role of melatonin against many stressful situations has also been extensively studied by researchers (Arnao and Hernández-Ruiz 2019a). This molecule represents an economically different way to promote plant protection against biotic stress due to its environment-friendly nature (Moustafa-Farag et al. 2019). MAPK (mitogen-activated protein kinase) cascades are one of the most significant pathways in eukaryotes, which deal with the exterior signaling and communicate them with interior interactions. These signaling pathways work quite efficiently against the pathogen-caused biotic stress (Sinha et al. 2011). As a response to biotic stress, MAPK cascades get activated in plants. MAPK cascades have several intermediates, such as MAPK1, MAPK2, MAPK3, MAPK4, MAPK5, MAPK6, and MAPK7, which are, in turn, activated by MAPKK (Pitzschke et al. 2009).

In plants, pathogen resistance is boosted by melatonin signals only by the activation of MAPKK intermediates, for example, MAPK4/5/7/9 in the upstream (Rasmussen et al. 2012). However, MAPK3/MAPK6 regulates the responses of defense-mechanism-interfering pathogens. On one side, the signaling pathways of MAPK3, MAPK4, and MAPK6 activate one of the most renowned molecular patterns (i.e. microbe-linked) of flagellin-extracted flg22 and hairpin peptide, while, on the other hand, MAPK4 acts as a negative regulator against resistance mechanisms (Bigeard et al. 2015). Some researchers exhibited that the pathogen does not depend on any signal molecules, such as kinase protein, calcium, hormones, jasmonic acid, salicylic acid, or ethylene during the treatment of flg22 with MAPKs (Lee and Back 2016). In *Arabidopsis thaliana*, melatonin functions as a triggering signal molecule against the pathogenic invasion. The commonly treated pathogen was Pst DC3000, and it was treated with gene expression (Lee et al. 2014). Moreover, when banana plants were treated with melatonin, it showed up regulation of nine MaHSp90s transcripts, because of which the banana plants ultimately became resistant to *Fusarium oxysporum* cubens (Wei et al. 2017).

1.5a. Antiviral mode of melatonin in plants: Very few studies have described the antiviral action of melatonin in plants. Some related examples are listed in Table 19.1. One report has shown reduction in TMV (tobacco mosaic virus) RNA and its concentration in the damaged seedlings of *Nicotiana glutinosa* and *Solanum lycopersicum* by the treatment of exogenous melatonin. This resulted to an increase in the concentration of SA (salicylic acid) in a NO-dependent pathway (Zhao et al. 2019a). Exogenous melatonin application to infected in vitro shoots of apple cultivar "Gala" destroyed ASGV (apple stem grooving virus) very efficiently. During the experiment, the infected apple shoots were grown in a shoot-proliferating medium which contained melatonin. After the incubation period, they noticed a greater number of virus-free apple shoots grown in the melatonin

TABLE 19.1
Antiviral Activity of Melatonin in a Variety of Plant Species

Name of crop	Name of biotic agent (virus)	Concentration of melatonin applied	Process of melatonin application	Effects on adding melatonin to the crops	References
Nicotiana glutinosa and *Solanum lycopersicum*	Tobacco Mosaic virus or TMV virus	100 μM	Root irrigation	Reduction in the viral concentration in the damaged seedlings	Zhao et al. (2019a)
Apple cultivar "Gala"	ASGV or apple stem grooving virus	15 μM	Added along with the growing medium	Higher rate of shoot regrowth	Chen et al. (2019a)
Oryza sativa (Rice)	Rice stripe virus (RSV)	10 μM	Pre-treatment of plants	Lower incidence of the virus	Lu et al. (2019)

medium compared to the shoots regrown in the medium without melatonin, and the rate of shoot regrowth was quite higher in the shoot-proliferating medium containing melatonin (Chen et al. 2019a). Further studies are certainly required to attribute the effect of melatonin on plant-virus relation. According to another study, melatonin has been added alongside NO, and this co-application has shown less viability and occurrence of the rice strip virus in *Oryza sativa* seedlings. On top of that, it also modified the expression of some genes, such as *OsPR1b* and *OsWRKY45*, which are associated with the SAR pathway (Lu et al. 2019).

1.5b. Antibacterial mode of melatonin in plants: Melatonin has exhibited many productive antibacterial activities against plant pathogens. For example, exogenous melatonin addition to rice leaves is able to reduce BLS or bacterial leaf streak caused by the bacterium *Xanthomonas oryzae pv. Oryzicola* (Chen et al. 2019b). In some cases, melatonin has been used in the supplemented form. For instance, treatment with supplemented melatonin could change the negative disorders of *Candidatus Liberibacter* asiaticus (CLas), the causal agent of devastating citrus greening disease huanglongbing, on its insect vector. In addition to that, it could also lower the CLas population inside the vector psyllids (Nehela and Killiny 2018). Another study suggested that melatonin plays an effective antibacterial role against CLas bacteria and all the gram-negative bacteria too (Nehela and Killiny 2020). By means of activation of melatonin biosynthesis genes, it can resist bacterial blight disease in cassava plant. The transcriptional activators of melatonin biosynthesis genes, such as *MeRAV2* and *MeRAV1* genes, are required for such defensive action (Wei et al. 2018b). MAPK cascades can be stimulated by melatonin in *Arabidopsis thaliana* infected with *Pseudomonas syringae*-DC3000 or (Pst)-DC3000, which eventually up regulate isochorismate synthase 1 (ICS1), the gene of SA biosynthesis (Lee et al. 2014; Lee and Back 2016; Lee and Back 2017). According to another study, nitric oxide (NO) supports the melatonin-triggered defense in *Arabidopsis* against *Pseudomonas syringae* DC3000 infection (Shi et al. 2016).

Researchers have found in (Pst)-DC3000-infected *Arabidopsis thaliana* that the high cell wall invertase (CWI) action within melatonin-applied *Arabidopsis* shows advancement in cell wall strengthening and callose-depositing factors, such as cellulose, galactose, and xylose (Zhao et al. 2015). Melatonin therapy is useful to promote the main enzyme for ethylene biosynthesis, 1-aminocyclopropane-1-carboxylate synthase 6 (ACS6). Thereafter, it induces PDF1.2 expression (Lee et al. 2014). According to transcriptomic data, ETI-associated and PTI-associated genes are activated in watermelon and *Arabidopsis* by exogenous treatment of melatonin (Mandal et al. 2018; Weeda et al. 2014). A report has shown that the sugars such as glucose, fructose, and cellulose and glycerol have significant role in the resistant pathways promoted by melatonin (Qian et al. 2015). The level of endogenous melatonin increases due to (Pst)-DC3000 infection in *Arabidopsis* plants and results in a remarkable accumulation of sugars (glucose, fructose, and cellulose) and glycerol. Moreover, it shows an improvement in the expression of downstream resistance genes (Mathur and Pramanik 2020). All the antibacterial actions of melatonin in plants are listed in Table 19.2.

Addition of 20 μM melatonin to *Arabidopsis thaliana* in nutrient medium activates defense-related genes to improve disease resistance against the bacterium *Pseudomonas syringae* (Shi et al.

TABLE 19.2
Antibacterial Activity of Melatonin in a Variety of Plant Species

Name of crop	Name of biotic agent (bacteria)	Concentration of melatonin applied	Process of melatonin application	Effects on adding melatonin to the crops	References
Oryza sativa (Rice)	*Xanthomonas oryzae pv. oryzicola*	200 μg/L	Spraying on the leaves	Reduces bacterial leaf streak symptom	Chen et al. (2019b)
Arabidopsis thaliana	*Pseudomonas syringae pv. Tomato DC3000*	50 μM	Added with growth media	High CWI action, advancement in cell wall strengthening and callose-depositing factors	Zhao et al. (2015)
Arabidopsis thaliana and *Nicotiana tabacum*	*Pseudomonas syringae pv. Tomato DC3000*	1 mM	Spraying on the leaves	Stimulates MAPK cascades	Lee and Back (2016)
Arabidopsis thaliana	*Pseudomonas syringae*	20 μM	Added with nutrient solution	Both the NO and melatonin levels increase in leaves, activation of defense-related genes, improvement in the disease-resistance response	Shi et al. (2015)
Arabidopsis thaliana, Nicotiana benthamiana	*Pseudomonas syringae*	1 μM	Foliar spray treatment	Promotes PR genes through MAPK signaling cascades	Lee and Back (2016)
Arabidopsis thaliana, Nicotiana benthamiana	*Pseudomonas syringae*	1 or 10 μM	Foliar spray treatment	Expresses defense-related genes, such as PR1, PR5, and PDF1, restricts the pathogen spread	Lee et al. (2014)
Arabidopsis thaliana	*Pseudomonas syringae*	20 μM	Added with nutrient solution	Induces sugars such as glucose, fructose, and cellulose and glycerol to strengthen disease resistance	Qian et al. (2015)
Arabidopsis thaliana	*Pseudomonas syringae*	10 μM	Foliar spray treatment	Stabilizes the signaling between SA and ET, enhances disease resistance against *Pseudomonas Syringae*	Lee et al. (2015)

2015). Foliar treatment of melatonin to *Arabidopsis thaliana* and *Nicotiana benthamiana* activates PR1, PR5, and PDF1.2, the defense-related genes, and induces PR genes through MAPK signaling cascades against the pathogen *Pseudomonas syringae* (Lee et al. 2014). Another study says that 10 μM dose of melatonin applied on the leaves of *Arabidopsis thaliana* is very effective against the encounter of *Pseudomonas syringae*. This results in balancing the signaling between salicylic acid (SA) and ethylene (ET) to create disease resistance (Lee et al. 2015).

1.5c. Antifungal mode of melatonin in plants: Melatonin plays a crucial role in defending against *Botrytis cinerea* in tomato by the synchronization of H_2O_2 production and jasmonic acid signaling technique (Liu et al. 2019). Melatonin can enhance fungicide vulnerability, induce the sensitivity of *Phytophthora infestans* to various environmental stresses, and trigger fungicide treatment efficacy against late blight of potato (Zhang et al. 2017). An enhancement in the amount of melatonin results in more defensive action in watermelon and other cucurbits against some leaf-borne pathogens, such as powdery mildew and soil-borne oomycetes. Such defensive movement is supported by the changes in gene expression connected to ETI- and PAMP-mediated resistance. One mmol/L dose of melatonin sprayed on the watermelon leaves was very effective against *Podosphaera xanthii* and *Phytophthora capsica* (Mandal et al. 2018). Application of 100 μM of melatonin showed significant reduction in the powdery mildew symptom on cucurbits by turning on antioxidant enzymes and antioxidant-related gene expression (Sun et al. 2019). *Phytophthora nicotianae* is responsible for black shank disease on tobacco. Combined application of melatonin and ethylicin, an oomycete fungicide, restricted both the in vitro and in vivo growth of *Phytophthora nicotianae*. Some detailed study depicted that nearly no sign of mycelial growth of *P. nicotianae* was found when tobacco was kept under potato dextrose agar (PDA) medium containing melatonin along with ethylicin. In addition to that, melatonin was also able to reduce the virulence of this particular pathogen, which became conspicuous by observing the low viability of the cell (Zhang et al. 2018). On the contrary, some related studies revealed that melatonin could resist the mycelial growth of *Phytopthora infestans* under in vitro culture, but during the transcriptomic studies with *Phytopthora infestans* hyphae, melatonin rather affected the genes responsible for metabolism of amino acid, stress endurance, and pathogenicity of the fungus (Zhang et al. 2017).

Reports have supported the exogenous application of melatonin in the replant soil, as it shows an increased growth of apple seedlings, improvement in the potassium level in the soil, and photosynthesis. All these together results in subsequent comfort from replant disease in apple (Li et al. 2018). Melatonin has shown similar successful defensive consequences against many fungi, such as *Botrytis sp., Penicillium sp., Fusarium sp., Penicillium nicotianae, and Alternaria sp.* (Arnao and Hernández-Ruiz 2015; Li et al. 2018; Zhang et al. 2018). It suppressed the growth of mold, yeast, and bacteria in the apple juice incubated for four to 12 hours (Zhang et al. 2018). According to some reports, melatonin, with its exogenous application, showed some fruitful results to develop and improve the gene expression of nine *HSP90* genes, which eventually diminished the severity of *Fusarium* wilt disease in banana. Among these nine *HSP90* genes, seven *HSP90* genes proved to be responsible in giving protection to *Fusarium* wilt (Wei et al. 2017). Melatonin addition to fruits showed significant difference in the fungal growth on them. When fruits treated with melatonin were compared to the fruits without any melatonin application, it was found that melatonin-treated fruits exhibited less growth of fungal hyphae than the fruits with no melatonin addition. When melatonin was applied to the fruits, it triggered the POD activity and also some defensive enzymatic activity, such as chitinases, glucanases, phenylalanine ammonia lyase (PAL), and 4-coumarate-coenzyme A ligase (4CL). Furthermore, the secondary metabolite production was induced by PAL and 4CL activity, which finally resulted in the protection of fruits by activating the defensive mode against fungal attack (Lin et al. 2019).

Exogenous application of melatonin to the apple plant through roots promoted the enzymatic activities of POD, CAT, APX, and PAL that can resist the destructive results of ROS during biotic stress (Yin et al. 2013). According to another study, the genes associated with SA (salicylic acid), JA (jasmonic acid), and ET (ethylene) signaling pathways in plants are altered and regulated by melatonin. It is incredible that melatonin provides resistance against gray mold to the cherry tomatoes after harvesting. Hyphal growths were observed in both melatonin-treated and melatonin-untreated samples. Less mycelial growth was present on the melatonin-dipped fruits in comparison with the fruits with no melatonin application. Moreover, melatonin resulted in enhanced activities of the defense-linked enzymes (Li et al. 2019). A different study showed the combination of melatonin with arbuscular mycorrhizal fungi (AMF). Such combination resulted in less severity of the disease symptom in cucumber against the pathogen *Fusarium sp.* This unique symbiosis showed enhancement in the photosynthetic rate, rise in antioxidant production, followed by improvement in disease tolerance in plants against infection (Ahammed et al. 2020). All the antifungal action of melatonin in plants are listed in Table 19.3.

TABLE 19.3
Antifungal Activity of Melatonin in a Variety of Plant Species

Name of crop	Name of biotic agent (fungus)	Concentration of melatonin applied	Process of melatonin application	Effects on adding melatonin to the crops	References
Musa acuminata (Banana)	*Fusarium oxysporum f. sp. cubense*	100 μM	Roots were pretreated with melatonin	Improved the gene expression of nine *HSP90* genes	Wei et al. (2017)
Solanum tuberosum (Potato)	*Phytophthora infestans*	3–10 mM	Applied as foliar spray	Enhanced fungicide vulnerability Induced the sensitivity of *Phytophthora infestans* to various environmental stress Triggered fungicide treatment efficacy against late blight of potato	Zhang et al. (2017)
Citrullus lanatus (Watermelon)	*Podosphaera xanthii* *Phytophthora capsici*	1 mmol/L	Spray on the leaves	Resisted pathogen attack	Mandal et al. (2018)
Nicotiana tabacum (Tobacco)	*Phytophthora nicotianae*	2 mmol/L	Leaves were pretreated with melatonin	Reduced the virulence of this particular pathogen	Zhang et al. (2018)
Citrus reticulate (Citrus)	*Penicillium digitatum*	50 μM	Applied on fruits	Activated the defensive mode against fungal attack Protected the fruits	Lin et al. (2019)
Cucumis sativus (cucumber)	*Pseudoperonospora cubensis*	100	–	Turned on antioxidant enzymes and antioxidant-related gene expression Significant reduction in the powdery mildew symptom in cucurbits	Sun et al. (2019)
Malus domestica (apple)	*Cylindrocarpon destructans*, *Phytophthora cactorum*, *R. solani* AG-5, and *Pythium* sp.	–	Exogenous application in the replant soil	Increased the growth of apple seedlings Improved the potassium level in the soil, and photosynthesis Provided subsequent comfort from replant disease	Li et al. (2018)

(*Continued*)

TABLE 19.3 *(Continued)*
Antifungal Activity of Melatonin in a Variety of Plant Species

Name of crop	Name of biotic agent (fungus)	Concentration of melatonin applied	Process of melatonin application	Effects on adding melatonin to the crops	References
Malus prunifolia (Willd.)	*Diplocarpon mali*	–	Applied through roots	Promoted the enzymatic activities of POD, CAT, APX, phenylalanine ammonia lyase (PAL) Resisted the destructive results of ROS during biotic stress Improved resistance to *Marssonina* apple blotch symptom	Yin et al. (2013)
Cherry tomato	*Botrytis cinerea*	0.1 mM	Fruits were dipped in melatonin solution	Inhibited gray mold development caused by the fungus	Li et al. (2019)
Cucumber	*Fusarium sp.*	–	Root application	Increased photosynthetic rate, Rise in antioxidant production Improved disease tolerance against infection	Ahammed et al. (2020)

19.6 FUTURE PERSPECTIVE

A number of studies have established the importance of melatonin in enhancing plant growth and development. It is discussed in detail how melatonin-associated pathways and mechanisms play a significant role in alleviating biotic stress in plants. Reports have also revealed the contribution of melatonin in controlling the deleterious effects of ROS in plants and the ways to overcome it. Most of these experiments have been performed in vitro in a controlled environment. However, the maximum utilization of these experiments lies with their application in the real-life case, such as in the agricultural fields in which the crops are mostly grown. Therefore, it must directly be practiced in the field instead of laboratory application, to get the real benefits in the future. It is more important to understand the endogenous host-pathogen relationship, and based on that understanding, the focus must be on how melatonin can be used to resist the specific pathogen. The exact mechanisms and signaling pathways of melatonin to deliberately defend pathogen attack is still unclear and unknown. Therefore, collaborative research is needed in the course of time to establish the contribution of melatonin in the field of crop development and food security (Mathur and Pramanik 2020).

19.7 CONCLUSION

In the recent time, melatonin has earned reasonable importance due to its tremendous advantages in the field of plant growth and development, especially in enhancing the potency against several stress factors. Considering large-scale research, there is a huge development in the thought process about this metabolite to elucidate its positive results in plants. However, more research and exposition are required to have a detailed and coherent knowledge about the melatonin-associated defensive signaling network (Lee et al. 2014; Qian et al. 2015; Shi et al. 2015). After a prolonged period of research, the discovery of AtPMTR1 in *Arabidopsis*, the first receptor of melatonin in plants, was a great finding (Wei et al. 2018a) that created an excellent opportunity to represent melatonin also as a phytohormone (Arnao and Hernández-Ruiz 2019c). Scientists need to explore more about several other melatonin receptor-related signaling pathways to understand its importance and contribution for the plant to withstand unfavorable biotic stresses (Arnao and Hernández-Ruiz 2019a; Zhan et al. 2019).

There has been extensive study related to the useful contribution of plant growth promotion rhizobacteria (PGPR) in protecting the plants against several biotic encounters and promoting the overall growth and development of plants in several ways, such as nitrogen fixation, solubilization of phosphorus, and accumulation of phytohormones (Almoneafy et al. 2013; Almoneafy et al. 2014; Kakar et al. 2014). This is, moreover, supported by many other studies in which it is reported that PGPR can generate melatonin. This melatonin produced from PGPR eventually increases its endogenous levels in various plant organs (Jiao et al. 2016; Liu et al. 2016; Ma et al. 2017). Researchers need to search and investigate more about the results of combined applications of PGPR and melatonin and how they can be beneficial for plants in defending against pathogen attacks and gaining more powerful resistance ability.

Melatonin is involved in performing various functions in plants. It includes promoting several physiological processes, such as growth and development in plant parts, germination of seed, flowering, fruit ripening, regulating the circadian cycle, and many more. It can withstand biotic stress in plants. Melatonin makes the plants resistant to various pathogen attacks and also improves disease-tolerating strength in them. By some unique qualities, such as delay in fruit ripening process, reducing chilling damage, and resisting the decay process, melatonin showed beneficial effects in the food industry. These extraordinary features of melatonin were helpful for the postharvest storage of fruits and vegetables and stretching out their shelf life (Arnao and Hernández-Ruiz 2021; Jayarajan and Sharma 2021; Lorente-Mento et al. 2021). Some major factors of melatonin that have been discussed earlier are (1) the pathways by which melatonin is produced in plants, (2) the role of melatonin receptors, CAND2/PMTR1, and (3) its contribution to induce stress tolerance in plants against biotic pathogen encounters. More intensive research is required for the further development of defense against biotic stress.

REFERENCES

Ahammed, G. J., Mao, Q., Yan, Y., Wu, M., Wang, Y., Ren, J., . . . & Chen, S. (2020). Role of melatonin in arbuscular mycorrhizal fungi-induced resistance to Fusarium wilt in cucumber. *Phytopathology*, *110*(5), 999–1009.

Almoneafy, A. A., Kakar, K. U., Nawaz, Z., Li, B., Chun-lan, Y., & Xie, G. L. (2014). Tomato plant growth promotion and antibacterial related-mechanisms of four rhizobacterial Bacillus strains against Ralstonia solanacearum. *Symbiosis*, *63*(2), 59–70.

Almoneafy, A. A., Ojaghian, M. R., Seng-fu, X., Ibrahim, M., Guan-Lin, X., Yu, S., . . . & Bin, L. (2013). Synergistic effect of acetyl salicylic acid and DL-Beta-aminobutyric acid on biocontrol efficacy of Bacillus strains against tomato bacterial wilt. *Tropical Plant Pathology*, *38*, 102–113.

Apel, K., & Hirt, H. (2004). Reactive oxygen species: Metabolism, oxidative stress, and signaling transduction. *Annual Review of Plant Biology*, *55*, 373.

Arnao, M. B., & Hernández-Ruiz, J. (2007). Melatonin promotes adventitious-and lateral root regeneration in etiolated hypocotyls of Lupinus albus L. *Journal of Pineal Research*, *42*(2), 147–152.

Arnao, M. B., & Hernández-Ruiz, J. (2015). Functions of melatonin in plants: A review. *Journal of Pineal Research*, *59*(2), 133–150.

Arnao, M. B., & Hernández-Ruiz, J. (2018). Melatonin and its relationship to plant hormones. *Annals of Botany*, *121*(2), 195–207.

Arnao, M. B., & Hernández-Ruiz, J. (2019a). Melatonin and reactive oxygen and nitrogen species: A model for the plant redox network. *Melatonin Research*, *2*(3), 152–168.

Arnao, M. B., & Hernández-Ruiz, J. (2019b). Melatonin: A new plant hormone and/or a plant master regulator? *Trends in Plant Science*, *24*(1), 38–48.

Arnao, M. B., & Hernández-Ruiz, J. (2019c). Melatonin as a chemical substance or as phytomelatonin rich-extracts for use as plant protector and/or biostimulant in accordance with EC legislation. *Agronomy*, *9*(10), 570.

Arnao, M. B., & Hernández-Ruiz, J. (2021). Melatonin as a plant biostimulant in crops and during post-harvest: A new approach is needed. *Journal of the Science of Food and Agriculture*, *101*(13), 5297–5304.

Back, K., Tan, D. X., & Reiter, R. J. (2016). Melatonin biosynthesis in plants: Multiple pathways catalyze tryptophan to melatonin in the cytoplasm or chloroplasts. *Journal of Pineal Research*, *61*(4), 426–437.

Bailey-Serres, J., & Mittler, R. (2006). The roles of reactive oxygen species in plant cells. *Plant physiology*, *141*(2), 311–311.

Bigeard, J., Colcombet, J., & Hirt, H. (2015). Signaling mechanisms in pattern-triggered immunity (PTI). *Molecular Plant*, *8*(4), 521–539.

Bolwell, G. P., Davies, D. R., Gerrish, C., Auh, C. K., & Murphy, T. M. (1998). Comparative biochemistry of the oxidative burst produced by rose and French bean cells reveals two distinct mechanisms. *Plant Physiology*, *116*(4), 1379–1385.

Byeon, Y., & Back, K. (2014). An increase in melatonin in transgenic rice causes pleiotropic phenotypes, including enhanced seedling growth, delayed flowering, and low grain yield. *Journal of Pineal Research*, *56*(4), 408–414.

Chen, L., Wang, M. R., Li, J. W., Feng, C. H., Cui, Z. H., Zhao, L., & Wang, Q. C. (2019a). Exogenous application of melatonin improves eradication of apple stem grooving virus from the infected in vitro shoots by shoot tip culture. *Plant Pathology*, *68*(5), 997–1006.

Chen, X., Hernández, C., Laborda, P., He, Y., Zhao, Y., Li, C., & Liu, F. J. P. P. (2019b). Melatonin treatments reduce the pathogenicity and inhibit the growth of Xanthomonas oryzae pv. oryzicola. *Plant Pathology*, *68*(2), 288–296.

Chen, Z., Xie, Y., Gu, Q., Zhao, G., Zhang, Y., Cui, W., . . . & Shen, W. (2017). The AtrbohF-dependent regulation of ROS signaling is required for melatonin-induced salinity tolerance in Arabidopsis. *Free Radical Biology and Medicine*, *108*, 465–477.

Choudhury, F. K., Rivero, R. M., Blumwald, E., & Mittler, R. (2017). Reactive oxygen species, abiotic stress and stress combination. *The Plant Journal*, *90*(5), 856–867.

Das, K., & Roychoudhury, A. (2014). Reactive oxygen species (ROS) and response of antioxidants as ROS-scavengers during environmental stress in plants. *Frontiers in Environmental Science*, *2*, 53.

Debnath, B., Islam, W., Li, M., Sun, Y., Lu, X., Mitra, S., . . . & Qiu, D. (2019). Melatonin mediates enhancement of stress tolerance in plants. *International Journal of Molecular Sciences*, *20*(5), 1040.

Erland, L. A., & Saxena, P. K. (2018). Melatonin in plant morphogenesis. *In Vitro Cellular & Developmental Biology-Plant*, *54*(1), 3–24.

Fan, J., Xie, Y., Zhang, Z., & Chen, L. (2018). Melatonin: A multifunctional factor in plants. *International Journal of Molecular Sciences*, *19*(5), 1528.

Foyer, C. H., Lopez-Delgado, H., Dat, J. F., & Scott, I. M. (1997). Hydrogen peroxide-and glutathione-associated mechanisms of acclimatory stress tolerance and signalling. *Physiologia Plantarum*, *100*(2), 241–254.

Fridovich, I. (1995). Superoxide radical and superoxide dismutases. *Annual Review of Biochemistry*, *64*(1), 97–112.

Gong, B., Yan, Y., Wen, D., & Shi, Q. (2017). Hydrogen peroxide produced by NADPH oxidase: A novel downstream signaling pathway in melatonin-induced stress tolerance in Solanum lycopersicum. *Physiologia Plantarum*, *160*(4), 396–409.

Grant, J. J., Yun, B. W., & Loake, G. J. (2000). Oxidative burst and cognate redox signalling reported by luciferase imaging: Identification of a signal network that functions independently of ethylene, SA and Me-JA but is dependent on MAPKK activity. *The Plant Journal*, *24*(5), 569–582.

Gu, Q., Chen, Z., Yu, X., Cui, W., Pan, J., Zhao, G., . . . & Shen, W. (2017). Melatonin confers plant tolerance against cadmium stress via the decrease of cadmium accumulation and reestablishment of microRNA-mediated redox homeostasis. *Plant Science*, *261*, 28–37.

Gu, Q., Wang, C., Xiao, Q., Chen, Z., & Han, Y. (2021). Melatonin confers plant cadmium tolerance: An update. *International Journal of Molecular Sciences*, *22*(21), 11704.

Gu, Q., Xiao, Q., Chen, Z., & Han, Y. (2022). Crosstalk between melatonin and reactive oxygen species in plant abiotic stress responses: An update. *International Journal of Molecular Sciences*, *23*(10), 5666.

Hardeland, R. (2016). Melatonin in plants: Diversity of levels and multiplicity of functions. *Frontiers in Plant Science*, *7*, 198.

Hardeland, R. (2017). Taxon-and site-specific melatonin catabolism. *Molecules*, *22*(11), 2015.

Hernández, I. G., Gomez, F. J. V., Cerutti, S., Arana, M. V., & Silva, M. F. (2015). Melatonin in Arabidopsis thaliana acts as plant growth regulator at low concentrations and preserves seed viability at high concentrations. *Plant Physiology and Biochemistry, 94*, 191–196.

Hernández-Ruiz, J., & Arnao, M. B. (2018). Relationship of melatonin and salicylic acid in biotic/abiotic plant stress responses. *Agronomy, 8*(4), 33.

Hernàndez-Ruiz, J., Cano, A., & Arnao, M. B. (2004). Melatonin: A growth-stimulating compound present in lupin tissues. *Planta, 220*(1), 140–144.

Hernández-Ruiz, J., Cano, A., & Arnao, M. B. (2005). Melatonin acts as a growth-stimulating compound in some monocot species. *Journal of Pineal Research, 39*(2), 137–142.

Huang, B., Chen, Y. E., Zhao, Y. Q., Ding, C. B., Liao, J. Q., Hu, C., . . . & Yuan, M. (2019a). Exogenous melatonin alleviates oxidative damages and protects photosystem II in maize seedlings under drought stress. *Frontiers in Plant Science, 10*, 677.

Huang, H., Ullah, F., Zhou, D. X., Yi, M., & Zhao, Y. (2019b). Mechanisms of ROS regulation of plant development and stress responses. *Frontiers in Plant Science, 10*, 800.

Jasso-Robles, F. I., Gonzalez, M. E., Pieckenstain, F. L., Ramírez-García, J. M., de la Luz Guerrero-González, M., Jiménez-Bremont, J. F., & Rodríguez-Kessler, M. (2020). Decrease of Arabidopsis PAO activity entails increased RBOH activity, ROS content and altered responses to Pseudomonas. *Plant Science, 292*, 110372.

Jayarajan, S., & Sharma, R. R. (2021). Melatonin: A blooming biomolecule for postharvest management of perishable fruits and vegetables. *Trends in Food Science & Technology, 116*, 318–328.

Jiao, J., Ma, Y., Chen, S., Liu, C., Song, Y., Qin, Y., . . . & Liu, Y. (2016). Melatonin-producing endophytic bacteria from grapevine roots promote the abiotic stress-induced production of endogenous melatonin in their hosts. *Frontiers in Plant Science, 7*, 1387.

Kakar, K. U., Duan, Y. P., Nawaz, Z., Sun, G., Almoneafy, A. A., Hassan, M. A., . . . & Xie, G. L. (2014). A novel rhizobacterium Bk7 for biological control of brown sheath rot of rice caused by Pseudomonas fuscovaginae and its mode of action. *European Journal of Plant Pathology, 138*(4), 819–834.

Kang, K., Kong, K., Park, S., Natsagdorj, U., Kim, Y. S., & Back, K. (2011). Molecular cloning of a plant N-acetylserotonin methyltransferase and its expression characteristics in rice. *Journal of Pineal Research, 50*(3), 304–309.

Kang, K., Lee, K., Park, S., Byeon, Y., & Back, K. (2013). Molecular cloning of rice serotonin N-acetyltransferase, the penultimate gene in plant melatonin biosynthesis. *Journal of Pineal Research, 55*(1), 7–13.

Kanwar, M. K., Yu, J., & Zhou, J. (2018). Phytomelatonin: Recent advances and future prospects. *Journal of Pineal Research, 65*(4), e12526.

Kolář, J., Johnson, C. H., & Machácčková, I. (2003). Exogenously applied melatonin (N-acetyl-5-methoxytryptamine) affects flowering of the short-day plant Chenopodium rubrum. *Physiologia Plantarum, 118*(4), 605–612.

Kołodziejczyk, I., Bałabusta, M., Szewczyk, R., & Posmyk, M. M. (2015). The levels of melatonin and its metabolites in conditioned corn (Zea mays L.) and cucumber (Cucumis sativus L.) seeds during storage. *Acta Physiologiae Plantarum, 37*(6), 1–11.

Kong, M., Sheng, T., Liang, J., Ali, Q., Gu, Q., Wu, H., . . . & Gao, X. (2021). Melatonin and its homologs induce immune responses via receptors trP47363-trP13076 in Nicotiana benthamiana. *Frontiers in Plant Science, 12*, 1197.

Lee, H. Y., & Back, K. (2016). Mitogen-activated protein kinase pathways are required for melatonin-mediated defense responses in plants. *Journal of Pineal Research, 60*(3), 327–335.

Lee, H. Y., & Back, K. (2017). Melatonin is required for H 2 O 2-and NO-mediated defense signaling through MAPKKK 3 and OXI 1 in A rabidopsis thaliana. *Journal of Pineal Research, 62*(2), e12379.

Lee, H. Y., Byeon, Y., & Back, K. (2014). Melatonin as a signal molecule triggering defense responses against pathogen attack in Arabidopsis and tobacco. *Journal of Pineal Research, 57*(3), 262–268.

Lee, H. Y., Byeon, Y., Tan, D. X., Reiter, R. J., & Back, K. (2015). Arabidopsis serotonin N-acetyltransferase knockout mutant plants exhibit decreased melatonin and salicylic acid levels resulting in susceptibility to an avirulent pathogen. *Journal of Pineal Research, 58*(3), 291–299.

Li, C., Zhao, Q., Gao, T., Wang, H., Zhang, Z., Liang, B., . . . & Ma, F. (2018). The mitigation effects of exogenous melatonin on replant disease in apple. *Journal of Pineal Research, 65*(4), e12523.

Li, S., Xu, Y., Bi, Y., Zhang, B., Shen, S., Jiang, T., & Zheng, X. (2019). Melatonin treatment inhibits gray mold and induces disease resistance in cherry tomato fruit during postharvest. *Postharvest Biology and Technology, 157*, 110962.

Li, Y. R., & Trush, M. (2016). Defining ROS in biology and medicine. *Reactive Oxygen Species*, *1*(1), 9–21.

Limson, J., Nyokong, T., & Daya, S. (1998). The interaction of melatonin and its precursors with aluminium, cadmium, copper, iron, lead, and zinc: An adsorptive voltammetric study. *Journal of Pineal Research*, *24*(1), 15–21.

Lin, Y., Fan, L., Xia, X., Wang, Z., Yin, Y., Cheng, Y., & Li, Z. (2019). Melatonin decreases resistance to postharvest green mold on citrus fruit by scavenging defense-related reactive oxygen species. *Postharvest Biology and Technology*, *153*, 21–30.

Liu, C., Chen, L., Zhao, R., Li, R., Zhang, S., Yu, W., . . . & Shen, L. (2019). Melatonin induces disease resistance to Botrytis cinerea in tomato fruit by activating jasmonic acid signaling pathway. *Journal of Agricultural and Food Chemistry*, *67*(22), 6116–6124.

Liu, T., Zhao, F., Liu, Z., Zuo, Y., Hou, J., & Wang, Y. (2016). Identification of melatonin in Trichoderma spp. and detection of melatonin content under controlled-stress growth conditions from T. asperellum. *Journal of Basic Microbiology*, *56*(7), 838–843.

Lorente-Mento, J. M., Guillén, F., Castillo, S., Martínez-Romero, D., Valverde, J. M., Valero, D., & Serrano, M. (2021). Melatonin treatment to pomegranate trees enhances fruit bioactive compounds and quality traits at harvest and during postharvest storage. *Antioxidants*, *10*(6), 820.

Lu, R., Liu, Z., Shao, Y., Sun, F., Zhang, Y., Cui, J., . . . & Zhou, T. (2019). Melatonin is responsible for rice resistance to rice stripe virus infection through a nitric oxide-dependent pathway. *Virology Journal*, *16*(1), 1–8.

Ma, Y., Jiao, J., Fan, X., Sun, H., Zhang, Y., Jiang, J., & Liu, C. (2017). Endophytic bacterium Pseudomonas fluorescens RG11 may transform tryptophan to melatonin and promote endogenous melatonin levels in the roots of four grape cultivars. *Frontiers in Plant Science*, *7*, 2068.

Manchester, L. C., Coto-Montes, A., Boga, J. A., Andersen, L. P. H., Zhou, Z., Galano, A., . . . & Reiter, R. J. (2015). Melatonin: An ancient molecule that makes oxygen metabolically tolerable. *Journal of Pineal Research*, *59*(4), 403–419.

Mandal, M. K., Suren, H., Ward, B., Boroujerdi, A., & Kousik, C. (2018). Differential roles of melatonin in plant-host resistance and pathogen suppression in cucurbits. *Journal of Pineal Research*, *65*(3), e12505.

Mandal, S., Zheng, X., Reiter, R. J., Feng, S., Wang, Y., Liu, S., . . . & Ren, M. (2017a). Melatonin attenuates potato late blight by disrupting cell growth, stress tolerance, fungicide susceptibility and homeostasis of gene expression in Phytophthora infestans. *Frontiers in Plant Science*, *8*, 1993.

Mathur, P., & Pramanik, S. (2020). Prospective role of melatonin in signaling and alleviation of stress in plants. In *Neurotransmitters in plant signaling and communication* (pp. 213–240). Cham: Springer.

Mittler, R. (2017). ROS are good. *Trends in Plant Science*, *22*(1), 11–19.

Moustafa-Farag, M., Almoneafy, A., Mahmoud, A., Elkelish, A., Arnao, M. B., Li, L., & Ai, S. (2019). Melatonin and its protective role against biotic stress impacts on plants. *Biomolecules*, *10*(1), 54.

Murch, S. J., Campbell, S. S., & Saxena, P. K. (2001). The role of serotonin and melatonin in plant morphogenesis: Regulation of auxin-induced root organogenesis in in vitro-cultured explants of St. John's wort (Hypericum perforatum L.). *In Vitro Cellular & Developmental Biology-Plant*, *37*(6), 786–793.

Nawaz, K., Chaudhary, R., Sarwar, A., Ahmad, B., Gul, A., Hano, C., . . . & Anjum, S. (2020). Melatonin as master regulator in plant growth, development and stress alleviator for sustainable agricultural production: Current status and future perspectives. *Sustainability*, *13*(1), 294.

Nehela, Y., & Killiny, N. (2018). Infection with phytopathogenic bacterium inhibits melatonin biosynthesis, decreases longevity of its vector, and suppresses the free radical-defense. *Journal of Pineal Research*, *65*(3), e12511.

Nehela, Y., & Killiny, N. (2020). Melatonin is involved in citrus response to the pathogen huanglongbing via modulation of phytohormonal biosynthesis. *Plant Physiology*, *184*(4), 2216–2239.

Pardo-Hernández, M., López-Delacalle, M., & Rivero, R. M. (2020). ROS and NO regulation by melatonin under abiotic stress in plants. *Antioxidants*, *9*(11), 1078.

Paredes, S. D., Korkmaz, A., Manchester, L. C., Tan, D. X., & Reiter, R. J. (2009). Phytomelatonin: A review. *Journal of Experimental Botany*, *60*(1), 57–69.

Park, S., Byeon, Y., & Back, K. (2013). Functional analyses of three ASMT gene family members in rice plants. *Journal of Pineal Research*, *55*(4), 409–415.

Pitzschke, A., Schikora, A., & Hirt, H. (2009). MAPK cascade signalling networks in plant defence. *Current Opinion in Plant Biology*, *12*(4), 421–426.

Prasad, T. K., Anderson, M. D., Martin, B. A., & Stewart, C. R. (1994). Evidence for chilling-induced oxidative stress in maize seedlings and a regulatory role for hydrogen peroxide. *The Plant Cell*, *6*(1), 65–74.

Qian, Y., Tan, D. X., Reiter, R. J., & Shi, H. (2015). Comparative metabolomic analysis highlights the involvement of sugars and glycerol in melatonin-mediated innate immunity against bacterial pathogen in Arabidopsis. *Scientific Reports*, *5*(1), 1–11.

Rasmussen, M. W., Roux, M., Petersen, M., & Mundy, J. (2012). MAP kinase cascades in Arabidopsis innate immunity. *Frontiers in Plant Science*, *3*, 169.

Reiter, R. J., Richardson, B. A., Johnson, L. Y., Ferguson, B. N., & Dinh, D. T. (1980). Pineal melatonin rhythm: Reduction in aging Syrian hamsters. *Science*, *210*(4476), 1372–1373.

Reiter, R. J., Tan, D. X., Rosales-Corral, S., Galano, A., Zhou, X. J., & Xu, B. (2018). Mitochondria: Central organelles for melatonin's antioxidant and anti-aging actions. *Molecules*, *23*(2), 509.

Salehi, B., Sharopov, F., Fokou, P. V. T., Kobylinska, A., Jonge, L. D., Tadio, K., . . . & Iriti, M. (2019). Melatonin in medicinal and food plants: Occurrence, bioavailability, and health potential for humans. *Cells*, *8*(7), 681.

Samanta, S., Banerjee, A., & Roychoudhury, A. (2021a). Exogenous melatonin regulates endogenous phytohormone homeostasis and thiol-mediated detoxification in two indica rice cultivars under arsenic stress. *Plant Cell Reports*, *40*, 1585–1602.

Samanta, S., Banerjee, A., & Roychoudhury, A. (2021b). Melatonin application differentially modulates the enzymes associated with antioxidative machinery and ascorbate glutathione cycle during arsenate exposure in indica rice varieties. *Plant Biology*, *23*(S1), 193–201.

Samanta, S., Banerjee, A., & Roychoudhury, A. (2022). Arsenic toxicity is counteracted by exogenous application of melatonin to different extents in arsenic-susceptible and arsenic-tolerant rice cultivars. *Journal of Plant Growth Regulation*, *41*, 2210–2231.

Sarropoulou, V. N., Therios, I. N., & Dimassi-Theriou, K. N. (2012). Melatonin promotes adventitious root regeneration in in vitro shoot tip explants of the commercial sweet cherry rootstocks CAB-6P (Prunus cerasus L.), Gisela 6 (P. cerasus× P. canescens), and MxM 60 (P. avium× P. mahaleb). *Journal of Pineal Research*, *52*(1), 38–46.

Sharif, R., Xie, C., Zhang, H., Arnao, M. B., Ali, M., Ali, Q., . . . & Li, Y. (2018). Melatonin and its effects on plant systems. *Molecules*, *23*(9), 2352.

Shi, H., Chen, K., Wei, Y., & He, C. (2016). Fundamental issues of melatonin-mediated stress signaling in plants. *Frontiers in Plant Science*, *7*, 1124.

Shi, H., Chen, Y., Tan, D. X., Reiter, R. J., Chan, Z., & He, C. (2015). Melatonin induces nitric oxide and the potential mechanisms relate to innate immunity against bacterial pathogen infection in Arabidopsis. *Journal of Pineal Research*, *59*(1), 102–108.

Sinha, A. K., Jaggi, M., Raghuram, B., & Tuteja, N. (2011). Mitogen-activated protein kinase signaling in plants under abiotic stress. *Plant Signaling & Behavior*, *6*(2), 196–203.

Sun, Q., Zhang, N., Wang, J., Zhang, H., Li, D., Shi, J., . . . & Guo, Y. D. (2015). Melatonin promotes ripening and improves quality of tomato fruit during postharvest life. *Journal of Experimental Botany*, *66*(3), 657–668.

Sun, Y., Liu, Z., Lan, G., Jiao, C., & Sun, Y. (2019). Effect of exogenous melatonin on resistance of cucumber to downy mildew. *Scientia Horticulturae*, *255*, 231–241.

Tan, D. X., Hardeland, R., Manchester, L. C., Korkmaz, A., Ma, S., Rosales-Corral, S., & Reiter, R. J. (2012). Functional roles of melatonin in plants, and perspectives in nutritional and agricultural science. *Journal of Experimental Botany*, *63*(2), 577–597.

Tan, D. X., Manchester, L. C., Esteban-Zubero, E., Zhou, Z., & Reiter, R. J. (2015). Melatonin as a potent and inducible endogenous antioxidant: Synthesis and metabolism. *Molecules*, *20*(10), 18886–18906.

Torres, M. A., & Dangl, J. L. (2005). Functions of the respiratory burst oxidase in biotic interactions, abiotic stress and development. *Current Opinion in Plant Biology*, *8*(4), 397–403.

Tousi, S., Zoufan, P., & Ghahfarrokhie, A. R. (2020). Alleviation of cadmium-induced phytotoxicity and growth improvement by exogenous melatonin pretreatment in mallow (Malva parviflora) plants. *Ecotoxicology and Environmental Safety*, *206*, 111403.

Tripathy, B. C., & Oelmüller, R. (2012). Reactive oxygen species generation and signaling in plants. *Plant Signaling & Behavior, 7*(12), 1621–1633.

Tsukagoshi, H., Busch, W., & Benfey, P. N. (2010). Transcriptional regulation of ROS controls transition from proliferation to differentiation in the root. *Cell, 143*(4), 606–616.

Wang, L. F., Li, T. T., Zhang, Y., Guo, J. X., Lu, K. K., & Liu, W. C. (2021a). CAND2/PMTR1 is required for melatonin-conferred osmotic stress tolerance in Arabidopsis. *International Journal of Molecular Sciences, 22*(8), 4014.

Wang, L. F., Lu, K. K., Li, T. T., Zhang, Y., Guo, J. X., Song, R. F., & Liu, W. C. (2021b). Maize phytomelatonin receptor1 functions in plant osmotic and drought stress tolerance. *Journal of Experimental Botany, 73*(17), 5961–5973.

Weeda, S., Zhang, N., Zhao, X., Ndip, G., Guo, Y., Buck, G. A., . . . & Ren, S. (2014). Arabidopsis transcriptome analysis reveals key roles of melatonin in plant defense systems. *PloS One, 9*(3), e93462.

Wei, J., Li, D. X., Zhang, J. R., Shan, C., Rengel, Z., Song, Z. B., & Chen, Q. (2018a). Phytomelatonin receptor PMTR 1-mediated signaling regulates stomatal closure in Arabidopsis thaliana. *Journal of Pineal Research, 65*(2), e12500.

Wei, Y., Chang, Y., Zeng, H., Liu, G., He, C., & Shi, H. (2018b). RAV transcription factors are essential for disease resistance against cassava bacterial blight via activation of melatonin biosynthesis genes. *Journal of Pineal Research, 64*(1), e12454.

Wei, Y., Hu, W., Wang, Q., Zeng, H., Li, X., Yan, Y., . . . & Shi, H. (2017). Identification, transcriptional and functional analysis of heat-shock protein 90s in banana (M usa acuminata L.) highlight their novel role in melatonin-mediated plant response to F usarium wilt. *Journal of Pineal Research, 62*(1), e12367.

Wolf, K., Kolář, J., Witters, E., van Dongen, W., van Onckelen, H., & Macháčková, I. (2001). Daily profile of melatonin levels in Chenopodium rubrum L. depends on photoperiod. *Journal of Plant Physiology, 158*(11), 1491–1493.

Xie, H. T., Wan, Z. Y., Li, S., & Zhang, Y. (2014). Spatiotemporal production of reactive oxygen species by NADPH oxidase is critical for tapetal programmed cell death and pollen development in Arabidopsis. *The Plant Cell, 26*(5), 2007–2023.

Yao, J. W., Ma, Z., Ma, Y. Q., Zhu, Y., Lei, M. Q., Hao, C. Y., . . . & Huang, X. (2021). Role of melatonin in UV-B signaling pathway and UV-B stress resistance in Arabidopsis thaliana. *Plant, Cell & Environment, 44*(1), 114–129.

Yin, L., Wang, P., Li, M., Ke, X., Li, C., Liang, D., . . . & Ma, F. (2013). Exogenous melatonin improves M alus resistance to M arssonina apple blotch. *Journal of Pineal Research, 54*(4), 426–434.

Yu, Y., Lv, Y., Shi, Y., Li, T., Chen, Y., Zhao, D., & Zhao, Z. (2018). The role of phyto-melatonin and related metabolites in response to stress. *Molecules, 23*(8), 1887.

Zafra, A., Rodríguez-García, M. I., & Alché, J. D. D. (2010). Cellular localization of ROS and NO in olive reproductive tissues during flower development. *BMC Plant Biology, 10*(1), 1–14.

Zhan, H., Nie, X., Zhang, T., Li, S., Wang, X., Du, X., . . . & Song, W. (2019). Melatonin: A small molecule but important for salt stress tolerance in plants. *International Journal of Molecular Sciences, 20*(3), 709.

Zhang, H., Liu, X., Chen, T., Ji, Y., Shi, K., Wang, L., . . . & Kong, J. (2018). Melatonin in apples and juice: Inhibition of browning and microorganism growth in apple juice. *Molecules, 23*(3), 521.

Zhang, N., Zhang, H. J., Sun, Q. Q., Cao, Y. Y., Li, X., Zhao, B., . . . & Guo, Y. D. (2017). Proteomic analysis reveals a role of melatonin in promoting cucumber seed germination under high salinity by regulating energy production. *Scientific Reports, 7*(1), 1–15.

Zhang, N., Zhao, B., Zhang, H. J., Weeda, S., Yang, C., Yang, Z. C., . . . & Guo, Y. D. (2013). Melatonin promotes water-stress tolerance, lateral root formation, and seed germination in cucumber (Cucumis sativus L.). *Journal of Pineal Research, 54*(1), 15–23.

Zhao, D., Wang, H., Chen, S., Yu, D., & Reiter, R. J. (2021). Phytomelatonin: An emerging regulator of plant biotic stress resistance. *Trends in Plant Science, 26*(1), 70–82.

Zhao, D., Yu, Y., Shen, Y., Liu, Q., Zhao, Z., Sharma, R., & Reiter, R. J. (2019b). Melatonin synthesis and function: Evolutionary history in animals and plants. *Frontiers in Endocrinology, 10*, 249.

Zhao, H., Xu, L., Su, T., Jiang, Y., Hu, L., & Ma, F. (2015). Melatonin regulates carbohydrate metabolism and defenses against Pseudomonas syringae pv. tomato DC 3000 infection in Arabidopsis thaliana. *Journal of Pineal Research, 59*(1), 109–119.

Zhao, L., Chen, L., Gu, P., Zhan, X., Zhang, Y., Hou, C., . . . & Wang, Q. C. (2019a). Exogenous application of melatonin improves plant resistance to virus infection. *Plant Pathology*, *68*(7), 1287–1295.
Zhu, Y., Gao, H., Lu, M., Hao, C., Pu, Z., Guo, M., . . . & Huang, X. (2019). Melatonin-nitric oxide crosstalk and their roles in the redox network in plants. *International Journal of Molecular Sciences*, *20*(24), 6200.
Zhu, Y., Guo, M. J., Song, J. B., Zhang, S. Y., Guo, R., Hou, D. R., . . . & Huang, X. (2021). Roles of endogenous melatonin in resistance to Botrytis cinerea infection in an Arabidopsis model. *Frontiers in Plant Science*, 1031.

20 Priming with and Exogenous Application of Melatonin in Improving Plant Adaptation against Unfavourable Conditions

Purabi Banerjee, Oksana Mandal, Parijat Bhattacharya, Kousik Atta, Pabitra Kumar Ghosh, Rajib Nath and Subhasis Mondal

20.1 INTRODUCTION

Melatonin (*N-acetyl*-5-methoxytryptamine) is a plant growth regulator containing an indolic ring structure (Arnao and Hernàndez-Ruiz 2018; Cherono et al. 2021), isolated from the essential amino acid named tryptophan (Back et al. 2016). Presence of melatonin in plants was first registered through radioimmunoassay and high-performance liquid chromatography (HPLC) in 1995 (Dubbels et al. 1995; Hattori et al. 1995). Afterwards, melatonin has been identified to possess diversified roles on plant growth, physiology and metabolic processes (Hardeland 2016; Arnao and Hernàndez-Ruiz 2019a). In fact, it has a great capability to act like a biostimulator under normal as well as under stress situations for crop plants, facilitating regular growth, photosynthetic process, flowering and economic yield (Nawaz et al. 2015; Fan et al. 2018; Debnath et al. 2019). The antioxidative role of melatonin has already been documented regarding the development of abiotic stress tolerance in several crop plants including barley (Arnao and Hernàndez-Ruiz 2009), cucumber (Li et al. 2013), apple (Wang et al. 2013), wheat (Turk et al. 2014), rice (Liang et al. 2015), soybean (Wei et al. 2015) and maize (Huang et al. 2019), etc.

Melatonin acts as a signaling hormone in case of drought stress when it prevents oxidative damage through reactive oxygen species (ROS) scavenging mechanisms by means of enhancement of enzymatic activities, thereby maintaining membrane integrity, chlorophyll biosynthesis and rate of net photosynthesis in crop plants (Ye et al. 2016; Nabi et al. 2019; Wang et al. 2022). Besides, it also modulates the synthesis and activities of enzymatic as well as non-enzymatic antioxidants (Li et al. 2015; Shafi et al. 2021). Moreover, several respective genes of plants undergo transcriptional up regulation in the presence of melatonin, which, in turn, mediate the enzymatic functions and redox potential, specifically under abiotic stress (Banerjee and Roychoudhury 2017; Li et al. 2018). Basically, melatonin produces a powerful cascade of metabolites, that is, free radical scavengers including N-acetyl 5-methoxykynuramine and N-acetyl-N-formyl-5-methoxyknuramine, as well as can easily cross the cellular membranes other than any antioxidants due to its amphiphilic character, rendering it to be a potent antioxidant as compared to the known antioxidant compounds in the world (Galano et al. 2013; Jafari et al. 2022). Numerous studies have demonstrated the significant functional aspects of melatonin in alleviating the extremities of unfavourable environmental stresses including heat and drought (Zhang et al. 2013; Liu et al. 2015; Liang et al. 2018), cold (Lei et al. 2004; Yang et al. 2018), salinity (Cen et al. 2020; Li et al. 2022), fluoride (Banerjee and

 DOI: 10.1201/9781003343752-20

Roychoudhury 2019) and heavy metal/metalloid accumulation (Tan et al. 2007; Posmyk et al. 2008; Samanta et al. 2022) in different crops.

Seed priming is a major alternative to apply any material related to plant nourishment, which exclusively encourages seed germination and early plant stand (Banerjee et al. 2022). Similar to seed priming, the exogenous foliar spray technique has already been established as an excellent way to facilitate plant growth under stressful situations (Banerjee et al. 2021). Priming technique implemented with melatonin has been reported to be a substantial cost-effective intervention, enabling the crop plants to withstand the adversities of unfavourable environmental conditions (Cao et al. 2019). In fact, melatonin priming in crop seeds has been found to be associated with the regulation of initial metabolic processes of the crop plant during its germination process under abiotic stresses (Marthandan et al. 2020). For example, priming of crop seeds with melatonin has been reported to promote rapid germination under heavy metal (chromium) stress in wheat (Lei et al. 2021). Besides, several studies have reported the functional aspects of melatonin priming in different crops in terms of improvement in osmotic regulation in rapeseed under moisture stress (Khan et al. 2019), regulation of starch metabolism in maize under cold stress (Cao et al. 2019) and stimulation in antioxidative properties in basil under salt stress (Bahcesular et al. 2020). Application of melatonin through foliar spray facilitates internal stress tolerance in crop plants by activating antioxidant defense (Tang et al. 2018; Zhang et al. 2019, 2020), which, in turn, ensures plant survival under abiotic stress conditions with special reference to heavy metal toxicity (Li et al. 2016; Gu et al. 2017). Interestingly, several plant genes related to stress relief are up regulated through exogenous melatonin (Arnao and Hernàndez-Ruiz 2021). This chapter emphasizes on promotional aspects of melatonin application through priming and exogenous foliar spray techniques, explaining its significance in improving crop production through activation of antioxidative defense mechanism against adversities of abiotic stress.

20.2 FUNCTIONAL ASPECTS OF MELATONIN IN CROP PHYSIOLOGY

The mechanistic principles of melatonin to improve physiological efficiency while inducing stress tolerance in crop plants have been elucidated in Figure 20.1. Eventually, seed priming as well as

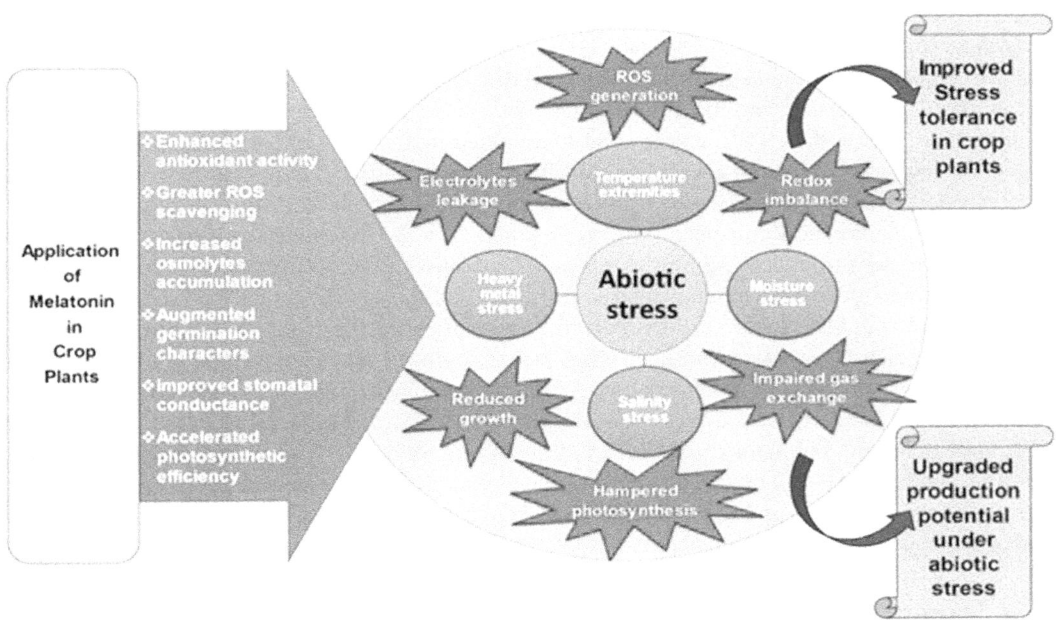

FIGURE 20.1 Impact of melatonin application on crop plants exposed to abiotic stresses.

exogenous application of plant growth regulators like melatonin has been reported several times to enhance the tolerance to abiotic stress in plants (Li et al. 2012; Arnao and Hernàndez-Ruiz 2018; Cherono et al. 2021).

Detailed functional aspects of melatonin with respect to different abiotic stresses have been documented as follows.

20.3 TEMPERATURE STRESS AND MELATONIN

An indolic substance generated from tryptophan, melatonin plays a crucial part in the general functions of plants, such as germination, photosynthesis, circadian rhythms, flowering, senescence and stress reduction. In many environments and in agriculture, the performance, yield and quality of plants are impacted by water scarcity. Additionally, the survival of drought stress depends on the intensity and duration of the drought period.

Under both favourable and unfavourable climatic conditions, seed priming boosts crop output and quality, which is possibly related to activating stress memory. Numerous advantages are purportedly offered by using melatonin administration as a part of agricultural techniques. More specifically, it has been demonstrated that melatonin seed priming enhances seed germination and increases seed resistance in challenging environments.

The principal role of melatonin in plants is to scavenge reactive oxygen species, which subsequently improves a plant's ability to withstand various stresses. The fundamental issue with the use of melatonin in crops, however, continues to be potential chemical alterations in the substance under various environmental conditions and with various formulations. One such technique is seed priming with melatonin, which controls plant metabolic activity during the first phases of germination under stress (Marthandan et al. 2020). Melatonin performs a number of distinct tasks for the growth and development of plants, particularly for the defense of plants under various abiotic stresses. By modifying subcellular antioxidant systems and abscisic acid concentrations in barley, Li et al. (2016) discovered that exogenous melatonin treatment improves the priming-induced cold tolerance. According to Lazar et al. (2013), in addition to chlorophyll (Chl), photosynthetic proteins in general are protected against ROS by melatonin. Melatonin is primarily produced in plant chloroplasts, which are also where photosynthesis takes place and oxygen free radicals are produced in large quantities. Under low-temperature stress, melatonin administration enhanced the sugar and acid content of Bermuda grass and increased Chl fluorescence (Fan et al. 2015). According to Liu et al. (2015), pre-treating tomato seedlings with melatonin considerably lessened the impact of drought stress on their PSII activity and maintained high photosynthetic capacity. Exogenous melatonin improved ability of cucumber seedlings to withstand cold temperature by speeding the ascorbate-glutathione cycle and controlling the photosynthetic electron flux (Zhao et al. 2016). Exogenous melatonin also enhanced the photosynthetic activity of tomato seedlings exposed to salt stress; the mechanism underlying this effect may be related to the positive effects of melatonin on the redox control of photosynthetic electron transport and the synthesis of D1 protein (Zhou et al. 2016). Low-temperature stress was less harmful to tomato seedlings when melatonin priming was applied to the leaves or the roots. It also markedly improved leaf photosynthetic capacity. As a preliminary investigation of the regulatory role of melatonin priming on photosynthetic capacity, spraying on the leaves was superior to root application; however, additional research is required to fully understand the regulatory mechanism of melatonin on the photosynthetic electron transport chain (Yang et al. 2018). The use of melatonin is a crucial agricultural executive contrivance that increases crop resistance and reduces light inhibition, although further investigation into the function of the melatonin priming mechanism is still required.

20.4 MOISTURE STRESS AND MELATONIN

Water stress (drought and waterlogging) comes under abiotic stress and is detrimental to plant growth and development. Melatonin (*N-acetyl*-5-methoxytryptamine), a bioactive plant hormone, can be used to reduce drought stress in several plant species like fruit crops (Lin et al. 2019; Xia et al. 2020), vegetable crops (Karaca and Cekic 2019; Kaya and Doganlar 2019), field crops (Cao et al. 2019; Su et al. 2019), ornamental and medicinal plants (Campos et al. 2019; Li et al. 2019) and model plants (Back and Lee 2019; Yang et al. 2019). Its application can be done in different ways such as pretreatment of roots (Li et al. 2019), direct soil treatment (Ding et al. 2019), seed coating (Wei et al. 2015), seed priming (Khan et al. 2019), foliar application (Li et al. 2019), in nutrient solutions and hydroponic systems (Zou et al. 2019) as well as it can be supplemented with irrigation, as stated by Xia et al. (2020).

The role of melatonin under drought stress is nothing short of a multifunctional regulator of plant status, in terms of both anatomical and physiological aspects. The anatomical changes induced by melatonin occur within the various plant organs, such as reduction of chloroplast ultrastructural damage and preservation of its system (Meng et al. 2014; Zhang et al. 2013), less membrane damage (Yan et al. 2016), more intact grana lamella of the chloroplast (Cui et al. 2017), protection of the mitochondrial structure (Shi et al. 2015), maintenance of cell expansion (Khan et al. 2019), better leaf thickness, stomatal size, spongy tissue (Meng et al. 2014; Khan et al. 2019), promotion of wax accumulation (Li et al. 2019) and cuticle formation (Ding et al. 2018).

In terms of the physiological aspects, melatonin plays a role in the antioxidant system, as stated by Sharma et al. (2020) and Xia et al. (2020), in reducing the oxidative damage so that there is less accumulation of reactive oxygen and nitrogen species (ROS and RNS) (Cao et al. 2019; Antoniou et al. 2017), the easing of toxic substance content (Liu et al. 2015), lower relative conductivity (Niu et al. 2019), less electrolyte leakage (Lee and Back 2016), lower lipid peroxidation (Xia et al. 2020; Ma et al. 2018), limitation on cellular redox disruption (Antoniou et al. 2017) and enhanced ascorbate (AsA)—glutathione (GSH) cycle capacity as in higher GSH and AsA contents (Liang et al. 2018). All of these beneficial effects are the results of regulation of enzymatic activity involving catalase (CAT), and superoxide dismutase (SOD), peroxidase (POD), ascorbate peroxidase (APX) and also non-enzymatic antioxidants and osmoprotectants (proline and others) (Kabiri et al. 2018; Ding et al. 2019; Ahmad et al. 2019) as well as secondary metabolites such as flavonoids, phenolics and phenylalanine ammonia lyase (Hossain et al. 2020). As per Fleta-Soriano et al. (2017), melatonin has also proven to improve photosystem II efficiency, thereby enhancing photoprotection. Being multifunctional, melatonin helps in the accumulation of soluble sugars and proline (Li et al. 2018) and thus regulates the osmotic potential of the cell (Ye et al. 2016). It also enhances plant resistance via higher cell turgor and water-holding capacity (Ma et al. 2018), higher stomatal conductance (Ye et al. 2016) and regulation of stomatal opening (Wei et al. 2018). This results in better growth and flowering (Wei et al. 2015; Zhang et al. 2019), visual quality (Ma et al. 2018), reduced leaf senescence (Liang et al. 2018), better efficiency in seed germination (Li et al. 2019), seed yield and root generation vitality (Zhang et al. 2013).

Melatonin promotes energy production through regulation of glycolytic protein expression and electron transport in the respiratory chain. It is a master regulator as it regulates many plant hormones, making the plants more tolerant when irrigation water is limited. Drought stimulates abscisic acid (ABA) biosynthesis, which results in stomatal closure and contributes to leaf senescence. With regard to this, melatonin maintains the ABA homeostasis (low to moderate concentrations) by positive regulation of its biosynthetic genes and negative regulation of the catabolic genes (Li et al. 2019; Bai et al. 2020; Li et al. 2015; Sharma et al. 2020). Gibberellins regulate photosynthesis, stomatal movement, seed germination and leaf senescence. As per Krugman et al. (2011) and Bai et al. (2020), drought stress inhibits gibberellin biosynthesis, which is further enhanced in response to melatonin treatment, causing drought tolerance (Sharma et al. 2020). Melatonin interacts with gibberellins (GAs) via GA-signalling, which further controls the biosynthesis of auxins (Yang et al. 2019; Sharma et al. 2020). Treatment with melatonin also up regulates cytokinin levels and some related signaling factors, which, in turn, improves the photosynthetic efficiency and drought-induced tolerance (Sharma et al. 2020; Ma et al. 2018; Xu et al. 2019).

Zhang et al. (2019) studied the impact of melatonin pretreatment on alfalfa under waterlogging stress and stated that melatonin could reduce the damage caused by the stress and increase plant growth, chlorophyll content and PSII efficiency. Waterlogging stress resulted in a twofold to fivefold increase of endogenous melatonin levels compared to unstressed plants. Melatonin reduces ethylene production via the down regulation of ethylene synthesis-associated genes that, in turn, reduces waterlogging-caused growth inhibition, premature senescence and chlorosis. Melatonin then increases polyamines levels by promoting the gene expression of the involved enzymes in polyamine metabolism, resulting in more stable cell membranes, less ethylene-responsive senescence and improved photosynthesis.

20.5 SALINITY STRESS AND MELATONIN

Salinization is one of the greatest threats in the agriculture field limiting the growth and productivity of crops. Plant molecular, biochemical and physiological processes are all directly impacted by soil salinization. Soil salinity continues to be one of the most harmful variables in agriculture and affects about 20% of irrigated land despite substantial efforts and breakthroughs in abiotic stress studies (Negrao et al. 2017). Evaluation of the physiological, biochemical and molecular effects of salinity is challenging because of the numerous complex physiological responses of plants to salinity stress. By utilizing intricate physiological features, metabolic pathways and molecular or gene networks, the plants develop a variety of tolerance mechanisms to deal with salinity stress.

The potential for priming techniques to be used as a "stress reliever" in agricultural crop production, however, can be greatly expanded due to their economic viability and simplicity of use. Melatonin seed priming is a straightforward inexpensive approach that improves germination and seedling establishment by triggering a number of physiological and metabolic processes. In order to speed up stress and cross tolerance, priming modulates molecular pathways by up regulating the expression of several stress-related genes and proteins. Altering important signaling molecules, transcription and chromatin states, which are vital for the second shock priming memory and epigenetic changes, enables the plants to tolerate salinity stress. As a result, priming can mediate stress tolerance as well as start a generalized stress tolerance to a variety of stimuli that further alter gene expression and improve crop yield.

The main necessity for growing crops has always been high-quality seeds. For field effectiveness, seedling morphogenesis must be established. Melatonin-primed seedlings performed better under salt stress than unprimed seedlings, indicating that seed priming could also increase tolerance to salt in cotton (Zhang et al. 2021). Since it was first found in plants, melatonin has undergone extensive research, and as a positive regulator under abiotic stress, it is used to reduce the effects of salt stress (Chen et al. 2020). Melatonin can accumulate in plants in response to biotic and abiotic challenges, which could lower ROS levels and lessen the impact of stress on plants (Li et al. 2019a, 2019b).

Exogenous melatonin may control genes associated to phytohormones, which control phytohormone levels and aid plant growth in salt stress (Wei et al. 2015; Arnao and Hernàndez-Ruiz 2018; Xu et al. 2019). Exogenous melatonin was used as a seed priming solution to boost SOD, POD and CAT activity while lowering MDA buildup and preventing salt damage. Melatonin-primed seedlings outgrew unprimed seedlings in terms of growth and phenotypic. As a result, modifications in unprimed seedlings, such as leaf yellowing and shrinkage, significantly impacted the outcomes of photosynthetic efficiency. Additionally, the seedlings primed with melatonin displayed the highest performance as seen by their photosynthetic indexes, which were in line with the phenotypic findings.

Melatonin-based seed priming demonstrated beneficial effects on plant growth and defense and has potential for wide-scale use in agricultural production. Furthermore, it lays the groundwork for the use of melatonin priming in cotton production in salinized land, which had a great significance for efficiently using the limited land resources (Zhang et al. 2021). By modulating the following mechanisms, priming with melatonin can dramatically increase the salt tolerance of seedling plants:

(1) Melatonin priming controlled the photosynthesis. In particular, melatonin priming dramatically reduced the expression of genes involved in photosynthesis to first adjust against salt stress. After adaptation, primed seedlings had considerably better photosynthetic efficiency than the control.
(2) The mechanism for scavenging ROS was improved. As a result, the activity of antioxidant enzymes dramatically increased, while the expression of genes encoding antioxidant enzymes was up regulated.
(3) Melatonin priming coordinated the phytohormone signal pathways to express during salt stress. Abiotic stress in cotton may be effectively prevented by using melatonin seed priming method. Furthermore, it greatly affects how salinized land is used and increases the supply of natural fiber for the textile sector.

Melatonin significantly and positively altered morphological, physiological and biochemical parameters in oat plants under salinity stress (Varghese et al. 2019). Additionally, melatonin triggered modifications in the expression of the genes responsible for producing the core PSII proteins. In the future, melatonin might be used to create transgenic crops that are more stress-tolerant and capable of efficient photosynthetic growth.

20.6 HEAVY METAL STRESS AND MELATONIN

Metals with specific gravity more than 5 gm/cc are generally characterized as heavy metals. But nowadays, the term "heavy metal" is used to indicate metals and metalloids that are toxic to the environment (Hawkes 1997; Chai et al. 2021). Environmental repercussions of heavy metals have become a huge matter of concern globally. Some of the heavy metals like copper, zinc, manganese and iron are essential elements for plant. When these elements are present in excessive amounts, it can severely impact the biochemical and physiological processes of plants. However, cadmium, mercury and lead, which are not categorized as essential elements, also pose severe toxic effects on crops (Gupta et al. 2013; Ghori et al. 2019; Briffa et al. 2020).

Seed priming is an important technique to relieve the plants from abiotic stresses. Heavy metal toxicity is often regarded as one of the most catastrophic abiotic stresses (Hoque et al. 2021). Like that of temperature, moisture and salinity stress, seed priming has been extensively used for remediation of heavy metal-induced stresses (Rhaman et al. 2020; Bhatia and Gupta 2022). Bio stimulants like melatonin play vital roles in imparting tolerance to heavy metal stress in an eco-friendly manner (Hoque et al. 2021; Sun et al. 2021).

Melatonin-mediated heavy metal stress tolerance is basically dependent upon reactive species detoxification, shielding against metabolite and nutrient imbalances as well as adjustment of osmotic imbalances (Hoque et al. 2021). ROS, like H_2O_2, $O_2^{\cdot}$, and NO are fundamentally used as signaling molecules at lower concentrations. However, during heavy metal stress, ROS concentration rapidly increases and severely impacts crop growth. Melatonin scavenges the excessive ROS for detoxification and relieves the crop from heavy metal stress (Nawaz et al. 2018; Nabaei and Amooaghaie 2019; Pardo-Hernández et al. 2020). Exposure to heavy metals leads to metabolite imbalances such as increased or decreased production of soluble protein, carbohydrate and enzymes which have devastating effects on crop growth. Exogenous application of melatonin has been reported to mask metabolite imbalances in crops (Zhang et al. 2017; Qiao et al. 2019; Hoque et al. 2021). Excessive heavy metals hamper osmotic balance inside plants through alteration of root morphology, which, in turn, results into decreased water uptake and eventual dehydration of plants. Pretreatment with melatonin through seed priming can enhance water retention in cells by enhancing the accumulation of proline by up regulating carbohydrate metabolism (Malar et al. 2014; Al-Huqail et al. 2020). Uptake of essential plant nutrients is indispensable for growth of crops. Excessive heavy metal accumulation leads to nutritional imbalances in plants through modification in root structure. Application of melatonin plays a vital role in nutrient homeostasis by modulating root architecture and thereby enhancing root activities (Liang et al. 2018; Jahan et al. 2020). Improvement in photosynthetic attributes as well as activation of stress-tolerant defensive genes also add to melatonin-induced heavy metal stress management (Sarropoulou et al. 2012; Zhao et al. 2017).

Thus, seed priming with melatonin can be an absolute breakthrough in mitigation of heavy metal–induced crop stress and which ultimately manifests into agricultural sustainability.

20.7 WAY FORWARDS

Abiotic stresses create various barriers at physiological as well as molecular levels in crop plants, which, in turn, bring about significant reduction in production potential worldwide. Application of growth regulators, especially through seed priming as well as with foliar spray, is a substantially efficient, convenient and affordable technique to boost resilience of crop plants under different abiotic stresses by preserving the anti-stress metabolites for counteraction of environmental adversities. Melatonin is one of these potent bioactive antioxidants having excellent capacity of ROS scavenging (Reiter et al. 2015; Kul et al. 2019). Notable research efforts have been devoted in recent times towards detailed functional aspects of melatonin in plant physiology. Apart from this, molecular regulation of melatonin in relation to numerous genes also deserves a great scope in varietal evolution of sustainable agriculture regarding stress alleviation (Sun et al. 2021; Nawaz et al. 2021). On the other side, exogenous foliar melatonin can also relieve oxidative stress by reducing lipid peroxidation as well as minimizing the accumulation of H_2O_2 in addition to priming, thereby increasing economic yield and extending the shelf lives of numerous crop plants (Xin et al. 2017; Asif et al. 2019).

The foundation of developing transgenic crops through utilization of the benefits from melatonin have already been established in the light of mutation breeding technology with the genes related to melatonin biosynthesis (Rajora et al. 2022). Multi-omics approaches covering proteomics, transcriptomics and metabolomics are much critical in validating the effect of melatonin in this aspect (Liang et al. 2015). In this regard, more research is necessary in the future to clarify the metabolic pathways along with genetic mechanisms involved in stress alleviation with melatonin (Kołodziejczyk et al. 2016). Further studies are also required using different phytohormone mutants to elucidate the interaction between melatonin and the concerned phytohormones and signaling molecules in terms of different physiological and biochemical mechanisms (Rehaman et al. 2021). Notably, melatonin seems to be a biodegradable and non-toxic compound of natural origin which may also support the progress of organic farming (Janas et al. 2013). Above all, using such a cheap source of plant growth promoter like synthetic melatonin or phytomelatonin extracts may be a key innovative strategy for the poor farmers in developing tolerance in crop plants against the prevailing abiotic stress (Arnao and Hernández-Ruiz 2019b; Pérez-Llamas et al. 2020).

20.8 CONCLUSION

Melatonin, the exclusive plant growth regulator or biostimulator, enhances stress tolerance in crop plants, besides improvising regular developmental processes as well as productivity. Seed priming with melatonin can positively influence the seed germination characteristics, initial plant stands, seedling establishment and physiological mechanisms under unfavorable environmental conditions. Besides, foliar application of exogenous melatonin has the ability to accelerate overall plant growth along with maintaining osmo-regulation and photosynthetic efficiencies of crop plants under different abiotic stresses. Irrespective of the application techniques, melatonin-mediated relaxation from oxidative stress is acquired by activating enzymatic and non-enzymatic antioxidants as well as by directly scavenging the reactive oxygen species to a greater extent. Not only do the accumulation levels of anti-stress osmolytes get elevated, but also, the functional aspects of the genes related to stress relief in plants are up regulated. However, there are several unexplored areas requiring much attention for further researches regarding synthesis of melatonin and its diversified roles in plant development under abiotic stress. Enriched with extensive knowledge in this aspect, the application of melatonin through seed priming and foliar spray may be considered as a promising way out to fight against the adversities of abiotic stress in a more eco-friendly and economic manner.

REFERENCES

Ahmad, S., M. Kamran, R. Ding, X. Meng, H. Wang, I. Ahmad, S. Fahad, and Q. Han. 2019. Exogenous melatonin confers drought stress by promoting plant growth, photosynthetic capacity and antioxidant defense system of maize seedlings. *Peer-Reviewed Journal* 7:e7793.

Al-Huqail, A. A., M. N. Khan, H. M. Ali, M. H. Siddiqui, A. A. Al-Huqail, F. M. AlZuaibr, and L. A. Al-Humaid. 2020. Exogenous melatonin mitigates boron toxicity in wheat. *Ecotoxicology and Environmental Safety* 201:110822.

Antoniou, C., G. Chatzimichail, R. Xenofontos, J. J. Pavlou, E. Panagiotou, A. Christou, and V. Fotopoulos. 2017. Melatonin systemically ameliorates drought stress-induced damage in Medicago sativa plants by modulating nitro-oxidative homeostasis and proline metabolism. *Journal of Pineal Research* 62:e12401.

Arnao, M. B., and J. Hernàndez-Ruiz. 2009. Protective effect of melatonin against chlorophyll degradation during the senescence of barley leaves. *Journal of Pineal Research* 46:58–63. https://doi.org/10.1111/j.1600-079X.2008.00625.x

Arnao, M. B., and J. Hernàndez-Ruiz. 2015. Functions of melatonin in plants: A review. *Journal of Pineal Research* 2:133–150. https://doi.org/10.1111/jpi.12253

Arnao, M. B., and J. Hernàndez-Ruiz. 2018. Melatonin and its relationship to plant hormones. *Annals of Botany* 121:195–207. https://doi.org/10.1093/aob/mcx114

Arnao, M. B., and J. Hernàndez-Ruiz. 2019a. Melatonin: A new plant hormone and/or a plant master regulator? *Trends in Plant Science* 24:38–48. doi: 10.1016/j.tplants.2018.10.010

Arnao, M. B., and J. Hernández-Ruiz. 2019b. Melatonin as a chemical substance or as phytomelatonin rich-extracts for use as plant protector and/or biostimulant in accordance with EC legislation. *Agronomy* 9:570.

Arnao, M. B., and J. Hernández-Ruiz. 2021. Melatonin as a regulatory hub of plant hormone levels and action in stress situations. *Plant Biology* 23:7–19.

Asif, M., A. Pervez, and R. Ahmad. 2019. Role of melatonin and plant-growth-promoting rhizobacteria in the growth and development of plants. *Clean Soil Air Water* 47:1800459.

Back, K., and H. J. Lee. 2019. 2-Hydroxymelatonin confers tolerance against combined cold and drought stress in tobacco, tomato, and cucumber as a potent anti-stress compound in the evolution of land plants. *Melatonin Research* 2:35–46.

Back, K., D. X. Tan, and R. J. Reiter. 2016. Melatonin biosynthesis in plants: Multiple pathways catalyze tryptophan to melatonin in the cytoplasm or chloroplasts. *Journal of Pineal Research* 61:426–437.

Bahcesular, B., E. D. Yildirim, M. Karaçocuk, M. Kulak, and S. Karaman. 2020. Seed priming with melatonin effects on growth, essential oil compounds and antioxidant activity of basil (*Ocimum basilicum* L.) under salinity stress. *Industrial Crops and Products* 146:112165.

Bai, Y., S. Xiao, Z. Zhang, Y. Zhang, H. Sun, K. Zhang, X. Wang, Z. Bai, C. Li, and L. Liu. 2020. Melatonin improves the germination rate of cotton seeds under drought stress by opening pores in the seed coat. *Peer-Reviewed Journal* 8:e9450.

Banerjee, A., and A. Roychoudhury. 2017. Melatonin as a regulator of abiotic stress tolerance in plants. In: Singh V. P., Singh S., Prasad S. M. (Eds.) *Mechanisms Behind Phytohormonal Signaling and Crop Abiotic Stress Tolerance*, New York: Nova Science Publishers, pp. 47–60.

Banerjee, A., and A. Roychoudhury. 2019. Melatonin application reduces fluoride uptake and toxicity in rice seedlings by altering abscisic acid, gibberellin, auxin and antioxidant homeostasis. *Plant Physiology and Biochemistry* 145:164–173.

Banerjee, P., V. K. Venugopalan, R. Nath, Y. S. Althobaiti, A. Gaber, H. Al-Yasi, and A. Hossain. 2021. Physiology, growth and productivity of spring: Summer slack gram (*Vigna mungo* L. Hepper) as influenced by heat and moisture stresses in different dates of sowing and nutrient management conditions. *Agronomy* 11:1–24.

Banerjee, P., V. K. Venugopalan, R. Nath, P. K. Chakraborty, A. Gaber, W. F. Alsanie, B. M. Raafat, and A. Hossain. 2022. Seed priming and foliar application of nutrients influence the productivity of relay grass pea (*Lathyrus sativus* L.) through accelerating the photosynthetically active radiation (PAR) use efficiency. *Agronomy* 12:1–18.

Bhatia, P., and M. Gupta. 2022. Micronutrient seed priming: New insights in ameliorating heavy metal stress. *Environmental Science and Pollution Research*:1–17.

Briffa, J., E. Sinagra, and R. Blundell. 2020. Heavy metal pollution in the environment and their toxicological effects on humans. *Heliyon* 6(9):e04691.

Campos, C. N., R. G. Avila, K. R. Dazio de Souza, L. M. Azevedo, and J. D. Alves. 2019. Melatonin reduces oxidative stress and promotes drought tolerance in young *Coffea arabica* L. plants. *Agricultural Water Management* 211:37–47.

Cao, L., X. J. Jin, and Y. X. Zhang. 2019. Melatonin confers drought stress tolerance in soybean (*Glycine max* L.) by modulating photosynthesis, osmolytes, and reactive oxygen metabolism. *Photosynthetica* 57:812–819.

Cao, Q., G. Li, Z. Cui, F. Yang, X. Jiang, L. Diallo, and F. Kong. 2019. Seed priming with melatonin improves the seed germination of waxy maize under chilling stress via promoting the antioxidant system and starch metabolism. *Scientific Reports* 9:1–12. https://doi.org/10.1038/s41598-019-51122-y

Cen, H., T. Wang, H. Liu, D. Tian, and Zhang, Y. 2020. Melatonin application improves salt tolerance of alfalfa (*Medicago sativa* L.) by enhancing antioxidant capacity. *Plants (Basel)* 9(2):220.

Chai, W. S., Cheun, J. Y., Kumar, P. S., Mubashir, M., Majeed, Z., Banat, F., and Show, P. L. 2021. A review on conventional and novel materials towards heavy metal adsorption in wastewater treatment application. *Journal of Cleaner Production* 296:126589.

Chen, L., L. Liu, B. Lu, T. Ma, D. Jiang, J. Li, K. Zhang, H. Sun, Y. Zhang, Z. Bai, and C. Li. 2020. Exogenous melatonin promotes seed germination and osmotic regulation under salt stress in cotton (*Gossypium hirsutum* L.). *PLoS One* 15. https://doi.org/10.1371/journal.pone.0228241

Cherono, S., C. Ntini, M. Wassie, M. D. Mollah, M. A. Belal, C. Ogutu, and Y. Han. 2021. Exogenous application of melatonin improves drought tolerance in coffee by regulating photosynthetic efficiency and oxidative damage. *Journal of the American Society for Horticultural Science* 146(1):24–32. https://doi.org/10.21273/JASHS04964-20

Cui, G., F. Sun, X. Gao, K. Xie, C. Zhang, S. Liu, and Y. Xi. 2018. Proteomic analysis of melatonin-mediated osmotic tolerance by improving energy metabolism and autophagy in wheat (*Triticum aestivum* L.). *Planta* 248:69–87.

Cui, G., X. Zhao, S. Liu, F. Sun, C. Zhang, and Y. Xi. 2017. Beneficial effects of melatonin in overcoming drought stress in wheat seedlings. *Plant Physiology and Biochemistry* 118:138–149.

Debnath, B., W. Islam, M. Li, Y. Sun, and X. Lu. 2019. Melatonin mediates enhancement of stress tolerance in plants. *International Journal of Molecular Sciences* 20:1040. doi: 10.3390/ijms20051040

Ding, F., G. Wang, M. Wang, and S. Zhang. 2018. Exogenous melatonin improves tolerance to water deficit by promoting cuticle formation in tomato plants. *Molecules* 23:1605.

Ding, Z., C. Wu, W. Tie, Y. Yan, G. He, and W. Hu. 2019. Strand-specific RNA-seq based identification and functional prediction of lncRNAs in response to melatonin and simulated drought stresses in cassava. *Plant Physiology and Biochemistry* 140:96–104.

Dubbels, R., R. J. Reiter, E. Klenke, A. Goebel, E. Schnakenberg, and C. Ehlers. 1995. Melatonin in edible plants identified by radioimmunoassay and by high performance liquid chromatography mass spectrometry. *Journal of Pineal Research* 18:28–31. doi: 10.1111/j.1600-079X.1995.tb00136.x

Fan J., Z. Hu, and Y. Xie. 2015. Alleviation of cold damage to photosystem II and metabolisms by melatonin in *Bermuda grass*. *Frontiers in Plant Science* 6:925.

Fan, J., Y. Xie, Z. Zhang, and L. Chen. 2018. Melatonin: A multifunctional factor in plants. *International Journal of Molecular Sciences* 19:1528. doi: 10.3390/ijms19051528

Fleta-Soriano, E., L. Díaz, E. Bonet, and S. Munné-Bosch. 2017. Melatonin may exert a protective role against drought stress in maize. *Journal of Agronomy and Crop Science* 203:286–294.

Galano, A., D. X. Tan, and R. J. Reiter. 2013. On the free radical scavenging activities of melatonin's metabolites, AFMK and AMK. *Journal of Pineal Research* 54:245–257.

Ghori, N. H., T. Ghori, M. Q. Hayat, S. RImadi, A. Gul, V. Altay, and M. Ozturk. 2019. Heavy metal stress and responses in plants. *International Journal of Environmental Science and Technology* 16(3):1807–1828.

Gu, Q., Z. Chen, X. Yu, W. Cui, J. Pan, and G. Zhao. 2017. Melatonin confers plant tolerance against cadmium stress via the decrease of cadmium accumulation and reestablishment of micro RNA-mediated redox homeostasis. *Plant Science* 261:28–37. doi: 10.1016/j.plantsci.2017.05.001

Gupta, D. K., Corpas, F. J., and Palma, J. M. (Eds.). 2013. *Heavy Metal Stress in Plants*, Berlin, Germany: Springer, p. 240.

Hardeland, R. 2016. Melatonin in plants: Diversity of levels and multiplicity of functions. *Frontiers in Plant Science* 7:198. doi: 10.3389/fpls.2016.00198

Hattori, A., H. Migitaka, M. Iigo, M. Itoh, K. Yamamoto, R. Ohtani Kaneko, M. Hara, T. Suzuki, and R. J. Reiter. 1995. Identification of melatonin in plants and its effects on plasma melatonin levels and binding to melatonin receptors in vertebrates. *Biochemistry and Molecular Biology International* 35:627–634.

Hawkes, S. J. 1997. What is a "heavy metal"? *Journal of Chemical Education* 74(11):1374.

Hoque, M. N., M. Tahjib-Ul-Arif, A. Hannan, N. Sultana, S. Akhter, M, Hasanuzzaman, F. Akter, M. S. Hossain, M. A. Sayed, M. T. Hasan, and M. Skalicky. 2021. Melatonin modulates plant tolerance to heavy metal stress: Morphological responses to molecular mechanisms. *International Journal of Molecular Sciences* 22(21):11445.

Hossain, M. S., J. Li, A. Sikdar, M. Hasanuzzaman, F. Uzizerimana, I. Muhammad, Y. Yuan, C. Zhang, C. Wang, and B. Feng. 2020. Exogenous melatonin modulates the physiological and biochemical mechanisms of drought tolerance in tartary buckwheat (*Fagopyrum tataricum*(L.) Gaertn). *Molecules* 25:282.

Huang, B., Y. E. Chen, Y. Q. Zhao, C. B. Ding, J. Q. Liao, C. Hu, L. J. Zhou, Z. W. Zhang, S. Yuan, and M. Yuan. 2019. Exogenous melatonin alleviates oxidative damages and protects photosystem II in maize seedlings under drought stress. *Frontiers in Plant Science* 10:677. doi: 10.3389/fpls.2019.00677

Jafari, M., A. R. Shahsavar, M. Talebi, and M. Hesami. 2022. Exogenous melatonin protects lime plants from drought stress-induced damage by maintaining cell membrane structure, detoxifying ROS and regulating antioxidant systems. *Horticulturae* 8(3):257.

Jahan, M. S., S. Guo, A. R. Baloch, J. Sun, S. Shu, Y. Wang, and R. Roy. 2020. Melatonin alleviates nickel phytotoxicity by improving photosynthesis, secondary metabolism and oxidative stress tolerance in tomato seedlings. *Ecotoxicology and Environmental Safety* 197:110593.

Janas, K. M., and M. M. Posmyk. 2013. Melatonin, an underestimated natural substance with great potential for agricultural application. *Acta Physiologiae Plantarum* 35:3285–3292.

Kabiri, R., A. Hatami, H. Oloumi, M. Naghizadeh, F. Nasibi, and Z. Tahmasebi. 2018. Foliar application of melatonin induces tolerance to drought stress in Moldavian balm plants (*Dracocephalummoldavica*) through regulating the antioxidant system. *Folia Horticulturae* 30:155–167.

Karaca, P., and F. Ö. Cekic. 2019. Exogenous melatonin-stimulated defense responses in tomato plants treated with polyethylene glycol. *International Journal of Vegetable Science* 25:601–609.

Kaya, A., and Z. B. Doganlar. 2019. Melatonin improves the multiple stress tolerance in pepper (*Capsicum annuum*). *Scientia Horticulturae* 256:108509.

Khan, M. N., J. Zhang, T. Luo, J. Liu, M. Rizwan, S. Fahad, Z. Xu, and L. Hu. 2019. Seed priming with melatonin coping drought stress in rapeseed by regulating reactive oxygen species detoxification: Antioxidant defense system, osmotic adjustment, stomatal traits and chloroplast ultrastructure perseveration. *Industrial Crops & Products* 140:111597 https://doi.org/10.1016/j.indcrop.2019.111597

Kołodziejczyk, I., K. Dzitko, R. Szewczyk, and M. M. Posmyk. 2016. Exogenous melatonin expediently modifies proteome of maize (*Zea mays* L.) embryo during seed germination. *Acta Physiologiae Plantarum* 38:1–16. doi: 10.1007/s11738-016-2166-y

Krugman, T., Z. Peleg, L. Quansah, V. Chague, A. B. Korol, E. Nevo, Y. Saranga, A. Fait, B. Chalhoub, and T. Fahima. 2011. Alteration in expression of hormone-related genes in wild emmer wheat roots associated with drought adaptation mechanisms. *Functional & Integrative Genomics* 11:565–583.

Kul, R., A. Esringü, E. Dadasoglu, Ü. Sahin, M. Turan, S. Örs, M. Ekinci, G. Agar, and E. Yildirim. 2019. Melatonin: Role in increasing plant tolerance in abiotic stress conditions. *Abiotic and Biotic Stress in Plants* 1:19.

Lazar, D., S. J. Murch, and M. J. Beilby. 2013. Exogenous melatonin affects photosynthesis in characeae *Charaaustralis*. *Plant Signaling & Behavior* 8:e23279.

Lee, H. J. and K. Back. 2016. 2-Hydroxymelatonin promotes the resistance of rice plant to multiple simultaneous abiotic stresses (combined cold and drought). *Journal of Pineal Research* 61:303–316.

Lei, K., S. Sun, K. Zhong, S. Li, H. Hu, C. Sun, Q. Zheng, Z. Tian, T. Dai, and J. Sun. 2021. Seed soaking with melatonin promotes seed germination under chromium stress via enhancing reserve mobilization and antioxidant metabolism in wheat. *Ecotoxicology and Environmental Safety* 220:112241.

Lei, X. Y., R. Y. Zhu, G. Y. Zhang, and Y. R. Dai. 2004. Attenuation of cold-induced apoptosis by exogenous melatonin in carrot suspension cells: The possible involvement of polyamines. *Journal of Pineal Research* 36:126–131. doi: 10.1046/j.1600-079x.2003.00106.x

Li, B., T. Sang, L. Z. He, J. Sun, J. Li, and S. R. Guo. 2013. Exogenous spermidine inhibits ethylene production in leaves of cucumber seedlings under NaCl stress. *Journal of the American Soiety for Horticultural Science* 138:108–113. doi: 10.21273/Jashs.138.2.108

Li, C., Q. He, F. Zhang, J. Yu, C. Li, T. Zhao, Y. Zhang, Q. Xie, B. Su, L. Mei, S. Zhu, and J. Chen. 2019a. Melatonin enhances cotton immunity to *Verticillium* wilt via manipulating lignin and gossypol biosynthesis. *The Plant Journal* 100:784–800. https://doi.org/10.1111/tpj.14477

Li, C., D. X. Tan, D. Liang, C. Chang, D. Jia, and F. Ma. 2015. Melatonin mediates the regulation of ABA metabolism, free-radical scavenging, and stomatal behaviour in two *Malus* species under drought stress. *Journal of Experimental Botany* 66:669–680.

Li, C., P. Wang, Z. Wei, D. Liang, C. Liu, L. Yin, D. Jia, M. Fu, and F. Ma. 2012. The mitigation effects of exogenous melatonin on salinity induced stress in *Malus hupehensis*. *Journal of Pineal Research* 53:298–306. doi: 10.1111/j.1600-079X.2012.00999.x

Li, D., D. Zhang, H. Wang, H. Li, S. Song, H. Li, and R. Li. 2019. Effects of melatonin on germination and amino acid content in different wheat varieties seeds under polyethylene glycol stress. *BioRxiv* 2019:710954.

Li, H., Y. Mo, Q. Cui, X. Yang, Y. Guo, C. Wei, J. Yang, Y. Zhang, J. Ma, and X. Zhang. 2019. Transcriptomic and physiological analyses reveal drought adaptation strategies in drought-tolerant and-susceptible watermelon genotypes. *Plant Science* 278:32–43.

Li, J., J. Liu, T. Zhu, C. Zhao, L. Li, and M. Chen. 2019b. The role of melatonin in salt stress responses. *International Journal of Molecular Sciences* 20:1735. https://doi.org/10.3390/ijms20071735

Li, J., Y. Liu, M. Zhang, X. Hualing, K. Ning, B. Wang, and M. Chen. 2022. Melatonin increases growth and salt tolerance of *Limonium bicolor* by improving photosynthetic and antioxidant capacity. *BMC Plant Biology* 22:1–14, https://doi.org/10.1186/s12870-021-03402-x

Li, J., Y. Yang, K. Sun, Y. Chen, X. Chen, and X. Li. 2019. Exogenous melatonin enhances cold, salt and drought stress tolerance by improving antioxidant defense in tea plant (*Camellia sinensis* (L.) O. Kuntze). *Molecules* 24:1826.

Li, J., L. Zeng, Y. Cheng, G. Lu, G. Fu, H. Ma, Q. Liu, X. Zhang, X. Zou, and C. Li. 2018. Exogenous melatonin alleviates damage from drought stress in *Brassica napus* L. (rapeseed) seedlings. *Acta Physiologiae Plantarum* 40:43.

Li, M. Q., M. K. Hasan, C. X. Li, G. J. Ahammed, X. J. Xia, and K. Shi 2016. Melatonin mediates selenium-induced tolerance to cadmium stress in tomato plants. *Journal of Pineal Research* 61:291–302. doi: 10.1111/jpi.12346

Li, X., D. X. Tan, and D. Jiang. 2016. Melatonin enhances cold tolerance in drought-primed wild-type and abscisic acid deficient mutant barley. *Journal of Pineal Research* 61:328–339.

Li, X., D. X. Tan, D. Jiang, and F. Liu. 2016. Melatonin enhances cold tolerance in drought-primed wild-type and abscisic acid-deficient mutant barley. *Journal of Pineal Research* 61:328–339.

Li, X., J. P. Wei, E. R. Scott, J. W. Liu, S. Guo, Y. Li, L. Zhang, and W. Y. Han. 2018. Exogenous melatonin alleviates cold stress by promoting antioxidant defense and redox homeostasis in *Camellia sinensis* L. *Molecules* 23:1–13. doi: 10.3390/molecules23010165

Liang, B., C. Ma, Z. Zhang, Z. Wei, T. Gao, Q. Zhao, F. Ma, and C. Li. 2018. Long-term exogenous application of melatonin improves nutrient uptake fluxes in apple plants under moderate drought stress. *Environmental and Experimental Botany* 155:650–661. doi: 10.1016/j.envexpbot.2018.08.016

Liang, C., G. Zheng, W. Li, Y. Wang, B. Hu, H. Wang, H. Wu, Y. Qian, X. G. Zhu, and D. X. Tan. 2015. Melatonin delays leaf senescence and enhances salt stress tolerance in rice. *Journal of Pineal Research* 59:91–101. https://doi.org/10.1111/jpi.12243

Liang, W., X. Ma, P. Wan, and L. Liu. 2018. Plant salt-tolerance mechanism: A review. *Biochemical and Biophysical Research Communications* 495(1):286–291.

Lin, Z., Y. Wang, H. Xia, and D. Liang. 2019. Effects of exogenous melatonin and abscisic acid on the antioxidant enzyme activities and photosynthetic pigment in 'Summer Black' grape under drought stress. *IOP Conference Series: Earth & Environmental Science* 295:012013.

Liu, J., W. Wang, L. Wang, and Y. Sun. 2015. Exogenous melatonin improves seedling health index and drought tolerance in tomato. *Plant Growth Regulation* 77:317–326.

Ma, X., J. Zhang, P. Burgess, S. Rossi, and B. Huang. 2018. Interactive effects of melatonin and cytokinin on alleviating drought-induced leaf senescence in creeping bentgrass (*Agrostis stolonifera*). *Environmental and Experimental Botany* 145:1–11.

Malar, S., R. Manikandan, P. J. Favas, S. V. Sahi, and P. Venkatachalam 2014. Effect of lead on phytotoxicity, growth, biochemical alterations and its role on genomic template stability in Sesbania grandiflora: A potential plant for phytoremediation. *Ecotoxicology and Environmental Safety* 108:249–257.

Marthandan, V., R. Geetha, K. Kumutha, V. G. Renganathan, A. Karthikeyan, and J. Ramalingam. 2020. Seed priming: A feasible strategy to enhance drought tolerance in crop plants. *International Journal of Molecular Sciences* 21:8258.

Meng, J. F., T. F. Xu, Z. Z. Wang, Y. L. Fang, Z. M. Xi, and Z. W. Zhang. 2014. The ameliorative effects of exogenous melatonin on grape cuttings under water-deficient stress: Antioxidant metabolites, leaf anatomy, and chloroplast morphology. *Journal of Pineal Research* 57:200–212.

Nabaei, M., and R. Amooaghaie. 2019. Nitric oxide is involved in the regulation of melatonin-induced antioxidant responses in Catharanthus roseus roots under cadmium stress. *Botany* 97(12):681–690.

Nabi, R. B. S., R. Tayade, A. Hussain, K. P. Kulkarni, Q. M. Imran, B. G. Mun, and B. W. Yun. 2019. Nitric oxide regulates plant responses to drought, salinity, and heavy metal stress. *Environmental and Experimental Botany* 161:120–133. doi: 10.1016/j.envexpbot.2019.02.003

Nawaz, K., R. Chaudhary, A. Sarwar, B. Ahmad, A. Gul, C. Hano, B. H. Abbasi, and S. Anjum. 2021. Melatonin as master regulator in plant growth, development and stress alleviator for sustainable agricultural production: Current status and future perspectives. *Sustainability* 13:294.

Nawaz, M. A., Y. Huang, Z. Bie, W. Ahmed, R. J. Reiter, and M. Niu. 2016. Melatonin: Current status and future perspectives in plant science. *Frontiers in Plant Science* 6:1230. doi: 10.3389/fpls.2015.01230

Nawaz, M. A., Y. Jiao, C. Chen, F. Shireen, Z. Zheng, M. Imelatoniniaz, and Y. Huang. 2018. Melatonin pretreatment improves vanadium stress tolerance of watermelon seedlings by reducing vanadium concentration in the leaves and regulating melatonin biosynthesis and antioxidant-related gene expression. *Journal of Plant Physiology* 220:115–127.

Negrao, S., S. M. Schmockel, and M. Tester. 2017. Evaluating physiological responses of plants to salinity stress. *Annals of Botany* 119:1–11.

Niu, X., C. Deqing, and D. Liang. 2019. Effects of exogenous melatonin and abscisic acid on osmotic adjustment substances of 'summer black' grape under drought stress. IOP Conf. Ser. *Earth & Environmental Science* 295:012012.

Pardo-Hernández, M., M. López-Delacalle, and R. M. Rivero. 2020. ROS and NO regulation by melatonin under abiotic stress in plants. *Antioxidants* 9(11):1078.

Pérez-Llamas, F., J. Hernández-Ruiz, A. Cuesta, S. Zamora, and M. B. Arnao. 2020. Development of a phytomelatonin-rich extract from cultured plants with excellent biochemical and functional properties as an alternative to synthetic melatonin. *Antioxidants* 9:158.

Posmyk, M. M., H. Kuran, K. Marciniak, and K. M. Janas. 2008. Pre-sowing seed treatment with melatonin protects red cabbage seedlings against toxic copper ion concentrations. *Journal of Pineal Research* 45:24–31. doi: 10.1111/j.1600–079X.2007.00552.x

Qiao, Y., L. Yin, B. Wang, Q. Ke, X. Deng, and S. Wang 2019. Melatonin promotes plant growth by increasing nitrogen uptake and assimilation under nitrogen deficient condition in winter wheat. *Plant Physiology and Biochemistry* 139:342–349.

Rajora, N., S. Vats, G. Raturi, V. Thakral, S. Kaur, V. Rachappanavar, M. Kumar, A. K. Kesarwani, H. Sonah, T. R. Sharma, and R. Deshmukh. 2022. Seed priming with melatonin: A promising approach to combat abiotic stress in plants. *Plant Stress* 4:1–10. https://doi.org/10.1016/j.stress.2022.100071

Rehaman, A., A. K. Mishra, A. Ferdose, T. S. Per, M. Hanief, A. T. Jan, and M. Asgher. 2021. Melatonin in plant defense against abiotic stress. *Forests* 12:1404. https://doi.org/10.3390/f12101404

Reiter, R., D. X. Tan, Z. Zhou, M. Cruz, L. Fuentes-Broto, and A. Galano. 2015. Phytomelatonin: Assisting plants to survive and thrive. *Molecules* 20:7396–7437. doi: 10.3390/molecules20047396

Rhaman, M. S., S. Imran, F. Rauf, M. Khatun, C. C. Baskin, Y. Murata, and M. Hasanuzzaman. 2020. Seed priming with phytohormones: An effective approach for the mitigation of abiotic stress. *Plants* 10(1):37.

Samanta, S., A. Banerjee, and A. Roychoudhury. 2022. Arsenic toxicity is counteracted by exogenous application of melatonin to different extents in arsenic-susceptible and arsenic-tolerant rice cultivars. *Journal of Plant Growth Regulation* 41:2210–2231.

Sarropoulou, V., K. Dimassi-Theriou, I. Therios, and M. Koukourikou-Petridou. 2012. Melatonin enhances root regeneration, photosynthetic pigments, biomass, total carbohydrates and proline content in the cherry rootstock PHL-C (*Prunus avium* × *Prunus cerasus*). *Plant Physiology and Biochemistry* 61:162–168.

Shah, A., A. K. Singh, and I. Zahoor. 2021. Melatonin: Role in abiotic stress resistance and tolerance. In: Aftab T., Hakeem K. R. (Eds.) *Plant Growth Regulators*, Springer: Cham, Switzerland, pp. 239–273.

Sharma, A., J. Wang, D. Xu, S. Tao, S. Chong, D. Yan, Z. Li, H. Yuan, and B. Zheng. 2020. Melatonin regulates the functional components of photosynthesis, antioxidant system, gene expression, and metabolic pathways to induce drought resistance in grafted *Caryacathayensis* plants. *Science of the Total Environment* 713:136675.

Shi, H., C. Jiang, T. Ye, D. X. Tan, R. J. Reiter, H. Zhang, R. Liu, and Z. Chan. 2015. Comparative physiological, metabolomic, and transcriptomic analyses reveal mechanisms of improved abiotic stress resistance in bermudagrass [*Cynodondactylon* (L). Pers.] by exogenous melatonin. *Journal of Experimental Botany* 66:681–694.

Su, X., X. Fan, R. Shao, J. Guo, Y. Wang, J. Yang, Q. Yang, and L. Guo. 2019. Physiological and iTRAQ-based proteomic analyses reveal that melatonin alleviates oxidative damage in maize leaves exposed to drought stress. *Plant Physiology and Biochemistry* 142:263–274.

Sun, C., L. Liu, L. Wang, B. Li, C. Jin, and X. Lin. 2021. Melatonin: A master regulator of plant development and stress responses. *Journal of Integrative Plant Biology* 63(1):126–145.

Tan, D. X., L. C. Manchester, P. Helton, and R. J. Reiter. 2007. Phytoremediative capacity of plants enriched with melatonin. *Plant Signaling & Behavior* 2:514–516. doi: 10.4161/psb.2.6.4639.

Tang, Y., L. Lin, Y. Xie, J. Liu, G. Sun, H. Li, et al. 2018. Melatonin affects the growth and cadmium accumulation of *Malachiumaquaticum* and *Galinsoga parviflora*. *International Journal of Phytoremediation* 20:295–300. doi: 10.1080/15226514.2017.1374341

Turk, H., S. Erdal, M. Genisel, O. Atici, Y. Demir, and D. Yanmis. 2014. The regulatory effect of melatonin on physiological, biochemical and molecular parameters in cold-stressed wheat seedling. *Plant Growth Regulation* 74:139–52. https://doi.org/10.1007/s10725-014-9905-0

Varghese, N., O. Alyammahi, S. Nasreddine, A. Alhassani, and M. A. Gururani. 2019. Melatonin positively influences the photosynthetic machinery and antioxidant system of *Avena sativa* during salinity stress. *Plants* 8(12):610. https://doi.org/10.3390/plants8120610

Wang, M., J. Gong, C. Song, Z. Wang, S. Song, J. Jiao, M. Wang, X. Zhang, and T. Bai. 2022. Exogenous melatonin alleviated growth inhibition and oxidative stress induced by drought stress in apple rootstock. *Biocell* 46(7):1763–70. doi: 10.32604/biocell.2022.018934

Wang, P., X. Sun, C. Li, Z. W. Wei, D. Liang, and F. W. Ma. 2013. Long-term exogenous application of melatonin delays drought-induced leaf senescence in apple. *Journal of Pineal Research* 54:292–302. https://doi.org/10.1111/jpi.12017

Wei, J., D. X. Li, J. R. Zhang, C. Shan, Z. Rengel, Z. B. Song, and Q. Chen. 2018. Phytomelatonin receptor PMELATONINR1-mediated signaling regulates stomatal closure in *Arabidopsis thaliana*. *Journal of Pineal Research* 65:e12500.

Wei, W., Q. T. Li, Y. N. Chu, R. J. Reiter, X. M. Yu, D. H. Zhu, W. K. Zhang, B. Ma, Q. Lin, J. S. Zhang, and S. Y. Chen. 2015. Melatonin enhances plant growth and abiotic stress tolerance in soybean plants. *Journal of Experimental Botany* 66:695–707. https://doi.org/10.1093/jxb/eru392

Xia, H., Z. Ni, R. Hu, L. Lin, H. Deng, J. Wang, Y. Tang, G. Sun, X. Wang, and H. Li. 2020. Melatonin alleviates drought stress by a non-enzymatic and enzymatic antioxidative system in kiwifruit seedlings. *International Journal of Molecular Sciences* 21:852.

Xin, D., J. Si, and L. Kou. 2017. Postharvest exogenous melatonin enhances quality and delays the senescence of cucumber. *Acta HorticulturaeSinica* 44:891–901.

Xu, L., G. Xiang, Q. Sun, Y. Ni, Z. Jin, S. Gao, and Y. Yao. 2019. Melatonin enhances salt tolerance by promoting *MYB108A*-mediated ethylene biosynthesis in grapevines. *Horticultural Research* 6:114. https://doi.org/10.1038/s41438-019-0197-4

Yan, W., L. Hongyan, M. Xuejiao, W. Xuejuan, and Z. Yuanbing. 2016. Effect of foliar spraying exogenous melatonin on physiological and biochemical characteristics of dendranthema morifolium. *Acta Botanica Boreali-OccidentaliaSinica* 36:2241–2246.

Yang, W. J., Y. T. Du, Y. B. Zhou, J. Chen, Z. S. Xu, Y. Z. Ma, M. Chen, and D. H. Min. 2019. Overexpression of Tacomelatonin improves melatonin production and enhances drought tolerance in transgenic arabidopsis. *International Journal of Molecular Sciences* 20:652.

Yang, X. L., H. Xu, and D. Li. 2018. Effect of melatonin priming on photosynthetic capacity of tomato leaves under low-temperature stress. *Photosynthetica* 56:884–892.

Yang, X. L., H. Xu, D. Li, X. Gao, T. L. Li, and R. Wang. 2018. Effect of melatonin priming on photosynthetic capacity of tomato leaves under low-temperature stress. *Photosynthetica* 56:884–892. https://doi.org/10.1007/s11099-017-0748-6

Ye, J., S. Wang, X. Deng, L. Yin, B. Xiong, and X. Wang. 2016. Melatonin increased maize (*Zea mays* L.) seedling drought tolerance by alleviating drought-induced photosynthetic inhibition and oxidative damage. *Acta Physiologiae Plantarum* 38:48.

Zhang, J., B. Zeng, Y. Mao, X. Kong, X. Wang, Y. Yang, and Q. Chen. 2017. Melatonin alleviates aluminium toxicity through modulating antioxidative enzymes and enhancing organic acid anion exudation in soybean. *Functional Plant Biology* 44(10):961–968.

Zhang, M., S. He, Y. Zhan, B. Qin, X. Jin, M. Wang, Y. Zhang, G. Hu, Z. Teng, and Y. Wu. 2019. Exogenous melatonin reduces the inhibitory effect of osmotic stress on photosynthesis in soybean. *PLoS ONE* 14(12):e0226542. https://doi.org/10.1371/journal.pone.0226542

Zhang, N., B. Zhao, H. J. Zhang, S. Weeda, C. Yang, Z. C. Yang, S. Ren, and Y. D. Guo. 2013. Melatonin promotes water-stress tolerance, lateral root formation, and seed germination in cucumber (*Cucumis sativus L.*). *Journal of Pineal Research* 54:15–23. doi: 10.1111/j.1600-079X.2012.01015.x

Zhang, T., Z. Shi, X. Zhang, S. Zheng, J. Wang, and J. Mo. 2020. Alleviating effects of exogenous melatonin on salt stress in cucumber. *Scientia Horticulturae* 262:109070.

Zhang, Q., X. Liu, Z. Zhang, N. Liu, D. Li, and L. Hu. 2019. Melatonin improved waterlogging tolerance in alfalfa (*Medicago sativa*) by reprogramming polyamine and ethylene metabolism. *Frontiers in Plant Science* 10:44.

Zhang, Y., X. Zhou, Y. Dong, F. Zhang, Q. He, J. Chen, S. Zhu, and T. Zhao. 2021. Seed priming with melatonin improves salt tolerance in cotton through regulating photosynthesis, scavenging reactive oxygen species and coordinating with phytohormone signal pathways. *Industrial Crops & Products* 169:113671.

Zhao, D., R. Wang, J. Meng, Z. Li, Y. Wu, and J. Tao. 2017. Ameliorative effects of melatonin on dark-induced leaf senescence in gardenia (*Gardenia jasminoides* Ellis): Leaf morphology, anatomy, physiology and transcriptome. *Scientific Reports* 7(1):1–19.

Zhao, H., L. Ye, and Y. Wang. 2016. Melatonin increases the chilling tolerance of chloroplast in cucumber seedlings by regulating photosynthetic electron flux and the ascorbate-glutathione cycle. *Frontiers in Plant Science* 7:1814.

Zheng, X., J. Zhou, D. X. Tan, N. Wang, L. Wang, D. Shan, and J. Kong. 2017. Melatonin improves waterlogging tolerance of *Malus baccata* (Linn.) borkh. seedlings by maintaining aerobic respiration, photosynthesis and ROS migration. *Frontiers in Plant Science* 8:483.

Zhou, X., H. Zhao, and K. Cao. 2016. Beneficial roles of melatonin on redox regulation of photosynthetic electron transport and synthesis of D1 protein in tomato seedlings under salt stress. *Frontiers in Plant Science* 7:1823.

Zou, J. N., X. J. Jin, Y. X. Zhang, C. Y. Ren, M. C. Zhang, and M. X. Wang. 2019. Effects of melatonin on photosynthesis and soybean seed growth during grain filling under drought stress. *Photosynthetica* 57:512–520.

21 Melatonin-Regulated Proteome and Metabolome Analysis during Optimal and Suboptimal Conditions

Shilpy Singh, Pooja Bhadrecha, and Jyoti Saini

21.1 INTRODUCTION

Melatonin is a pleiotropic chemical which serves a variety of purposes in plants. The biosynthesis of melatonin from tryptophan is discussed here, along with phytomelatonin bioavailability to the plants. Melatonin regulates growth where it controls enzyme expression, promoters, as well as receptors of hormones required for responses including rooting and shooting of plants. Melatonin is well documented to have a significant role in overall functioning of plants, including photosynthesis and seed germination, and also acts for floral transition, disease resistance, stress reduction, and plant development in its capacity as a stimulant or inhibitor. Plants can be protected from harmful and harsh environmental circumstances by melatonin, which functions as an effective antioxidant (Banerjee and Roychoudhury, 2017). Examples of abiotic stressors include drought, salinity, low and high ambient temperatures, heavy metals, and herbicides; and the biotic stress is caused by pathogenic microorganisms, insects, and competitive plants. The potential of melatonin has been examined against *Mycobacterium tuberculosis*, various Gram-positive and Gram-negative bacteria resistant to numerous drugs. Melatonin reduced tobacco mosaic virus in plant-virus interaction models and effectively eliminated the apple stem grooving virus (ASGV) from apple plant, in vitro. Melatonin has been linked to increased fungicide susceptibility, decreased *Phytophthora infestans* stress tolerance, and elevated resistance against *Botrytis cinerea*, according to models of plant-fungi interactions. Melatonin effects on the plant proteome and metabolome under stress conditions are discussed here as components in signaling cascades, transcriptional factors, interactions with other biomolecules providing protection, transferring energy or scavenge free radicals, and interaction with other molecules. Utilizing electrophoretic and chromatographic equipment with hyphenated mass spectrometric techniques is the primary method for identifying proteins and metabolites. Investigating the structural characterization of the investigated compounds using MS techniques requires the use of this information. Abiotic stress reactions in plants are extremely complicated phenomena with unique traits for each species, and they gradually hamper plant growth and development. It is well-known that abiotic stressors cause a multigene response that alters proteins and result in buildup of primary and secondary metabolites (Roychoudhury et al., 2008). Melatonin roles are studied in reference to a range of various organisms like plants, mammals, amphibians, microbes, as well as birds. Melatonin is reported to be discovered within the dinoflagellate *Gonyaulax polyhedra* and has since been discovered in other organisms, including algae, fungi, bacteria, plants, etc. (Arnao and Hernández-Ruiz, 2019a). Pineal glands from bovine species were found to contain melatonin in 1958 (Lerner et al., 1958). In 1995, two separate teams led by Dubbles and Hattori discovered melatonin in edible plants (Dubbels et al., 1995; Hattori et al., 1995). The synthesis route for melatonin from tryptophan was discovered shortly once it

DOI: 10.1201/9781003343752-21

was fully recognized as *N-acetyl*-5-methoxytryptamine in 1959 (Axelrod and Weissbach, 1960) and was fully found in 1995 in higher plants, including most edible plant species (Dubbels et al., 1995). Melatonin is reported to be directly connected to varied physiological processes that are crucial for the plants to respond to external stimuli, including protecting tissues in developmental phases from damage and stress reactions along with rerouting plant growth as well as development (Arnao and Hernández-Ruiz, 2019b). Melatonin is found in at least 20 different plant families and is highly conserved throughout all biological kingdoms. According to multiple studies (Arnao and Hernández-Ruiz, 2019c), melatonin serves majorly in numerous processes. Lately it was proposed that melatonin might be a phytohormone when the initial potential receptor for plant melatonin was discovered in *Arabidopsis thaliana* (Arnao and Hernández-Ruiz, 2020, 2019a). Along with regulating circadian rhythms, germination of seeds, growth and development of roots, leaf senescence, and plant responses to stressful environmental conditions, melatonin also exhibits a wide range of hormone-like activities (Sun et al., 2020). Billions of years ago, organisms like *Rhodospirillum rubrum* (Manchester et al., 1995), *Arthrospira platensis* (syn. *Spirulina platensis*) (Majima et al., 1999), and others developed the ability to produce melatonin against reactive oxygen species (Kurland et al., 2000). Latest studies demonstrate its role in signaling, during various stress conditions (biotic as well as abiotic) (Sun et al., 2021), which eventually regulates plant defense system against various pathogens, along with elevating tolerance to abiotic stresses (Sun et al., 2021). More significantly, melatonin, whether exogenously given or endogenously produced, increases plant resilience to several stresses (Sun et al., 2021). By boosting antioxidant enzyme activities, postponing senescence, and safeguarding the photosynthetic system, melatonin improves plant ability to withstand environmental stress (Chang et al., 2021; Samanta et al., 2022). Additionally, investigations on transcriptomes demonstrate phyto-hormones gibberellic acid (GA) and jasmonic acid (JA) to interact during melatonin signaling, which gradually controls plant responses in stress conditions (Tan et al., 2019; Tan and Reiter, 2020; Samanta et al., 2021). Additionally, studies have reported melatonin to play a crucial role as a phyto-protectant via reduction of stress effects on plants, including drought (Sharma and Zheng, 2019), heat (Buttar et al., 2020), salinity (Altaf et al., 2020, 2021), and cold (Sharaf et al., 2019). The most notable property of melatonin is its ability to scavenge reactive oxygen species (ROS) and reactive nitrogen species (RNS) in stress conditions. According to reports, overabundance of these ROS and RNS under stress can cause cell membrane damage, protein and DNA denaturation, lipid peroxidation, carbohydrate oxidation, pigment degradation, as well as abnormal activities showcased by enzymes (Tiwari et al., 2020a). In times of stress, melatonin, a potent antioxidant, neutralizes these free radicals and preserves ionic equilibrium. Melatonin not only neutralizes ROS but also activates antioxidant defenses (Arnao and Hernández-Ruiz, 2015). Numerous analysis have shown melatonin-mediated up-regulation of genes encoding SOD (superoxide dismutase), APX (ascorbate peroxidase), CAT (catalase), GPX (glutathione peroxidase), or leading to synthesis of ASA (ascorbic acid) and GSH (glutathione) in stress conditions (Das and Roychoudhury, 2014; Sun et al., 2019). Additionally, it has been recorded that the genes involved in ASA-GSH cycle are up regulated to protect plants from stress (Roychoudhury and Basu, 2012; Tiwari et al., 2021). Different kinds of proteins may be taken into consideration, such as heat-shock protein, osmolyte biosynthesis-related enzymes, or ROS-scavenging systems. The content of the plant metabolome is significantly altered by the majority of critical abiotic stresses like drought, salt, as well as high temperature (Ruan et al., 2013; Obata and Fernie, 2012). The development of new crop species depends on understanding the role played by secondary metabolites, as well as the primary metabolites, for stress tolerance processes. Additionally, metabolites (such as hormones and their interactions) are essential for plant defense against pathogens under biotic stress. In this chapter, we have made an attempt to discuss about the melatonin-regulated proteome and metabolome analysis during optimal and suboptimal conditions.

A variety of biosynthetic isozyme transcripts are often up regulated by stresses, enhancing the generation of endogenous melatonin in plants (Ahn et al., 2021). As per latest records, transgenic rice contains all of the genes for the melatonin biosynthesis-related enzymes (Byeon et al., 2013).

This information may help to describe the specific intricacies related to the plant melatonin biosynthetic pathway. Melatonin also is subjected to breakdown in addition to its biosynthesis. An enzyme-dependent or non-enzymatic process is used to break down the secondary metabolite AFMK(N1-acetyl-N2-formyl-5-methoxykynuramine) in vascular plants (Tan et al., 2016). In rice, 2-hydroxymelatonin (99 percent) and 4-hydroxymelatonin (0.05 percent), both of which are present in much lower amounts, are known to be melatonin-dependent metabolites (Byeon, et al., 2015).

21.2 DETECTION OF MELATONIN

Melatonin can be detected and measured using a variety of analytical techniques in raw as well as processed food products. Few additional elements are also present along with melatonin (Ramakrishna et al., 2011). A general approach to prepare samples is either by liquid-liquid extraction, solid-phase extraction, or microextraction with the help of a tight-packed sorbent. Melatonin can be detected in plants using a variety of techniques, including ELISA (enzyme-linked immunosorbent assay), RIA (radioimmunoassay), and GC-MS (gas chromatographic-mass spectrometric), as shown in Figure 21.1. RIA is not a trustworthy approach for detecting melatonin in plants. HPLC in conjunction with mass spectrometer (HPLC-MS) or an electrochemical detector (HPLC-ECD) has preferably been used to conduct their analyses (Ye et al., 2017; Özen and Ekşi, 2016; Ramakrishna et al., 2012).

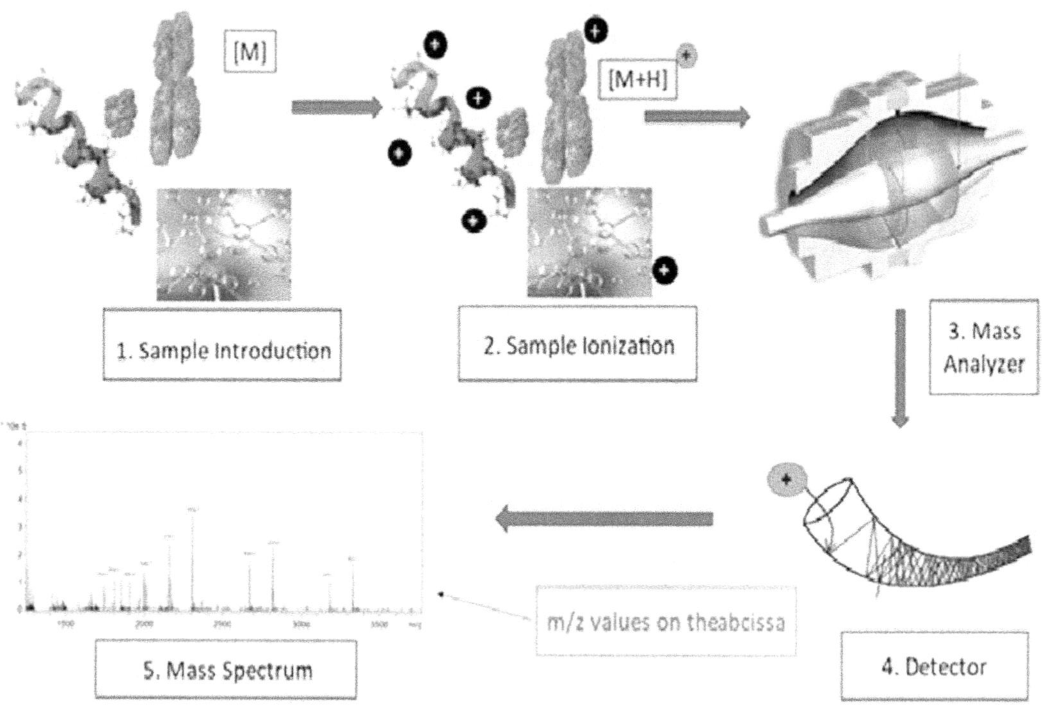

FIGURE 21.1 The introduction and ionization of the sample mark the first step in the MS analysis. For example, molecules (M) or atoms that need to be transformed into positive ionic species by protonation (and are therefore represented as M+ or [M + H]+ or negative ionic species produced by deprotonation (and are therefore represented by [M—H]-) are then run through a mass analyzer where they are separated according to m/z and are then detected and counted. In the mass spectrum, the m/z value and ion abundance are represented as a peak.

Different extraction solvents have also been utilized for various techniques. Melatonin was measured by GC-MS and extracted using an ultrasound-assisted solid-liquid extraction technique by Oladi et al., 2014.

21.3 MELATONIN INTERACTIONS WITH PLANT HORMONES

Melatonin may have diverse effects on plant growth. Due to the structural resemblance of melatonin with auxin IAA (indole-3-acetic acid), botanists have carried out investigations on the potential of melatonin function as a regulator of normal plant growth and development (Arnao and Hernández-Ruiz, 2007). Melatonin works in terms of a plant growth regulator, promotes growth, enhances development, and also helps plants fight against various challenges (Arnao and Hernández-Ruiz, 2019c). Melatonin participates in safeguarding plants through participating in different physiological and metabolic pathways, as summarized in Table 21.1. Stress influences plant growth and subsequently results in reduced yield (Sadak et al., 2020). Melatonin is a potent antioxidant capable of reducing levels of ROS and RNS along with detoxification of a variety of chemical contaminants. Cell membranes are stabilized by melatonin, which also makes them more fluid and controls gene expression. In plants, melatonin controls gene expression destined for producing essential enzymes, hormones, promoters, receptors, etc., involved in growth responses like branching and rooting, and hence serves an exclusive role for regulating growth (Debnath et al., 2019). Furthermore, by altering phenotypic responses, melatonin might raise their levels of IAA. Phytomelatonin may also be a stimulator in plants which encourages germination of seeds, growth of the plant, protective mechanisms in stress conditions which increase plant tolerance to adverse environmental conditions, or as a growth regulator which influences the process of flowering and fruit ripening (Tan and Reiter, 2020). Through the salicylic acid– and NO-dependent mechanism, melatonin promotes the assemblage of carbohydrates that raises the NO level (endogenous) and stimulates defense in case of any bacterial infections. Both of these mechanisms alter hormone levels by increasing or decreasing activity of those enzymes which are a part of hormone production or catabolism. NO and melatonin can activate genes involved in auxin signal transduction. For instance, melatonin treatment in plants has been documented as a negative impact on the production of ABA by up-regulating ABA-catabolism genes and down-regulating NCED3 (9-cis-epoxycarotenoid dioxygenase 3), which is an important biosynthetic gene (Osakabe et al., 2014).

TABLE 21.1
Regulation of Biotic Stress by Melatonin and Hormonal Cross-Talk

Hormone	Plant name	Pathogen	Molecular mechanism	Reference
Salicylic acid (SA)	*Arabidopsis thaliana*	*Pseudomonas syringae*	Defense-related genes are activated	Shi et al. (2015a); Zhao et al. (2015)
	Tobacco (*Nicotiana tabaccum*)	Tobacco mosaic virus	Defense genes PR1 and PR5 are upregulated	Zhao et al. (2019)
Jasmonic acid	Tomato (*Solanum lycopersicum*)	*Botrytis cineraria*	AOC, LoxD, and PI II are up-regulated, SlMYC2 and JZD 1 are down-regulated	Sharma et al. (2020a)
Indole acetic acid (IAA)	Banana (*Musa paradisica*)	*Fusarium oxysporum*cubense	(HSP90)-mediated heat shock protein–induced protection	Wei et al. (2017)
	Apple (*Malus prunifolia*)	Apple stem grooving virus	Lack of investigation	Chen et al. (2019)
Ethylene	Cucumber (*Cucumis sativus*)	*Podosphaeraxanthii*	Ethylene biosynthesis genes are up-regulated	Weeda et al. (2014)

Due to active involvement of ABA in reaction to water stress, melatonin therapy enables plants to deal with this particular kind of abiotic stress. Plants that are exposed to drought benefit from the ability of melatonin to boost the expression of cytokinin genes such as histidine kinases (HKs), histidine phosphotransferases (HPs), and response regulators (Type-A RRs, Type-B RRs). Melatonin has the ability to simultaneously up-regulate and down-regulate the cytokinin zeatin (ZT), which is produced mostly under conditions of biotic stress (Sharma et al., 2020b). On the contrary, melatonin also activates a number of biosynthetic genes that positively affect the production of jasmonic acid (JA) and brassinosteroids (24-EBL). These methods improve the defenses against various abiotic stresses, where the 24-EBL and JA play key roles (Shahzad et al., 2018).

21.4 BIOAVAILABILITY OF PHYTOMELATONIN IN PLANTS

21.4.1 Phytomelatonin: A Phytohormone

21.4.1.1 Plant Growth: Melatonin as a Stimulant or Inhibitor

Exogenous melatonin treatment of plants or seeds exhibited a stimulating and inhibitory influence on root and shoot systems, as well as the yield, which is dependent on the concentration (Hernández-Ruiz and Arnao, 2008). Melatonin is produced as a stress response, which eventually leads to plant protection, involving several physiological processes affecting plants growth and result in a reduced yield (Sadak et al., 2020). However, this chemical appears to be inhibitory when the concentrations are high, and exhibit auxin-like action (Chen et al., 2009). In particular, it was discovered that the growth-promoting effects of melatonin and IAA were 63 percent equivalent (Hernández-Ruiz et al., 2004). On several plant species, this auxin-like effect has been noted by a number of authors. For instance, nanomolar doses promoted growth in oats, wheat, barley, and canary grass. Inhibition of root growth in canary grass and wheat roots were observed when the concentrations were in micromolar (56 percent to 86 percent with respect to IAA) and wheat roots (to 86 percent) (Hernàndez-Ruiz et al., 2005). Similar effects were also reported in the mustard roots, when 0.1 M melatonin treatment stimulated root growth, while 100 M caused an inhibitory response (Chen et al., 2009).

21.4.1.2 Control of Photosynthesis in the Presence of Abiotic Stress

Photosynthesis in plants is exclusive to the leafy green sections of the plant (Pan et al., 2012). Multiple metabolic processes are controlled by photosynthesis, which transforms solar to chemical energy (Chen et al., 2018a). In the chloroplast, which is extremely vulnerable to abiotic stressors, photosynthesis takes place. Melatonin treatment has been found to boost photosynthetic activity in chloroplasts, even during abiotic stress (Han et al., 2017). Melatonin enhances photosynthesis and preserves the redox state in a dose-dependent manner in case of *Brassica juncea* (Mir et al., 2020). As per the data reported, melatonin is essential for both photosynthesis and photoprotection. In the macroalga *Ulva* sp. under salt stress (Tal et al., 2011) and in the freshwater *Chara australis* (L.) (Lazar et al., 2013), melatonin stimulates the overall biomass and content of chloroplast and boosts photosystem II efficiencies. Stomata typically close in stressed plants to minimize water loss, which reduces photosynthesis and affects stomatal conductance (Bhat et al., 2021). Melatonin boosts the final rate of photosynthesis in salt stress by increasing stomatal function and allowing the stomata to reopen (Ye et al., 2016). Additionally, melatonin increased the degree of gene transcription for those products involved in photosynthesis during salt stress, and preserved the photosynthetic system (Shi et al., 2014). According to reports, under conditions of drought stress, melatonin scavenges ROS and enhances enzymes involved in antioxidant activity, which eventually removes any oxidative damages to the cells brought about by drought stress (Zhang et al., 2014).

21.4.1.3 Impact of Melatonin on Seed Germination

Seed priming is a useful technique for enhancing the sowing medium, which, in turn, increases the viability of the seed and the potential strength of the future plant. One technique for controlling metabolism in plants is priming the seeds with melatonin (Marthandan et al., 2020). Pretreatment of the seeds with various growth-promoting substances and/or providing them with varied developmental factors, like temperature or light, for stimulating metabolism, is the fundamental idea behind priming of seeds (Pawar and Laware, 2018). By reducing imbibition time, turning on germination inducers, repairing DNA breaks, and controlling moisture levels, the technique of seed priming ensures germination in a uniform manner (Marthandan et al., 2020). Exogenous melatonin has been reportedly applied for pre-sowing the seeds, a therapy that enhanced germination of seeds and the growth and vigor of seedlings. Magorzata et al. noticed elevated germination of seeds and increased seedling weight, as compared to the control and hydro-primed seeds after, melatonin (25–50 M)-treatment of *Zea mays* seeds, applied in osmopriming (Posmyk et al., 2009b). Similar outcomes were seen when melatonin was administered via hydropriming to seeds of *Cucumis sativus* and *Brassica oleracea* (Posmyk et al., 2009a). Additionally, studies have reported the value of melatonin priming of seeds in various plant species, such as wheat, where germination of seeds was encouraged during chromium stress (Lei et al., 2021), basil, where antioxidant activity was increased under salinity stress (Bahcesular et al., 2020), maize, where starch metabolism was controlled in cold (chill) stress (Cao et al., 2019), and rapeseed (Khan et al., 2019). As seen among the cucumber seeds, they had altered germination of seeds, followed by varying viability (Posmyk et al., 2009a, 2009b), high melatonin levels (> 500 mM) specifically triggered oxidative alterations in the protein amino acidic sequence. Field tests using *Cucumis sativus*, *Zea mays*, and *Vigna radiate* seeds which were melatonin-primed resulted in higher percentage of mature plants which also reflected delayed onset of foliar senescence along with elevated production yields (Szafranska et al., 2013). As seen in tomato (*Solanum lycopersicum*) plants, melatonin treatment caused enhanced transcription of ethylene signal transduction genes in plants. This hormone was produced, perceived, and signaled, which accelerated the ripening and softening of fruit and improved color and flavor (Sun et al., 2015)

21.4.1.4 Impact on Disease Resistance

Plants are at serious risk from diseases caused by viruses, fungi, and bacteria since they are frequently contagious and sometimes fatal. Recently, melatonin-treated plants were found to have a wide range of beneficial properties. According to Yin et al., melatonin pretreatment increased the resilience of apple plants against the *Marssonina* apple blotch. Melatonin therapy improved the activity of plant defense-related enzymes, antioxidants, and photosystem efficiency (Yin et al., 2013). Melatonin may be a cost-effective way to safeguard plants against pathogen infestations because it is an environmentally benign substance. Melatonin may also function as a defense-signaling molecule in the fight against the pathogenic bacterium *Pseudomonas syringae* (Pst) DC3000 in *Arabidopsis* (Lee et al., 2014).

21.4.1.5 Role of Phytomelatonin in the Floral Transition

The timing of blooming has been identified as a new function for melatonin in regulating the growth (Arnao and Hernàndez-Ruiz, 2020). In a preliminary investigation, it was discovered that disrupting blooming was caused by signaling in the leaves of *Chenopodium rubrum* (Kolár et al., 2003; Machácčková and Krekule, 2002). This disruption may have been caused by alterations in the cycle of melatonin levels. In fact, the circadian regulating function for melatonin has been reported to be discovered in *C. rubrum* for the first time (Wolf et al., 2001). In transgenic rice plants containing sheep SNAT, restricted blooming was noticed (Byeon and Back, 2014). Though melatonin does not affect the flowering stage, it does regulate some crucial floral transitional stages. Shi et al. (2016) postulated the mechanism behind this impact and went on further describing the significance of melatonin during development in *Arabidopsis*. A model has been developed to explain how melatonin suppresses floral transition via influencing stability of DELLA proteins and subsequently disrupting transcription factors essential for flowering (Arnao and Hernández-Ruiz, 2020).

21.5 ANTIOXIDANT CAPABILITIES AND POTENTIAL MECHANISMS FOR BIOTIC AND ABIOTIC STRESS TOLERANCE

Plants can be shielded from harsh environmental circumstances by melatonin, an effective antioxidant molecule (Ianas et al., 1991). In nature, melatonin is widely dispersed and interacts with several free radicals. The ability of up-regulating several genes involved in response to stress conditions, such as HSFA3 (heat stress transcription factor A-3), AWPM (ABA-induced wheat plasma membrane polypeptide), CYTC2 (cytochrome c-2), SAD (stearoyl-acyl carrier protein desaturase), catalase, CNX1 (Calnexin 1), APX, MAPK (mitogen-activated protein kinase), bZIP60 (basic Leucine Zipper Domain 60), BIP2 (luminal binding protein 2), BIP3 (luminal binding protein 3), along with the luminal binding protein 4, as well as for down-regulation of stress-related genes CDPK1 (calcium-dependent protein kinase 1), MAPK1, TSPMS (thermos spermine synthase), ERF4 (ethylene-responsive transcription factor 4), HSP80 (heat shock protein 80), and ERD15 (protein early responsive to dehydration 15) (Lee and Back, 2018), results in an anti-stress effect that is primarily linked to antioxidant and scavenger activity. In terms of both direct and indirect processes, melatonin can effectively function as an antioxidant. When compared to ASA, it has a more effective direct mechanism for scavenging free radicals, ROS, and RNS. Melatonin is known to up-regulate other antioxidant enzymes at the cellular level, in addition to directly scavenging ROS, and also functions in signaling (Shi et al., 2015b).

21.5.1 Antioxidant Capabilities of Melatonin also Serve as a Mediator in the Defense Against Biotic and Abiotic Stress

Melatonin can occasionally operate like the mediator of many antioxidant pathways, such as CAT, SOD, and the glutathione ascorbate cycle, and it protects the plant in stress conditions. Resistance to numerous stimuli, including dehydration, oxidation, salt, chemical, heavy metal, and radiation stress, can be achieved through the generation of melatonin or exogenous therapy (Zhang et al., 2015). Vulnerability to exogenous melatonin has been reported to enhance ion channel gene expression, when exposed to saline stress, which was recorded in leaves which may conserve ion homeostasis and hence elevate resistance (Li et al., 2018). Similar to stress caused by freezing, it can also increase the regulation of the DREBs (drought response element-binding factors)/C-repeat binding factors (CBFs), the COR15a (cryoprotective protein) responsible for resistance to freezing and drought, the CAMTA1 transcription factor involved in resistance to drought, as well as the ZAT10 and ZAT12 genes that offer resistance to cold (Bajwa et al., 2014). Numerous studies have recorded that abiotic stress results in the endogenous production of melatonin, indicating the chemical as a protectant for plant cells against the damage caused by oxidative stress. Chemical stress–treated plants such as barley (Arnao and Hernández-Ruiz, 2009) and lupin (Arnao and Hernández-Ruiz, 2013) demonstrated enhanced endogenous melatonin levels. Additionally, compared to untreated control, the plants showed enhanced vegetative development and survival after receiving exogenous melatonin treatment prior to chemical stress exposure. Similar findings were made with cabbage seedlings (Posmyk et al., 2008) and peas (*Pisum sativum*) (Tan et al., 2007) which were grown in the presence of harmful copper concentrations and displayed enhanced vegetative development and survival after melatonin treatment.

21.6 MELATONIN'S ROLE AS A PLANT PROTECTOR IN STRESS

21.6.1 Biotic Stress

The term "biotic stress" refers to harm that multiple living species, such as weeds, insects, parasitic nematodes, bacteria, viruses, fungus, and other native or cultivated plants, cause to crops. Fungi and bacteria can affect many plant organs and can result in symptoms including spotting in leaves, vascular wilts, cankers, etc. (Agrios et al., 2005). ROS and RNS are very well-protected against oxidation by melatonin, which also acts as a protective agent for plants in a variety of stressful situations, which is one of the most researched uses for this hormone (Arnao, 2014). The excessive production of ROS and RNS as a result of the altered redox homeostasis is primarily to blame for "oxidative burst", which has been a common plant defense and signaling response in stress (Tiwari et al., 2020b), as shown in Figure 21.2. The early recognition response to a pathogen attack seeks to stop the spread of the virus at the point of entry by inducing localized programmed cell death (PCD), which results in a phenomenon known as a "hypersensitive response" (Zhao et al., 2021).

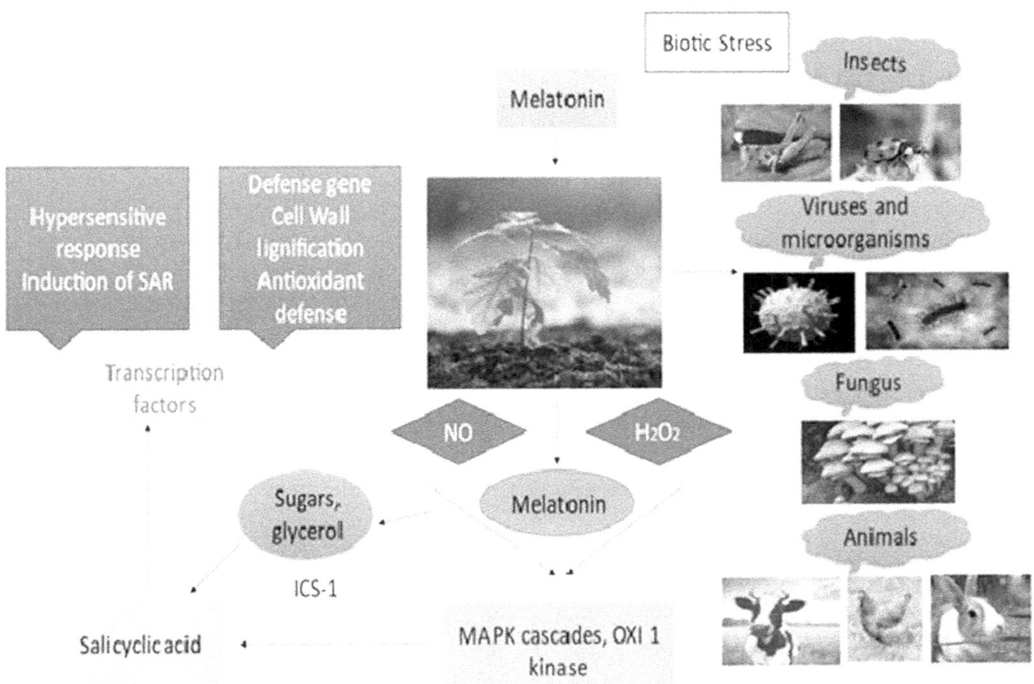

FIGURE.21.2 Melatonin involvement in preventing biotic stress in plants: signal transduction and mechanism.

Source: Melatonin serves as a signaling molecule for defense after microbial invasion. Melatonin-triggered innate immunity signal transduction is mediated by the interaction of hydrogen peroxide and nitric oxide (NO), mitogen-activated protein kinase (MAPK) cascades, and oxidative signal inducible 1 (OXI1) kinase. As a result of this mechanism, the gene encoding the enzyme isochorismate synthase 1 (ICS1), which is necessary for the production of salicylic acid (SA), is upregulated, raising the levels of SA.

For suppressing stress conditions of biotic origin, for example those caused by pests, a variety of control techniques have been used. These include chemical, genetic, biological, and agricultural controls, along with management of pests, that combine employing various pest control techniques which may eventually change suitability of pest microenvironment as well as affect their prevalence (Wyckhuys et al., 2019), as shown in Table 21.2.

21.6.1.1 Antibacterial Agent

Melatonin has been shown to be efficient against several Gram-positive and Gram-negative bacteria which are resistant to multiple drugs (Wei et al., 2018). An intense inhibitory effect on *Mycobacterium tuberculosis* caused by melatonin has been demonstrated post-application (10 μM). Cell wall instability, reduction of substrates in the cells, and modulation of cAMP and Ca^{2+} in bacteria were the main documented mechanisms of antibacterial activity. Melatonin (200 g mL^{-1}) is reported to have a direct inhibitory effect on the bacteria *X. oryzae* pv. *oryzae* as well as *oryzicola*, pathogenic to rice and responsible for causing leaf streaking or leaf blight (Chen et al., 2018a). Melatonin can enhance gene expression for isochorismate synthase 1 (ICS1), involved in biosynthesis of SA, in *A. thaliana* intruded by (Pst)-DC300 (Lee et al., 2018). Activity of invertase at the cell wall is elevated if *A. thalianais* is treated with melatonin, which eventually results in increased strength and callose deposit at the cell walls when infested by (Pst)-DC3000, according to research by Zhao et al. (2015). Furthermore, glycerol and sucrose actively participate in melatonin-related protection against (Pst)-DC3000 in SA- and NO-dependent mechanisms of *Arabidopsis* (Lee et al., 2014).

21.6.1.2 Antiviral Agent

Plant viruses provide a persistent hazard to the crop production system because there is no direct treatment for these infections (Kumar et al., 2017). Viral management can be via alternative strategies like defense induced by host and control of vector (Zhao et al., 2019). In relation to this, employing melatonin exogenously leads to a decrease in the levels of tobacco mosaic virus (TMV) RNA and virus-infected *Nicotiana glutinosa* and *Solanum lycopersicum* seedlings (Zhao et al., 2019). PR1 and PR5 genes were highly expressed, which caused SAR positively linked with lower virus load in plants treated with melatonin. According to a related study, apple stem grooving virus was entirely eradicated from previously infected apple explants upon addition of melatonin (Chen et al., 2019, 2018b). The increase in SA concentration in the NO-dependent pathway was thought to be the cause of beneficial effects (Zhao et al., 2019). This highlights the use of melatonin as an additive to the tissue culture mixture when producing virus-free plants.

21.6.1.3 Antifungal Agent

Melatonin promoted tomato fruit resistance to *Botrytis cinerea* in a plant-fungi interaction model by controlling H_2O_2 production and the jasmonic acid signaling pathway (Liu et al., 2019). Melatonin provided resistance against *Phytophthora infestans* and increased the effectiveness of fungicide treatment against potato late blight (Zhang et al., 2017b). In addition, phytomelatonin at a concentration of 100 M considerably decreased the growth of the plant pathogen *Colletotrichum gloeosporioides*, which otherwise leads to anthracnose in pepper (*Capsicum annum*), by 76 and 71 percent, respectively (Ali et al., 2021). Effectiveness of melatonin is well documented against several harmful fungi, including *Diplocarpon mali* (apples) (Yin et al., 2013), *Fusarium oxyspoum* (bananas) (Wei et al., 2017), and *Botrytis cineraria* (tomatoes) (Liu et al., 2019). In plants, the modulation rates of ROS and RNS serve as an indicator for numerous cellular as well as physiological responses stressors, both directly and indirectly (Arnao et al., 2019).

TABLE 21.2
Melatonin Functioning under Biotic Stress Conditions

Pathogen	Host	Melatonin dosage (µM)	Melatonin application	Mechanism	References
Bacterial agent					
Pseudomonas syringae	*Arabidopsis thaliana* *Nicotiana benthamiana*	10 µM	Foliar spray	PR1, PR5, and plant defensin defense genes are up-regulated	Lee et al. (2014)
Pseudomonas syringae	*Arabidopsis thaliana*	10 µM	Foliar spray	Defense activation via SA and ET signaling	Qian et al. (2015)
Pseudomonas syringae	*Arabidopsis thaliana*	20 µM	Added to the culture media	Signaling for defense via NO	Shi and Chan (2014)
Pseudomonas syringae	*Arabidopsis thaliana*	50 µM	Added to the culture media	Callose formation and cell wall strengthening	Zhao et al. (2015)
Pseudomonas syringae	*Arabidopsis thaliana*	1 µM	Foliar spray	Cascade of MAPK is activated	Lee and Back (2017)
Xanthomonas oryzae pv. oryzae	Rice (*Arabidopsis thaliana*)	100 µM	Foliar spray	Reduction of bacterial virulence and metabolic genes	Chen et al. (2018a)
P. syringae	*A. thaliana, N. benthamiana*	1µM	Leaf treatment	PR gene induction via MAPK signaling cascades	Lee et al. (2018)
P. syringae	*A. thaliana*	20µM	Added to nutrient solution	Sugars and glycerol participate in melatonin-mediated innate immunity through SA- and NO-dependent pathways	Qian et al. (2015)
Viral pathogens					
Tobacco mosaic virus (TMV)	Tomato (*Solanum lycopersicum*)	100 µM	Applied to Roots	PR1 and PR5 expressions that are elevated	Zhao et al. (2019)
Apple stem grooving virus (ASGV)	Apple (*Malus domestica*)	15 µM	Addition to nutrient media	Virus removal from infected shoots (unknown mechanism)	Chen et al. (2019)
Fungal pathogens					
Phytophthora infestans	Potato (*Solanum tubersum*)	10 mM	Spray on leaves and tuber slices	Mycelia and spore disruption increases fungicide susceptibility	Zhang et al. (2017b)

(*Continued*)

TABLE 21.2 ***(Continued)***
Melatonin Functioning under Biotic Stress Conditions

Pathogen	Host	Melatonin dosage (μM)	Melatonin application	Mechanism	References
Podosphaera xanthii and *Phytophthora capsici*	Cucurbit (*Citrullus lanatus*)	100 μm	Leaf treatment	Control over PTI and ETI	Mandal et al. (2018)
Pseudoperonospora cubensis	Cucumber (*Cucumis sativus*)	100 μm	Leaf treatment	Antioxidant gene activation	Sun et al. (2019)
Fusarium oxysporum	Banana (*Musa acuminata*)	100 μm	Root and leaf spray	Control of heat shock protein expression	Wei et al. (2017)
Diplocarpon mali	Apple (*Malus prunifolia*)	50–500 μm	Root spray	Enzymes involved in plant defence activity	Yin et al. (2013)
Botrytis cinerea, *Rhizopus stolonifer*	Strawberry (*Fragaria ananassa*)	100 μm	Leaf treatment	Keep intracellular H2O2 constant	Aghdam and Fard (2017), Liu et al. (2019)
Verticillium dahliae	Cotton (*Gossypium* spp.)	10 mM	Foliar treatments	Increased resistivity and encouraged the buildup of lignin and gossypol	Li et al. (2019)

21.6.2 Abiotic Stress

Resistance to numerous stimuli, including dehydration, oxidation, salt, chemical, heavy metal, and radiation stress, can be achieved through melatonin synthesis or exogenous therapy (Zhang et al., 2015). Additionally, abiotic stresses are evolving more and more under the current conditions of global climatic change (Nadeem et al., 2019) and are a serious danger to the sustainability of agriculture. Exogenous administration of melatonin may be a way to help plants thrive under difficult circumstances. Melatonin treatment to the soil and to the leaves of maize boosted drought tolerance, according to Ye et al., 2016. Similar investigations have revealed that melatonin administration improved stress tolerance in plants. One such technique is seed priming with melatonin, which controls metabolism of the plant at the initial time during germination under stress (Marthandan et al., 2020). Figure 21.3 displays a generalized model for functions of melatonin in defending plants from abiotic stress.

We also go into the mechanistic details of regulation transcriptional modifications in plants by melatonin, under stress, enlisting how melatonin modulates effects caused by several abiotic stresses. Melatonin has been demonstrated to mitigate or protect plants from a variety of challenges, which are covered in more detail later (Table 21.3).

21.6.2.1 Drought Stress Tolerance Is Improved by Melatonin

In particular, in arid and semi-arid environments, drought is a severe worry that reduces the performance, quality, and productivity of plants. (Yang et al., 2019). The primary stressor that changes physiology, biochemistry, and molecular pathways is drought (Ahmed et al., 2021). Melatonin

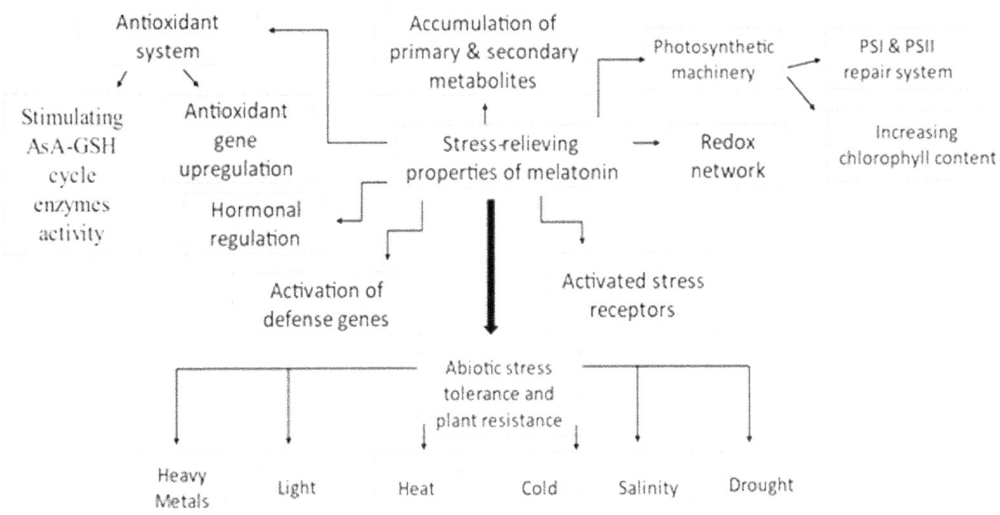

FIGURE 21.3 Several processes allow plants to withstand abiotic stress via melatonin. Under stressful situations, melatonin boosts antioxidant levels in plants and activates antioxidant enzymes to scavenge reactive oxygen species (ROS). Melatonin mediates stomatal movement through interactions with abscisic acid (ABA), which prevents chlorophyll from degrading and boosts photosynthetic efficiency in challenging conditions. Toxic substances including heavy metals and Na^+ are reduced in concentration when melatonin is present. Defense genes are directly regulated and up-regulated by melatonin. Under stress, ability of plants to synthesize secondary metabolites is improved by melatonin. Melatonin also alters the elements of cell walls, reducing cellular damage.

TABLE 21.3
Melatonin Functioning under Abiotic Stress Conditions

Type of stress	Plant species	Melatonin dose	Response	Reference(s)
Drought	*Arabidopsis*	50 μM	mRNA expression control of several stress-responsive genes, including COR15A, RD22, and KIN1, sucrose accumulation	Shi et al. (2015c)
	Cynodon dactylon L.	4, 20, and 100 μM	High concentrations of amino acids, organic acids, sugars, and sugar alcohols, decreased reactive oxygen species (ROS), increased antioxidant metabolism	Shi et al. (2014)
	Malus	100 μM	Increased photosynthetic rate, reduced electrolyte leakage, and improved water conservation in leaves	Li et al. (2012)
	Zea mays	50 μM melatonin	Increases in photosynthetic capacity and plant growth, the activity of antioxidant enzymes rises	Ahmad et al. (2020)
	Agrostis stolonifera L.	20 μM	Melatonin-treated ipt-transgenic plants exhibited improved photochemical efficiency, chlorophyll content, up-regulation of dehydrated genes, and substantial up-regulation of cytokinin-signaling and production genes	Ma et al. (2018)
Salinity stress	*Gossypium hirsutum* L.	20 μM	Starch concentration, hypocotyl length, and germination rate have all increased; melatonin reduced dormancy by controlling the expression of ABA and GA genes in the plant signal transduction pathway	Chen et al. (2020)
	Arabidopsis thaliana	10 μM	Increased starch content, hypocotyl length, and germination rate; melatonin controls gene expression in ABA and GA transduction pathway	Chen et al. (2020)
	Triticum aestivum (L.)	70 μM	Increased tissue growth, root and shoot weight, salinity tolerance; melatonin and SA induce nitrogen metabolism, melatonin and SA increase the content of Spd, Spm, and PAs	Talaat et al. (2021)

TABLE 21.3 *(Continued)*
Melatonin Functioning under Abiotic Stress Conditions

Type of stress	Plant species	Melatonin dose	Response	Reference(s)
	Brassica napus (L.)	1 µM	Decreased lipid peroxidation, hydrogen peroxide content, redox and ion homeostasis maintenance	Zhao et al. (2018)
	Brassica juncea	1 µM	Increased plant height, leaf length/width, and stem diameter; increased gas exchange parameters, relative water content; reduced abscisic acid and increased salicylic acid content	Park et al. (2021)
	Oryza sativa	10 µM	Decreasing the sodium content to maintain Na/K homeostasis, decreased membrane lipid oxidation, increased chlorophyll content	Xie et al. (2020)
	Limonium bicolor (Bunge)	50, 100, 200, 500 µM	Higher seed germination, high content of melatonin, high gibberellic acid (GA), low content of abscisic acid (ABA)	Li et al. (2019)
High temperature	*Triticum aestivum*	20 µM	Improved photosynthesis, decreased oxidative stress markers	Iqbal et al. (2021c)
	Lolium perenne (L.)	100 µM	Regulate CK biosynthesis genes, down-regulation of ABA biosynthesis genes, melatonin supplementation enhanced the endogenous melatonin and CK content with reduced content of ABA under heat stress	Zhang et al. (2017a)
	Solanum lycopersicum	100 µM	Enhanced PA levels, melatonin increased endogenous NO content, decreased oxidative stress parameters	Jahan et al. (2019)
Cold stress	*Cucumis sativus*	100 µM	Increased antioxidants provide cold tolerance, increased PsbA (D1) protein of PSII	Feng et al. (2021)
	Citrullus lunatus	150 µM	Decreased oxidative stress markers such as electrolyte leakage, MDA, and ROS accumulation	Li et al. (2021)
Cold Stress 4 ◦C	*Citrullus lanatus*	150 µM	Increased accumulation of MeJA and hydrogen peroxide, increased Fv/Fm	Li et al. (2021)
	Solanum lycopersicum	100 µmol	Increased Fv/Fm, increased NPQ	Ding et al. (2017)

(Continued)

TABLE 21.3 *(Continued)*
Melatonin Functioning under Abiotic Stress Conditions

Type of stress	Plant species	Melatonin dose	Response	Reference(s)
Heavy metals				
Pb	*Nicotiana tabacum*	200 nM	Prevents DNA damage, efficacious antioxidant, and decreased superoxide radical accumulation	Kobylinska et al. (2017)
	Zea mays	0.1 mM	Melatonin in combination with cPTIO resulted in an elevation of hydrogen peroxide, MDA contents, and electrolyte leakage parameters; cPTIO in combination with melatonin was ineffective in changing the activities of enzyme	Okant and Kaya (2019)
Cd	*Triticum aestivum*	100 µM	Enhanced growth attributes, decreased oxidative stress, reduced endogenous NO levels	Kaya et al. (2019)
Herbicides				
Pendimethaline (Pend) (N-(1-ethylpropyl)-3,4-dimethyl-2,6-dinitrobenzenamine)	*Capsicum annuum* (L.)	8, 16, and 32 mM	Chlorophyll content and relative water content reduced; increase in carotenoid, proline, and glutathione	Kaya et al. (2019)
Paraquat	*Pisum sativum* (L.)	200 µM	Increased photosynthetic pigments, improved functioning of the photosynthetic apparatus, increased water content	Szafranska et al. (2017)
Thermotolerance	*Arabidopsis*	20 µM	Up-regulation of heat-shock factors (HSFA1s)	Shi et al. (2014)
	Zea mays (L).	(100 µmol/L)	Increased photosynthesis, reduced oxidative damage	Ye et al. (2016)
pH stress	*Malus hupehensis*	5 µM	Enhanced antioxidant and PA content and its gene, increased MDA	Li et al. (2012)

reduces ROS generation and oxidative damage via heightened activity of several antioxidant enzymes and levels of non-enzymatic antioxidants (Zhang et al., 2013). By controlling nitro-oxidative and osmoprotective homeostasis in *Medicago sativa*, melatonin may act like the priming factor which increases tolerance to dry circumstances (Antoniou et al., 2017).

21.6.2.2 Improved Endurance to High Temperature Due to Melatonin

Accumulation of reactive oxygen species (ROS) may cause peroxidative damage in membranes and can also lead to alterations in redox status. Few studies have proved that melatonin treatment (exogenous) reverses the damages caused due to germination and growth of plants in stress conditions (Wang et al., 2018). Protein stability, enzyme activities, and alterations in biomolecules associated with ROS production are severely affected by heat. Suppressed photosynthesis, growth, and yield is caused by higher temperature, which also disturb the fluidity of membrane and enzyme activity (Iqbal et al., 2021a, 2021b). Low temperature, like other stressors, lead to significant oxidative damage in wheat seedlings, although melatonin administration boosted SOD, CAT, APX and GR activity, thereby lowering ROS levels and lipid peroxidation (Turk et al., 2014).

21.6.2.3 Melatonin-Induced Improved Salt Tolerance

The major environmental factors which restrict plant growth and jeopardize sustainable output in many regions of the world is salinity (Zhu et al., 2021). Saline soil limits crop productivity by causing excessive ion uptake, which interferes with their ion homeostasis and causes oxidative stress (Sairam et al., 2002). Salinity is the major environmental factor which restricts growth and jeopardizes sustainable output at various regions of throughout world (Zhu, 2001). Saline soil limits crop productivity by causing excessive ion uptake, which interferes with their ion homeostasis and causes oxidative stress (Sairam et al., 2002). Melatonin lowered sodium concentration for maintenance of sodium-potassium balance, decreased oxidation of lipids, as well as boosted synthesis of chlorophyll to improve salt tolerance, based on analyses of the metabolite abundances in plants treated with salt and metabolites from metabolome profile.

21.6.2.4 Melatonin in the Presence of Heavy Metals

The environment has been impacted by rising amounts of heavy metal pollution. Plants can benefit from protective effects of melatonin against heavy metal stress. For instance, melatonin pretreatment of seeds prior to sowing the seedlings protected red cabbage against harmful concentration of copper ion, and foliar melatonin spraying of tobacco provided cadmium tolerance (Wang et al., 2019). Melatonin helps plants thrive by enhancing their antioxidant defences and lowering heavy metal toxicity by chelating metals and preventing their buildup at the rhizospheric surface (Li and Yan, 2021). According to a recent study, *Eruca vesicaria* (L.) plants exposed to Pb stress exhibited improved control over the photosynthetic enzymes that support growth (Mohamed et al., 2021).

21.6.2.5 Herbicide Stress Tolerance Improved by Melatonin

The herbicides persist in the soil for a very long time, harming subsequent crops as well as plants which are employed as "green manure". Chemical herbicides are exclusively used in agricultural field to control weed growth and development, and their use at high concentration inhibits physiological function of plants (Lukatkin, 2013). The effects of pendimethaline (herbicide) and stress caused by drought conditions on pepper were investigated by Kaya and Doganlar. By increasing the antioxidant potential, photosynthetic pigments, and mRNA, melatonin at 50 M reduced the negative impacts of 10 percent PEG and herbicides (Kaya and Doganlar, 2019).

21.7 EFFECT OF MELATONIN ON PLANT PROTEOME AND METABOLOME UNDER STRESS CONDITIONS

Plant functional characterization is increasingly using proteome analysis as an effective technique. Proteomics is a crucial area of applied biology that aids in the investigation of proteins under various situations and characteristics (Chaturvedi et al., 2016; Zargar et al., 2015). Various techniques involved in proteome analysis evaluates functioning of plants at several stages, since nucleotide sequence information is readily available, and because mass spectrometry has made progress in sensitive and quick protein identification (Canovas, 2004), because (1) proteins directly contribute to the development of novel plant phenotypes by regulating physiological traits to adapt to environmental changes, and (2) proteins have a major role in stress management and response. The functions that proteins perform are extremely diverse. They interact with other molecules, function as enzymes or transcriptional factors, participate in signaling cascades, transfer energy, or scavenge free radical, as part of the protective mechanisms. Various chemicals, along with numerous proteins produced later in embryogenesis and ROS-scavenging mechanisms, could be taken into account (Simpson et al., 2021).

In many leguminous crops, such as soybean (Yu et al., 2016; Khan and Komatsu, 2016), chickpea (Gayen et al., 2019; Parankusam et al., 2017), and common bean (Zadraznik et al., 2017), proteomic studies assisted in identification of drought and/or heat proteins, expressed in a differential manner. Additionally, the next sections address how proteomics can be used to analyze how proteins are expressed over time and in diverse patterns in various legumes, under drought and high-temperature conditions.

The application of mass spectrometric techniques is the principal method used for metabolite identification. For functional genomics and systems biology, metabolomics is a crucial tool. Metabolites have a role in signaling, storage, development, scaffolding etc. for the entire plant under stress (Wen et al., 2015). Molecules that sustain cell turgor and stabilize proteins and cell membranes include amino acids, carbohydrates, polyols, and tertiary sulfonium. They also contribute in balancing redox by scavenging the ROS that could harm cells. The resistant species have higher levels of secondary metabolites that are connected to defensive systems, and the ability to maintain the levels of primary metabolic pathways and activating the opposing defensive systems appears to be crucial for survival, growth, and development of plants. Secondary metabolites play a function in the defense response in case of infections and operate as attractants or repellents to herbivores as well as insects, along with resisting detrimental effects caused by ultraviolet radiations.

21.8 CONCLUSION

Melatonin has recently drawn attention of researchers because of its direct benefit in controlling plant development and enhancing tolerance against various stress challenges. Melatonin is produced by plants in a variety of subcellular locations, helps in signaling pathways, regulates and promotes plant growth as well as development, and also helps plants fight against various challenges. Further research is required to investigate the global transcriptome, proteome, and metabolome profile regulated by melatonin during biotic and abiotic stresses, so as to have an overall idea regarding the entire signaling components that are controlled by melatonin, so as to increase crop resilience against stress.

REFERENCES

Aghdam, M. S., Fard, J. R. 2017. Melatonin treatment attenuates postharvest decay and maintains nutritional quality of strawberry fruits (*Fragariaanannasa* cv. Selva) by enhancing GABA shunt activity. *Food Chem*, 221:1650–1657. https://doi.org/10.1016/j.foodchem.2016.10.123

Agrios, G. N. 2005. Preface. In *Plant Pathology*, 5th ed., ed. G. N. Agrios. San Diego, CA: Academic Press.

Ahmad, S., Kamran, M., Zhou, X., et al. 2020. Melatonin improves the seed filling rate and endogenous hormonal mechanism in grains of summer maize. *Physiol Plant*, 172:1059–1072.

Ahmed, S., Kouser, S., Asgher, M., Gandhi, S. G. 2021. Plant aquaporins: A frontward to make crop plants drought resistant. *Physiol Plant*, 172:1089–1105.

Ahn, H. R., Kim, Y. J., Lim, Y. J., et al. 2021. Key genes in the melatonin biosynthesis pathway with circadian rhythm are associated with various abiotic stresses. *Plants*, 10:129.

Ali, M., Lamin-Samu, A. T., Muhammad, I., et al. 2021. Melatonin mitigates the infection of *Colletotrichumgloeosporioides*via modulation of the chitinase gene and antioxidant activity in *Capsicum annuum* L. *Antioxidants*, 10:1–25. https://doi.org/10.3390/antiox10010007

Altaf, M. A., Shahid, R., Ren, M. X., et al. 2020. Exogenous melatonin enhances salt stress tolerance in tomato seedlings. *Biol Plant*, 64:604–615. https://doi.org/10.32615/bp.2020.090

Altaf, M. A., Shahid, R., Ren, M. X., et al. 2021. Melatonin alleviates salt damage in tomato seedling: A root architecture system, photosynthetic capacity, ion homeostasis, and antioxidant enzymes analysis. *Sci Hortic*, 285:110145. https://doi.org/10.1016/j.scienta.2021.110145

Antoniou, C., Chatzimichail, G., Xenofontos, R., et al. 2017. Melatonin systemically ameliorates drought stress-induced damage in *Medicago sativa* plants by modulating nitro-oxidative homeostasis and proline metabolism. *J Pineal Res*, 62:e12401.

Arnao, M. B. 2014. Phytomelatonin: Discovery, content, and role in plants. *Adv Bot*, 2014:815769.

Arnao, M. B., Hernández-Ruiz, J. 2007. Melatonin promotes adventitious-and lateral root regeneration in etiolated hypocotyls of *Lupinusalbus* L. *J Pineal Res*, 42:147–152.

Arnao, M. B., Hernández-Ruiz, J. 2009. Chemical stress by different agents affects the melatonin content of barley roots. *J Pineal Res*, 46:295–299.

Arnao, M. B., Hernández-Ruiz, J. 2013. Growth conditions determine different melatonin levels in *Lupinusalbus* L. J. *Pineal Res*, 55:149–155.

Arnao, M. B., Hernández-Ruiz, J. 2015. Functions of melatonin in plants: A review. *J Pineal Res*, 59:133–150. https://doi.org/10.1111/jpi.12253

Arnao, M. B., Hernández-Ruiz, J. 2019a. Melatonin and reactive oxygen and nitrogen species: A model for the plant redox network. *Melatonin Res*, 2:152–168.

Arnao, M. B., Hernández-Ruiz, J. 2019b. Melatonin as a chemical substance or as phytomelatonin rich-extracts for use as plant protector and/or biostimulant in accordance with EC legislation. *Agronomy*, 9:570.

Arnao, M. B., Hernández-Ruiz, J. 2019c. Melatonin: A new plant hormone and/or a plant master regulator? *Trends Plant Sci*, 24:38–48. https://doi.org/10.1016/j.tplants.2018.10.010

Arnao, M. B., Hernández-Ruiz, J. 2020. Melatonin in flowering, fruit set and fruit ripening. *Plant Reprod*, 11:77–87. https://doi.org/10.1007/s00497-020-00388-8

Axelrod, J., Weissbach, H. 1960. Enzymatic O-methylation of N-acetylserotonin to melatonin. *Science*, 131:1312. https://doi.org/10.1126/science.131.3409.1312

Bahcesular, B., Yildirim, E. D., Karaçocuk, M., Kulak, M., Karaman, S. 2020. Seed priming with melatonin effects on growth, essential oil compounds and antioxidant activity of basil (*Ocimumbasilicum* L.) under salinity stress. *Ind Crops Prod*, 146:112165.

Bajwa, V. S., Shukla, M. R., Sherif, S. M., Murch, S. J., Saxena, P. K. 2014. Role of melatonin in alleviating cold stress in *Arabidopsis thaliana*. *J Pineal Res*, 56:238–245. https://doi.org/10.1111/jpi.12115

Banerjee, A., Roychoudhury, A. 2017. Melatonin as a regulator of abiotic stress tolerance in plants. In *Mechanisms Behind Phytohormonal Signaling and Crop Abiotic Stress Tolerance*, ed. V. P. Singh, S. Singh, S. M. Prasad, 47–60. New York: Nova Science Publishers.

Bhat, M. A., Mir, R. A., Kumar, V., et al. 2021. Mechanistic insights of CRISPR/Cas-mediated genome editing towards enhancing abiotic stress tolerance in plants. *Physiol Plant*, 172:1255–1268.

Buttar, Z. A., Wu, S. N., Arnao, M. B., Wang, C., Ullah, I., Wang, C. 2020. Melatonin suppressed the heat stress-induced damage in wheat seedlings by modulating the antioxidant machinery. *Plants*, 9:1–17. https://doi.org/10.3390/plants9070809

Byeon, Y., Back, K. 2014. An increase in melatonin in transgenic rice causes pleiotropic phenotypes, including enhanced seedling growth, delayed flowering, and low grain yield. *J Pineal Res*, 56: 408–414.

Byeon, Y., Lee, H. Y., Hwang, O. J., et al. 2015. Coordinated regulation of melatonin synthesis and degradation genes in rice leaves in response to cadmium treatment. *J Pineal Res*, 58:470–478. https://doi.org/10.1111/jpi.12232.

Byeon, Y., Park, S., Kim, Y. S. 2013. Microarray analysis of genes differentially expressed in melatonin-rich transgenic rice expressing a sheep serotonin N-acetyltransferase. *J Pineal Res*, 55:357–363. https://doi.org/10.1111/jpi.12077.

Canovas, F. M., Dumas-Gaudot, E., Recorbet, et al. 2004. Plant proteome analysis. *Proteomics*, 4(2):285–298. doi:10.1002/pmic.200300602. PMID:14760698.

Cao, Q., Li, G., Cui, Z., et al. 2019. Seed priming with melatonin improves the seed germination of waxy maize under chilling stress via promoting the antioxidant system and starch metabolism. *Sci Rep*, 9:1–12.

Chang, J., Guo, Y., Yan, J., et al. 2021. The role of watermelon cafeic acid O-methyltransferase (ClCOMT1) in melatonin biosynthesis and abiotic stress tolerance. *Hortic Res*, 8:210.

Chaturvedi, P., Ghatak, A., Weckwerth, W. 2016. Pollen proteomics: From stress physiology to developmental priming. *Plant Reprod*, 29(1–2):119–132.

Chen, L., Liu, L., Lu, B., et al. 2020. Exogenous melatonin promotes seed germination and osmotic regulation under salt stress in cotton (*Gossypium hirsutum* L.). *PLoS One*, 15:e0228241.

Chen, L., Wang, M. R., Li, J. W., et al. 2019. Exogenous application of melatonin improves eradication of apple stem grooving virus from the infected in vitro shoots by shoot tip culture. *Plant Pathol*, 68:997–1006. https://doi.org/10.1111/ppa.13018

Chen, Q., Qi, W. B., Reiter, R. J., Wei, W., Wang, B. M. 2009. Exogenously applied melatonin stimulates root growth and raises endogenous indoleacetic acid in roots of etiolated seedlings of *Brassica juncea*. *J Plant Physiol*, 166:324–328.

Chen, X., Sun, C., Laborda, P., et al. 2018a. Melatonin treatment inhibits the growth of *Xanthomonas Oryzae* Pv. Oryzae. *Front Microbiol*, 9:2280. https://doi.org/10.3389/fmicb.2018.02280

Chen, Y. E., Mao, J. J., Sun, L. Q., et al. 2018b. Exogenous melatonin enhances salt stress tolerance in maize seedlings by improving antioxidant and photosynthetic capacity. *Physiol Plant*, 164:349–363.

Das, K., Roychoudhury, A. 2014. Reactive oxygen species (ROS) and response of antioxidants as ROS-scavengers during environmental stress in plants. *Frontiers in Environmental Science*, 2:53.

Debnath, B., Islam, W., Li, M., et al. 2019. Melatonin mediates enhancement of stress tolerance in plants. *Int J Mol Sci*, 20:1040. https://doi.org/10.3390/ijms20051040.

Ding, F., Wang, M., Liu, B., Zhang, S. 2017. Exogenous Melatonin mitigates photoinhibition by accelerating non-photochemical quenching in tomato seedlings exposed to moderate light during chilling. *Front Plant Sci*, 8. https://doi.org/10.3389/fpls.2017.00244.

Dubbels, R., Reiter, R., Klenke, E., et al. 1995. Melatonin in edible plants identified by radioimmunoassay and by high performance liquid chromatography-mass spectrometry. *J Pineal Res*, 18:28–31.

Feng, Y., Fu, X., Han, L., et al. 2021. Nitric oxide functions as a downstream signal for melatonin-induced cold tolerance in cucumber seedlings. *Front Plant Sci*, 12:1432.

Gayen, D., Gayali, S., Barua, P., et al. 2019. Dehydration induced proteomic landscape of mitochondria in chickpea reveals large-scale coordination of key biological processes. *J Proteomics*, 192: 267–279.

Han, Q. H., Huang, B., Ding, C. B., et al. 2017. Effects of melatonin on anti-oxidative systems and photosystem II in cold-stressed rice seedlings. *Front Plant Sci*, 8:785.

Hattori, A., Migitaka, H., Masayaki, I., Itoh, M., Yamamoto, K., Ohtani-Kaneko, R. 1995. Identification of melatonin in plant seed its effects on plasma melatonin levels and binding to melatonin receptors in vertebrates. *Biochemistry and Molecular Biology International*, 35:627–634.

Hernández-Ruiz, J., Arnao, M. B. 2008. Melatonin stimulates the expansion of etiolated lupin cotyledons. *Plant Growth Regul*, 55:29–34.

Hernández-Ruiz, J., Cano, A., Arnao, M. B. 2004. Melatonin: A growth-stimulating compound present in lupin tissues. *Planta*, 220:140–144.

Hernández-Ruiz, J., Cano, A., Arnao, M. B. 2005. Melatonin acts as a growth-stimulating compound in some monocot species. *J Pineal Res*, 39:137–142.

Ianas, O., Olinescu, R., Badescu, I. 1991. Melatonin involvement in oxidative processes. *Endocrinologie*, 29:147–153.

Iqbal, N., Fatma, M., Gautam, H., et al. 2021a. The crosstalk of melatonin and hydrogen sulfide determines photosynthetic performance by regulation of carbohydrate metabolism in wheat under heat stress. *Plants*, 10:1778.

Iqbal, N., Umar, S., Khan, N. A., Corpas, F. J. 2021b. Crosstalk between abscisic acid and nitric oxide under heat stress: Exploring new vantage points. *Plant Cell Rep*, 40:1429–1450.

Iqbal, N., Umar, S., Khan, N. A., Corpas, F. J. 2021c. Nitric oxide and hydrogen sulfidecoordinately reduce glucose sensitivity and decrease oxidative stress via ascorbate-glutathione cycle in heat-stressed wheat (*Triticum aestivum*L.) plants. *Antioxidants*, 10:108.

Jahan, M. S., Shu, S., Wang, Y., et al. 2019. Melatonin alleviates heat-induced damage of tomato seedlings by balancing redox homeostasis and modulating polyamine and nitric oxide biosynthesis. *BMC Plant Biol*, 19:1–16.

Kaya, A., Doganlar, Z. B. 2019. Melatonin improves the multiple stress tolerance in pepper (*Capsicum annuum*). *Sci Hortic*, 256:108509.

Kaya, C., Okant, M., Ugurlar, F., Alyemeni, M. N., Ashraf, M., Ahmad, P. 2019. Melatonin-mediated nitric oxide improves tolerance to cadmium toxicity by reducing oxidative stress in wheat plants. *Chemosphere*, 225:627–638.

Khan, M. N., Komatsu, S. 2016. Proteomic analysis of soybean root including hypocotyl during recovery from drought stress. *J Proteomics*, 144:39–50.

Khan, M. N., Zhang, J., Luo, T., et al. 2019. Seed priming with melatonin coping drought stress in rapeseed by regulating reactive oxygen species detoxification—Antioxidant defense system, osmotic adjustment, stomatal traits and chloroplast ultrastructure perseveration. *Ind Crops Prod*, 140, 111597. https://doi.org/10.1016/j.indcrop.2019.111597.

Kobylinska, A., Reiter, R. J., Posmyk, M. M. 2017. Melatonin protects cultured tobacco cells against lead-induced cell death via inhibition of cytochrome c translocation. *FrontPlant Sci*, 8:1560.

Kolár, J., Johnson, C. H., Macháčková, I. 2003. Exogenously applied melatonin (N-acetyl-5-methoxytryptamine) affects flowering of the short-day plant *Chenopodium rubrum*. *Physiol Plant*, 118: 605–612.

Kumar, R., Bohra, A., Pandey, A. K., Pandey, M. K., Kumar, A. 2017. Metabolomics for plant improvement: Status and prospects. *FrontPlant Sci*, 8:1302.

Kurland, C. G., Andersson, S. G. E. 2000. Origin and evolution of the mitochondrial proteome. *Microbiol Mol Biol Rev*, 64:786–820.

Lazar, D., Murch, S. J., Beilby, M. J., Al Khazaaly, S. 2013. Exogenous melatonin affects photosynthesis in characeae *Charaaustralis*. *Plant Signal Behav*, 8:e23279.

Lee, H. Y., Back, K. 2017. Melatonin is required for H2O2- and NO mediated defensesignaling through MAPKKK3 and OXI1 in *Arabidopsis thaliana*. *J Pineal Res*, 62(2):e12379. https://doi.org/10.1111/jpi.12379

Lee, H. Y., Back, K. 2018. Melatonin plays a pivotal role in conferring tolerance against endoplasmic reticulum stress via mitogen activated protein kinases and bZIP60 in *Arabidopsis thaliana*. *Melatonin Res*, 1:94–108.

Lee, H. Y., Byeon, Y., Back, K. 2014. Melatonin as a signal molecule triggering defense responses against pathogen attack in *Arabidopsis* and tobacco. *J Pineal Res*, 57:262–268. https://doi.org/10.1111/jpi.12165

Lei, K., Sun, S., Zhong, K., et al. 2021. Seed soaking with melatonin promotes seed germination under chromium stress via enhancing reserve mobilization and antioxidant metabolism in wheat. *Ecotoxicol Environ Saf*, 220:112241.

Lerner, A. B., Case, J. D., Takahashi, Y., Lee, T. H., Mori, W. 1958. Isolation of melatonin, a pineal factor that lightens melanocytes. *J Am Chem Soc*, 80:2587. https://doi.org/10.1021/ja01543a060.

Li, C., He, Q., Zhang, F., et al. 2019. Melatonin enhances cotton immunity to *Verticillium wilt* via manipulating lignin and gossypol biosynthesis. *Plant J*, 100:784–800.

Li, C., Wang, P., Wei, Z., et al. 2012. The mitigation effects of exogenous melatonin on salinity-induced stress in *Malus hupehensis*. *J Pineal Res*, 53:298–306.

Li, H., Guo, Y. L., Lan, Z., et al. 2021. Methyl jasmonate mediates melatonin induced cold tolerance of gr grafted watermelon plants. *Hortic Res*, 8:57.

Li, L., Yan, X. 2021. Insights into the roles of melatonin in alleviating heavy metal toxicity in crop plants. *Phyton*, 90:1559–1572.

Li, X. N., Brestic, M., Tan, D. X., et al. 2018. Melatonin alleviates low PS I-limited carbon assimilation under elevated CO_2 and enhances the cold tolerance of offspring in chlorophyll b-defcient mutant wheat. *J Pineal Res*, 64:e12453.

Liu, C., Chen, L., Zhao, R., et al. 2019. Melatonin induces disease resistance to *Botrytis cinerea* in tomato fruit by activating jasmonic acid signaling pathway. *J Agric Food Chem*, 67:6116–6124. https://doi.org/10.1021/acs.jafc.9b00058

Lukatkin, A. S., Gar'Kova, A. N., Bochkarjova, A. S., Nushtaeva, O. V., da Silva, J. A. T. 2013. Treatment with the herbicide TOPIK induces oxidative stress in cereal leaves. *Pestic Biochem Physiol*, 105:44–49.

Ma, X. J., Zhang, J., Burgess, P., Rossi, S., Huang, B. 2018. Interactive effects of melatonin and cytokinin on alleviating drought-induced leaf senescence in creeping bentgrass (*Agrostis stolonifera*). *Environ Exp Bot*, 145:1–11.

Macháčková, I., Krekule, J. 2002. Sixty-five years of searching for the signals that trigger flowering. *Russ J Plant Physiol*, 49:451–459.

Majima, A., Hattori, A., Wada, M., Chiba, A., Aoki, K. 1999. Dynamic release of melatonin in cyanobacterium, *Spirulina platensis*. *Proc Jpn Soc Comp Endocrinol*, 14:50.

Manchester, L. C., Poeggeler, B., Alvares, F. L., Ogden, G. B., Reiter, R. J. 1995. Melatonin immunoreactivity in the photosynthetic prokaryote *Rhodospirillum rubrum*: Implications for an ancient antioxidant system. *Cell Mol Biol Res*, 41:391–395.

Mandal, M. K., Suren, H., Ward, B., Boroujerdi, A., Kousik, C. 2018. Differential roles of melatonin in plant-host resistance and pathogen suppression in cucurbits. *J Pineal Res*, 65(3):e12505. https://doi.org/10.1111/jpi.12505

Marthandan, V., Geetha, R., Kumutha, K., Renganathan, V. G., Karthikeyan, A., Ramalingam, J. 2020. Seed priming—A feasible strategy to enhance drought tolerance in crop plants. *Int J Mol Sci*, 21:8258.

Mir, S., Alam, H. 2020. Melatonin modulates photosynthesis, redox status, and elemental composition to promote growth of *Brassica juncea*-a dose-dependent effect. *Protoplasma*, 257:1685–1700.

Mohamed, H. A., Moussa, H. R., Selem, E., El-Deen Sayed Ragab, M. H. 2021. Does exogenous application of melatonin ameliorate lead toxicity in *Erucavesicaria* plants? *Egypt J Bot*, 61:33–40.

Nadeem, M., Li, J., Yahya, M., et al. 2019. Grain legumes and fear of salt stress: Focus on mechanisms and management strategies. *Int J Mol Sci*, 20:799.

Obata, T., Fernie, A. R. 2012. The use of metabolomics to dissect plant responses to abiotic stresses. *Cell Mol Life Sci*, 69:3225–3243.

Okant, M., Kaya, C. 2019. The role of endogenous nitric oxide in melatonin-improved tolerance to lead toxicity in maize plants. *Environ Sci Pollut Res*, 26:11864–11874.

Oladi, E., Mohamadi, M., Shamspur, T., Mostafavi, A. 2014. Spectro fluorimetric determination of melatonin in kernels of four different pistacia varieties after ultrasound-assisted solid—liquid extraction. *Spectrochimica Acta. Part A, Molecular and Biomolecular Spectroscopy*, 132:326–329. https://doi.org/10.1016/j.saa.2014.05.010

Osakabe, Y., Osakabe, K., Shinozaki, K., Tran, L. S. P. 2014. Response of plants to water stress. *Front Plant Sci*, 5:540.

Özen, İ. T., Ekşi, A. 2016. Melatonin and serotonin content of the main sour cherry varieties and commercially produced sour cherry concentrates. *Eur Int J Sci Tech*, 5(4):57–64.

Pan, J., Lin, S., Woodbury, N. W. 2012. bacteriochlorophyll excited-state quenching pathways in bacterial reaction centers with the primary donor oxidized. *J Phys Chem*, 116:2014–2022.

Parankusam, S., Bhatnagar-Mathur, P., Sharma, K. K. 2017. Heat responsive proteome changes reveal molecular mechanisms underlying heat tolerance in chickpea. *Environ Exp Bot*, 141:132–144.

Park, H. S., Kazerooni, E. A., Kang, S. M., Al-Sadi, A. M., Lee, I. J. 2021. Melatonin enhances the tolerance and recovery mechanisms in *Brassica juncea* (L.) Czern. under saline conditions. *Front Plant Sci*, 12:593717.

Pawar, V., Laware, S. 2018. Seed priming—A critical review. *Int J Sci Res Biol Sci*, 5:94–101.

Posmyk, M. M., Bałabusta, M., Wieczorek, M., Sliwinska, E., Janas, K. M. 2009a. Melatonin applied to cucumber (*Cucumis sativus* L.) seeds improves germination during chilling stress. *J Pineal Res*, 46:214–223.

Posmyk, M. M., Janas, K. M., Malgorzata, P. M., Marta, B., Krystyna, J. M. 2009b. Melatonin applied by osmopriming as a biostimulator improving cucumber (*Cucumis sativus* L.) seedling growth at abiotic stress conditions. In Proceedings of the International Symposium on Environmental Science and Technology (ISEST), Shanghai, China, 2–5 June.

Posmyk, M. M., Kuran, H., Marciniak, K., Janas, K. M. 2008. Presowing seed treatment with melatonin protects red cabbage seedlings against toxic copper ion concentrations. *J Pineal Res*, 45:24–31.

Qian, Y., Tan, D. X., Reiter, R. J., Shi, H. 2015. Comparative metabolomic analysis highlights the involvement of sugars and glycerol in melatonin-mediated innate immunity against the bacterial pathogen in *Arabidopsis*. *Sci Rep*, 5:1–11. https://doi.org/10.1038/srep15815

Ramakrishna, A., Giridhar, P., Ravishankar, G. A. 2011. Phytoserotonin: A review. *Plant Signal Behav*, 6:800–809. https://doi.org/10.4161/psb.6.6.15242

Ramakrishna, A., Giridhar, P., Sankar, K. U., Ravishankar, G. A. 2012. Endogenous profiles of indoleamines: Serotonin and melatonin in different tissues of *Coffeacanephora* P ex Fr. as analyzed by HPLC and LC-MS-ESI. *Acta Physiol Plant*, 34:393–396. https://doi.org/10.1007/s11738-011-0829-2

Roychoudhury, A., Basu, S. 2012. Ascorbate-glutathione and plant tolerance to various abiotic stresses. In *Oxidative Stress in Plants: Causes, Consequences and Tolerance*, ed. N. A. Anjum, S. Umar, A. Ahmad, 177–258. New Delhi: IK International Publishers.

Roychoudhury, A., Basu, S., Sarkar, S. N., Sengupta, D. N. 2008. Comparative physiological and molecular responses of a common aromatic indica rice cultivar to high salinity with non-aromatic indica rice cultivars. *Plant Cell Reports*, 27(8):1395–1410.

Ruan, C. J., Rumpunen, K., Nybom, H. 2013. Advances in improvement of quality and resistance in a multipurpose crop: Sea buckthorn. *Crit Rev Biotechnol*, 33:126–144.

Sadak, M. S., Abdalla, A. M., Elhamid, E. M. A., Ezzo, M. I. 2020. Role of melatonin in improving growth, yield quantity and quality of *Moringa oleifera* L. plant under drought stress. *Bull Natl Res Cent*, 44:1–13.

Sairam, R. K., Rao, K. V., Srivastava, G. 2002. Differential response of wheat genotypes to long term salinity stress in relation to oxidative stress, antioxidant activity and osmolyte concentration. *Plant Sci*, 163:1037–1046.

Samanta, S., Banerjee, A., Roychoudhury, A. 2021. Exogenous melatonin regulates endogenous phytohormone homeostasis and thiol-mediated detoxification in two indica rice cultivars under arsenic stress. *Plant Cell Reports*, 40:1585–1602.

Samanta, S., Banerjee, A., Roychoudhury, A. 2022. Arsenic toxicity is counteracted by exogenous application of melatonin to different extents in arsenic-susceptible and arsenic-tolerant rice cultivars. *J Plant Growth Regul*, 41:2210–2231.

Shahzad, B., Tanveer, M., Che, Z., et al. 2018. Role of 24-epibrassinolide (EBL) in mediating heavy metal and pesticide induced oxidative stress in plants: A review. *Ecotoxicol Environ Saf*, 147:935–944.

Sharaf, Y., Aghdam, M. S., Luo, Z., et al. 2019. Melatonin treatment promotes endogenous melatonin accumulation and triggers GABA shunt pathway activity in tomato fruits during cold storage. *Sci Hortic* (Amsterdam), 254:222–227. https://doi.org/10.1016/j.scienta.2019.04.056

Sharma, A., Wang, J., Xu, D., et al. 2020a. Melatonin regulates the functional components of photosynthesis, antioxidant system, gene expression, and metabolic pathways to induce drought resistance in grafted *Caryacathayensis* plants. *Sci Total Environ*, 713:136675.

Sharma, A., Zheng, B. 2019. Melatonin mediated regulation of drought stress: Physiological and molecular aspects. *Plants*, 8:190.

Sharma, P., Manchanda, R., Goswami, R., Chawla, S. 2020b. Biodiversity and therapeutic potential of medicinal plants. In *Environmental Concerns and Sustainable Development*, 27–44. Germany: Springer.

Shi, H., Chan, Z. 2014. The cysteine2/histidine2-type transcription factor zinc finger of *Arabidopsis thaliana* 6-activated C-repeat-binding factor pathway is essential for melatonin-mediated freezing stress resistance in *Arabidopsis*. *J Pineal Res*, 57:185–191.

Shi, H., Chen, Y., Tan, D. X., Reiter, R. J., Chan, Z., He, C. 2015a. Melatonin induces nitric oxide and the potential mechanisms relate to innate immunity against bacterial pathogen infection in *Arabidopsis*. *J Pineal Res*, 59:102–108. https://doi.org/10.1111/jpi.12244

Shi, H., Jiang, C., Ye, T., et al. 2014. Comparative physiological, metabolomic, and transcriptomic analyses reveal mechanisms of improved abiotic stress resistance in bermudagrass [*Cynodondactylon* (L). Pers.] by exogenous melatonin. *J Exp Bot*, 66: 681–694.

Shi, H., Reiter, R. J., Tan, D. X., Chan, Z. 2015b. Indole-3-acetic acid inducible 17 positively modulates natural leaf senescence through melatonin-mediated pathway in *Arabidopsis*. *J Pineal Res*, 58:26–33.

Shi, H., Tan, D. X., Reiter, R. J., Ye, T., Yang, F., Chan, Z. 2015c. Melatonin induces class A1 heat-shock factors (HSFA 1s) and their possible involvement of thermotolerance in *Arabidopsis*. *J Pineal Res*, 58:335–342.

Shi, H., Wei, Y., Wang, Q., Reiter, R. J., He, C. 2016. Melatonin mediates the stabilization of DELLA proteins to repress the floral transition in *Arabidopsis*. *J Pineal Res*, 60:373–379.

Simpson, J. P., Wunderlich, C., Li, X., Svedin, E., Dilkes, B., Chapple, C. 2021. Metabolic source isotopic pair labeling and genome-wide association are complementary tools for the identification of metabolite-gene associations in plants. *Plant Cell*, 33:492–510.

Sun, C., Liu, L., Wang, L., Li, B., Jin, C., Lin, X. 2021. Melatonin: A master regulator of plant development and stress responses. *J Integr Plant Biol*, 63:126–145.

Sun, C., Lv, T., Huang, L., Liu, X., Jin, C., Lin, X. 2020. Melatonin ameliorates aluminium toxicity through enhancing aluminium exclusion and re-establishing redox homeostasis in roots of wheat. *J Pineal Res*, 68:e12642.

Sun, Q., Zhang, N., Wang, J., et al. 2015. Melatonin promotes ripening and improves quality of tomato fruit during postharvest life. *J Exp Bot*, 66:657–668.

Sun, Y., Liu, Z., Lan, G., Jiao, C., Sun, Y. 2019. Effect of exogenous melatonin on resistance of cucumber to downy mildew. *Sci Hortic* (Amsterdam), 255:231–241. https://doi.org/10.1016/j.scienta.2019.04.057

Szafranska, K., Glinska, S., Janas, K. M. 2013. Ameliorative effect of melatonin on meristematic cells of chilled and re-warmed *Vigna radiata* roots. *Biol Plant*, 57:91–96.

Szafranska, K., Reiter, R. J., Posmyk, M. M. 2017. Melatonin improves the photosynthetic apparatus in pea leaves stressed by paraquat via chlorophyll breakdown regulation and its accelerated de novo synthesis. *Front Plant Sci*, 8:878.

Tal, O., Haim, A., Harel, O., Gerchman, Y. 2011. Melatonin as an antioxidant and its semi-lunar rhythm in green macroalga *Ulva* sp. *J Exp Botany*, 62:1903–1910.

Talaat, N. B. 2021. Polyamine and nitrogen metabolism regulation by melatonin and salicylic acid combined treatment as a repressor for salt toxicity in wheat (*Triticum aestivum* L.) plants. *Plant Growth Regul*, 165:1–15.

Tan, D. X., Hardeland, R., Back, K., Manchester, L. C., Alatorre-Jimenez, M. A., Reiter, R. J. 2016. On the significance of an alternate pathway of melatonin synthesis via 5-methoxytryptamine: Comparisons across species. *J Pineal Res*, 61:27–40.

Tan, D. X., Manchester, L. C., Helton, P. 2007. Phyto-remediative capacity of plants enriched with melatonin. *Plant Signaling & Behavior*, 2:514–516. https://doi.org/10.4161/psb.2.6.4639

Tan, D. X., Reiter, R. J. 2020. An evolutionary view of melatonin synthesis and metabolism related to its biological functions in plants. *J Exp Bot*, 71:4677–4689.

Tan, X., Long, W., Zeng, L., et al. 2019. Melatonin-induced transcriptome variation of rapeseed seedlings under salt stress. *Int J Mol Sci*, 20:5355.

Tiwari, R. K., Bashyal, B. M., Shanmugam, et al. 2021. Impact of *Fusarium* dry rot on physicochemical attributes of potato tubers during postharvest storage. *Postharvest BiolTechnol*, 181:111638. https://doi.org/10.1016/J.POSTHARVBIO.2021.111638

Tiwari, R. K., Lal, M. K., Kumar, R. 2020a. Mechanistic insights on melatonin-mediated drought stress mitigation in plants. *Physiol Plantarum*, 172:1212–1226. https://doi.org/10.1111/ppl.13307

Tiwari, R. K., Lal, M. K., Naga, K. C., et al. 2020b. Emerging roles of melatonin in mitigating abiotic and biotic stresses of horticultural crops. *Sci Hortic* (Amsterdam), 272:109592.

Turk, H., Erdal, S., Genisel, M., Atici, O., Demir, Y., Yanmis, D. 2014. The regulatory effect of melatonin on physiological, biochemical and molecular parameters in cold-stressed wheat seedlings. *Plant Growth Regul*, 74:139–152.

Wang, M., Duan, S., Zhou, Z., Chen, S., Wang, D. 2019. Foliar spraying of melatonin confers cadmium tolerance in *Nicotiana tabacum* L. *Ecotox Environ Safe*, 170:68–76.

Wang, Y., Reiter, R. J., Chan, Z. 2018. Phytomelatonin: A universalabiotic stress regulator. *J Exp Bot*, 69:963–974.

Weeda, S., Zhang, N., Zhao, X., et al. 2014. *Arabidopsis* transcriptome analysis reveals key roles of melatonin in plant defense systems. *PLoS One*, 9(3):e93462. https://doi.org/10.1371/journal.pone.0093462

Wei, Y., Chang, Y., Zeng, H., Liu, G., He, C., Shi, H. 2018a. RAV transcription factors are essential for disease resistance against cassava bacterial blight via activation of melatonin biosynthesis genes. *J Pineal Res*, 64:12454.

Wei, Y., Hu, W., Wang, Q., et al. 2017. Identifcation, transcriptional and functional analysis of heat-shock protein 90s in banana (*Musa acuminata* L.) highlight their novel role in melatonin-mediated plant response to *Fusarium* wilt. *J Pineal Res*, 62(1):e12367. https://doi.org/10.1111/jpi.12367

Wei, Y., Jin, J., Jiang, S., et al. 2018b. Quantitative response of soybean development and yield to drought stress during different growth stages in the Huaibei Plain, China. *Agronomy*, 8(7):97.

Wen, W., Li, K., Alseekh, S., et al. 2015. Genetic determinants of the network of primary metabolism and their relationships to plant performance in a maize recombinant inbred line population. *Plant Cell*, 27:1839–1856.

Wolf, K., Kolář, J., Witters, E., van Dongen, W., van Onckelen, H., Macháčková, I. 2001. Daily profile of melatonin levels in *Chenopodium rubrum* L. depends on photoperiod. *J Plant Physiol*, 158:1491–1493.

Wyckhuys, K. A. G., Hughes, A. C., Buamas, C. 2019. Biological control of an agricultural pest protects tropical forests. *Commun Boil*, 2:10.

Xie, Z., Wang, J., Wang, W., et al. 2020. Integrated analysis of the transcriptome and metabolome revealed the molecular mechanisms underlying the enhanced salt tolerance of rice due to the application of exogenous melatonin. *Front Plant Sci*, 11:618680.

Yang, W. J., Du, Y. T., Zhou, Y. B. 2019. Overexpression of TaCOMT improves melatonin production and enhances drought tolerance in transgenic *Arabidopsis*. *Int J Mol Sci*, 20:652.

Ye, J., Wang, S., Deng, X., Yin, L., Xiong, B., Wang, X. 2016. Melatonin increased maize (*Zea mays* L.) seedling drought tolerance by alleviating drought-induced photosynthetic inhibition and oxidative damage. *Acta Physiol Plant*, 38:1–13.

Ye, T., Hao, Y. H., Yu, L., Shi, H., Reiter, R. J., Feng, Y. Q. 2017. A simple, rapid method for determination of melatonin in plant tissues by UPLC coupled with high resolution orbitrap mass spectrometry. *Front Plant Sci*, 8:1–10. https://doi.org/10.3389/fpls.2017.00064

Yin, L., Wang, P., Li, M., et al. 2013. Exogenous melatonin improves Malus resistance to Marssonina apple blotch. *J Pineal Res*, 54:426–434. https://doi.org/10.1111/jpi.12038

Yu, X., James, A. T., Yang, A., et al. 2016. A comparative proteomic study of drought-tolerant and drought-sensitive soybean seedlings under drought stress. *Crop Pasture Sci*, 67(5):528–540.

Zadraznik, T., Egge-Jacobsen, W., Meglic, V., et al. 2017. Proteomic analysis of common bean stem under drought stress using in-gel stable isotope labeling. *J Plant Physiol*, 209:42–50.

Zargar, S. M., Fujiwara, M., Inaba, S., et al. 2015. Correlation analysis of proteins responsive to Zn, Mn, or Fe deficiency in *Arabidopsis* roots based on iTRAQ analysis. *Plant Cell Rep*, 34(1):157–166.

Zhang, H. J., Zhang, N., Yang, R. C., et al. 2014. Melatonin promotes seed germination under high salinity by regulating antioxidant systems, ABA and GA 4interaction in cucumber (*Cucumis sativus* L.). *J Pineal Res*, 57:269–279.

Zhang, J., Shi, Y., Zhang, X., et al. 2017a. Melatonin suppression of heat-induced leaf senescence involves changes in abscisic acid and cytokinin biosynthesis and signaling pathways in perennial ryegrass (*Loliumperenne* L.). *Environ Exp Bot*, 138:36–45.

Zhang, N., Sun, Q., Zhang, H., et al. 2015. Roles of melatonin in abiotic stress resistance in plants. *J Exp Bot*, 66:647–656.

Zhang, N., Zhao, B., Zhang, H. J., et al. 2013. Melatonin promotes water-stress tolerance, lateral root formation, and seed germination in cucumber (*Cucumis sativus* L.). *J Pineal Res*, 54:15–23.

Zhang, S., Zheng, X., Reiter, R. J., et al. 2017b. Melatonin attenuates potato late blight by disrupting cell growth, stress tolerance, fungicide susceptibility and homeostasis of gene expression in *Phytophthora infestans*. *Front Plant Sci*, 8:1993. https://doi.org/10.3389/fpls.2017.01993

Zhao, D., Wang, H., Chen, S., Yu, D., Reiter, R. J. 2021. Phytomelatonin: An emerging regulator of plant biotic stress resistance. *Trends Plant Sci*, 26:70–82. https://doi.org/10.1016/j.tplants.2020.08.009

Zhao, D., Yu, Y., Shen, Y., et al. 2019. Melatonin synthesis and function: Evolutionary history in animals and plants. *Front Endocrinol*, 10:249.

Zhao, G., Zhao, Y., Yu, X., et al. 2018. Nitric oxide is required for melatonin-enhanced tolerance against salinity stress in rapeseed (*Brassica napus* L.) *Seedlings*. *Int J Mol Sci*, 19:1912.

Zhao, H., Xu, L., Su, T., Jiang, Y., Hu, L., Ma, F. 2015. Melatonin regulates carbohydrate metabolism and defenses against *Pseudomonas syringae* pv. tomato DC3000 infection in *Arabidopsis thaliana*. *J Pineal Res*, 59:109–119. https://doi.org/10.1111/jpi.12245

Zhao, L., Chen, L., Gu, P., et al. 2019. Exogenous application of melatonin improves plant resistance to virus infection. *Plant Pathol*, 68:1287–1295. https://doi.org/10.1111/ppa.13057

Zhu, J. K. 2001. Plant salt tolerance. *Trends Plant Sci*, 6:66–71.

Zhu, W., Han, H., Liu, A., et al. 2021. Combined ultraviolet and darkness regulation of medicinal metabolites in *Mehoniabealei* revealed by proteomics and metabolomics. *J Proteom*, 233:104081.

Index

R

S

T

U

V

W

Y

Z

Taylor & Francis Group
an **informa** business

Taylor & Francis eBooks

www.taylorfrancis.com

A single destination for eBooks from Taylor & Francis with increased functionality and an improved user experience to meet the needs of our customers.

90,000+ eBooks of award-winning academic content in Humanities, Social Science, Science, Technology, Engineering, and Medical written by a global network of editors and authors.

TAYLOR & FRANCIS EBOOKS OFFERS:

- A streamlined experience for our library customers
- A single point of discovery for all of our eBook content
- Improved search and discovery of content at both book and chapter level

REQUEST A FREE TRIAL

support@taylorfrancis.com

Routledge
Taylor & Francis Group

CRC Press
Taylor & Francis Group

For Product Safety Concerns and Information please contact our EU
representative GPSR@taylorandfrancis.com
Taylor & Francis Verlag GmbH, Kaufingerstraße 24, 80331 München, Germany

www.ingramcontent.com/pod-product-compliance
Ingram Content Group UK Ltd.
Pitfield, Milton Keynes, MK11 3LW, UK
UKHW052218190926
475702UK00006B/1087
* 9 7 8 1 0 3 2 3 8 1 5 5 8 *